T0238294

Lecture Notes in Computer Sci

Edited by G. Goos and J. Hartmanis

Lecture Notes in Computer Science

Edited by G. Goos and J. Hartmanis

Advisory Board: W. Brauer D. Gries J. Stoer

Gérard Cohen Teo Mora Oscar Moreno (Eds.)

Applied Algebra, Algebraic Algorithms and Error-Correcting Codes

10th International Symposium, AAECC-10
San Juan de Puerto Rico, Puerto Rico,
May 10-14, 1993
Proceedings

Springer-Verlag
Berlin Heidelberg New York
London Paris Tokyo
Hong Kong Barcelona
Budapest

Series Editors

Gerhard Goos
Universität Karlsruhe
Postfach 69 80
Vincenz-Priessnitz-Straße 1
W-7500 Karlsruhe, FRG

Juris Hartmanis
Cornell University
Department of Computer Science
4130 Upson Hall
Ithaca, NY 14853, USA

Volume Editors

Gérard Cohen
Ecole Nationale Supérieure des Télécommunications
46, rue Barrault, F-75634 Paris Cedex 13, France

Teo Mora
Università di Genova, Dipartimento di Matematica
Via L. B. Alberti 4, I-16132 Genova, Italy

Oscar Moreno
University of Puerto Rico, Department of Mathematics
Rio Pedras, Puerto Rico 00931

CR Subject Classification (1991): E.3-4, I.1, G.2, F.2

ISBN 3-540-56686-4 Springer-Verlag Berlin Heidelberg New York
ISBN 0-387-56686-4 Springer-Verlag New York Berlin Heidelberg

Printed in Germany

Typesetting: Camera ready by author/editor
Printing and binding: Druckhaus Beltz, Hemsbach/Bergstr.
45/3140-543210 - Printed on acid-free paper

Preface

The AAECC Symposia Series was started ten years ago by Alain Poli (Toulouse), who organized, together with R. Desq, D. Lazard and P. Camion, the first conference in the series (Toulouse, June 1983) and was in charge of most of the following editions.

AAECC (the acronym has shifted its meaning over the years before stabilizing as "Applied Algebra, Algebraic Algorithms and Error Correcting Codes") aims to attract high-level research papers and to encourage cross-fertilization among different areas which share the use of algebraic methods and techniques for applications in the sciences of computing, communications, and engineering.

Algebra, in its broader sense, has always been viewed as a frame to describe in a formal setting both the properties of the objects giving mathematical models of reality and the rules under which they can be manipulated. Its importance for applications has grown in recent years with the introduction of technological areas (related to signal processing, error correcting codes, information processing, software engineering, etc.) in which the symbolic nature of the objects studied make the techniques of calculus and numerical analysis inapplicable. For these areas, algebra provides both a theoretical framework for the development of theories and algorithmic techniques for the concrete manipulation of objects.

While in principle covering any area related to applications of algebra to communication and computer sciences, by their previous history the AAECC Symposia are mainly devoted to research in coding theory and computer algebra.

The theory of error-correcting codes deals with the transmission of information in the presence of noise. Coding is the systematic use of redundancy in the formation of the messages to be sent so as to enable the recovery of the information present originally after it has been corrupted by (not too much) noise in the transmission over the channel. There has been a great deal of theoretical and applied work in this subject since the famous paper of Shannon in 1949. Applications of coding range from the lowly Hamming codes used in dynamic memories to the sophisticated Reed-Solomon codes used in compact disks and in many commercial and military systems. There are also convolutional codes widely used in satellite systems.

Computer algebra is devoted to the investigation of algorithms, computational methods, software systems and computer languages, oriented to scientific computations performed on exact and often symbolic data, by manipulating formal expressions by means of the algebraic rules they satisfy. It studies such problems from three different but confluent viewpoints: a) development and analysis of algebraic algorithms (both from the viewpoint of practical performance and of theoretical complexity); b) design and analysis of software systems for symbolic manipulation; c) applications of scientific and/or technological systems. It is important to stress that the mathematical theories to which computer algebra applies are not necessarily only the algebraic ones: polynomial equations, algebraic geometry, commutative algebra and group theory have a well-established research activity using symbolic computation techniques, but the same is equally true for analytic theories, e.g. differential equations, as shown by a couple of papers in these proceedings. Computer algebra views algebra more as a method than as an object of research.

In the past, coding has interacted with group theory, combinatorics and finite geometries (the proof of the non-existence of a projective plane of order 10 by a coding approach is a recent example). More recently it has developed remarkable and unexpected connections with algebraic geometry and number theory (Goppa's algebraic geometric codes, Serre's improvement on Weil's bound for number of points of curves over finite fields, the p-adic Serre bound, improvements on Ax and Chevalley-Warning Theorems, etc.). This connection is creating links between the two major areas represented in AAECC, coding theory and computer algebra, e.g. by the use of Gröbner bases for decoding algebraic geometric codes or other algebraic codes.

Questions of complexity are naturally linked with the computational issues of both coding theory and computer algebra and represent an important share of the area which AAECC aims to cover; the same holds for cryptography where algebraic techniques are gaining relevance.

Finally let us mention the area of sequence design or spread spectrum multiple access, represented here by an invited contribution: originally developed in the Second World War for communications in a hostile enviroment where the enemy tries to jam one's message, it now includes non-military applications such as mobile radio, cellular telephony, and wireless computer communications.

Except for AAECC 1 (*Discrete Mathematics*, **56**,1985) and AAECC 7 (*Discrete Applied Mathematics*, **33**,1991), the proceedings of all the symposia are published in Springer *Lecture Notes in Computer Sciences*, Vols. 228, 229, 307, 356, 357, 508, 539.

It is a policy of AAECC to maintain a high scientific standard, comparable to that of a journal, and at the same time a fast publication of the proceedings. This is made possible only thanks to the cooperation of a large body of referees.

We aimed to have each submission evaluated by at least three referees, and we failed only in 9 cases. We had 147 independent reports from 105 referees on the 47 submissions. Of these, 6 were withdrawn during the procedure, 12 were rejected, 7 accepted for oral presentation only, 22 accepted for oral presentation and inclusion in the proceedings. The proceedings also contain six invited contributions; a seventh, by G. Lachaud, was not received in time for inclusion in the proceedings.

The conference was organized by the University of Puerto Rico and sponsored by the Army Research Office Cornell MSI project and by the NSF EPSCoR of Puerto Rico project.

We express our thanks to the staff of the Gauss Laboratory of the University of Puerto Rico and especially to Tita Santos, for handling the local organization, and to the Springer-Verlag staff and especially to A. Hofmann for their help in the preparation of these proceedings.

February 1993 G. Cohen, T. Mora, O. Moreno

Conference Board

Gerard Cohen (Paris), Teo Mora (Genova), Oscar Moreno (Puerto Rico)

Conference Committee

T. Beth (Karlsruhe), J. Calmet (Karlsruhe), G. Cohen (Paris), M. Giusti (Palaiseau), J. Heintz (Buenos Aires), H. Imai (Yokohama), H. Janwa (Bombay), R. Kohno (Yokohama), H. F. Mattson (Syracuse), A. Miola (Roma), T. Mora (Genova), O. Moreno (Puerto Rico), A. Poli (Toulouse), T. R. N. Rao (Lafayette, LA), S. Sakata (Toyohashi)

Referees

G. Attardi, T. Beth, E. Biglieri, D. Bini, M. Blaum, M. Bronstein, D. Bruschi, G. Butler, P. Camion, J.F. Canny, H. Chabanne, I. Chakravarti, A. Chan, P. Charpin, M. Clausen, A. Cohen, G. Cohen, D. Coppersmith, J. Davenport, J.L. Dornstetter, L.A. Dunning, A. Duval, D. Duval, T. Etzion, H.J. Fell, G.L. Feng, R. Froeberg, G. Gallo, Z. Ge, W. Geiselmann, M. Giusti, S. Goldwasser, D. Gordon, M. Goresky, R. Grossman, S. Harari, J. Heintz, T. Helleseth, H. Imai, H. Janwa, J. Justesen, M. Kalkbrener, E. Kaltofen, A. Kerber, N. Koblitz, R. Kohno, G. Lachaud, W. Lassner, D. Lazard, D. Lebrigand, S. Litsyn, A. Logar, I. Luengo, H.S. Maini, H.F. Mattson, H.M. Möller, T. Mora, R. Morelos-Zaragoza, O. Moreno, J. Moulin-Ollagnier, D. Mundici, H. Niederreiter, A.M. Odlyzko, F. Ollivier, G. Persiano, K.T. Phelps, V. Pless, A. Poli, C. Pomerance, T.R.N. Rao, T. Recio, J.J. Risler, L. Robbiano, F. Rodier, M.F. Roy, S. Sakata, R. Safavi-Naini, B.D. Saunders, W. Schmidt, R. Schoof, C. Scovel, J. Seberry, W. Seiler, N. Sendrier, A. Sgarro, K. Shirayanagi, M. Singer, P. Sole, H.J. Stetter, D.R. Stinson, B. Sturmfels, M. Sweedler, M. Szegedy, A. Tietavainen, H.C.A. van Tilborg, J. van Tilburg, L. Tolhuizen, C. Traverso, K. Tzeng, U. Vaccaro, F. Vatan, S. Vladut, J. Wolfmann, H. Yoshida, G. Zemor.

Contents

Sequence Based Methods for Data Transmission and Source Compression

A. R. Calderbank[1], P. C. Fishburn[1] and A. Rabinovich[2]

[1] Mathematical Sciences Research Center,
AT&T Bell Laboratories, Murray Hill, NJ 07974
[2] Statistics Department, Stanford University, Palo Alto, CA 94305

Abstract. In the last 10 years the invention of trellis coded modulation has revolutionized communication over bandlimited channels and is starting to be used in magnetic storage. Part of the reason is that sophisticated signal processing systems involving finite state machines can now be fabricated inexpensively. This paper discusses new developments in the performance analysis of finite state machines.
This is the extended abstract of an invited lecture to be given at the 10th International Symposium on Applied Algebra, Algebraic Algorithms, and Error-Correcting Codes, Puerto Rico, May 10–14, 1993.

1 Introduction

This paper surveys recent work on sequence based methods for data transmission and source compression. We shall focus on new methods for analyzing the expected and worst-case performance of finite state machines. We suppose that every state transition in the finite state machine is associated with a symbol or set of symbols. The output of a finite state machine is then a set of symbol sequences or codewords. This set can be searched efficiently to find the optimum codeword with respect to any nonnegative measure that can be calculated on a symbol by symbol basis. The search algorithm is dynamic programming, that is to say the Viterbi algorithm.

The most familiar application of the Viterbi algorithm is in the decoding of channel outputs that have been corrupted by noise. A more recent application is the trellis coded quantization work of Marcellin and Fischer [14] where the measure is mean squared error (mse). We shall begin by describing a new graphical method for analyzing the covering properties of binary convolutional codes. This may be viewed as trellis coded quantization of a binary source, since the squared Euclidean distance $d^2(x, y)$ between two binary vectors x, y is just the Hamming distance $d_H(x, y)$. For complete details see [1, 2]. These are the first papers to define covering radius of a convolutional code and to describe a procedure for calculating this quantity.

The evolution of the Viterbi algorithm is determined by vectors of path metrics. The set of possible vectors forms the decoder state space. Bounds on the differences between path metrics are of practical importance in digital implementations of the Viterbi algorithm. Section 3 describes an example taken from magnetic recording where it was possible to completely determine the decoder state space.

Section 4 considers the problem of finding the closest convolutional codeword to a sequence of source samples drawn from a uniform source on [0, 1]. The mean squared error per dimension can be interpreted as the second moment of a Voronoi region of an infinite lattice. This quantity is of importance in data transmission and vector quantization.

2 Graphical Analysis of the Covering Properties of Convolutional Codes

In this section we consider quantization of equiprobable binary data using a decoder for a binary convolutional code. Given an arbitrarily long binary sequence we wish to calculate the expected and worst-case Hamming distortion per dimension. This normalization gives the fraction of bits that we need to change in order to reach a codeword in the convolutional code.

The key observation is that a convolutional code with 2^v states gives 2^v approximations to a given source sequence and that these approximations do not differ very much. If we subtract the smallest path metric from the others then we obtain a vector with bounded entries. These vectors determine the evolution of the Viterbi algorithm. Hence the possible one-step trajectories of the Viterbi algorithm determine a finite directed graph on these vectors.

We shall briefly describe the new graphical method by means of a simple but representative example, the rate 1/2 convolutional code C with generator matrix $[1 + D^2, 1 + D + D^2]$. The encoder state diagram is shown below in Fig. 1.

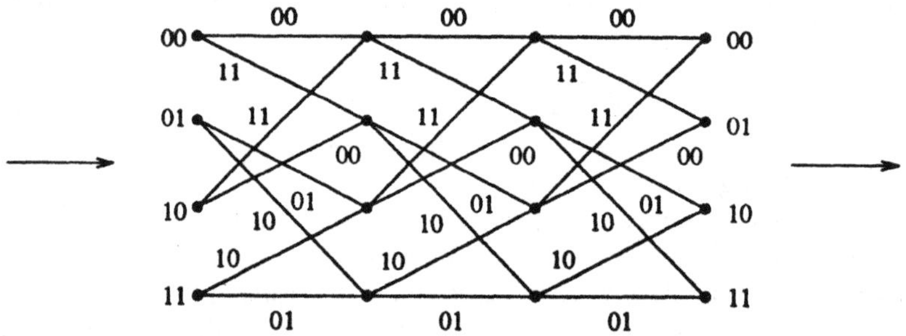

Fig. 1. Trellis diagram for the convolutional code with generator matrix $[1 + D^2, 1 + D + D^2]$.

The decoder has a copy of the trellis shown in Fig. 1. In every signaling interval, the decoder calculates and stores that path terminating in a given state that is closest to the binary data sequence. The decoder also calculates the *path metric* which measures distance from the data sequence to the codeword corresponding to the most likely path.

The one-step trajectories of the Viterbi decoding algorithm are determined by vectors of 4 path metrics. Accordingly we define decoder states as 4-tuples of individual path metrics. More formally, if Γ denotes the encoder state space of the convolutional encoder, then the decoder states are *metric functions* $\nu : \Gamma \to \mathbb{R}^+$.

Next we define a *metric state* to be the translation of a metric function v by a function that is constant on v, so that the minimum value of the translated function is 0. We make this definition because it is the *differences* between individual path metrics that determine the one-step trajectories of the Viterbi algorithm.

Let C_{ab} be the set of codewords terminating in encoder state ab, and let $d^2(C_{ab}, C_{a'b'})$ be the minimum distance between the sets C_{ab} and $C_{a'b'}$. The quantity $d^2(C_{ab}, C_{a'b'})$ depends on the free distance of the convolutional code, and in this case it is easy to verify that $d^2(C_{ab}, C_{a'b'}) \leq 3$. Since the sets C_{ab} are cosets of a linear space, it follows that for each sequence $x \in C_{ab}$ there exists a sequence $x' \in C_{a'b'}$ such that $d^2(x, x') = d^2(C_{ab}, C_{a'b'})$.

Now let z be a binary data sequence. It follows from the triangle inequality that

$$|d^2(z, C_{ab}) - d^2(z, C_{a'b'})| \leq d^2(C_{ab}, C_{a'b'}) \,. \quad (1)$$

Hence the difference between individual path metrics is bounded as claimed above.

2.1 Amalgamating Metric States

The one-step Viterbi trajectories determine a directed graph on metric states. We label the edges of this graph with aggregate incremental mean squared error (mse) as shown in Fig. 2 below.

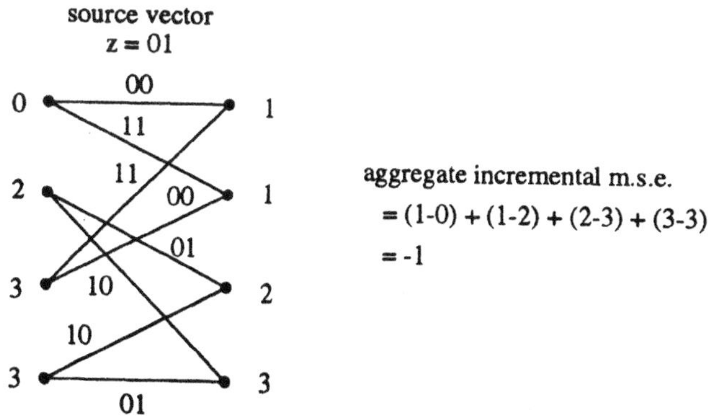

Fig. 2. An aggregate incremental m.s.e. calculation.

Next we collapse the directed graph by amalgamating metric states. A vertex labelled $[a, b, c, d]$ in the collapsed graph represents the set $\{(abcd), (badc),$

$(cdab)$, $(dcba)\}$ of metric states. An edge in the collapsed graph represents many edges in the original graph, but it is easy to verify that all these edges have the same aggregate incremental Hamming distortion (see Fig. 3). The vertices in the collapsed graph are called *m-states.*

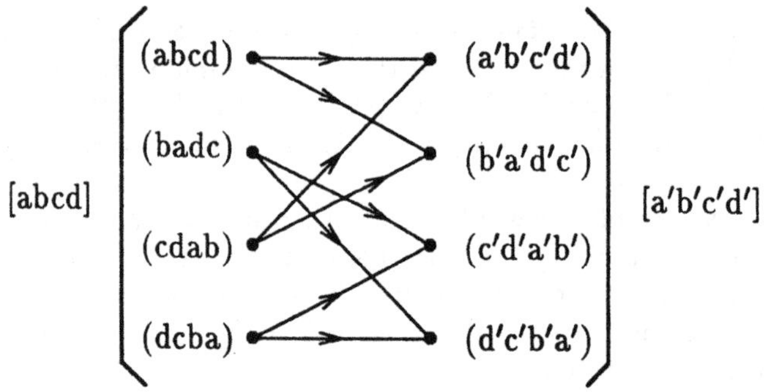

Fig. 3. Edges between metric states that collapse to the same edge between m-states.

For our convolutional code C, there are 31 metric states and 9 m-states. The directed graph on m-states is shown in Fig. 4, together with the vector of steady state probabilities. Double lines represent transitions with probability 1, and single lines represent transitions with probability 0.5.

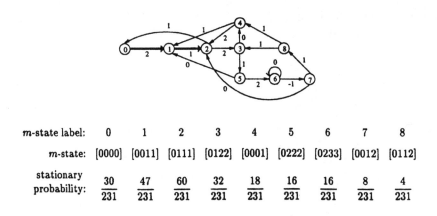

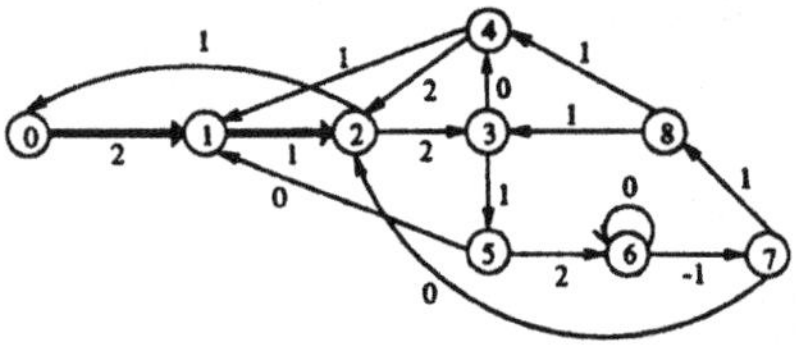

m-state label:	0	1	2	3	4	5	6	7	8
m-state:	[0000]	[0011]	[0111]	[0122]	[0001]	[0222]	[0233]	[0012]	[0112]
stationary probability:	$\frac{30}{231}$	$\frac{47}{231}$	$\frac{60}{231}$	$\frac{32}{231}$	$\frac{18}{231}$	$\frac{16}{231}$	$\frac{16}{231}$	$\frac{8}{231}$	$\frac{4}{231}$

Fig. 4. The directed graph on m-states together with associated stationary probabilities.

For a general convolutional code, there is an algebraic method for amalgamating metric states. We observe that addition of a codeword from the convolutional code may be viewed as a symmetry of the infinite encoder trellis. At time i this symmetry permutes encoder states by adding some fixed binary vector to each state label. This determines an elementary abelian group of order 2^v where v is the memory of the code. The m-states are the orbits of metric states under the action of this group. The approximation of a binary source vector by a convolutional codeword is represented as a path in the directed graph on m-states. This path is constant on cosets of the convolutional code. Thus Fig. 4 is a picture of the way in which the coset space of our convolutional code C evolves with time.

Figure 4 also makes it very easy to calculate expected and worst-case performance.

The expected performance μ is the expected incremental mse per dimension over an arbitrarily long sequence of equiprobable source samples. For each m-state u, the expected one-step aggregate gain out of u is

$$g(u) = \sum_{u' \in U} P(u, u')g(u, u') , \tag{2}$$

where $P(u, u')$ is the probability of the transition from u to u', and $g(u, u')$ is the aggregate incremental m.s.e. associated with this transition. Since we aggregated over 4 states, and since there are 2 source samples per step, it follows that

$$\mu = \left(\sum_{u \in U} \pi(u)g(u) \right) \Big/ (2 \times 4) = \frac{32}{231} = 0.138528\ldots . \tag{3}$$

The covering radius of this convolutional code is the worst case incremental m.s.e. per dimension for a particular arbitrarily long sequence of source samples. This quantity is determined by the cycles in Fig. 4. If A is such a cycle, then the average aggregate gain per edge of the cycle is

$$g(A) = \frac{1}{|A|} \sum_{(u,u') \in A} g(u, u') , \tag{4}$$

where $|A|$ is the number of edges in the cycle. The covering radius R is determined by the cycles for which this average aggregate gain per edge is maximum; formally

$$R = \sup_{\text{cycles } A} \{g(A)/(4 \times 2)\} , \tag{5}$$

and again the product 4×2 in the denominator takes account of the aggregation over 4 states and 2 source samples. For the convolutional code [101, 111] the covering radius $R = 1/6$, and a worst-case cycle is ⓪ $\xrightarrow{2}$ ① $\xrightarrow{1}$ ② $\xrightarrow{1}$ ⓪. This cycle corresponds to the periodic source sequence $\ldots 00, 00, 01, 00, 00, 01, \ldots$ as is shown in Fig. 5, where states are labelled by the path metrics, and the source samples appear in parentheses.

It is interesting to compare this approach to the covering properties of convolutional codes with recent work on the covering radius of block codes (see [5] and [10] for an introduction to this subject). Very little of the latter work concerns expected performance because this involves calculation of the complete coset weight distribution which is quite hard.

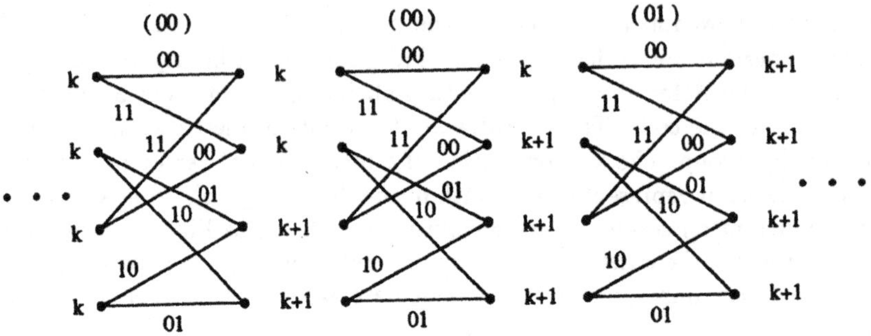

Fig. 5. A periodic sequence at maximal distance from the code.

2.2 Duality Questions

Figure 6 shows a different representation of the encoder state diagram which may be used to calculate the distance spectrum of the convolutional code. Codewords correspond to paths in this graph, and the weight of this codeword is the sum of the edge labels. Generating function techniques may be used to calculate the number of codewords with Hamming weight i of length N.

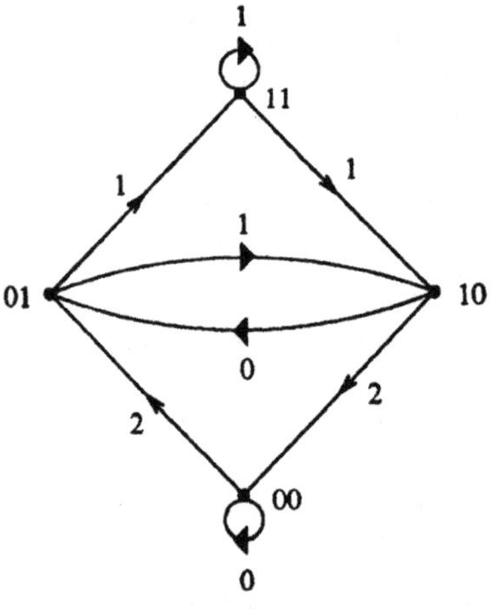

Fig. 6. Standard representation of the encoder state diagram for the convolutional code with generator matrix $[1 + D^2, 1 + D + D^2]$.

A question of some interest to us is whether the duality between packing and covering can be expressed as a mathematical relation between Fig. 4 and Fig. 6. We are also interested in any connections with the algebraic theory of convolutional codes developed by Forney [7].

3 Magnetic Recording

In Section 2, source samples were drawn from a finite set and we were able to calculate expected and worst case performance by analyzing a finite directed graph. When the source samples are drawn from a continuous distribution, the feasible metric functions describe a bounded region $\mathcal{R}$ of real Euclidean space (this follows from Equation (1)).

Digital implementation of the Viterbi decoding algorithms can be simplified given bounds on the size of path metric differences. A deeper, and more difficult problem, is to describe the probability distribution induced on the region $\mathcal{R}$ as a function of the statistics of the noise. These problems have been looked at recently in the context of finite state modulation codes for the magnetic recording channel. The survey article by Marcus, Siegel and Wolf [15] is a wonderful introduction to this subject. The first problem of computing accurate bounds for path metric differences has been addressed in two recent papers by Hekstra [12] and by Siegel et al. [17]. The insights obtained in this second paper were used to simplify the design of a 30 MHz trellis codec chip for partial response channels applicable to magnetic recording (see Shung et al. [16] for details). The calculation of the bounds by Siegel et al. is formulated as a series of linear programming problems.

A recent paper by Calderbank, Fishburn and Siegel [3] solves the problem of determining exactly the region R for Even-Mark-Modulation (EMM) trellis-coded duobinary (class 1) partial response. The encoder trellis is shown in Fig. 7 below. In this paper the noisy sample region is taken to be $[-2, 2]$, and all path metrics in the initial state x_0 are taken to be zero. The region R is the set of all decoder states that can be reached from x_0 in one or more steps. It is convenient to parametrize the decoder state space in terms of pairs (a, b) where $a = DM(3,1)$ and $b = DM(2,3)$. The region R then becomes a polygon in the plane. Exact determination of this polygon (which is shown in Fig. 8) gave stronger bounds on individual path metrics than were obtained via linear programming. It would be interesting to understand how properties of this polygon (for example the area) depend on the distance properties of the code and on the noise sample region.

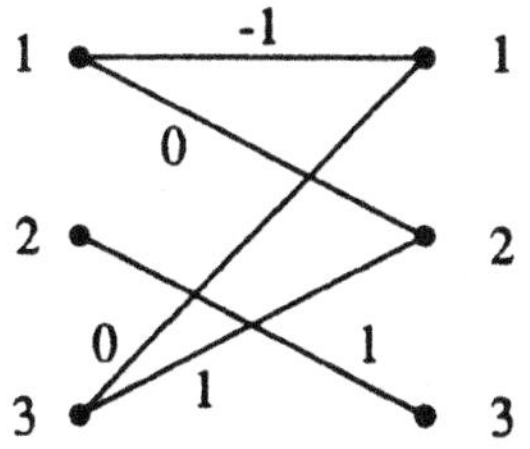

Fig. 7. The encoder trellis for EMM-coded duobinary.

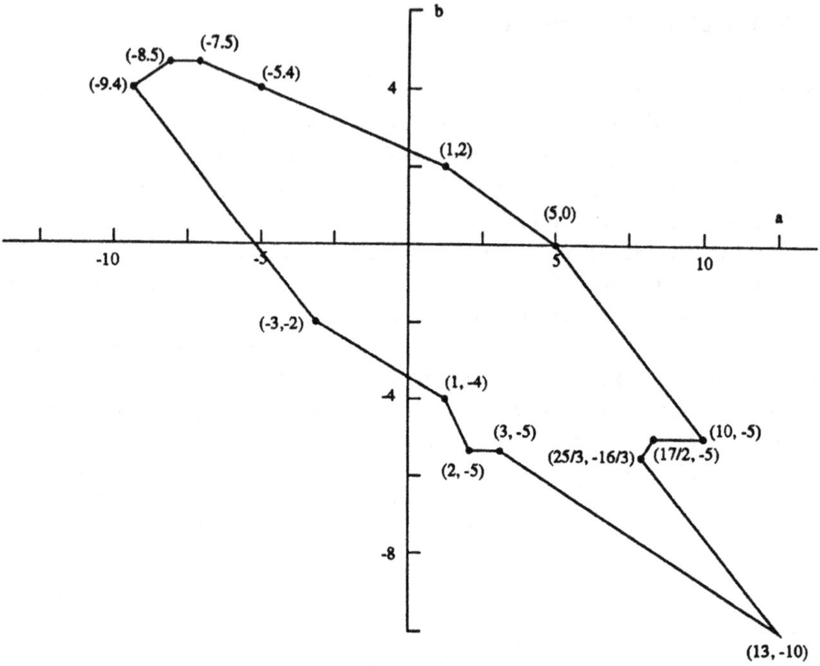

Fig. 8. The decoder state space for EMM-coded duobinary.

4 Trellis Shaping and Quantization of Uniform Sources

In this section we briefly describe trellis shaping, a technique invented by Forney [9] for decreasing the average transmitted power in a communications system. We shall describe in greater detail a second and closely related application, which is the compression of a uniform source using the trellis coded quantization technique proposed by Marcellin and Fischer [14].

Both applications involve representing a source sequence x as the sum of a codeword c and an error sequence $e = (e_i)$. In quantization, the objective is the codeword c, and the expected value $E(e_i^2)$ is the mean squared error (per dimension). In trellis shaping the objective is the error sequence e. The signal constellation will be the error sequences e that result from a suitably chosen discrete set of source sequences x. Here the expected value $E(e_i^2)$ will determine the extent to which average transmitted signal power is reduced.

The mathematical objective in this analysis is calculation of the circumradius and second moment of the Voronoi region of an infinite dimensional lattice. In source compression, these Voronoi regions are the quantization cells, the second moment gives the expected mean squared error per symbol, and the circumradius gives the worst case error per symbol. In data transmission, the signal constellation is uniformly distributed over the Voronoi region, and the second moment gives the average transmitted signal power. We begin with some basic definitions and notation.

A *lattice* is an additive subgroup of $\mathbb{R}^N$. If C is an $[N,k]$ binary linear code, then the *binary lattice* $\Lambda(C)$ is defined by

$$\Lambda(C) = \{x \in \mathbb{Z}^N | x \equiv c \pmod 2 \text{ for some } c \in C\}.$$

It is easy to verify that $\Lambda(C)$ is closed under addition so that it is indeed a lattice. The lattice $\Lambda(C)$ lies between $2\mathbb{Z}^N$ and $\mathbb{Z}^N$; the index of $2\mathbb{Z}^N$ in C is 2^k, and the index of $\Lambda(C)$ in $\mathbb{Z}^N$ is 2^{N-k}. A binary convolutional code C determines an infinite binary lattice $\Lambda(C)$ consisting of all sequences with integer entries that are congruent to some codeword modulo 2.

If $\Lambda_s \subseteq \mathbb{R}^N$ is a lattice, and $y \in \Lambda_s$ is a lattice point, then the *Voronoi region* $\mathcal{R}(y)$ consists of those points in $\mathbb{R}^N$ that are at least as close to y as to any other $y' \in \Lambda_s$. Thus

$$\mathcal{R}(y) = \left\{x \in \mathbb{R}^N \,\middle|\, \|x-y\| \le \|x-y'\| \text{ for all } y' \in \Lambda_s\right\}.$$

The interiors of different Voronoi regions are disjoint though two neighboring Voronoi regions may share a face. These faces lie in the hyperplanes midway between two neighboring lattice points. Translation by $y \in \Lambda_s$ maps the Voronoi region $\mathcal{R}(w)$ to the Voronoi region $\mathcal{R}(w+y)$. Thus all Voronoi regions are congruent, and we may speak of the Voronoi region $\mathcal{R}(\Lambda_s) = \mathcal{R}(0)$.

The application to data transmission involves finite-dimensional signal constellations, where signal points are taken from a lattice Λ. The constellation Ω consists of all lattice points that lie within a region $\mathcal{R}$.

Shaping gains are savings in average power that are obtained by manipulating the shape of the region $\mathcal{R}$. Clearly spherical regions maximize shape gain. However, an efficient algorithm is required to address signal points. Conway and Sloane [6] circumvent the problem of addressing spherical constellations by considering instead constellations bounded by the Voronoi region of a sublattice Λ_s of Λ. This enables them to use a decoder or vector quantizer for the lattice Λ_s to address the signal constellation. It must be remarked that this approach to constellation design has not prevailed in the discussions for the CCITT standard for voiceband modems operating at 19.2 kb/s and above. The approach of Laroia, Favardin and Tretter [13] (which builds on earlier work of Calderbank and Ozarow [4]) provides an efficient algorithm for addressing an "almost spherical" region and has been preferred.

A Voronoi constellation based on the lattice partition Λ/Λ_s is the set of points in some translate $\Lambda + a$ that falls within the Voronoi region $\mathcal{R}(\Lambda_s) = \mathcal{R}(0)$. Different methods have been suggested for choosing the translate a, and for resolving ties when a coset of Λ_s has more than one minimum norm point (see Conway and Sloane [6], Forney [8] and Headley [11]). The Voronoi region $\mathcal{R}(0)$ contains one and only one point from each equivalence class of $\mathbb{R}^N$ modulo Λ_s. Hence the lattice points $v \in \Lambda$ for which $v + a \in \mathcal{R}(0)$ form a complete set of coset representatives for Λ_s in Λ. Conway and Sloane [6] observed that any complete set Γ of coset representatives for Λ_s in Λ suffice to generate the Voronoi constellation Ω. If Q is the lattice vector quantizer for Λ_s then Ω is simply the set of error vectors $x + a - Q(x + a)$ as x ranges over Γ.

The points in the Voronoi constellation are uniformly distributed throughout the Voronoi region $\mathcal{R}(0)$, and the savings in average power (over a signal constellation of the same size bounded by the unit cube $[-1/2, 1/2]^N$ suitably scaled) can be quantified using the continuous approximation. This approximation replaces the average signal power of a constellation bounded by a region $\mathcal{R}$ with the average power of a probability distribution that is uniform within $\mathcal{R}$ and zero elsewhere.

The second moment $G(\mathcal{R})$ results from taking the average squared distance from a point in $\mathcal{R}$ to the centroid 0, and normalizing to obtain a dimensionless quantity. Formally

$$G(\mathcal{R}) = \frac{\int_{\mathcal{R}} ||x||^2 dv}{NV(\mathcal{R}_{(0)})^{1+2/N}} ,$$

where

$$V(\mathcal{R}) = \int_{\mathcal{R}} dv$$

is the volume of the region $\mathcal{R}$. The ***shape gain*** $\gamma(\Lambda_s)$ is the ratio of the second moment of the cube $[-1/2, 1/2]^N$, which is 1/12, to $G(R(\Lambda_s))$. The shape gain in N-dimensional space is bounded above by the shape gain of the N-sphere. The limiting gain is $\pi e/6 = 1.53$ dB.

In the theory of quantization, this ratio of second moments measures the reduction in mean squared error (compared with uniform scalar quantization) that comes from the shape of the quantization cell (the ***cell shape gain***). For binary lattices $\Lambda(C)$ the problem of calculating the second moment of the Voronoi region $\mathcal{R}$ simplifies considerably. For simplicity let C be an $[N, k]$ binary linear code. Then the map $\varphi : [0, 2]^N \rightarrow \mathcal{R}$ given by $\varphi(\omega) = \omega - Q(\omega)$ produces a uniform covering of $\mathcal{R}$ in which every vector is covered 2^k times. Thus a uniform distribution on $[0, 2]$ will produce a uniform distribution on $\mathcal{R}$.

To simplify further, we need the observation that if $r = (r_1, \ldots, r_N) \in [0, 1]^N$, then the closest point to r in the lattice $\Lambda(C)$ is just the closest codeword in C. If $\omega = (\omega_1, \ldots, \omega_N)$, then let $S(\omega) = \{i | 1 < \omega_i < 2\}$. Define $\omega' = (\omega'_1, \ldots, \omega'_N)$ by setting $\omega'_i = \omega_i$ if $i \notin S(\omega)$ and $\omega' = 2 - \omega'_i$ if $i \in S(\omega)$. Now the closest lattice point to ω' is a codeword in C, say $c' = (c'_1, \ldots, c'_N)$. Define a lattice point $c \in \Lambda(C)$ by setting $c_i = c'_i$ if $i \notin S(\omega)$, and $c_i = 2 - c'_i$ if $i \in S(\omega)$. Then c is the closest lattice point to ω. Observe that $\omega_i - c_i = \omega'_i - c'_i$ if $i \notin S(\omega)$ and $\omega_i - c_i = -(\omega'_i - c'_i)$ if $i \in S(\omega)$. Hence we may create a random variable uniformly distributed over the Voronoi region $\mathcal{R}$ by passing an iid source uniformly distributed over $[0, 1]$ through a maximum likelihood decoder for C, and then changing signs by tossing a fair coin. In particular, to calculate the second moment $G(\mathcal{R})$ it suffices to consider an iid source uniformly distributed over $[0, 1]$.

We define the ***normalized second moment*** μ of a binary lattice $\Lambda(C)$ to be the mean squared error per dimension when an iid source S uniformly distributed over $[0, 1]$ is quantized by a maximum likelihood decoder for C. Quantization of

S at rate 1 using the points 0 and 1 gives mse equal to 1/12. The binary code reduces the rate from 1 to R, and interpolation predicts that mse will increase by a factor $4^{(1-R)}$. The *cell shape gain* $\gamma(\Lambda(C))$ is given by $\gamma(\Lambda(C)) = [4^{(1-R)}/12]/\mu$ and is expressed in dB.

Calderbank and Fishburn [1] were able to calculate the normalized second moment exactly for the rate 1/2 convolutional code with generator matrix $[1, 1+D]$. As the memory of the convolutional code grows, exact calculation of this quantity quickly becomes intractable. This is because calculating the steady state distribution on the region representing feasible metric functions requires the solution of a multidimensional integral equation. Calderbank, Fishburn and Rabinovich [2] subsequently obtained intriguing formulae for the normalized second moment of rate $1/n$ convolutional codes that depend on the distribution of differences between a single pair of path metrics. These formulae were used to calculate the normalized second moments listed in Table 1. For comparison the normalized second moments of the Gosset Lattice E_8 and the Leech lattice are 0.65 dB and 1.03 dB respectively. It would be interesting to have a more extensive table of normalized second moments for convolutional codes.

Table 1. Normalized Second Moments of Convolutional Codes.

Rate	# States	Generator Matrix (octal)	Normalized Second Moment μ	Shape Gain (dB.)
1/4	4	$[5,7,7,7]$	.198	0.7570
	8	$[54,64,64,74]$	.189	0.9596
1/3	4	$[5,7,7]$	.1717	0.8742
	8	$[54,64,75]$	.1623	1.118
1/2	2	$[2,6]$	.14661	0.5569
	4	$[5,7]$	.1332	0.9734
	8	$[64,74]$	.1304	1.066
	16	$[46,72]$	.1287	1.123
	32	$[65,57]$	.1273	1.170
	64	$[554,744]$	.126	1.215

References

1. A. R. Calderbank and P. C. Fishburn, "The normalized second moment of the binary lattice determined by a convolutional code," preprint 1991.
2. A. R. Calderbank, P. C. Fishburn, and A. Rabinovich, Covering properties of convolutional codes and associated lattices, preprint, 1992.
3. A. R. Calderbank, P. C. Fishburn, and P. H. Siegel, "State space characterization of Viterbi detector path metric differences," preprint, 1992.
4. A. R. Calderbank and L. H. Ozarow, "Nonequiprobable signaling on the Gaussian channel," *IEEE Trans. Inform. Theory*, vol. IT-36, pp. 726–740, 1990.
5. G. D. Cohen, M. G. Karpovsky, H. F. Mattson Jr., and J. R. Schatz, "Covering radius — Survey and recent results," *IEEE Trans. Inform. Theory*, vol. IT-31, pp. 328–343, 1985.
6. J. H. Conway and N. J. A. Sloane, "A fast encoding method for lattice codes and quantizers," *IEEE Trans. Inform. Theory*, vol. IT-29, pp. 820–824, 1983.
7. G. D. Forney, Jr., "Structural analysis of convolutional codes via dual codes," *IEEE Trans. Inform. Theory*, vol. IT-19, pp. 512–518, 1973.
8. G. D. Forney, Jr., "Multidimensional constellations — Part II: Voronoi constellations," *IEEE J. Select. Areas Commun.*, vol. SAC-7, pp. 941–958, 1989.
9. G. D. Forney, Jr., "Trellis Shaping," *IEEE Trans. Inform. Theory*, vol. IT-38, pt. 1, pp. 281–300, 1992.
10. R. L. Graham and N. J. A. Sloane, "On the covering radius of codes," *IEEE Trans. Inform. Theory*, IT-31, pp. 385–401, 1985.
11. P. Headley, "A counterexample to a Voronoi constellation conjecture," *IEEE Trans. Inform. Theory*, vol. IT-37, pp. 1665–1667, 1991.
12. A. Hekstra, "An Alternative to Metric Rescaling in Viterbi Decoders," *IEEE Trans. Commun.*, vol. 37, pp. 1220–1222, November 1989.
13. R. Laroia, N. Favardin, and S. Tretter, "On SVQ shaping of multidimensional constellations — High-rate large-dimensional constellations," preprint, 1992.
14. M. W. Marcellin and T. R. Fischer, "Trellis coded quantization of memoryless and Gauss-Markov sources," *IEEE Trans. Commun.*, vol. 38, pp. 82–93, 1990.
15. B. H. Marcus, P. H. Siegel, and J. K. Wolf, "Finite-state modulation codes for data storage," *IEEE J. Select. Areas Commun.*, vol. 10, pp. 5–37, 1992.
16. C. B. Shung, R. Karabed, P. H. Siegel and H. K. Thapar, "A 30 MHz Trellis Codec Chip for Partial Response Channels," IBM Technical Report RJ 7771 (71888), October 24, 1991.
17. P. Siegel, C. Shung, T. Howell, and H. Thapar, "Exact Bounds for Viterbi Detector Path Metric Differences," *Proceedings of 1991 IEEE Int. Conf. on Acoustics, Speech, and Signal Processing* (*ICASSP'91*), Toronto, May 14–17, 1991.

On the Apparent Duality of the Kerdock and Preparata Codes *

A. Roger Hammons, Jr.[1], P. Vijay Kumar[2], A. R. Calderbank[3], N.J.A. Sloane[3], Patrick Solé[4]

[1] Hughes Aircraft Company,
8433 Fallbrook Avenue, Canoga Park, CA 91304-0445 U.S.A.
[2] Communication Sciences Institute,
EE-Systems, University of Southern California,
Los Angeles, CA 90089-2565 U.S.A.
[3] Mathematical Sciences Research Center
AT&T Bell Laboratories, Murray Hill, NJ 07974 U.S.A.
[4] CNRS-I3S, 250 rue A. Einstein, bâtiment 4
Sophia - Antipolis, 06560 Valbonne, France

Abstract. The Kerdock and extended Preparata codes are something of an enigma in coding theory since they are both Hamming-distance invariant and have weight enumerators that are MacWilliams duals just as if they were dual linear codes. In this paper, we explain, by constructing in a natural way a Preparata-like code $\mathcal{P}_L$ from the Kerdock code $\mathcal{K}$, why the existence of a distance-invariant code with weight distribution that is the McWilliams transform of that of the Kerdock code is only to be expected. The construction involves quaternary codes over the ring $\mathbb{Z}_4$ of integers modulo 4. We exhibit a quaternary code $\mathcal{Q}$ and its quaternary dual $\mathcal{Q}^\perp$ which, under the Gray mapping, give rise to the Kerdock code $\mathcal{K}$ and Preparata-like code $\mathcal{P}_L$, respectively. The code $\mathcal{P}_L$ is identical in weight and distance distribution to the extended Preparata code. The linearity of $\mathcal{Q}$ and $\mathcal{Q}^\perp$ ensures that the binary codes $\mathcal{K}$ and $\mathcal{P}_L$ are distance invariant, while their duality as quaternary codes guarantees that $\mathcal{K}$ and $\mathcal{P}_L$ have dual weight distributions. The quaternary code $\mathcal{Q}$ is the $\mathbb{Z}_4$-analog of the first-order Reed-Muller code. As a result, $\mathcal{P}_L$ has a simple description in the $\mathbb{Z}_4$-domain that admits a simple syndrome decoder. At length 16, the code $\mathcal{P}_L$ coincides with the Preparata code.

1 Introduction

Recently, a family of nearly optimal four-phase sequences of period $N = 2^r - 1$, r odd, with alphabet $\{1, j, -1, -j\}$, $j = \sqrt{-1}$, was discovered first by Solé [17] and later independently by Boztaş , Hammons, and Kumar [2] (see also [18]). After replacing each complex fourth root-of-unity j^a by its exponent $a \in \{0, 1, 2, 3\}$, this family may be viewed as a linear quaternary code over the ring $\mathbb{Z}_4$ of

* This work was supported in part by the National Science Foundation under Grant NCR-9016077 and by Hughes Aircraft Company under its Ph.D. fellowship program.

integers modulo 4. Since the family has low correlation values, it also possesses large minimum Euclidean distance and thus the potential for excellent error-correcting capability.

A 2-adic (*i.e.*, base 2) expansion of the four-phase sequences is contained in [2]. Interestingly, this bore a striking resemblance to the original expression [12] for the nonlinear binary Kerdock code. A second connection with the Kerdock code arose during attempts to construct good binary codes from the four-phase sequence family using the Gray map. This was a logical step to pursue as the Gray map translates a quaternary code with large minimum Euclidean distance into a binary code of twice the length having large minimum Hamming distance. The codes that resulted were nonlinear and had the same parameters as shortened versions of the Kerdock code.

In exploring these connections, it was discovered that the original quaternary code could be enlarged in a natural way, to a linear quaternary code $\mathcal{Q}$ whose image under the Gray map is *precisely* the Kerdock code. It was only natural to consider whether the interesting link between the Kerdock code and a linear quaternary code could also be used to explain the apparent duality of the Kerdock and Preparata codes.

The new perspective does indeed provide an explanation, although not the one that might first be suspected. We show that the binary images $\mathcal{G}(\mathcal{C})$ and $\mathcal{G}(\mathcal{C}^{\perp})$ under the Gray map of a linear quaternary code $\mathcal{C}$ and its $\mathbb{Z}_4$-dual are always Hamming distance invariant. Furthermore, these binary codes have the property that their weight distributions are always dual under the MacWilliams transform. As a consequence, the Kerdock code possesses a *natural* "quaternary-dual" code $\mathcal{P}_L = \mathcal{G}(\mathcal{Q}^{\perp})$, identical in size, weight, and distance to the extended Preparata code. Although the Preparata and Preparata-like ($\mathcal{P}_L$) codes have similar finite field transform descriptions, they are in general not the same.

Interestingly, at length 16, the Preparata and Preparata-like codes do coincide. In fact, the Kerdock code, the extended Preparata code, and the Preparata-like code all coincide with the Nordstrom-Robinson code $\mathcal{N}_{16}$. Thus, the Nordstrom-Robinson code can be generalized in one way to get the extended Preparata codes, in another way to get the Kerdock codes, and in yet another way to get the Preparata-like codes!

From the standpoint of decoding, it is not necessary to distinguish between the binary codes and their quaternary parents. An important advantage in working in the $\mathbb{Z}_4$-domain, where the codes are linear, is that it is meaningful to speak of syndromes. Moreover, the codes $\mathcal{Q}$ and $\mathcal{Q}^{\perp}$ are $\mathbb{Z}_4$-analogs of the binary first-order Reed-Muller code $RM(1,r)$ and its dual $RM(r-2,r)$. This connection makes decoding of the Kerdock and Preparata codes, at least conceptually, easier.

Our discussion of these results is organized in the following manner. In Sect. 2, we present the general theory of linear quaternary codes, their duals, and their images as binary codes under the Gray map. We show that the curious linear-like properties of the Kerdock and Preparata codes are true in general for this new class of binary codes. We demonstrate that the Kerdock code is a member of this class and provide a finite field transform characterization of its Preparata-like pseudo-dual. In Sect. 3, we present simple Galois ring theoretic descriptions of the codes. In Sect. 4, we consider the impact of the new quaternary interpretation on decoding algorithms. Finally, Sect. 5 provides an update on related results by the authors.

2 Theory of Quaternary and Related Binary Codes

2.1 Quaternary Codes

The interest here is in linear block codes over the ring $\mathbb{Z}_4$ of integers modulo 4. By a linear (n, M) quaternary code, we shall mean a set $\mathcal{C}$, consisting of M $\mathbb{Z}_4$-valued n-tuples, that is closed under addition modulo 4. Algebraically, $\mathcal{C}$ is a $\mathbb{Z}_4$-module, *i.e.* a module over the ring $\mathbb{Z}_4$.

Like linear block codes over finite fields, linear quaternary codes possess natural duals. If $\mathcal{C}$ is a linear (n, M) quaternary code, its dual $\mathcal{C}^{\perp}$ is defined to be the set of all $\mathbb{Z}_4$-valued n-tuples $\underline{v} = (v_1, v_2, \ldots, v_n)$ satisfying

$$\underline{u} \cdot \underline{v} = \sum_{t=1}^{n} u_t v_t = 0, \qquad \forall \underline{u} = (u_1, u_2, \ldots, u_n) \in \mathcal{C} \ . \tag{1}$$

To each $\mathbb{Z}_4$-valued codeword $\underline{c} = (c(t) : t \in I)$, we associate the equivalent complex roots-of-unity sequence $\underline{s} = \omega^{\underline{c}} = (\omega^{c(t)} : t \in I)$, where $\omega = i = \sqrt{-1} = e^{2\pi i/4}$. Then, given a set $\mathcal{C}$ of quaternary vectors, we let

$$\Omega\ (\mathcal{C}) = \{\omega^{\underline{c}} : \underline{c} \in \mathcal{C}\} \tag{2}$$

denote the corresponding set of complex sequences. When $\mathcal{C}$ is regarded as a set of quaternary CDMA signature sequences, its effectiveness depends upon the complex correlations (or inner products) of the sequences in $\Omega\ (\mathcal{C})$. When $\mathcal{C}$ is regarded as a quaternary code, its error-correcting capability depends upon the Euclidean distance properties of $\Omega\ (\mathcal{C})$. Note that if $\underline{c}_0 = (c_0(t) : t = 1, 2, \ldots, n)$ and $\underline{c}_1 = (c_1(t) : t = 1, 2, \ldots, n)$ are quaternary vectors with associated complex vectors $\underline{s}_0 = \Omega\ (\underline{c}_0)$ and $\underline{s}_1 = \Omega\ (\underline{c}_1)$, then

$$||\underline{s}_1 - \underline{s}_0||^2 = ||\underline{s}_1||^2 + ||\underline{s}_0||^2 - 2\ \mathrm{Re}\{\underline{s}_1^{\mathrm{H}} \underline{s}_0\} \tag{3}$$

$$= 2n - 2\mathrm{Re}\{\zeta(\underline{c}_1 - \underline{c}_0)\}, \tag{4}$$

where ^{H} denotes the Hermitian inner product and

$$\zeta(\underline{c}_1 - \underline{c}_0) = \sum_{t=1}^{n} \omega^{c_1(t) - c_0(t)} \tag{5}$$

is the *complex correlation* of $\underline{c}_0$ and $\underline{c}_1$. Note that ζ depends only on the difference $\underline{c} = \underline{c}_1 - \underline{c}_0$. By the above, if the nontrivial correlations of $\Omega\ (\mathcal{C})$ are low in magnitude, then the set also possesses large minimum Euclidean distance.

2.2 Equivalent Binary Codes under the Gray Map

In communication systems employing quadrature phase-shift keying (QPSK), the preferred assignment of two information bits to the four possible phases is the one in which adjacent phases differ by only one binary digit. This mapping is called *Gray encoding* and has the advantage that, when a quaternary codeword is transmitted across the additive white Gaussian noise channel, the errors most likely to occur are those causing a single erroneously decoded information bit.

For $c \in \mathbb{Z}_4$, it is natural from an algebraic viewpoint to consider the 2-adic expansion

$$c = 2a + b \tag{6}$$

where $a, b \in \{0, 1\}$. If u, v denote the two binary digits assigned to the complex number ω^c by the Gray encoder, then the pairs (a, b) and (u, v) are related by

$$u = a \tag{7}$$

$$v = a \oplus b. \tag{8}$$

where $\oplus$ denotes addition modulo 2.

Consider the map $g : \Omega(\mathbb{Z}_4{}^n) \to (\mathbb{Z}_2)^{2n}$ in which $g(\omega^{\underline{c}}) = (\underline{u} \mid \underline{v}) = (\underline{a} \mid \underline{a} \oplus \underline{b})$. It is easy to see that g maps complex vectors into binary vectors of twice the length in such a way that the squared Euclidean distance between two complex vectors is proportional to the Hamming distance between the corresponding binary vectors.

Thus, given a quaternary code $\mathcal{C}$ whose associated complex sequence set has good Euclidean distance, it is natural to examine the binary code $\mathcal{B}$ obtained by mapping the quaternary codeword $\underline{c} = 2\underline{a} + \underline{b}$ in $\mathcal{C}$ to the binary codeword $(\underline{a} \mid \underline{a} \oplus \underline{b})$. In general, if $\mathcal{C}$ is a code of length n, then $\mathcal{B}$ is a nonlinear code of length $2n$. We will use $\mathcal{G} : (\mathbb{Z}_4)^n \to (\mathbb{Z}_2)^{2n}$ to denote this *Gray map* transformation, so that $\mathcal{G}(\mathcal{C}) = \mathcal{B}$.

2.3 Weight and Distance Properties

In this section, we discuss the weight and distance properties of the quaternary codes $\mathcal{C}$ and $\mathcal{C}^{\perp}$ and their respective associated binary codes $\mathcal{B} = \mathcal{G}(\mathcal{C})$ and $\mathcal{B}_{\perp} = \mathcal{G}(\mathcal{C}^{\perp})$. The two principal results to be derived here are the following:

1. $\mathcal{B}$ and $\mathcal{B}_{\perp}$ are distance invariant.
2. The weight distributions of $\mathcal{B}$ and $\mathcal{B}_{\perp}$ are MacWilliams transforms of one another.

A binary code $\mathcal{B}$ is said to be *distance invariant* if the Hamming weight distributions of its translates $\underline{c} + \mathcal{B}$ are the same for all $\underline{c} \in \mathcal{B}$.

Theorem 1. *If $\mathcal{C}$ is a linear quaternary code, then its binary Gray representation $\mathcal{G}(\mathcal{C})$ is distance invariant.*

Proof. When $\mathcal{C}$ is a linear quaternary code, the complex signal set $\Omega(\mathcal{C})$ is Euclidean distance invariant, *i.e.* the set of relative Euclidean distances

$$\{||\underline{s}-\underline{s}_0|| \ : \ \underline{s} \in \Omega(\mathcal{C})\} \tag{9}$$

does not depend on the choice of $\underline{s}_0 \in \Omega(\mathcal{C})$. This is clear from (3) and (4). Since Hamming distances in $\mathcal{G}(\mathcal{C})$ are proportional to squared-Euclidean distances in $\Omega(\mathcal{C})$, $\mathcal{G}(\mathcal{C})$ must be Hamming distance invariant. □

Next, we recall from [15], (pp. 141–145), that the *complete weight enumerator* for a binary code $\mathcal{B}$ of block length n is a polynomial in two variables given by

$$W_{\mathcal{B}}(z_0, z_1) = \sum_{\underline{c}=[c_1,c_2,\ldots,c_n]\in\mathcal{B}} z_0^{k_0(\underline{c})} z_1^{k_1(\underline{c})} , \tag{10}$$

where $k_i(\underline{c})$ is the number of times the value i appears as an entry in the vector $\underline{c}$. If $\mathcal{B}$ is a linear binary code with dual $\mathcal{B}^{\perp}$, the weight enumerators of the two codes are related by the MacWilliams transform:

$$W_{\mathcal{B}^{\perp}}(z_0, z_1) = \frac{1}{|\mathcal{B}|} W_{\mathcal{B}}(z_0 + z_1, z_0 - z_1) \ . \tag{11}$$

Similarly, the complete weight enumerator for a quaternary code $\mathcal{C}$ is defined by

$$W_{\mathcal{C}}(z_0, z_1, z_2, z_3) = \sum_{\underline{c}=[c_1,c_2,\ldots,c_n]\in\mathcal{C}} z_0^{k_0(\underline{c})} z_1^{k_1(\underline{c})} z_2^{k_2(\underline{c})} z_3^{k_3(\underline{c})} \ . \tag{12}$$

The relationships between the complete weight enumerators of $\mathcal{C}$, its dual $\mathcal{C}^{\perp}$, and their binary representations $\mathcal{G}(\mathcal{C})$ and $\mathcal{G}(\mathcal{C}^{\perp})$ are given by the following theorems.

Theorem 2 (MacWilliams Transform for Codes over $\mathbb{Z}_4$). *If $\mathcal{C}$ and $\mathcal{C}^{\perp}$ are dual quaternary codes, then*

$$W_{\mathcal{C}^{\perp}}(z_0, z_1, z_2, z_3) = \frac{1}{|\mathcal{C}|} W_{\mathcal{C}}\Big(\sum_{i=0}^{3} z_i\omega^{0\cdot i}, \sum_{i=0}^{3} z_i\omega^{1\cdot i}, \sum_{i=0}^{3} z_i\omega^{2\cdot i}, \sum_{i=0}^{3} z_i\omega^{3\cdot i}\Big) \ . \tag{13}$$

Proof. The proof is essentially the same as in [15], (Thm. 10, pp. 141–145) for codes over $GF(4)$. □

Lemma 3. *If $\mathcal{C}$ is a quaternary code and $\mathcal{G}(\mathcal{C})$ is its binary Gray representation, then*

$$W_{\mathcal{G}(\mathcal{C})}(z_0, z_1) = W_{\mathcal{C}}(z_0^2, z_0 z_1, z_1^2, z_0 z_1) \ . \tag{14}$$

Proof. Immediate from the definitions. □

Theorem 4. *If $\mathcal{C}$ and $\mathcal{C}^{\perp}$ are dual quaternary codes, then the weight distributions of the nonlinear binary codes $\mathcal{G}(\mathcal{C})$ and $\mathcal{G}(\mathcal{C}^{\perp})$ are related by the binary MacWilliams transform.*

Proof. By the above theorems,

$$W_{\mathcal{G}(\mathcal{C}^\perp)}(z_0, z_1) = W_{\mathcal{C}^\perp}(z_0^2, z_0 z_1, z_1^2, z_0 z_1) \tag{15}$$

$$= \frac{1}{|\mathcal{C}|} W_{\mathcal{C}}((z_0+z_1)^2, z_0^2 - z_1^2, (z_0-z_1)^2, z_0^2 - z_1^2) \tag{16}$$

$$= \frac{1}{|\mathcal{G}(\mathcal{C})|} W_{\mathcal{G}(\mathcal{C})}(z_0 + z_1, z_0 - z_1), \tag{17}$$

which is the desired result. □

2.4 Quaternary View of the Kerdock and Preparata-Like Codes

A monic polynomial in $\mathbb{Z}_4\,[x]$ is a ***primitive basic irreducible*** if its projection modulo 2 is a primitive, irreducible polynomial in $\mathbb{Z}_2\,[x]$. Let $f(x) = f_0 + f_1 x + f_2 x^2 + \ldots + f_{r-1} x^{r-1} + x^r$ be a primitive basic irreducible dividing $x^N - 1$ in $\mathbb{Z}_4\,[x]$, where $N = 2^r - 1$. Let $\mathcal{Q}^-(r)$ denote the set of all $\mathbb{Z}_4$-valued sequences that satisfy the linear recurrence over $\mathbb{Z}_4$ whose characteristic polynomial is $f(x)$. Specifically, the $\mathbb{Z}_4$-valued sequence $c(t)$ is in $\mathcal{Q}^-(r)$ iff

$$c(t) = -\sum_{i=0}^{r-1} f_i\, c(t-i) \qquad \forall t \geq r\ . \tag{18}$$

It is not hard to show that all the sequences in $\mathcal{Q}^-$ have common period N and thus may be regarded as elements of $\mathbb{Z}_4{}^N$. Then, it is clear by construction that $\mathcal{Q}^-(r)$ is a cyclic, linear quaternary code of length N, rank r, and size 4^r.

Let $\mathcal{A}$ denote the family of cyclically distinct sequences obtained from $\mathcal{Q}^-$ by deleting the all-zero sequence and failing to distinguish between a sequence and any of its cyclic shifts. The corresponding collection $\Omega\,(\mathcal{A})$ of complex-valued sequences has been studied in [2] [17][18] as a family of asymptotically optimal CDMA signature sequences (referred to as Family $\mathcal{A}$ in [2]). Since the sequences of $\Omega\,(\mathcal{A})$ have low values of auto- and cross-correlation, the set also has large minimum Euclidean distance.

A little experimentation shows that $\mathcal{Q}^-(r)$ can be enlarged while preserving the minimum Euclidean distance of the corresponding complex sequence set. Let $\mathcal{Q}(r)$ denote the following enlarged code:

1. Extend the block length of the code from $N = 2^r - 1$ to $N' = 2^r$ by introducing a leading "0" to each $\mathcal{Q}^-$ codeword.
2. Increase the size of the code from 4^r to 4^{r+1} by adding an arbitrary $\mathbb{Z}_4$-valued constant vector to each lengthened codeword.

This is analogous to the process by which one enlarges the binary maximum-length code to obtain the first-order Reed-Muller code.

Let $\underline{c}^- = (c^-(t) : t \in I)$ denote a $\mathcal{Q}^-$ codeword where $I = \{0, 1, 2, \ldots, N-1\}$. Then, the codewords of $\mathcal{Q}$ are of the form

$$\underline{c} = (c(t) : t \in I^\star)\ , \tag{19}$$

where $I^\star \doteq I \cup \{\star\}$, $c(\star) = \delta$ is the arbitrary $\mathbb{Z}_4$-valued constant, and $c(t) = c^-(t) + \delta$ for all $t \in I$.

Theorem 5. *When r is an odd integer, the binary code $\mathcal{G}(\mathcal{Q}(r))$ derived via the Gray map is the Kerdock code $\mathcal{K}(r+1)$.*

Proof. Let $tr(x) = \sum_{i=1}^{r-1} x^{2^i}$ denote the usual binary trace mapping, and define $Q(x) = \sum_{i=1}^{s} tr(x^{1+2^i})$. One may show as in [2] that, when $r = 2s+1$, each codeword of $\mathcal{Q}(r)$ has 2-adic components

$$a(t) = tr(\eta\alpha^t) \oplus Q(\gamma\alpha^t) \oplus A \tag{20}$$

$$b(t) = tr(\gamma\alpha^t) \oplus B \tag{21}$$

$\forall (t \in I^\star)$, where α is a primitive element in the finite field $GF(2^r)$, $\alpha^\star \doteq 0$, and the elements $\eta, \gamma \in GF(2^r)$ and $A, B \in GF(2)$ are arbitrary.

From Kerdock's original construction [12], one finds that $a(t)$ and $a(t) \oplus b(t)$ are also the left and right halves of the codewords of $\mathcal{K}(r+1)$. Thus, $\mathcal{G}(\mathcal{Q}(r)) = \mathcal{K}(r+1)$ as claimed. □

In light of Theorem 5, it is natural to think of the Kerdock codes as linear quaternary codes and wonder about their quaternary duals. Let $\mathcal{P}_L$ denote the binary code that results from applying the Gray map to the quaternary code $\mathcal{Q}^\perp$, *i.e.*, $\mathcal{P}_L = \mathcal{G}(\mathcal{Q}^\perp)$. Then $\mathcal{P}_L$ is a nonlinear binary code of length $2(N+1) = 2^{r+1}$ and size $4^{2^r-(r+1)}$. From Theorems 1 and 4, we have the following corollary.

Corollary 6. *$\mathcal{P}(r)$ is a nonlinear, distance invariant, binary code whose weight distribution is related to that of the Kerdock code $\mathcal{K}(r+1)$ by the MacWilliams transform.*

Thus, $\mathcal{P}(r)$ is a nonlinear code of length $2(N+1) = 2^{r+1}$, size 4^{2^r-r-1}, and minimum Hamming distance 6. Like the extended Preparata code $\mathcal{P}(r+1)$, it is an optimal code capable of simultaneously correcting up to two-bit errors and detecting all three-bit errors. We shall refer to $\mathcal{P}_L$ as the *Preparata-like* code. As we shall see in the next section, the two codes $\mathcal{P}$ and $\mathcal{P}_L$ also share similar finite field transform descriptions.

2.5 Comparing the Preparata and Preparata-Like Codes

We define the finite field transform $\underline{\hat{a}} = (\hat{a}(\lambda) : \lambda \in I^\star)$ of the binary vector $\underline{a} = (a(t) : t \in I^\star)$ as follows:

$$\hat{a}(\lambda) = \sum_{t \in I^\star} a(t)\alpha^{\lambda t}, \qquad \forall \lambda \in I^\star , \tag{22}$$

where α is a fixed primitive element in $GF(2^r)$. We define the *half-convolution* of the sequence $\underline{\hat{a}}$ at lag λ by the equation

$$\mathcal{H}(\underline{\hat{a}}, \lambda) = \sum_{\substack{\lambda_1, \lambda_2 \in I \\ \lambda_1 \leq \lambda_2 \\ \lambda_1 + \lambda_2 = \lambda}} \hat{a}(\lambda_1)\hat{a}(\lambda_2) . \tag{23}$$

The summation is a half, rather than full, convolution because we exclude the cases $\lambda_1 > \lambda_2$.

Theorem 7. *The Preparata-like code $\mathcal{P}_L$ consists of all vectors $(\underline{a} \mid \underline{a} \oplus \underline{b})$ for which $\underline{a}, \underline{b} \in (\mathbb{Z}_2)^{N+1}$ satisfy*

$$\hat{b}(0) + b(\star) = 0 \tag{24}$$

$$\hat{b}(1) = 0 \tag{25}$$

$$\hat{a}(0) + a(\star) = \hat{b}(0)b(\star) + \mathcal{H}(\hat{\underline{b}}, 0) \tag{26}$$

$$\hat{a}(1) = \mathcal{H}(\hat{\underline{b}}, 1). \tag{27}$$

The proof of this theorem requires some knowledge of Galois ring theory, and may be found in [9]. It should be emphasized that this somewhat complicated characterization of the Preparata-like code is given primarily for comparison with a similar description of the extended Preparata code. In practice, one would prefer to work in the $\mathbb{Z}_4$-domain where the code $\mathcal{P}_L$ has a more natural description. Indeed, the dual code $\mathcal{Q}^\perp$ has a parity check matrix very similar to that of the even-weight subcode of the binary extended Hamming code. This is discussed in detail in Sect. 3.

For comparison with the Preparata-like code, a transform theoretic characterization of the extended Preparata codes $\mathcal{P}(r+1)$ can be readily derived from the simple description of the codes given by Baker, Van Lint, and Wilson [1]. In particular, the vector $(\underline{a} \mid \underline{a} \oplus \underline{b})$ is in $\mathcal{P}(r+1)$ iff the binary components $\underline{a} = (a(t) : t \in I^\star)$ and $\underline{b} = (b(t) : t \in I^\star)$ have transforms satisfying

$$\hat{b}(0) + b(\star) = 0 \tag{28}$$

$$\hat{b}(1) = 0 \tag{29}$$

$$\hat{a}(0) + a(\star) = 0 \tag{30}$$

$$(\hat{a}(1))^3 = \hat{b}(3). \tag{31}$$

The similarity in descriptions of the two codes $\mathcal{P}$ and $\mathcal{P}_L$ is evident. Interestingly, in one particular case, the two codes are the same!

Theorem 8. *For $r = 3$, $\mathcal{P}(r) = \mathcal{P}(r+1)$.*

Remark. When $r = 3$, both $\mathcal{K}(4)$ and $\mathcal{P}(4)$ coincide with $\mathcal{N}_{16}$, the Nordstrom-Robinson $(16, 256, 6)$ code. The discovery of $\mathcal{N}_{16}$ was published by Nordstrom and Robinson in 1967. Preparata's analysis of the distance properties of the code and his generalization to the codes that bear his name were both published in 1968. Kerdock's generalization of $\mathcal{N}_{16}$ was published in 1972. Thus, the Nordstrom-Robinson code can be generalized in one way to get the extended Preparata codes, in another way to get the Kerdock codes, and in yet another way to get our Preparata-like codes that are dual (in the quaternary sense) to the Kerdock codes!

3 Galois Ring Theoretic Descriptions

This section has two parts. The first provides a description of the internal structure of the Galois ring, of the trace function acting on the Galois ring, and of a Fourier-like transform defined on n-tuples over the Galois ring. The second part applies this Galois ring theory to derive binary and quaternary descriptions of $\mathcal{Q}$ and $\mathcal{Q}^{\perp}$ (and thus the Kerdock and Preparata-like codes). For a more detailed mathematical development of Galois rings, the reader is referred to [14].

3.1 Galois Rings

A polynomial $f(x) \in \mathbb{Z}_4\,[x]$ is called a *basic irreducible* if its reduction modulo 2 is an irreducible polynomial $\bar{f}(x) \in \mathbb{Z}_2\,[x]$. Similarly, $f(x)$ is a *primitive* basic irreducible if $\bar{f}(x)$ is primitive. The Galois ring $GR(4,r)$ denotes a Galois extension of dimension r over the integer ring $\mathbb{Z}_4$. Every such Galois ring is isomorphic to a quotient ring $\mathbb{Z}_4\,[x]/(f(x))$, where $f(x)$ is a basic irreducible polynomial in $\mathbb{Z}_4\,[x]$. It is convenient, although not essential, to assume that $f(x)$ is a primitive basic irreducible dividing $x^N - 1$ in $\mathbb{Z}_4\,[x]$, where as usual $N = 2^r - 1$. This is analogous to preferring to construct a finite field using a polynomial that is primitive as well as irreducible. Clearly, one can extend the reduction modulo 2 mapping defined on $\mathbb{Z}_4\,[x]$ to a surjective ring homomorphism $\mu : GR(4,r) \rightarrow GF(2^r)$.

Let $R = GR(4,r)$ and $K = GF(2^r)$. The Galois ring R is a local ring whose zero divisors form the unique maximal ideal $2R$. The units of R may be expressed [14] as a direct product of groups $R^* = G_1 \times G_2$, where G_1 is a cyclic group of order N and G_2 is a direct product of r cyclic groups of order 2. Let β be a root of $f(x)$. Then, β has multiplicative order N, cyclically generates G_1, and has modulo 2 projection $\alpha = \mu(\beta)$ that is primitive in K. As a $\mathbb{Z}_4$-module, $R = < 1, \beta, \beta^2, \ldots, \beta^{r-1} >$. It is not hard to show that the Frobenius mapping $\sigma_R : \beta \longrightarrow \beta^2$ generates the Galois automorphism group of R over $\mathbb{Z}_4$. For a more detailed discussion of these facts, see Sect. 3 of [2].

Let σ_K denote the corresponding Galois automorphism of K that maps α to α^2, and let $tr : K \rightarrow \mathbb{Z}_2$ denote the finite field trace mapping. Similarly, let $T : R \rightarrow \mathbb{Z}_4$ denote the Galois ring trace mapping defined by $T(\gamma) = \sum_{i=1}^{r} \sigma_R^i(\gamma)$.

3.2 Galois Ring Transforms

In analogy with finite field transform theory, we define the *Galois ring transform* $\underline{\hat{c}} = (\hat{c}(\lambda) : \lambda \in I^\star)$ of a quaternary sequence $\underline{c} = (c(t) : t \in I^\star)$ as follows:

$$\hat{c}(\lambda) = \sum_{t \in I^\star} c(t) \beta^{\lambda t} \ , \tag{32}$$

for all $\lambda \in I^\star$, where as usual $\beta^\star = 0$.

The inversion formula

$$c(t) = -\sum_{\lambda \in I^*} \hat{c}(\lambda)\beta^{-\lambda t} . \tag{33}$$

follows in the usual way from the fact that

$$\sum_{\lambda \in I} \beta^{\lambda} = \frac{1-\beta^N}{1-\beta} = 0 . \tag{34}$$

Let $c(t)$ have binary 2-adic components $a(t)$ and $b(t)$ satisfying $c(t) = 2a(t)+b(t)$. Since $a(t)$ and $b(t)$ are $\{0,1\}$-valued, we may talk of either their finite field transforms or their Galois ring transforms. In contexts where both are possible, we shall use $\hat{\cdot}$ for the Galois ring transform and $\tilde{\cdot}$ for the finite field transform. Note that $\hat{c}(\lambda) = 2\hat{a}(\lambda) + \hat{b}(\lambda)$.

Theorem 9. *Every sequence in Q^- has a unique representation as $s_\gamma(t) = T(\gamma\beta^t)$ for some element $\gamma \in R$, where $\beta \in R$ is a fixed root of the characteristic polynomial of the linear recurrence defining Q^- and $0 \le t \le N-1$. Conversely, every sequence of this form is a member of Q^-.*

Proof. See [2]. □

This trace characterization leads to a simple Galois ring theoretic description of the Kerdock code.

Theorem 10. *The quaternary code Q consists precisely of the set of all $\mathbb{Z}_4$-valued vectors $\underline{c} = (c(t) : t \in I^*)$ satisfying*

$$c(t) = T(\gamma\beta^t) + A , \tag{35}$$

where $\gamma \in R$ and $A \in \mathbb{Z}_4$ are arbitrary parameters.

Theorem 11. *The quaternary dual code $Q^\perp$ consists of all $\mathbb{Z}_4$-valued sequences $\underline{c} = (c(t) : t \in I^*)$ whose Galois ring transform satisfies*

$$\hat{c}(0) + c(\star) = 0 \tag{36}$$

$$\hat{c}(1) = 0. \tag{37}$$

Based on Theorems 10 and 11, the codes Q and $Q^\perp$ may be regarded as quaternary generalizations of the first-order and $(r-2)^{\text{th}}$-order Reed-Muller codes. In fact, all of the Reed-Muller codes have natural quaternary analogs (for more on this, see [9]).

4 Quaternary Decoding Algorithms

Although in the theoretical development, we have made a distinction between the quaternary codes $\mathcal{Q}$ and $\mathcal{Q}^{\perp}$ and their associated nonlinear binary codes $\mathcal{K}$ and $\mathcal{P}_L$, from a practical viewpoint, they are really two different descriptions of the same codes.

It is natural to expect that the simpler quaternary descriptions of the Kerdock and Preparata-like codes would lead to simpler decoding algorithms that work in the $\mathbb{Z}_4$-domain. This is indeed the case for the Preparata-like codes. There is an optimal syndrome decoder that provides correction of all two-bit errors and detection of all three-bit errors. In the case of the Kerdock codes, the quaternary viewpoint leads to a soft-decision decoding algorithm that is comparable in complexity, to previously known binary techniques, derived in a similar fashion from the binary descriptions of the codes.

5 Update

The discoveries about the Kerdock and Preparata codes are in a paper [7] presented by Hammons and Kumar at the IEEE International Symposium on Information Theory (San Antonio January 1993, but submitted in June 1992), in Hammons' dissertation [8], and in a manuscript submitted in early November 1992 to the *IEEE Transactions on Information Theory*, but now replaced by [9]. Hammons and Kumar also realized in June 1992 that the $\mathbb{Z}_4$ Kerdock and 'Preparata' codes could be generalized to give the quaternary Reed-Muller codes $QRM(r, m)$ described in [9].

In late October 1992, Calderbank, Sloane and Solé submitted a research announcement (now replaced by [3]) to the *Bulletin of the American Mathematical Society*, also containing the discoveries about the Kerdock and Preparata codes, as well as results concerning the existence of quaternary versions of Reed-Muller, Golay and Hamming codes. They discovered the quaternary versions of the Goethals and Delsarte-Goethals codes in early November.

The two teams (Hammons-Kumar and Calderbank-Sloane-Solé) worked independently until the middle of November 1992, when discovering the considerable overlap between their work, they decided to join forces. The common starting point for the two independent discoveries was the formula (page 1107 of [2]) providing a base-2 expansion for the four-phase sequence family, Family $\mathcal{A}$. Ref. [9] is a compositum of the results obtained by the 5 authors.

The discovery that the Nordstrom-Robinson code is a quaternary version of the octacode was made by Forney, Sloane and Trott in early October 1992 and is described in [5]. It can be shown [4] that several of the well-known binary nonlinear single-error-correcting codes also have a simpler description as codes over $\mathbb{Z}_4$ (although here, the corresponding $\mathbb{Z}_4$-codes are nonlinear). Large sequence families for code-division multiple-access (CDMA) that are supersets of the near-optimum four-phase families described above, and which are related to the Delsarte-Goethals codes are investigated in [13]. Some related constructions for lattices may be found in [19].

References

1. R. D. Baker, J. H. Van Lint, and R. M. Wilson: On the Preparata and Goethals Codes, IEEE Trans. on Inform. Theory **IT-29, 3** (May 1983), 342–345.
2. S. Boztaş, A. R. Hammons, and P. V. Kumar: 4-phase Sequences with Near Optimum Correlation Properties, IEEE Trans. on Inform. Theory, **IT-38, 3**, (May 1992), 1101–1113.
3. A. R. Calderbank, A. R. Hammons, P. V. Kumar, N. J. A. Sloane and P. Solé, A Linear Construction for Certain Kedock and Preparata Codes, Bull. Amer. Math. Soc., submitted.
4. J. H. Conway and N. J. A. Sloane, Quaternary Constructions for the Binary Codes of Julin, Best and Others, preprint.
5. G. D. Forney, Jr., N. J. A. Sloane and M. D. Trott, The Nordstrom-Robinson Code is the Binary Image of the Octacode, Proc. DIMACS/IEEE Workshop on Coding and Quantization, 1993, to appear.
6. A. R. Hammons and P. V. Kumar: A Linear Quadriphase Interpretation of the Binary Kerdock Code, Tech. Rep. **CSI-92-04-01**, (April 1992), Comm. Sciences Inst., Univ. of South. Calif., Los Angeles, CA.
7. A. R. Hammons and P. V. Kumar, On the Apparent Duality of the Kerdock and Preparata Codes, IEEE Int. Symp. on Inform. Theory, San Antonio, TX January 1993.
8. A. R. Hammons and P. V. Kumar, On four-phase sequences with low correlation and their relation to Kerdock and Preparata Codes, Ph.D. Diss., Univ. of South. Calif., Los Angeles, CA Nov. 1992.
9. A. R. Hammons, P. V. Kumar, A.R. Calderbank, N.J.A. Sloane and P. Solé, The $\mathbb{Z}_4$-Linearity of Kerdock, Preparata, Goethals and Related Codes, submitted to the IEEE Trans. on Inform. Theory.
10. T. W. Hungerford: Algebra, Grad. Texts in Math., **73**, Springer-Verlag, New York, 1974.
11. W. M. Kantor, On the Inequivalence of Generalized Preparata Codes, IEEE Trans. on Inform. Theory, **IT-29, 3**, May 1983, 345–348.
12. A. M. Kerdock, A Class of Low-Rate Nonlinear Binary Codes, Inform. and Control, **20**, 1972, 182–187.
13. P. V. Kumar, A. R. Hammons and T. Helleseth, Large Sequence families for CDMA, IEEE Trans. on Inform. Theory, to be submitted.
14. B. R. MacDonald: Finite Rings with Identity, Marcel Dekker, Inc., New York, 1974.
15. F. J. MacWilliams and N. J. A. Sloane: The Theory of Error-Correcting Codes, North-Holland Publishing Company, Amsterdam, 1977.
16. F. P. Preparata: A Class of Optimum Nonlinear Double-Error-Correcting Codes, Inform. and Control, **13**, 1968, 378–400.
17. P. Solé, A Quarternary Cyclic Code and a Family of Quadriphase Sequences with Low Correlation Properties, Lec. Notes in Comp. Science. **388**, 1989, 193–201.
18. P. Udaya and M. U. Siddiqi: Large Linear Complexity Sequences over $\mathbb{Z}_4$ for Quadriphase Modulated Communication Systems Having Good Correlation Properties, IEEE Int. Symp. on Inform. Theory, Budapest, Hungary, June 1991.
19. K. Yang, A. R. Hammons, and P. V. Kumar: Constructions for Lattices Based on 4-Phase Sequences, Technical Report **CSI-92-06-02**, (June 1992), Comm. Sciences Inst., Univ. of South. Calif., Los Angeles, CA, June 1992.

Bounds for Codes as Solutions of Extremum Problems for Systems of Orthogonal Polynomials

Vladimir Levenshtein

Keldysh Institute for Applied Mathematics
Russian Academy of Sciences
Miusskaya Sq. 4, Moscow, 125047, Russia
FAX: 7-095-9720737

Abstract. In this survey paper codes in finite and infinite polynomial metric spaces with given values of parameters are considered. We are especially interested in such parameters as the minimal distance and the maximal strength of the design generated by a code. We discuss some general bounds on the size of the codes, that is bounds satisfied by any code. For this reason they can be used to prove that too good codes do not exist. These bounds are obtained as solutions of certain extremum problems for the system of orthogonal polynomials which corresponds to the polynomial metric space. As applications we have the best known bounds for some problems in geometry, coding theory, number theory and complexity theory.

1. Introduction. Some Properties of Polynomial Metric Spaces.

The notion of a polynomial metric space is one of the main achievements of the Delsarte theory [6]. It allows him to introduce a general concept of a design which brings to light the unified nature of different combinatorial objects and many known bounds for them. Furthermore this notion gives a possibility to describe all nonnegative-definite matrices (for infinite case, all nonnegative-definite degenerate kernels) $F(x,y)$ which depend only on distance $d(x,y)$ by using a system of orthogonal polynomials. As a result a method of obtaining general bounds on the size of codes and designs is reduced to some extremum problems for a system of orthogonal polynomials. To describe solutions of these problems and corresponding bounds is the main goal of the paper.

In this paper we call *polynomial metric spaces* both finite polynomial metric spaces, which are P- and Q-polynomial association schemes [6,3,5,36], and infinite ones [43,7,9,15,32,25,13,21] which are the connected two-point homogeneous spaces totally classified by Wang [43] as the Euclidean unit spheres, the real, complex and quaternionic projective spaces and the Cayley projective plane. There is no complete classification for the finite case, but we mention the most important examples:

(a) The Hamming space $H(n,r)$ consists of all n-tuples with components from the alphabet $\{0,1,\dots,r-1\}$, and the Hamming distance between two elements $x,y \in H(n,r)$ equals the number of positions where they differ.

(b) The Johnson space $J(n,w)$ is the set of all w-subsets of a n-set, and the distance between two elements $x,y \in J(n,w)$ is defined as $w - |x \cap y|$.

(c) The Grassmann space $J(n,w,q)$ is the set of all w-dimensional subspaces of the vector space F_q^n over the finite field F_q, and the distance between two elements $x,y \in J(n,w,q)$ is $w - dim\ (x \cap y)$.

Every polynomial metric space $\mathfrak{M}$ is a compact metric space with certain additional properties. The metric defined on $\mathfrak{M}$ and the diameter of $\mathfrak{M}$ will be denoted by $d(x,y)$ and by $D=D(\mathfrak{M}) = \max_{x,y\in\mathfrak{M}} d(x,y)$. In the finite case $\mathfrak{M}$ is a (Q-polynomial) distance-regular graph [6,3,5] and the distance function $d(x,y)$ takes values $0,1,\dots,D$.

Each polynomial metric space $\mathfrak{M}$ is endowed with a unique normalized measure $\mu(\ )$ so that the measure of an arbitrary closed metric sphere of radius d, $0\leq d\leq D$, doesn't depend on its center and so can be denoted simply by $\mu(d)$, in particular $\mu(D)=1$ (we hope that using $\mu(\ )$ both as a function of subsets of $\mathfrak{M}$ and as one of a real d does not cause a confusion). In the finite case μ can be the usual normalized counting measure, and $\mu(d)=|\mathfrak{M}|^{-1}\sum_{i\leq d} k_i$, where k_i is the number of points at distance i from an arbitrary fixed point of $\mathfrak{M}$, i.e. $\mu(d)$ is a step function having $D+1$ jumps. $\mu(d)$ is absolutely continuous in the infinite case. In the particular, for the Euclidean unit sphere $S^{n-1}= \{x=(x_1,\dots,x_n)\in\mathbb{R}^n: \sum_{i=1}^{n} x_i^2\}$,

$$\mu(d)=\frac{\Gamma(n/2)}{\Gamma((n-1)/2)\Gamma(1/2)}\int_{1-d^2/2}^{1}(1-v^2)^{(n-3)/2}dv.$$

Consider the Hilbert space $\mathcal{L}_2(\mathfrak{M},\mu)$ of complex-valued quadratic integrable functions with the usual inner product

$$<u,v> = \int_{\mathfrak{M}} u(x)\,\overline{v(x)}\,d\mu(x).$$

For any polynomial metric space, $\mathcal{L}_2(\mathfrak{M},\mu)$ decomposes into a direct sum of mutually orthogonal finite dimensional subspaces V_i, $i=0,1,\ldots$, where V_0 is the one-dimensional subspace of constant functions and the decomposition consists of exactly $D+1$ members if $\mathfrak{M}$ is finite. Moreover there exist polynomials $Q_i(t)$ of degree i, $i=0,1,\ldots$, of a real variable t and a continuous real-valued function $t(d)$ taking different values at the different values of $d(x,y)$ such that for all $x,y \in \mathfrak{M}$ and $i=0,1,\ldots$

$$Q_i(t(d(x,y))) = \frac{1}{r_i} \sum_{j=1}^{r_i} v_{i,j}(x)\overline{v_{i,j}(y)} \tag{1}$$

holds, where $\dim V_i = r_i$, $i=0,1,\ldots$, and the system $\{v_{i,j}(x)$, $j=1,2,\ldots,r_i\}$ constitutes an orthonormal basis of V_i. Sometimes $\mathfrak{M}$ is said to a polynomial space with substitution $t(d)$.

Throughout this paper, we restrict ourselves to the case of polynomial metric spaces with *monotone substitutions* $t(d)$. Since $t(d)$ can be chosen up to a linear transformation we can assume without loss of generality that it satisfies

$$t(D(\mathfrak{M}))=-1 \le t(d(x,y)) \le t(0)=1 \text{ for all } x,y\in\mathfrak{M}. \tag{2}$$

Then $t(d)$ will be referred to as a *standard substitution* for $\mathfrak{M}$. The inverse of $t(d)$ will be denoted by $t^{-1}(\quad)$, i.e. $t^{-1}(t)=d$ if and only if $t=t(d)$. For example, in the case of the Hamming space $H(n,r)$ the corresponding system of polynomials $Q_i(t)$ are obtained from the Krawtchouk polynomials by the standard substitution $t(d)=1-2d/n$, and for the Euclidean sphere $\mathbf{S}^{n-1}$, the standard substitution $t(d)=1-d^2/2$ transforms the Euclidean distance between points into inner product and the polynomials $Q_i(t)$ are equal to the Gegenbauer polynomials.

Now we recall several fundamental facts about the polynomial system $Q_i(t)$ related to a polynomial metric space. From (1) and from our assumption (2) follows that for all $x,y \in$

$\mathfrak{M}$ and $i,j=0,1,\ldots$,

$$r_i \int_{\mathfrak{M}} Q_i(t(d(x,y)))\, Q_j(t(d(y,z)))\, d\mu(y) = \delta_{i,j} Q_i(t(d(x,z)))$$

holds ($\delta_{i,j}$ is the Kronecker symbol). Taking $x=z$ and defining $\nu(t)$ by $\nu(t(d))=1-\mu(d)$ we can see that the polynomials $Q_i(t)$ satisfy the following condition of orthogonality and normalization

$$r_i \int_{-1}^{1} Q_i(t)Q_j(t)d\nu(t) = \delta_{i,j}, \quad Q_i(1)=1, \quad i=0,1\ldots , \tag{3}$$

where $\nu(t)$ is a nondecreasing leftcontinuous function and the integral is taken in the Lebesgue-Stieltjes sense. We notice that for a finite $\mathfrak{M}$, $\nu(t)$ is a step function with $D+1$ jumps and (3) turns into

$$r_i |\mathfrak{M}|^{-1} \sum_{d=0}^{D} Q_i(t(d))Q_j(t(d))k_d = \delta_{i,j}, \quad Q_i(1)=1, \quad i=0,1\ldots,D \tag{4}$$

where k_d is the number of points at distance d from a fixed one. In infinite case $\nu(t)$ is absolutely continuous and (3) takes the form

$$r_i \int_{-1}^{1} Q_i(t)Q_j(t)w(t)dt = \delta_{i,j}, \quad Q_i(1)=1, \quad i=0,1\ldots , \tag{5}$$

where $w(t)=\nu'(t)$. For a polynomial metric space, the expression

$$f(t)=\sum_{i=0}^{l} f_i Q_i(t) \text{ with } f_i \geq 0, \ i=1,\ldots,l \ , \ l=0,1,\ldots \tag{6}$$

yields a general form of a continuous function $f(t)$ such that $f(t(d(x,y)))$ is a nonnegative-definite matrix on $\mathfrak{M}$ (in finite case) or a nonnegative-definite degenerate kernel on $\mathfrak{M}$ (in infinite case), i.e. can be represented as $\sum_{i=1}^{r} u_i(x)\overline{u_i(y)}$ with some complex functions $u_1(x),\ldots,u_r(x)$ on $\mathfrak{M}$. Since $r_0=1$ and

$$f_0 = \int_{-1}^{1} f(t)d\nu(t) = \int_{\mathfrak{M}}\int_{\mathfrak{M}} f(t(d(x,y)))d\mu(x)d\mu(y),$$

for any finite subset $W \subseteq \mathfrak{M}$ and any polynomial (6) from (1) follows the "mean inequality"

$$\frac{1}{|W|^2}\sum_{x,y\in W} f(t(d(x,y))) \geq \int_{\mathfrak{M}}\int_{\mathfrak{M}} f(t(d(x,y)))d\mu(x)d\mu(y). \qquad (7)$$

2. Code Parameters and Extremum Problems for the System of Orthogonal Polynomials.

A finite nonempty subset W of a metric space $\mathfrak{M}$ is said to be a *code.* The code distance, i.e. the minimal distance between distinct elements of W, and the diameter of W, i.e. the maximal distance between elements of W, will be denoted by $d(W)$ and $D(W)$ respectively. Moreover $l(W)$ denotes the number of distinct distances between code elements. A code $W \subseteq \mathfrak{M}$ with $d(W) \geq d$ will be referred to as a *d-code.* Denote by $A(\mathfrak{M},d)$ the maximum cardinality of a d-code in $\mathfrak{M}$. The covering radius of W is denoted by $\rho(W)$ and it is defined by $\rho(W) = \max_{y\in\mathfrak{M}} \min_{x\in W} d(x,y)$. In other words, $\rho(W)$ is the minimum number ρ such that the metric balls of radius ρ with centers in elements of W cover the whole space $\mathfrak{M}$.

For a polynomial metric space $\mathfrak{M}$ with substitution $t(d)$ we introduce the following denotations. A nonzero polynomial $f(t)$ is called *annihilating* for a code $W \subseteq \mathfrak{M}$ if $f(t(d(x,y)))=0$ for every pair $x,y\in W$ $(x\neq y)$. The annihilating polynomial of minimal degree for W (i.e. of degree $l(W)$) is denoted by $f_W(t)$. If $y\in\mathfrak{M}\setminus W$ then let $f_{y,W}(t)$ be a nonzero polynomial of minimal degree satisfying $f_{y,W}(t(d(x,y)))=0$ for all $x\in W$. Clearly, $f_W(t)$ and $f_{y,W}(t)$ are determined up to a constant factor. Therefore we assume afterwards that $f_W(t)=f(t)$ (and respectively $f_{y,W}(t)=f(t)$) means only that the sets of zeroes of these polynomials coincide).

A code $W \subseteq \mathfrak{M}$ is called a τ*-design* (τ = 0,1, ...) if

$$\sum_{x\in W} v(x) = 0 \text{ for all } v(x) \in \bigcup_{i=1}^{\tau} V_i. \qquad (8)$$

Let $\tau(W)$ be the maximal number τ for which W constitutes a τ-design. This definition of τ-design [6,8-10,14] is a generalization of the classical ones [4]. Indeed, a subset W of the Johnson space $J(n,w)$ is a τ-design if every τ-subset of the basic n-set is contained in the same number of members of W. In the Grassmann space $J(n,w,q)$ a subset W forms a τ-design if every τ-dimensional linear subspace of the vector space F_q^n is contained in the same number of elements of W. A τ-design W in the Hamming space $H(n,r)$ is nothing other than an orthogonal array of strength τ, i.e. if we associate W with an array whose rows are all the n-tuples from W then every τ of its columns contain all ordered τ-tuples exactly $|W|r^{-\tau}$ times. Finally we notice that a finite subset $W \subset S^{n-1}$ forms a τ-design [9] if and only if the cubature formula

$$\int_{S^{n-1}} f(x)\, d\mu(x) \cong \frac{1}{|W|} \sum_{x \in W} f(x)$$

is exact for every polynomial $f(x)$ of degree at most τ in variables $x_1, x_2, \ldots, x_n$, where $x=(x_1,x_2,\ldots,x_n) \in S^{n-1}$.

By (1) the property (8) is equivalent to the condition that for any polynomial $f(t)$ in a real t of degree at most τ there holds the "mean equality"

$$\frac{1}{|W|^2} \sum_{x,y \in W} f(t(d(x,y)))=f_0= \int_{\mathfrak{M}}\int_{\mathfrak{M}} f(t(d(x,y)))d\mu(x)d\mu(y). \qquad (9)$$

This shows that, for uniform distributions, a τ-design is a "good approximation" of the whole space $\mathfrak{M}$ while the parameter τ characterizes the degree of such approximation. This is a base of using τ-designs in randomized algorithms. The problem is to construct a τ-design of the minimum size.

Let $W \subseteq \mathfrak{M}$ be a code and let $f(t)=\sum_{i=0}^{l} f_i Q_i(t)$ be an arbitrary polynomial of some degree l ($l \leq D$ for finite $\mathfrak{M}$). Then from (1) and (2) follows the equality

$$|W|f(1)+ \sum_{\substack{x,y \in W \\ x \neq y}} f(t(d(x,y)))=|W|^2 f_0+\sum_{i=1}^{l} \frac{f_i}{r_i} \sum_{j=1}^{r_i} |\sum_{x \in W} v_{i,j}(x)|^2 \qquad (10)$$

which proved to be fundamental [7] for the investigation of cardinality of codes and designs in polynomial metric spaces.

If $W \subseteq \mathfrak{M}$ and $\tau(W) \geq \tau$, then for any polynomial $f(t)=\sum_{i=0}^{l} f_i Q_i(t)$ such that

$$f_0>0,\ f_i \leq 0 \text{ for } i=\tau+1,...,l, \tag{11}$$

$$f(1)>0,\ f(t) \geq 0 \quad \text{for } -1 \leq t < 1, \tag{12}$$

(9) yields the Delsarte inequality for τ -designs

$$|W| \geq \Omega(f) = \frac{f(1)}{f_0}. \tag{13}$$

This means that any polynomial satisfying (11)-(12) gives rise to some lower bound on the size of a τ-design. So for the system $\{Q_i(t)\}$ of orthogonal polynomials we have

τ-design problem: *Find the supremum of the ratio $\Omega(f)$ over the class of polynomials $f(t)$ satisfying the properties (11)-(12).*

If $W \subseteq \mathfrak{M}$ is a d-code and $t(d)=s$, then for any polynomial $f(t)=\sum_{i=0}^{l} f_i Q_i(t)$ such that

$$f_0>0,\ f_i \geq 0 \quad \text{for } i=1,...,l, \tag{14}$$

$$f(1)>0,\ f(t) \leq 0 \quad \text{for } -1 \leq t \leq s, \tag{15}$$

(9) yields the Delsarte inequality for d-codes

$$|W| \leq \Omega(f) = \frac{f(1)}{f_0}. \tag{16}$$

Now any polynomial satisfying (14)-(15) gives rise to some upper bound on the size of a d-code. In particular, the polynomial t-s of the first degree for different polynomial metric spaces leads to the bounds known as Plotkin, Johnson, Rankin , Grey-Rankin bounds [23]. Considerable progress was made in 1971-1974 by Sidelnikov [28-30] who, for different spaces, proved and used the "mean inequality" (7) for polynomials of the form t^l (the power sum method). It allowed him for the first time to obtain a asymptotic improvement of the Bassalygo-Elias bound on rate of codes in the Hamming space with correction of a given fraction of errors and the Rogers bound on logarithm of the highest packing density of n-dimensional Euclidean space [29, 30]. The next significant step was made in 1977 when McEliece, Rodemich, Rumsey and

Welch [24] suggested polynomials satisfying (14)-(15) which yield stronger bounds than the power sum method for the Hamming and Johnson spaces. In 1978 Kabatiansky and Levenshtein [15] used the same polynomials for the Euclidean spheres and other polynomial metric spaces. So for the system $\{Q_i(t)\}$ of orthogonal polynomials there is

***d*-code problem**: *Find the infimum of the ratio* $\Omega(f)$ *over the class of polynomials* $f(t)$ *satisfying the properties* (14)-(15).

Afterwards we describe a solution (Levenshtein [17], 1978) of the *d*-code problem over class of polynomials degree of which is limited from above by a certain function of *d*.

One more extremum problem for the system $\{Q_i(t)\}$ is connected with evaluation of the covering radius of τ-designs. Notice that by (10) and (8) a code W a τ-design in $\mathfrak{M}$ if and only if for every polynomial $f(t)$ of degree at most τ and for every $y\in\mathfrak{M}$

$$\frac{1}{|W|}\sum_{x\in W} f(t(d(x,y))) = f_0.$$

(compare with (9)). Therefore if a polynomial $f(t)$ possesses the properties

$$\deg f(t)\le\tau,\ f_0>0,\ f(t)\le 0 \text{ for } -1\le t\le s, \tag{17}$$

then for any τ-design W and any $y\in\mathfrak{M}$ there exists $x\in W$ such that $f(t(d(x,y)))>0$ and hence $t(d(x,y))>s$. This implies that $\rho(W) < t^{-1}(s)$ and leads to

τ-design covering problem: *Find the supremum of numbers* s, $-1\le s<1$, *such that there exists a polynomial* $f(t)$ *satisfying the properties* (17).

3. General Bounds for Codes and Designs. To formulate known results for codes and designs, following [21], we consider certain adjacent orthonormal systems $\{Q_i^{a,b}(t),\ i=0,1,\dots\}$ $(a,b=0,1,\dots)$ of the system $\{Q_i(t),\ i=0,1,\dots\}$. First we introduce the weight functions $\nu^{a,b}(t)$ as follows: If $\nu(t)$ is a step function (case of finite $\mathfrak{M}$) then let $\nu^{a,b}(t)$ be also a step function obtained from $\nu(t)$ by multiplying its jumps k_i in the points $t(i)$ $(i=0,1,\dots,D)$ by $c^{a,b}(1-t(i))^a(1+t(i))^b$ and if $\nu(t)$ is absolutely continuous then let $\nu^{a,b}(t)$ be also an absolutely continuous function satisfying $(\nu^{a,b}(t))' = c^{a,b}(1-t)^a(1+t)^b w(t)$. In both cases $c^{a,b}$ is a positive constant normalizing the Lebesgue-Stieltjes measure so that

$$\int_{-1}^{1} d\nu^{a,b}(t) = c^{a,b} \int_{-1}^{1} (1-t)^a(1+t)^b d\nu(t) = 1$$

holds. Now, the measure uniquely determines the sequences of positive constants $r_i^{a,b}$ and polynomials $Q_i^{a,b}(t)$ of degree i such that

$$r_i^{a,b} c^{a,b} \int_{-1}^{1} Q_i^{a,b}(t) Q_j^{a,b}(t) d\nu^{a,b}(t) = \delta_{i,j}, \quad Q_i^{a,b}(1) = 1, \quad i=0,1\ldots \tag{18}$$

If $\mathfrak{M}$ is finite then the system $Q_i^{a,b}(t)$ is finite as well. By (2), in this case it consists of $D - 1 + \delta_{a,0} + \delta_{b,0}$ elements. Moreover, it is clear that $Q_i^{0,0}(t) = Q_i(t)$ and $r_i^{0,0} = r_i$.

Let $t_i^{a,b}$ be the largest zero of the polynomial $Q_i^{a,b}(t)$. In [19] it was proved that for any $k=1,2,\ldots$

$$t_{k-1}^{1,1} < t_k^{1,0} < t_k^{1,1}, \text{ where } t_0^{1,1} = -1.$$

For the kernels

$$K_i^{a,b}(s,t) = \sum_{j=0}^{i} r_j^{a,b} Q_j^{a,b}(s) Q_j^{a,b}(t) \quad i=0,1,\ldots\ ,\ . \tag{19}$$

the following Christoffel-Darboux formula is valid

$$(s-t)K_i^{a,b}(s,t) = r_i^{a,b} m_i^{a,b} (Q_{i+1}^{a,b}(s) Q_i^{a,b}(t) - Q_i^{a,b}(s) Q_{i+1}^{a,b}(t)) \tag{20}$$

where $m_i^{a,b}$ denotes the the ratio of the highest coefficient of $Q_i^{a,b}(t)$ to that of $Q_{i+1}^{a,b}(t)$. From (18)-(20) follows

$$Q_i^{0,1}(t) = \frac{K_i^{0,0}(t,-1)}{K_i^{0,0}(1,-1)}, \quad Q_i^{1,0}(t) = \frac{K_i^{0,0}(t,1)}{K_i^{0,0}(1,1)}, \quad Q_i^{1,1}(t) = \frac{K_i^{0,1}(t,1)}{K_i^{0,1}(1,1)} \tag{21}$$

for all i for which the functions in (21) are defined [21].

Bound for t-designs. (Delsarte [6], Dunkl [11]). *For every* $W\subseteq\mathfrak{M}$ *holds*

$$|W|\geq\begin{cases}\sum_{i=0}^{k} r_i & \text{if } \tau(W)=2k, \qquad (22)\\[2ex] \left(1-\dfrac{Q_k^{1,0}(-1)}{Q_{k+1}(-1)}\right)\sum_{i=0}^{k} r_i & \text{if } \tau(W)=2k+1. \qquad (23)\end{cases}$$

Moreover the bounds are attained if and only if $f_W(t)=Q_k^{1,0}(t)$ *and* $f_W(t)=(t+1)Q_k^{1,1}(t)$ *respectively.*

Delsarte [6] used the polynomial $f(t)=(Q_k^{1,0}(t))^2$ in (13) in order to obtain (22). Dunkl [11] used the polynomial $f(t)=(t+1)(Q_k^{1,1}(t))^2$ in (13) in order to obtain (23). As it will be explained later these polynomials are solutions of the τ-design problem over the class of polynomials of degree at most τ. τ-designs for which the bounds (22)-(23) are attained are called *tight* τ-designs in $\mathfrak{M}$.

Bound for d-codes. (Levenshtein [17-21]). *Let* $W\subset\mathfrak{M}$ *and* $s=t(d(W))$. *Then*

$$|W|\leq B(s)=\begin{cases}\left(1-\dfrac{Q_{k-1}^{1,0}(s)}{Q_k(s)}\right)\sum_{i=0}^{k-1} r_i & \text{if } t_{k-1}^{1,1}\leq s\leq t_k^{1,0}, \qquad (24)\\[2ex] \left(1-\dfrac{Q_k^{1,0}(s)}{Q_k^{0,1}(s)}\right)\sum_{i=0}^{k} r_i & \text{if } t_k^{1,0}<s<t_k^{1,1}; \qquad (25)\end{cases}$$

in particular

$$|W|\leq B(t_k^{1,1})=\left(1-\frac{Q_k^{1,0}(-1)}{Q_{k+1}(-1)}\right)\sum_{i=0}^{k} r_i \qquad \text{if } s=t_k^{1,1}, \qquad (26)$$

$$|W|\leq B(t_k^{1,0})=\sum_{i=0}^{k} r_i \qquad \text{if } s=t_k^{1,0}. \qquad (27)$$

Moreover $B(s)$ *is a continuous increasing function, and the bounds* (24)-(25) *are attained for a code* $W\subset\mathfrak{M}$ *if and only if*

$$f_W(t)=\begin{cases}(t-s)K^{1,0}_{k-1}(t,s) & if \quad t^{1,1}_{k-1}\le s\le t^{1,0}_{k},\\ (t+1)(t-s)K^{1,1}_{k-1}(t,s) & if \quad t^{1,0}_{k}<s<t^{1,1}_{k},\end{cases}$$

and

$$\tau(W)=\begin{cases}2k\text{-}1 & if \quad t^{1,1}_{k-1}\le s<t^{1,0}_{k},\\ 2k & if \quad t^{1,0}_{k}\le s<t^{1,1}_{k}.\end{cases}$$

The bounds for d-codes are obtained by using in (16) the polynomials

$$f^{(s)}(t)=\begin{cases}f^{(s)}_{2k-1}(t) & \text{if} \quad t^{1,1}_{k-1}\le s\le t^{1,0}_{k},\\ f^{(s)}_{2k}(t) & \text{if} \quad t^{1,0}_{k}<s<t^{1,1}_{k},\end{cases} \tag{28}$$

where $s=t(d)$ and

$$f^{(s)}_{2k-1}(t)=(t-s)(K^{1,0}_{k-1}(t,s))^2, \tag{29}$$

$$f^{(s)}_{2k}(t)=(t+1)(t-s)(K^{1,1}_{k-1}(t,s))^2, \tag{30}$$

i.e. $B(s)=\Omega(f^{(s)}(t))$. These polynomials satisfy (15), and for $t^{1,1}_{k-1}\le s\le t^{1,0}_{k}$ the polynomial $f^{(s)}_{2k-1}(t)$ satisfies (14) so that the bounds (24) and their significant special cases (26) and (27) are valid [20] in any polynomial metric space $\mathfrak{M}$ with a standard substitution. The property (14) for the polynomial $f^{(s)}_{2k}(t)$ where $t^{1,0}_{k}<s<t^{1,1}_{k}$ and the bounds (25) were proved for antipodal spaces [20], for all infinite polynomial metric spaces and for the so-called finite decomposable polynomial metric spaces. [21]. Notice that almost all classical distance-regular graphs, in particular $H(n,r)$, $J(n,w)$, $J(n,w,q)$ are decomposable. We notice also that the above mentioned bound is an improvement of

the bounds [24,15] obtained by using in (16) the polynomials $(t-s)(K_k(t,s))^2$ for $t_k<s<t_{k+1}$. Moreover, we shall see that for any polynomial $f(t)$ such that $f(t)\leq 0$ for $-1\leq t\leq s$ and $\deg f(t)\leq \deg f^{(s)}(t)$, the inequality $\Omega(f)\geq B(s)$ holds.

Bounds for Packing and Covering Radii of τ-designs. *For every code $W\subset\mathfrak{M}$*

$$d(W)\leq\begin{cases} t^{-1}(t_k^{1,0}) & \text{if } \tau(W)=2k, \quad (31)\\ t^{-1}(t_k^{1,1}) & \text{if } \tau(W)=2k+1, \quad (32)\end{cases}$$

$$\rho(W)\leq\begin{cases} t^{-1}(t_k^{0,1}) & \text{if } \tau(W)=2k, \quad (33)\\ t^{-1}(t_k^{0,0}) & \text{if } \tau(W)=2k+1. \quad (34)\end{cases}$$

Moreover the bounds (31) *and* (32) *are attained if and only if* $f_W(t)=Q_k^{1,0}(t)$ *and* $f_W(t)=(t+1)Q_k^{1,1}(t)$ *respectively, and the bounds* (33) *and* (34) *are attained if and only if there exists a point* $y\in\mathfrak{M}\backslash W$ *such that* $f_{y,W}(t)=(t+1)Q_k^{0,1}(t)$ *and* $f_{y,W}(t)=Q_k^{0,0}(t)$ *respectively.*

The bounds (31)-(32) are the consequence of the bounds for τ-designs and d-codes given above. The bounds (33)-(34) for the Hamming space are due to Tietavainen [39-41]. The bounds (33)-(34) in the general case and the necessary and sufficient conditions for their attainability were found by Fazekas and Levenshtein [12] as solutions of the τ-design covering problem. Notice that bounds (31)-(32) are attained for all tight designs. In [12] it was proved that the bound (33) is attained for any tight $2k$-design in the cases of the binary Hamming spaces and the Euclidean spheres.

4. Optimality of the Bounds in the Framework of the Method under Consideration.

A proof of optimality of the bounds for d-codes and τ-designs in a polynomial metric space is based on the following

Theorem 1. (Sidelnikov [31], Levenshtein [21]). *Let for any s, $-1\leq s<1$,*

$$h(s)=\begin{cases} 2k-1 & if\ t_{k-1}^{1,1}\le s<t_k^{1,0} \\ 2k & if\ t_k^{1,0}\le s<t_k^{1,1} \end{cases}.$$

Then all distinct zeroes $\alpha_1,\ldots,\alpha_{k(s)}$ *of the polynomial* $f^{(s)}(t)$, *defined by (28), belong to the interval* $[-1,s]$, *and for any polynomial* $f(t)=\sum_{i=0}^{l} f_i Q_i(t)$ *of degree* $l\le h(s)$ *the following equality holds:*

$$f_0=(B(s))^{-1}f(1)+\sum_{i=1}^{k(s)} \rho_i f(\alpha_i), \tag{35}$$

where $B(s)$ *is defined by (24)-(25) and all coefficients* ρ_i *are positive.*

From (35) follows the optimality of the polynomials $f^{(s)}(t)$ with $s=t(d)$ for the d-code problem over the class of polynomials of degree at most $h(s)$. In [21] it was shown that the polynomials $f^{(s)}(t)$ for $s=t_k^{1,0}$ and $s=t_k^{1,1}$ have the same set of distinct zeroes as $(Q_k^{1,0}(t))^2$ and $(t+1)(Q_k^{1,1}(t))^2$ respectively. Therefore from (35) follows also that these polynomials are optimal for the τ-design problem ($\tau=2k$ and $\tau=2k+1$ respectively) over the class of polynomials of degree at most τ. The latter result is due to Schoenberg and Szego [27]. It should be noted that numerous cases of attainability of the bounds for d-codes and τ-designs in different polynomial metric spaces (see [18-21]) serve as proofs of the fact that in these cases the corresponding polynomials are optimal over the class of polynomials of any degree. On the other hand, Odlyzko and Sloane [26] strengthened by a computer the bound for d-codes in the special cases of the Euclidean spheres S^{n-1}, $3\le n\le 24$, $n\ne 8$, $n\ne 24$, and $d=1$, considering polynomials of higher degree than $h(s)$, $s=t(1)=1/2$. Another very interesting improvement of the bounds (24)-(25) for certain range of values d is due to Tietavainen [37] for the Hamming space and to Astola [2] for the Euclidean sphere. Moreover, for sufficiently small d the bound for d-codes (24)-(25) is weaker than trivial general bounds for d-codes

$$|W| \leq (\mu(d/2-0))^{-1}, \qquad (36)$$

which is consequence of the fact that open metric balls of radius $d/2$ with centers in points of a d-code W do not intersect. In order to extend the bound (24)-(25) to the range of small values of d, the so-called "multiple packing" method is used. This method takes into account relationships between packing densities of a subspace and the whole space or of the same space with distinct packing radii. For example, the Bassalygo inequality

$$A(H(n,2),d) \leq A(J(n,w),2d)$$

and the inequality

$$A(S^{n-1},d) \leq \sqrt{2\pi n}\left(\frac{d'}{d}\right)^{n-1} A(S^{n},d') \text{ if } d<d'$$

were used in [24] and [15] respectively. Notice also that for finite polynomial metric spaces the conditions (12) and (15) can be relaxed and replaced by the corresponding conditions for integers. It gives a possibility to obtain one more general bound for d-codes similar to the Singleton bound [23]. Furthermore, by our definition, finite polynomial metric spaces are P-polynomial association schemes and hence there exist a system of polynomials $\{P_i(t)\}$ of degree i $(i=0,1,\ldots,D)$ and a (standard) substitution function $m(z)$ of a real z such that for any integers i and d $(i,d=0,1,\ldots,D)$, $Q_i(t(d))=P_d(m(i))$ and (compare with (4))

$$k_i|\mathfrak{M}|^{-1}\sum_{d=0}^{D} P_i(m(d))P_j(m(d))r_d=\delta_{i,j},\ P_i(1)=1,\ i=0,1,\ldots,D. \qquad (37)$$

It allows us to prove the Hammimng bound (36) by using the above given solution of the τ-design problem but only for the system $\{P_i(t)\}$ of orthogonal polynomials. The solution of the d-code problem for the system $\{P_i(t)\}$ gives rise to a new general bound on the size of τ-designs in a finite polynomial metric space. This bound for the Hamming space will be submitted by the author to IEEE Transaction on Information Theory.

A proof of optimality of the above given solution of τ-design covering problem is based on the following

Theorem 2. (Sidelnikov [31], Levenshtein [21], Fazekas and Levenshtein [12]). *If* $s \geq t_k^{0,0}$ ($s \geq t_k^{0,1}$) *then for any polynomial* $f(t)=\sum_{i=0}^{D} f_i Q_i(t)$ *of degree at most* $2k$-1 *(at most* $2k$*) such that* $f(t) \leq 0$ *for* $-1 \leq t \leq s$ *there holds the inequality* $f_0 \leq 0$, *with equality if* $f(t)$ *is defined by* (29) *for* $s=t_k^{0,0}$ *(respectively, by* (30) *for* $s=t_k^{0,1}$ *).*

5. Some Applications. The bound for d-codes (24)-(25) was written in the explicit form for the Euclidean spheres in [18], for projective spaces in [19] and for many polynomial metric spaces in the most detailed form in [20] (in Russian). Unfortunately the later publications of other authors (for example, [38,35,33,34,16,42]), where the power sum method or nonoptimal polynomials were used, show that this bound is really unknown to English readers. Now we point out some applications of the bound. The bound for d-codes on Euclidean spheres (see [18,20]) leads for sufficiently large n to the best known upper bound on the highest packing density of n-dimensional Euclidean space, defines exactly the kissing numbers for n=8 and n=24 and gives the best asymptotical upper bound for them, improves in a certain range the Shannon reliability function for the additive Gaussian channel with limited signal power (see also [15]). The bound for d-codes in projective spaces [19,20] gave a possibility to improve the known Sidelnikov-Welch bounds [28,44] on crosscorrelation of binary, r-ary, real and complex codes. Moreover this bound allows us to improve some lower bounds for modules of character sums of polynomials over finite fields known in the number theory. Recently certain problems connected with randomness properties of codes and designs were intensively investigated in the complexity theory [1]. The bounds presented in [19,20] give the best known bounds for some of these problems (see [22]).

References.

1. Alon N., Goldreich O., Hastad J., Peralta R., Simple constructions of almost *k*-wise independent random variables, Proc. of the 31st Annual Symposium on the Foundations of Computer Sciencs, 1991.
2. Astola J.T., The Tietavainen bound for spherical codes, Discr. Appl. Math. 7 (1984), 471-477.
3. Bannai E., Ito T., Algebraic Combinatorics I, Association Schemes, Bejamin/Cummings, London, 1984.
4. Beth Th., Jungnickel D., Lenz H., Design theory, Bibl Inst. Wissenschaftsverlag, Mannheim, 1985.
5. Brouwer A.E., Cohen A.M., Neumaier A., Distance-regular graphs, Springer-Verlag, Berlin, 1989.
6. Delsarte Ph., An algebraic approach to the association schemes of coding theory, Philips Res. Reports Suppl. 10 (1973).
7. Delsarte Ph., Goethals J.-M. and Seidel J.J., Bounds for systems of lines, and Jacobi polynomials, Philips Res.Reports 30 (1975), 91^*-105^*.
8. Delsarte Ph., Associations schemes and *t*-design in regular semilattices, J.Combin.Th.(A) 20 (1976), 230-243.
9 Delsarte Ph., Goethals J.-M. and Seidel J.J., Spherical codes and designs. Geometriae Dedicata 6 (1977), 363-388.
10. Delsarte Ph., Hahn polynomials, discrete harmonics, and *t*-designs, SIAM J.Appl.Math. 34 (1978), 157-166.
11. Dunkl C.F., Discrete quadrature and bounds on *t*-design, Mich.Math.J. 26 (1979), 81-102.
12. Fazekas G., Levenshtein V.I., On upper bounds for code distance and covering radius of designs in polynomial metric spaces, Fifth Joint Soviet-Swedish Intern. Workshop on Information Theory, Moscow, 1990, 65-68 (The full text is submitted to J. Combin. Th. (A)).
13. Godsil C.D., Polynomial spaces, Discrete Math. 73 (1988/89), 71-88.
14. Hoggar S.G., *t*-designs in projective spaces, Europ.J.Comb.3 (1982), 233-254.
15. Kabatiansky G.A.and Levenshtein V.I., Bounds for packings on a sphere and in space (in Russian), Problemy Peredachi

Informacii 14, N1 (1978), 3-25. English translation in Problems of Information Transmission 14, N1 (1978), 1-17.
16. Kumar P.V., Liu C.-M., On lower bounds to the maximum correlation of complex root-of-unity sequences, IEEE Trans. Inform. Theory 36 (1990), 633-640.
17. Levenshtein V.I., On choosing polynomials to obtain bounds in packing problems (in Russian), in Proc. Seventh All-Union Conf. on Coding Theory and Information Transmission, Part II, Moscow-Vilnius, 1978, p.103-108.
18. Levenshtein V.I., On bounds for packings in n-dimensional Euclidean space (in Russian), Dokl. Akad. Nauk SSSR 245, N6 (1979), 1299-1303. English translation in Soviet Math. Doklady 20, N2 (1979), 417-421.
19. Levenshtein V.I., Bounds oh the maximal cardinality of a code with bounded modules of the inner product (in Russian), Dokl. Akad. Nauk SSSR 263 N6 (1982); English translation, in Soviet Math. Doklady 25 N2 (1982), 526-531.
20. Levenshtein V.I., Bounds for packings of metric spaces and some their applications (in Russian), Problemy Kiberneticki, Issue 40, Moscow, "Nauka" Publishing House, 1983, p.43-110.
21. Levenshtein V.I., Designs as maximum codes in polynomial metric spaces, Acta Applicandae Mathematicae, 29 (1992), 1-82.
22. Levenshtein V.I., Bounds for self-complementary codes and their applications, submitted to Proc. of "Eurocode-92".
23. MacWilliams F.J and Sloane N.J.A., The theory of error-correcting codes, North Holland Publ.Co., Amsterdam, 1977.
24. McEliece R.J., Rodemich E.R., Rumsey H.,jr., and Welch L.R., New upper bounds on the rate of a code via the Delsarte-MacWilliams inequalities, IEEE Trans.Inform.Theory IT-23 (1977), 157-166.
25. Neumaier A., Combinatorial configurations in terms of distances, Memorandum 81-09 (Wiskunde), Eindhoven Univ. Technol., 1981.
26. Odlyzko A.M., Sloane N.J.A., New upper bounds on the number of units spheres that can touch a unit sphere in n dimensions, J. Combin. Th. (A) 26 (1979), 210-214.
27. Schoenberg I. and Szego G., An extremum problem for polynomials, Composito Math. 14 (1960), 260-268.
28. Sidelnikov V.M., On mutual correlation of sequences (in Russian), Problemy Kiberneticki 24 (1971), 15-42, "Nauka" Publishing House, Moscow; a short version in English in Soviet Math. Doklady 12 N1(1971), 197-201.
29. Sidelnikov V.M., On the densest packing of balls on the surface of an n-dimensional Euclidean sphere and the number of binary code vectors with a given code distance (in Russian), Dokl. Akad. Nauk SSSR 213 N5 (1973); English

translation, in Soviet Math. Doklady 14 N6 (1973), 1851-1855.
30. Sidelnikov V.M., New bounds for the density of sphere packings in an n-dimensional Euclidean space (in Russian), Mat. Sbornik 95 (1974). English translation, in Math. USSR Sbornik 24 (1974), 147-157.
31. Sidelnikov V.M., On extremal polynomials used to estimate the size of codes (in Russian), Problemy Peredachi Informacii 16, N3 (1980), 17-30. English translation in Problems of Information Transmission 16, N3 (1980), 174-186.
32. Sloane N.J.A., Recent bounds for codes, sphere packing and related problems obtained by linear programming and others methods, Contemporary Mathematics, 9 (1982), 153-185.
33. Sole P. and Mehrotra K., Generalization of the Norse bounds to codes of higher strength, IEEE Trans. Inform. Theory IT-37 (1991), 190-191.
34. Sole P., The covering radius of spherical designs, Graphs and Combinatorics, 5 (1991), 423-431.
35. Tarnanen H., On character sums and codes, Discrete Math.57 (1985), 285-295.
36. Terwilliger P., A characterization of P- and Q-polynomial schemes, J.Combin. Th.(A) 45 (1987), 8-26.
37. Tietavainen A., Bounds for binary codes just outside the Plotkin range, Inform. and Control, 47 N2 (1980), 85-93.
38. Tietavainen A., Lower bounds for the maximum moduli of certain character sums, J. London Math. Soc. (2) 29 (1984), 204-210.
39. Tietavainen A., Covering radius problems and character sums, in Proc. Fourth Joint Swedish-Soviet International Workshop on Information Theory, Gotland, Sweden, 1989, p.196-198.
40. Tietavainen A., An upper bound on the covering radius of codes as a function of the dual distance, IEEE Trans. Inform. Theory 36 (1990), 1472-1474.
41. Tietavainen A., Covering radius and dual distance, Designs, Codes and Cryptography 1 (1991), 31-46.
42. Tietavainen A., On the cardinality of sets of sequences with given maximum correlation, Discrete Math. 106/107 (1992), 471-477.
43. Wang H.-C., Two-point homogeneous spaces, Ann. Math. 55 (1952), 177-191.
44. Welch L.R., Lower bounds on the maximum correlation of signals, IEEE Trans. Inform. Theory 20 (1974), 397-399.

Systems of Algebraic Equations Solved by Means of Endomorphisms *

H. Michael Möller

FB Mathematik der FernUniversität Hagen, W-5800 Hagen, Germany

Abstract. Recently, several authors studied methods based on endomorphisms for localizing and computing the common zeros of systems of polynomial equations $f_i(x_1, \ldots, x_n) = 0,\ i = 1, \ldots, s$, in case the ideal $\mathcal{I}$ generated by $f_1, \ldots, f_s$ has dimension zero. The main idea is to consider the trace and the eigenvalues of the endomorphisms $\Phi_f : [g] \mapsto [g \cdot f]$, where $[\ .\ ]$ denotes equivalence classes modulo $\mathcal{I}$ in the polynomial ring. In this paper we give discuss some of these methods and combine them with the concept of dual bases for describing zero dimensional ideals.

1 Introduction

The interpretation of polynomial rings $\mathcal{P}$ and ideals $\mathcal{I} \subset \mathcal{P}$ as k-vector spaces has been fruitful for getting insight into the ideal structure and for improving existing methods. In computer algebra, Lazard investigated this connection early [La 77],[La 81], and Buchberger never failed in his development of Gröbner basis techniques to stress the connection to linear algebra e.g. [Bu 88], but also many other authors mentioned this connection and investigated ideals with linear techniques.

However, the multiplicative structure of $\mathcal{P}$ and $\mathcal{P}/\mathcal{I}$ has, at least implicitely, always been used. This was done by considering with a coefficient vector of a polynomial f the "shifted" coefficient vectors for power product multiples of f. And, mentioned just for curiosity, starting point for the development of Gröbner basis techniques was Gröbner's proposal to Buchberger to develop a method for computing the multiplication table of $\mathcal{P}/\mathcal{I}$, [Bu 65].

In recent years, the interest in the multiplicative structure has been renewed. By using the endomorphisms

$$\Phi_f : \mathcal{P}/\mathcal{I} \longrightarrow \mathcal{P}/\mathcal{I},\ \Phi_f([u]) := [f \cdot u]$$

where $[u]$ denotes the equivalence class modulo $\mathcal{I}$ generated by $u \in \mathcal{P}$, some new results or new interpretations of old results have been found. In this paper, we concentrate on zero dimensional ideals $\mathcal{I}$ and intend to present in a unified notation the method of Stetter [AS 88] for computing the set of all common zeros of the polynomials in $\mathcal{I}$ using eigenvectors of Φ_f, a method for computing this

* This work is supported in part by the CEC, ESPRIT Basic Research Action 6846 (PoSSo)

set of zeros by using minimal polynomials [YNT 92], and a real root isolating method based on a trace formula for quadratic forms in $\mathcal{P}/\mathcal{I}$, [PRS 92], [Be 91]. Using the concept of dual bases, [$M^3$91], [$M^3$92], we give new proofs, show the connection of the method by [FGLM] to the minimal polynomial computation, and discuss complexity aspects.

By the first two methods, systems of polynomial equations are solved directly; the real root isolating method can serve as preprocesor for a numerical calculation by Newton's method. Hence they all fit into the PoSSo project of solving systems of polynomial equations. An other very interesting method has been presented in a thesis by Cardinal (Université de Rennes), in which for zero dimensional ideals $\mathcal{I}$ generated by n polynomials the common zeros are computed. There, the space $\mathcal{P}/\mathcal{I}$ is equipped with an inner product structure allowing an elegant description of the Φ_f's. The computation of the zeros is then done by a method known in numerical analysis as the von-Mises-iteration. This thesis however became known to the author so recently, that the result can not included here in details.

2 Ideals and Dual Bases

In the following, k is always a field, $\mathcal{P} := k[x_1, \ldots, x_n]$, and $\mathcal{I} \subset \mathcal{P}$ is an ideal of dimension zero. Then $\mathcal{P}/\mathcal{I}$ is a finite dimensional k-vector space, i.e. $\mathcal{I}$ is a k-vector space of finite codimension. A basis of $\mathcal{P}/\mathcal{I}$ can be obtained by a Gröbner basis $\mathcal{G}$ of $\mathcal{I}$. Consider the set $\mathcal{B}$ of all power products $x_1^{i_1} \cdots x_n^{i_n}$ which are not divisible by the leading power product of a $g \in \mathcal{G}$. Then the corresponding equivalence classes $[x_1^{i_1} \cdots x_n^{i_n}]$ constitute a basis of $\mathcal{P}/\mathcal{I}$, see for instance [Bu 88]. We will denote this basis briefly by $[\mathcal{B}]$.

In this section, we resume some relevant parts of the concept of dual bases as described in [$M^3$92] or in the shorter version [$M^3$91], both based on Gröbner's exposition in [Gr 70]. Let $L_1, \ldots, L_s$ be functionals over $\mathcal{P}$, i.e. in $Hom_k(\mathcal{P}, k)$. They are linearly independent if and only if $q_1, \ldots, q_s \in \mathcal{P}$ exist, such that

$$L_i(q_j) = 0 \ \text{ if } \ i \neq j, \ L_i(q_i) = 1 \ . \tag{1}$$

Polynomials $q_1, \ldots, q_s$ satisfying (1) are called *biorthogonal to* $L_1, \ldots, L_s$.

Let $V \subset \mathcal{P}$ be a k-vector space of codimension s. Then there are s linearly independent functionals $L_1, \ldots, L_s$, such that $p \in V \Leftrightarrow L_1(p) = \ldots = L_s(p) = 0$. The set $\{L_1, \ldots, L_s\}$ is called a *dual basis of* V. Conversely, if s functionals L_i are linearly independent, then $\{p \in \mathcal{P} \mid L_1(p) = \ldots = L_s(p) = 0\}$ is a k-vector space of codimension s, i.e. every set of s linearly independent functionals is a dual basis.

If $\{L_1, \ldots, L_s\}$ is a dual basis of V, then V is a zero dimensional ideal if and only if the functionals

$$L_{ij} : p \mapsto x_i \cdot p, \ L_{ij} \in Hom_k(\mathcal{P}, k), \ i = 1 \ldots, n, j = 1, \ldots, s \ , \tag{2}$$

belong to $span_k\{L_1, \ldots, L_s\}$.

Two kinds of dual bases for zero dimensional ideals were considered in [$M^3$91], [$M^3$92]. The first one needs a Gröbner basis $\mathcal{G}$ of $\mathcal{I}$ with respect to an arbitrary admissible term ordering. Then the basis $[\mathcal{B}]$ of $\mathcal{P}/\mathcal{I}$ and the normalform mapping $f \mapsto Can(f, \mathcal{G})$ is available. This maps every $f \in \mathcal{P}$ to a linear combination $\sum c_t(f)t$ where the summation is extended over the finite set $\mathcal{B}$. This normalform mapping satisfies $f - \sum c_t(f)t \in \mathcal{I}$, see for instance [Bu 88]. Then $f \mapsto c_t(f)$ is a functional for every $t \in \mathcal{B}$; and $f \in \mathcal{I}$ if and only if $c_t(f) = 0$ for each $t \in \mathcal{B}$. This means, the functionals $f \mapsto c_t(f)$ constitute a dual basis of $\mathcal{I}$. By construction, $[\mathcal{B}]$ is biorthogonal to it.

This kind of bases allows the practical determination to what equivalence class in $\mathcal{P}/\mathcal{I}$ a $p \in \mathcal{P}$ belongs. This is based on the following obvious fact, which we will state explicitely because of its importance.

Proposition 1. *Let $\{L_1, \ldots, L_s\}$ be a dual basis of the zero dimensional ideal $\mathcal{I}$. Then for equivalence classes $[p]$ of $\mathcal{P}/\mathcal{I}$ the following holds*

$$q \in [p] \Longleftrightarrow L_i(q) = L_i(p), \ i = 1, \ldots, s \ .$$

Proof. $q \in [p] \Longleftrightarrow p - q \in \mathcal{I} \Longleftrightarrow L_i(q) = L_i(p), \ i = 1, \ldots, s$. □

The second type of dual bases gives insight into the primary decomposition of $\mathcal{I}$ by describing the behaviour of the polynomials in the points $y \in \bar{k}^n$, $\bar{k}$ the algebraic closure of k, which are the common zeros of all $p \in \mathcal{I}$. For describing this local behaviour, we define

$$\partial_t := \frac{1}{i_1! \cdots i_n!} \frac{\partial^{i_1 + \cdots i_n}}{\partial x_1^{i_1} \cdots \partial x_n^{i_n}} \quad \text{for power products } t = x_1^{i_1} \cdots x_n^{i_n}, \tag{3}$$

and $\partial_t := 0$ else.

Definition 2. A k-vector subspace V of $ID := span_k\{\partial_t \mid t \text{ a power product }\}$ is called closed, if it has finite dimension and if for every $\sum c_t \partial_t \in V$ the operators $\sum c_t \partial_{t/x_i}$ also belong to V for every $i = 1, \ldots, n$. (Here t/x_i is a power product if x_i divides t and $\partial_{t/x_i} = 0$ else.)

If $y \in k^n$, then the prime ideal $\mathcal{M}_y := \{p \in \mathcal{P} \mid p(y) = 0\}$ has obviously as dual basis $\{id_y\}$, $id_y : p \mapsto p(y)$. If $\{D_1, \ldots, D_s\}$ is a basis of a closed subspace of ID, then $\{id_y \circ D_1, \ldots, id_y \circ D_s\}$ is a dual basis of an $\mathcal{M}_y$-primary ideal $\mathcal{Q}$. We will call this dual basis a *local dual basis*. Its length is the length of the primary ideal. Each $\mathcal{M}_y$-primary ideal has such a local dual basis. If $y_1, \ldots, y_r \in k^n$ are pairwise different and U_i a local dual basis of an $\mathcal{M}_{y_i}$-primary ideal $\mathcal{Q}_i$, then the union of the U_i is a dual basis of the ideal $\bigcap \mathcal{Q}_i$, which will then also be called a local dual basis. Since in case $k = \bar{k}$, every zero dimensional ideal has a representation $\bigcap \mathcal{Q}_i$, the so called primary decomposition, we obtain in that case, that every zero dimensional ideal has a local dual basis. And the proposition applied to such a local basis describes the exact local conditions, which a polynomial has to satisfy to become member of the ideal. For zero dimensional ideals, the local Max Noether conditions and the Hilbert Nullstellensatz are consequences of these local conditions.

Example 1. Consider the univariate case $n = 1$, $\mathcal{P} = k[x]$. Then every closed subspace of *ID* contains with an operator of order m, $\sum_{i=0}^{m} c_i \partial_{x^i}$, $c_m \neq 0$, also operators of order $m-1, , m-2, \ldots, 0$, hence it contains $\partial_{x^0} = \partial_1$, and then also ∂_{x^i} for $i = 1, \ldots, m$. Thus, in the univariate case the only closed subspaces are $span_k\{\partial_1, \ldots, \partial_{x^j}\}$, $j \geq 0$.

Let $y_1, \ldots, y_r \in k$ be pairwise different. Then the functionals

$$L_{ij} : p \mapsto \partial_{x^j} p(y_i) = \frac{d^j}{dx^j} p(y_i),\ j = 0, \ldots, s_i,\ i = 1, \ldots, r\ , \tag{4}$$

constitute a local dual basis of a zero dimensional ideal $\mathcal{I}$, and in the algebraically closed case, $k = \bar{k}$, every ideal $\mathcal{I}$, $(0) \neq \mathcal{I} \neq (1)$ possesses such a local dual basis. Since $L_{ij}(p) = 0$, $j = 0, \ldots, s_i$, $i = 1, \ldots, r$, holds for exactly all $p \in \mathcal{I}$, the ideal is just the set of all solutions of a homogeneous Hermite interpolation problem. If a polynomial set $\{p_{ij} \mid j = 0, \ldots, s_i,\ i = 1, \ldots, r\}$ is biorthogonal to the set of L_{ij}'s, then the p_{ij} allow to solve the inhomogeneous Hermite interpolation problem

$$L_{ij}(p) = c_{ij},\ j = 0, \ldots, s_i,\ i = 1, \ldots, r\ . \tag{5}$$

A solution is given by $\sum_{i=1}^{r} \sum_{j=0}^{s_i} c_{ij} p_{ij}$, and the set of all solutions of this inhomogeneous problem constitute an equivalence class in $\mathcal{P}/\mathcal{I}$.

The last remark, which is an immediate consequence of the proposition applied to local dual bases, can be extended to multivariate Hermite interpolation. For an application of Gröbner basis techniques to multivariate Hermite interpolation problems, see also [BW 91].

3 The Endomorphisms

We consider endomorphisms Φ_f of $\mathcal{P}/\mathcal{I}$,

$$\Phi_f : [g] \mapsto [f \cdot g] \tag{6}$$

for arbitrary $f \in \mathcal{P}$. Since $[f][g] = [f \cdot g]$ holds for arbitrary polynomials $f, g \in \mathcal{P}$, we could also define $\Phi_{[f]} := \Phi_f$. Mainly for simplifying the notations, we prefer to write Φ_f instead of $\Phi_{[f]}$.

The endomorphisms Φ_f commute, since $\Phi_f \Phi_g = \Phi_{fg}$ and $\mathcal{P}$ is commutative. Assume a reduced Gröbner basis $\mathcal{G}$ of $\mathcal{I}$ and the basis $[\mathcal{B}] = \{[u_1], \ldots, [u_s]\}$ of $\mathcal{P}/\mathcal{I}$ is given. Then the matrix M_f corresponding to the endomorphism Φ_f can be computed by the following procedure.

First we compute by calculations as those described in [FGLM] or [$M^3$91] for each power product $t \notin \mathcal{B}$, with $t = x_i u_j$ for a suitable $u_j \in \mathcal{B}$ and $1 \leq i \leq n$, the normalform $Can(t, \mathcal{G})$. Then $t - Can(t, \mathcal{G}) \in \mathcal{I}$, and the set of all these polynomials constitute the so called border basis of $\mathcal{I}$. These computations require $\mathcal{O}(b \cdot s^2)$ additions/multiplications, where b is the cardinality of the border basis and $s = dim(\mathcal{P}/\mathcal{I})$. Using the rough bound $b + s \leq ns + 1$, we get $\mathcal{O}(n \cdot s^3)$ additions/multiplications, see [$M^3$91] or [$M^3$92]. Since $Can(u, \mathcal{G}) = u$ for every

$u \in \mathcal{B}$, we have now the normalforms for all $x_i u,\ u \in \mathcal{B}, 1 \leq i \leq n$. For fixed i, the coefficients $L_\mu(x_i u_j)$ of $Can(x_i u_j, \mathcal{G}) = \sum_\mu L_\mu(x_i u_j) u_\mu$ are the entries of the j-th row of the matrix M_{x_i} corresponding to Φ_{x_i}. Using $\Phi_{x_j}\Phi_u = \Phi_{x_j u}$ we get recursively for every $u_j \in \mathcal{B}$ the matrix M_{u_j} by a multiplication of two matrices, since these u_j are power products of $x_1, \ldots, x_n$. This means for every $u_i \in \mathcal{B}$ with $deg(u_i) > 1$ a matrix by matrix multiplication. Hence additional $\mathcal{O}(s^4)$ additions/multiplications. Once having the matrices $M_{u_1}, \ldots, M_{u_s}$, the matrix corresponding to Φ_f is $\sum_{i=1}^s c_i M_{u_i}$, if $[f] = \sum_{i=1}^s c_i [u_i]$. This computation requires an other $\mathcal{O}(s^3)$ set of additions/multiplications.

By means of the eigenvalues and eigenspaces of the endomorphisms Φ_f, informations on the zero-space $V(\mathcal{I}) := \{y \in \bar{k}^n \mid p(y) = 0\ \forall p \in \mathcal{I}\}$ can be obtained.

Lemma 3. *$\lambda \in k$ is an eigenvalue of Φ_f if and only if*

$$\mathcal{I} : (f - \lambda) \neq \mathcal{I}\ . \tag{7}$$

The eigenvector space to the eigenvalue λ is the set $\{[h] \mid h \in \mathcal{I} : (f - \lambda)\}$. If $V(\mathcal{I}) \subset k^n$, then the set of eigenvalues is $\{f(y) \mid y \in V(\mathcal{I})\}$.

Proof. The equivalence $\Phi_f([h]) = \lambda[h] \Longleftrightarrow (f - \lambda) \cdot h \in \mathcal{I}$ gives immediately (7) and the eigenvector spaces. The condition $\mathcal{I} : (f - \lambda) \neq \mathcal{I}$ is equivalent to $f - \lambda$ in a prime ideal associated to $\mathcal{I}$, see e.g.[Gr 70]. If $V(\mathcal{I}) \subset k^n$, then the associated primes are the maximal ideals $\mathcal{M}_y = \{p \in \mathcal{P} \mid p(y) = 0\},\ y \in V(\mathcal{I})$. □

A local dual basis of the ideal $\mathcal{I} : (f - \lambda)$ can be described in terms of a local dual basis of $\mathcal{I}$. This is simple for $f = x_i$ and algebraically closed k. We will describe this relation only for $x_i = x_1$. It will show that the eigenvectors contain still the main informations on the set $V(\mathcal{I})$, except a weakening of the differential conditions at the zero which corresponds to the eigenvalue.

Lemma 4. *Let $V(\mathcal{I}) \subset k^n$, and let $L_{ij} : p \longmapsto \sum c_t^{(ij)} \partial_t p(y_i),\ j = 0, \ldots, s_i, i = 1, \ldots, r$, be a local dual basis of $\mathcal{I}$. Then $\mathcal{I} : (x_1 - \lambda)$ has for a local dual basis a maximal linearly independent subset of $\{\tilde{L}_{ij} \mid j = 0, \ldots, s_i,\ i = 1, \ldots, r\}$, where $\tilde{L}_{ij} = L_{ij}$, if λ is not the first component of y_i, and otherwise*

$$\tilde{L}_{ij}(p) = \sum_t c_t^{(ij)} \partial_{t/x_1} p(y_i)\ .$$

Proof. Let $\mathcal{I} = \bigcap \mathcal{Q}_i$ be the primary decomposition of $\mathcal{I}$, $\mathcal{Q}_i$ $\mathcal{M}_{y_i}$-primary. Then $\mathcal{I} : (x_1 - \lambda) = \bigcap(\mathcal{Q}_i : (x_1 - \lambda))$. This is the primary decomposition of $\mathcal{I} : (x_1 - \lambda)$ with $\mathcal{M}_{y_i}$-primary $\mathcal{Q}_i : (x_1 - \lambda)$. Then $\mathcal{Q}_i : (x_1 - \lambda) = \mathcal{Q}_i$ if and only if $x_1 - \lambda \notin \mathcal{M}_{y_i}$. Therefore $L_{ij} = \tilde{L}_{ij}$ in that case. And if $x_1 - \lambda \in \mathcal{M}_{y_i}$, then λ equals the first component of y_i and the Leibniz rule shows then the equivalence of $L_{ij}((x_1 - \lambda) \cdot f) = 0$ to $\tilde{L}_{ij}(p) = 0$. Hence in that case $(x_1 - \lambda)p \in \mathcal{Q}_i$ if and only if $\tilde{L}_{ij}(p) = 0,\ j = 0 \ldots, s_i$. □

For an appropriate choice of the basis of $\mathcal{P}/\mathcal{I}$, the matrices corresponding to the endomorphisms $\Phi_f, f \in \mathcal{P}$, have upper triangular form. Using a set of polynomials which is biorthogonal to a local dual basis, we will present a constructive proof. We first need some definitions.

Definition 5. For a differential operator $0 \neq D := \sum c_t \partial_t \in ID$ the *order of* D, $ord(D)$, is defined as the maximal order ($deg(t)$) of ∂_t with nonzero c_t. A basis $D_1, \ldots, D_m$ of a subspace $V \subset ID$ is called *consistently ordered*, if every $0 \neq D \in V$ is already in $span_k\{D_i \mid ord(D_i) \leq ord(D)\}$. If $y_1, \ldots, y_r$ are pairwise different points of k^n, and if $\{D_{ij} \mid 1 \leq j \leq s_i\}$ is a consistently ordered basis of a closed subspace V_i of ID, $i = 1, \ldots, r$, then the local dual basis of all $id_{y_i} \circ D_{ij}$ is also called consistently ordered.

The existence of a consistenly ordered basis for every linear subspace $V \subset ID$ of finite dimension is easily proved by linear algebra techniques. Therefore, if k is algebraically closed, then $\mathcal{I}$ has a consistently ordered basis.

Proposition 6. *Let $\{L_1, \ldots, L_s\}$ be a consistently ordered local dual basis of $\mathcal{I}$ numbered in such a way, that $i < j$ if $L_i = id_{y_\mu} \circ D_{\mu\sigma}$ and $L_j = id_{y_\nu} \circ D_{\nu\tau}$, so that $\mu < \nu$ or $\mu = \nu$ and $ord(D_{\mu\tau}) > ord(D_{\mu\sigma})$. Let $q_1, \ldots, q_s$ be biorthogonal to $L_1, \ldots, L_s$. Then the following statements hold true.*

(A) $\{[q_1], \ldots, [q_s]\}$ *is a basis of* $\mathcal{P}/\mathcal{I}$.

(B) $\mathcal{I}_i := \{p \in \mathcal{P} \mid L_j(p) = 0 \; \forall j \geq i\}$ *is an ideal and* $\mathcal{I} = \mathcal{I}_1 \subset \mathcal{I}_2 \subset \ldots \subset \mathcal{I}_s$.

(C) $L_i(q_j \cdot q_l) = 0$ *if* $i > \min\{j, l\}$, *and* $L_i(q_i \cdot q_l) = q_l(y_\mu)$ *if* $L_i = id_{y_\mu} \circ D_{\mu\tau}$.

(D) $[q_i \cdot q_l] = q_l(y_\mu)[q_i] + \sum_{j<i} c_{ij}^{(l)}[q_j]$ *with some* $c_{ij}^{(l)} \in k$, *if* $L_i = id_{y_\mu} \circ D_{\mu\tau}$.

Proof. $p - \sum L_i(p) q_i \in \mathcal{I}$ for every $p \in \mathcal{P}$. Hence the s equivalence classes $[q_i]$ generate $\mathcal{P}/\mathcal{I}$, an s-dimensional vector space. This gives (A).

By the ordering of the L_i, the first s_1 functionals belong to an $\mathcal{M}_{y_1}$-primary ideal, $L_i = id_{y_1} \circ D_{1i}, i \leq s_1$, s_1 the length of this $\mathcal{M}_{y_1}$-primary component of $\mathcal{I}$; the next s_2 functionals belong to an $\mathcal{M}_{y_2}$-primary ideal etc. Then also the space spanned by all D_{1j}, $1 < j \leq s_1$, is closed because the order of $\sum c_t \partial_{t/x_i}$ is less than the order of $\sum c_t \partial_t$, and hence by the consistent ordering of $V_1 := span_k\{D_{11}, \ldots, D_{1s_1}\}$, the image of V_1 under $\sum c_t \partial_t \mapsto \sum c_t \partial_{t/x_i}$ is contained in $span_k\{D_{12}, \ldots, D_{1s_1}\}$. Therefore $\{L_2, \ldots, L_s\}$ is a local dual basis for an ideal $\mathcal{I}_2$. This ideal satisfies by construction $\mathcal{I}_1 \subset \mathcal{I}_2$. The assumptions of the proposition allow an inductive argument. This gives (B).

The polynomial q_j is contained in the ideal $\mathcal{I}_{j+1}$ by biorthogonality. Hence also $q_j \cdot q_l \in \mathcal{I}_{j+1}$. Therefore $L_i(q_j \cdot q_l) = 0$ for $i > j$. By symmetry also $L_i(q_j \cdot q_l) = 0$ for $i > l$. If $L_i = id_{y_\mu} \circ D_{\mu\tau}$, then by Leibniz rule $L_i(q_i \cdot q_l) = q_l(y_\mu) L_i(q_i) = q_l(y_\mu)$, because all other summands in the rule contain functionals with lower order differential operators evaluated at q_i which by closedness and consistent ordering can expressed as linear combination of $L_{i+1}(q_i), \ldots$ belonging to the same m_{y_μ}-primary ideal and $L_j(q_i) = 0$ for $j > i$. Therefore (C) holds.

$f := q_i \cdot q_l - q_l(y_\mu) \cdot q_i$ satisfies $L_j(f) = 0$ for $j \geq i$ by (C). Then $f^* := f - \sum_{j=1}^{i-1} L_j(f) q_j$ satisfies $L_j(f^*) = 0$ for $j = 1, \ldots, s$. Therefore $f^* \in \mathcal{I}$. Hence turning over to equivalence classes, (D) follows. □

Corollary 7. *Take as basis $p_1, \ldots, p_s$ with $p_i := q_{s+1-i}$, where the $q_1, \ldots, q_s$ are biorthogonal to the consistently ordered local dual basis of $\mathcal{P}/\mathcal{I}$ in proposition 6. Then every Φ_f has as corresponding matrix an upper triangular matrix with diagonal $(f(y_r), \ldots, f(y_r), f(y_{r-1}), \ldots, f(y_1))$. Here $f(y_i)$ is written s_i times, when s_i denotes the length of the $\mathcal{M}_{y_i}$-primary component of $\mathcal{I}$.*

Proof. The assertion follows immediately from proposition 6 for every $f = q_l$. Since every $f \in \mathcal{P}$ is modulo $\mathcal{I}$ a linear combination of $q_1, \ldots, q_s$ and all $p \in \mathcal{I}$ vanish at the zeros y_i, the assertion follows for arbitrary $f \in \mathcal{I}$. □

This corollary is useful for the determining of the common zeros $V(\mathcal{I})$, as we will see at latest in section 6. Knowing the upper triangular matrices at least for $f = x_i$, $i = 1, \ldots, n$, we read off the zeros including their multiplicities. This was already remarked by [YNT 92]. However, our construction of the basis $\{p_1, \ldots, p_s\}$ is only possible, if we already know the common zeros and the local differential conditions leading to the local dual basis of $\mathcal{P}/\mathcal{I}$. It is an interesting task to construct with other methods a basis, which transforms simultaneously n commuting matrices to upper triangular form. For this purpose, there is yet no standard numerical algorithm known to the author.

But for the computation of the common zeros using the eigenvectors of one mapping Φ_f, the exists a numerical method, which we will present in the next section.

4 The Method of Stetter

In this section, k is always $\mathbb{C}$. For numerical computations of eigenvalues and eigenvectors, it is useful to fix a basis of $\mathcal{P}/\mathcal{I}$ and to consider the matrix corresponding to Φ_f instead of Φ_f itself.

We fix the usual basis $[\mathcal{B}] = \{[u_1], \ldots, [u_s]\}$ of $\mathcal{P}/\mathcal{I}$, where $\mathcal{B} = \{u_1, \ldots, u_s\}$ is the set of power products, being no multiples of leading terms of Gröbner basis elements of $\mathcal{I}$. Then $1 \in \mathcal{B}$, since $\mathcal{I} \neq \mathcal{P}$. Let $u_1 = 1$. If there are $x_i, i \in \{1, \ldots, n\}$, such that $x_i \notin \mathcal{B}$, then let for simplicity be $x_j \in \mathcal{B}$ for $j = 1, \ldots, l$, and $x_j \notin \mathcal{B}$ for $j > l$. Then no power product u with $u \in \mathcal{B}$ is divisible by an x_j, $j > l$, and there are polynomials

$$f_j := x_j + \sum_{u \in \mathcal{B}} c_u^{(j)} \cdot u \in \mathcal{G}, \quad j = l+1, \ldots, n \ . \tag{8}$$

Let us denote the entires of M_f, the matrix corresponding to Φ_f, denote by $m_{ij}(f)$. Then the matrix equation applied to the basis $[\mathcal{B}]$ is

$$\begin{pmatrix} \Phi_f([u_1]) \\ \vdots \\ \Phi_f([u_s]) \end{pmatrix} = \begin{pmatrix} m_{11}(f) & \ldots & m_{1s}(f) \\ \vdots & & \vdots \\ m_{s1}(f) & \ldots & m_{ss}(f) \end{pmatrix} \begin{pmatrix} [u_1] \\ \vdots \\ [u_s] \end{pmatrix} \ . \tag{9}$$

Applying now the functionals id_{y_i} from the local dual basis of $\mathcal{I}$, we get

$$\begin{pmatrix} f(y_i)\cdot u_1(y_i) \\ \vdots \\ f(y_i)\cdot u_s(y_i) \end{pmatrix} = \begin{pmatrix} m_{11}(f) \dots m_{1s}(f) \\ \vdots \qquad\qquad \vdots \\ m_{s1}(f) \dots m_{ss}(f) \end{pmatrix} \begin{pmatrix} u_1(y_i) \\ \vdots \\ u_s(y_i) \end{pmatrix} . \tag{10}$$

Hence $f(y_i)$ is an eigenvalue of M_f with eigenvector $(u_1(y_i), \dots, u_s(y_i))^T$. Since $x_1, \dots, x_l$ are among the u_i, we can read off the first l components of the zero y_i. The remaining components can be found by inserting the first l components into $f_j = 0$, where f_j is given by (8). For the special instance $f = x_i$ the result was obtained by Stetter, presented first in a joint paper with Auzinger [AS 88], and investigated in forthcoming papers [St 93a],[St 93b]. By (10) and lemma 3, we get

Theorem 8. *Let M_f be the matrix corresponding to Φ_f. Then $\{f(y_j) \mid y_j \in V(\mathcal{I})\}$ is the set of eigenvalues of M_f. If every eigenspace has dimension one, then the eigenspace of $f(y_j)$ is spanned by $(u_1(y_j), \dots, u_s(y_j))^T$.*

In the case, if all eigenspaces of a matrix have dimension one, the matrix is called *non-derogatory.* Then, by using numerical software, all eigenvectors with first component 1 can be calculated approximately. From these, (approximations to) all members of $V(\mathcal{I})$ can be read off as described above. However, the condition that the matrix is non-derogatory, can not be weakened. This shows the following.

Example 2. Let $\mathcal{P} = \mathbb{C}[x, y]$ and $f_1 = x^2-2, f_2 = y^2-2xy+2$. Then $\mathcal{I} := (f_1, f_2)$ has $\{f_1, f_2\}$ as Gröbner basis and $V(\mathcal{I}) = \{(\sqrt{2}, \sqrt{2}), (-\sqrt{2}, -\sqrt{2}).\}$ Then $[\mathcal{B}] = \{[1], [x], [y], [xy]\}$. We take the endomorphism Φ_x. The matrix corresponding Φ_x is

$$M_x := \begin{pmatrix} 0 & 1 & 0 & 0 \\ 2 & 0 & 0 & 0 \\ 0 & 0 & 0 & 1 \\ 0 & 0 & 2 & 0 \end{pmatrix} .$$

Its eigenvalues are $\pm\sqrt{2}$, each with a two dimensional eigenspace. The eigenspace for $\sqrt{2}$ is spanned by

$$v_1 := (1, \sqrt{2}, 0, 0)^T,\ v_2 := (0, 0, 1, \sqrt{2})^T .$$

If we could read off the components of the common zero, they were at the second and third position. But every $v_1 + av_2$ is an eigenvector and the reading off would give then points $(\sqrt{2}, a)$, which are no zeros except of that one for $a = \sqrt{2}$.

The requirement that the matrix is non-derogatory is, at least for a radical $\mathcal{I}$, not very restrictive. If $\mathcal{I}$ is a radical, take for instance a linear polynomial f. If on every hyperplane $f = \lambda$ there is at most one point of $V(\mathcal{I})$, then all eigenvector spaces are one dimensional. But this means that the hyperplane $f = \lambda$ must not contain any line joining two points of $V(\mathcal{I})$. These are at most $\binom{s}{2}$ different lines, where $s = dim(\mathcal{P}/\mathcal{I}) = card(V(\mathcal{I}))$. Thus, for getting a non-derogatory matrix M_f, we have at most $\binom{s}{2}$ linear restrictions on the continuum of choices for f.

5 A Method Using the Minimal Polynomial

The definition and a construction of minimal polynomials for endomorphisms of finite dimensional k-vector spaces can be found in [Ga 77]. We will quote it briefly from [Ga 77]. In this context, we like to remind that we do not need that the field k is algebraically closed (in contrast to Stetters method).

Definition 9. If Φ is an endomorphism of a finite dimensional k-vector space U, then the minimal polynomial for Φ is the polynomial $\psi(t) = \sum_{i=1}^{m} c_i t^i$, of least degree with $c_m = 1$ such that $\sum_{i=1}^{m} c_i \Phi^i(v) = 0$ for all $v \in U$.

The minimal polynomial is uniquely determined by Φ. Its zeros are the eigenvalues of Φ, hence it divides the characteristic polynomial of Φ.

The procedure for computing the minimal polynomial for Φ reads for Φ_f as follows. Starting with an arbitrary $u \in \mathcal{P}$, the sequence $[u], \Phi_f([u]), \Phi_f^2([u]), \ldots,$ is calculated until the first $\Phi_f^m([u])$ depends linearly on its predecessors. Since the k-vector space is finite, the procedure terminates for arbitrary $u \in \mathcal{P}$ and gives constants $c_i(u) \in k$ with $f^m \cdot u - \sum_{i=0}^{m-1} c_i(u) f^i \cdot u \in \mathcal{I}$. Hence $\psi_u(t) := t^m - \sum_{i=0}^{m-1} c_i(u) t^i$ is the least degree polynomial with leading coefficient 1, such that $\psi_u(f) \in \mathcal{I} : u$. Because of $\mathcal{I} = \mathcal{I} : 1 \subseteq \mathcal{I} : u$, the minimal polynomial is therefore ψ_1. And the described construction is an FGLM-like procedure, see [$M^3$91].

The space $U_f := span_k\{\Phi_f^i([1]) \mid i \geq 0\}$ is a subspace of $\mathcal{P}/\mathcal{I}$ which is invariant under ψ_1. If we restrict Φ_f to U_f, then ψ_1 is simultaneously characteristic and minimal polynomial for this restricted mapping. The corresponding matrix is non-derogatory, i.e. a matrix with one dimensional eigenvectorspaces (and for each eigenvalue λ_i and corresponding eigenvector v_0 a sequence of principal eigenvectors $v_1, \ldots, v_m$, defined by $\Phi_f(v_i) = \lambda v_i + v_{i-1}, i \geq 1$).

If we consider Φ_{x_i} restricted to U_{x_i}, then we get the same eigenvalues as for the unrestricted Φ_{x_i}. By theorem 8, these are the i-th component of the points in $V(\mathcal{I})$. This could be done with every Φ_{x_i}, $1 \leq i \leq n$, and gives separately informations for every component of the zeros.

For combining these informations, U-resultant methods are proposed. An interesting approach has been given by [YNT 92], who summarized briefly U-resultant methods and presented an algorithm and variants of it based on these U-resultants. The main idea of this algorithm is to transform first the ideal $\mathcal{I}$ in general position, such that the (new) n-th components of all $y \in V(\mathcal{I})$ are pairwise different. Then they calculate the minimal polynomial for Φ_{x_n}, factor it (over $\mathcal{P}$) and calculate for every factor h the minimal polynomials of the maps Φ_{x_i}, $1 \leq i < n$, considered as endomorphism of $\mathcal{P}/(\mathcal{I} + (h))$. [YNT 92] showed that the minimal polynomial of such Φ_{x_i} is of type $x_i^j + \frac{1}{j} G(x_n) x_i^{j-1} + \cdots$, such that $G(a)$ is the i-th component of a $y \in V(\mathcal{I})$ with last component equal to a and $h(a) = 0$. A variant of the algorithm use sequences of polynomials obtained by gcd-computations instead of the irreducible factors h of the minimal polynomial.

Useful for solving systems of equations are so called triangular sets,

$$\begin{aligned} p_1(x_1) &= x_1^{d_1} + \textstyle\sum_{i=0}^{d_1-1} c_{1i} x_1^i \\ p_2(x_1, x_2) &= x_2^{d_2} + \textstyle\sum_{i=0}^{d_2-1} c_{2i}(x_1) x_2^i \\ &\vdots \\ p_n(x_1, \ldots, x_n) &= x_n^{d_n} + \textstyle\sum_{i=0}^{d_2-1} c_{ni}(x_1, \ldots, x_{n-1}) x_n^i \end{aligned} \tag{11}$$

where $c(x_1, \ldots, x_j)$ denotes a polynomial in $k[x_1, \ldots, x_j]$. It is an interesting problem, under what condition such $p_j(x_1, \ldots, x_j)$ can be interpreted as a minimal polynomial of an endomorphism. Because then, it can be computed by FGLM-like methods. If it is not in the ideal $\mathcal{I}$, then $\mathcal{I}$ can be replaced by $\tilde{\mathcal{I}} := \mathcal{I} + (p_j)$. This gives $V(\tilde{\mathcal{I}}) = V(\mathcal{I})$, but multiplicities of zeros might be reduced. A special instance of such a procedure is to take as p_j the minimal polynomial for the endomorphisms Φ_{x_j}. Plugging all n of them into $\mathcal{I}$ gives the radical of $\mathcal{I}$.

6 Counting the Number of Real Solutions

In this section, k is a real closed field and C its algebraic closure. Then the number of solutions, i.e. the cardinality of $V(\mathcal{I}) = \{y_j \in C^n \mid f(y_j) = 0 \,\forall f \in \mathcal{I}\}$ is equal to $s = dim(\mathcal{P}/\mathcal{I})$, if each y_j is counted with multiplicity s_j, where s_j denotes the length of the primary component of $\mathcal{I}$, or in terms of local dual bases, s_j the number of local dual basis elements of type $f \mapsto \sum c_t \partial_t f(y_j)$, with suitable $c_t \in k$.

For the count of solutions without multiplicities, for the count of the number of real solutions, i.e. the number of points in $V_k(\mathcal{I}) := \{y_j \in k^n \mid f(y_j) = 0 \,\forall y_j \in \mathcal{I}\}$, and the count of real solutions in a given semi-algebraic set $\#\{y_j \in V_k(\mathcal{I}) \mid h_1(y_j) > 0, \ldots, h_r(y_j) > 0\}$, there a method of Ch.Hermite can be applied, who counted real zeros by calculating signatures of appropriate quadratic forms. The interest in this procedure has recently been renewed. There is a thesis by Pedersen [Pe 91] and a forthcoming paper [PRS 92] in which it is showed how to perform effectively the needed operations by using Gröbner basis techniques. Main tool for the calculation is a trace formula, which has also been considered by [Be 91] and theoretical and algorithmical aspects have been discussed in [BW 92].

For each polynomial $h \in \mathcal{P}$, we can construct a quadratic form

$$Q_h : \mathcal{P}/\mathcal{I} \longrightarrow k, \; Q_h([f]) := Tr(\Phi_{f^2 h}) \; . \tag{12}$$

Here the map Φ_{f^2h} is the endomorphism for the polynomial f^2h, and Tr denotes the trace. Then the main theorem for this approach is the following.

Theorem 10. *For given $h \in \mathcal{P}$ the signature of Q_h, $\sigma(Q_h)$, and the rank of Q_h, $\rho(Q_h)$, satisfy*

$$\begin{aligned} \sigma(Q_h) &= \#\{y_j \in V_k(\mathcal{I}) \mid h(y_j) > 0\} - \#\{y_j \in V_k(\mathcal{I}) \mid h(y_j) < 0\} \; , \\ \rho(Q_h) &= \#\{y_j \in V(\mathcal{I}) \mid h(y_j) \neq 0\} \; . \end{aligned} \tag{13}$$

Proof. The trace of a map is invariant under basis transformations. If we take the basis described in corollary 7, then we immediately see that

$$Tr(\Phi_{f^2h}) = \sum_{j=1}^{r} s_j f^2(y_j)h(y_j) ,$$

where $V(\mathcal{I}) = \{y_1, \ldots, y_r\}$ and s_j the length of the $\mathcal{M}_{y_j}$-primary component of $\mathcal{I}$. This gives immediately the formula for the rank. For showing that $Tr(\Phi_{f^2h})$ is the sum of σ_1 squares of real numbers minus σ_2 squares of real numbers, we observe, that the y_i appear in pairs, to every $y_j \notin k^n$, there is a $\bar{y}_j \in V(\mathcal{I})$ with same length s_j of the corresponding primary components. Decomposing therefore the partial sum

$$S_j := s_j f(y_j)^2 h(y_j) + s_j f(\bar{y}_j)^2 h(\bar{y}_j)$$

by substituting real and imaginary parts of $f(y_j)$, $f(\bar{y}_j) = \overline{f(y_j)}$ and of the square roots of $h(y_j)$ and $h(\bar{y}_j) = \overline{h(y_j)}$, we get

$$\begin{aligned} S_j &= s_j[\Re f(y_j) + i\Im f(y_j)]^2[\Re\sqrt{h(y_j)} + i\Im\sqrt{h(y_j)}]^2 \\ &+ s_j[\Re f(y_j) - i\Im f(y_j)]^2[\Re\sqrt{h(y_j)} - i\Im\sqrt{h(y_j)}]^2 \end{aligned}$$

and after regrouping terms

$$\begin{aligned} S_j &= 2s_j[\Re f(y_j)\Re\sqrt{h(y_j)} - \Im f(y_i)\Im\sqrt{h(y_j)}]^2 \\ &- 2s_j[\Re f(y_j)\Im\sqrt{h(y_j)} + \Im f(y_i)\Re\sqrt{h(y_j)}]^2. \end{aligned}$$

Hence every S_j corresponding to a conjugate complex pair of points contributes a square for the positive and a square for the negative part. The signature as difference of σ_1 and σ_2 is hence not influenced by conjugate complex pairs and counts therefore only the number of those terms $s_j f(y_j)^2 h(y_j)$, $y_j \in k^n$, with positive $h(y_j)$ minus those with negative $h(y_j)$. □

Rank and signature of Q_1 gives therefore the number of all points in $V(\mathcal{I})$ and $V_k(\mathcal{I})$ resp. The cardinalities of

$$\begin{aligned} V_+^{(h)} &:= \{y_j \in V_k(\mathcal{I}) \mid h(y_j) > 0\}, \\ V_o^{(h)} &:= \{y_j \in V_k(\mathcal{I}) \mid h(y_j) = 0\}, \\ V_-^{(h)} &:= \{y_j \in V_k(\mathcal{I}) \mid h(y_j) < 0\}, \end{aligned}$$

can be found as solution of the linear system of equations

$$\begin{pmatrix} 1 & 1 & 1 \\ 0 & 1 & -1 \\ 0 & 1 & 1 \end{pmatrix} \begin{pmatrix} \#V_o^{(h)} \\ \#V_+^{(h)} \\ \#V_-^{(h)} \end{pmatrix} = \begin{pmatrix} \sigma(Q_1) \\ \sigma(Q_h) \\ \sigma(Q_{h^2}) \end{pmatrix} . \tag{14}$$

Two such identities $Ac = u$, $A'c' = u'$ may be combined with a tensor product $(A \otimes A')(c \otimes c') = u \otimes u'$ and the solution describes then the number of points in $V_k(\mathcal{I})$, where two polynomials h and h' have a fixed sign pattern. If this tensoring is iterated to find for every combination of signs the number of points in $V_k(\mathcal{I})$ such that r given polynomials $h_1, \ldots, h_r$ fulfill the sign conditions, then a $3^r \times 3^r$ matrix arises. There is the algorithm of Ben-Or, Kozen, and Reif, which reduces the problem of solving this $3^r \times 3^r$ system of equation into $r \log r$ systems, each with at most $\#V_k(\mathcal{I})$ equations in $\#V_k(\mathcal{I})$ unknowns. For a more detailed discussion, see [Pe 91] or [PRS 92].

The practical computation of trace and rank for Q_h can be done by using the basis $[\mathcal{B}] = \{[u_1], \ldots, [u_s]\}$ of $\mathcal{P}/\mathcal{I}$. Then the matrix corresponding to the quadratic form Q_h has as (i,j)-entry by calculating the corresponding matrices. We have already seen that the computation of all M_{u_i} costs in total $\mathcal{O}(s^4)$ multiplications/additions. The computation of M_h needs $\mathcal{O}(s^3)$ operations. The matrix corresponding to $\Phi_{u_i u_j h}$ is $M_{u_i} M_{u_j} M_h$. It requires s matrix multiplications $M_{u_i} M_h = M_{u_i h}$, and then for $1 \le i \le j \le s$ the multiplications $M_{u_j} M_{u_i h}$. Here the amount is $s + \binom{s+2}{2}$ matrix multiplications with $\mathcal{O}(s^3)$ operations each, i.e. in total $\mathcal{O}(s^5)$ arithmetical operations. Hence all matrices $M_{u_i u_j h}$ are calculated using $\mathcal{O}(s^5)$ multiplications/additions. The trace $Tr(\Phi_{u_i u_j h})$ is just the sum of the $M_{u_i u_j h}$-diagonal entries. Thus the matrix for Q_h is computed using $\mathcal{O}(s^5)$ arithmetical operations. This complexity can of course be reduced by using a more sophisticated algorithm for matrix multiplication. Then the total costs will become something like $\mathcal{O}(s^{4.7})$.

The main problem is the increase of length of the matrix entries and the storing of so many matrices. Let ℓ be an upper bound for the length of the entries in the matrices M_{x_i}, and also a bound for the coefficients c_i in $[h] = \sum c_i[u_i]$. (This ℓ may be quite large, because the M_{x_i} contain for instance the coefficient vectors of Gröbner basis elements as rows and these may be large compared to the coefficient size of an other basis of the ideal.) Following the proposed way of calculation, we see that the maximal entry length is in the worst case $s\ell$ for M_{u_i}, $2s\ell$ for M_h, and therefore $4s\ell$ for $M_{u_i u_j h}$. Even if there is an other way of arranging the succession of operations, this size of coefficients has to be expected and requires a very careful bookkeeping of which are the actually really needed matrices with long entries.

Up to now, we looked only for the costs of establishing the matrix for Q_h. An additional problem is the complexity of computing its rank and trace. This however can be done numerically and deserves therefore other arguments.

7 Conclusion

A naïve proposal for calculating the common real zeros could be as follows. Use first the real root counting method to find their number in semi-algebraic sets like balls and reduce the volume of these sets until the zeros are isolated. Then numerical methods like Newton's can then be applied for obtaining the zeros in arbitrary precision. Our discussion of the method in section 6 gave, that at most for low numbers s this procedure should be applied. But if the real root counting method is applied only after a preprocessing which reduces the original problem to one or a series of problems with small s, then it might become a useful tool.

The method of Stetter methods can be used as well for counting the number of real zeros or number of zeros in a given semi-algebraic subset of $I\!R^n$, because the numerical computation of eigenvectors gives approximations to the zeros and hence to their position. If the exact multiplicity of every zero is known, then by an eventually needed increase of the floating point accuracy, sufficiently precise statements concerning the number of real points (in a semi-algebraic set) can be made. We have seen, that Stetter's method needs a matrix computed with $\mathcal{O}(s^3)$ complexity and requires then a numerical algorithm for the eigenvector computation. With respect to the high complexity of establishing the matrices for Q_1, Q_h, and Q_{h^2}, and then the following rank and signature computations, Stetter's method may be a good alternative.

The only drawback for Stetter's method is the requirement that the givem matrix M_f is non-derogatory. Here, the method using minimal polynomials can perhaps be helpful. Further investigations are needed for deciding under what conditions what method is most favourable. Such investigations will be made within the PoSSo-project.

References

[AS 88] Auzinger, W., Stetter, H.J.: An elimination algorithm for the computation of all zeros of a system of multivariate polynomial equations. Birkhäuser International Series in Numerical Mathematics **86** (1988) 11 - 30.

[Be 91] Becker, E.: Sums of squares and trace forms in real algebraic geometry. Cahier du Séminaire d'Histoire des Mathématiques, 2e Série, Vol. 1, Université Pierre et Marie Curie, 1991.

[BW 91] Becker, T., Weispfenning, V.: The Chinese remainder problem, multivariate interpolation, and Gröbner bases. Proc. ISSAC'91 (ed.: S.M.Watt), acm press (1991) 64 - 69.

[BW 92] Becker, E., Wörmann, T.: On the trace formula for quadratic forms and some applications. Proc. RAGSQUAD'92 (ed.: T.Lam), to appear.

[Bu 65] Buchberger, B.: An algorithm for finding a basis for the residue class ring of a zero dimensional polynomial ideal (German). Ph.D. thesis, Univ. of Innsbruck, Austria, 1965.

[Bu 88] Buchberger, B.: Application of Gröbner bases in non-linear computational geometry. Springer Lecture Notes in Computer Sciences. **296** (1988) 52 - 80.

[FGLM] Faugère, J.C., Gianni, P., Lazard, D., Mora, T.: Efficient computation of zero-dimensional Gröbner bases by change of ordering. Unpublished manuscript 1989.

[Ga 77] Gantmacher, F.R.: The Theory of Matrices. vol.I, Chelsea Publ. Comp. 1977.

[Gr 70] Gröbner, W.: Algebraische Geometrie I/II. B.I.Hochschultaschenbücher **273, 737**, Bibliogr. Institut. 1970.

[La 77] Lazard, D.: Algèbre linéaire sur $K[X1,\ldots,Xn]$ et élimimation. Bull. Soc. Math. France **105** (1977), 165 - 190.

[La 81] Lazard, D.: Résolution des systèmes d'équations algébriques. Theoret. Comp. Sci. **15** (1981), 77 - 110.

[$M^3$91] Marinari, M.G., Möller, H.M., Mora, T.: Gröbner bases of ideals given by dual bases. Proc. ISSAC'91 (ed.: S.M.Watt), acm press (1991) 55 - 63.

[$M^3$92] Marinari, M.G., Möller, H.M., Mora, T.: Gröbner bases of ideals defined by functionals with an application to ideals of projective points. J. of AAECC (to appear).

[Pe 91] Pedersen, P.: Counting real zeros. Thesis, Courant institute, N.Y. University (1991).

[PRS 92] Pedersen, P., Roy, M.-F., Szpirglas, A.: Counting real zeros in the multivariate case. Proc. of MEGA'92 (ed.:A. Galligo), 1992 (to appear).

[St 93a] Stetter, H.J.: Verification in computer algebra systems. In: Validation Numerics – Theory and Applications Computing Suppl.**8** (1993) (to appear).

[St 93b] Stetter, H.J.: Multivariate polynomial equations as matrix eigenproblems. Preprint.

[YNT 92] Yokoyama, K., Noro, M., Takeshima, T.: Solutions of systems of algebraic equations and linear maps on residue class rings. J. Symb. Comp. **14** (1992) 399–417.

Criteria for Sequence Set Design in CDMA Communications

Robert A. Scholtz

Communication Sciences Institute, University of Southern California, Los Angeles, CA 90089-2565, USA

Abstract. This paper surveys some of the traditional sequence designs that have been considered for code-division multiple-access (CDMA) communications. The design criteria are based on periodic and partial-period cross-correlation properties of the sequence sets. Variations of the design criteria are suggested here which are related more directly to CDMA system performance.

1 A Discrete Communications Model

Code-division multiple-access is a signalling technique that is used in spread-spectrum communication systems (e.g., see Simon et al., 1985) to make signals separable by a receiver when they share the same frequency band. The receiver traditionally uses a synchronous correlation detector to extract the desired carrier, which is identified by a distinctive *spreading code* that modulates it.

A transmitter in the CDMA system contains a *pseudonoise (PN) generator* that produces a spreading code $\{b_n\}$ from a prescribed alphabet. The effect of sending phase-shift keyed (PSK) spreading-code and data modulation through a channel with delay τ, carrier frequency and phase shifts ω and ϕ respectively, and detecting it with a correlation receiver controlled by a spreading code $\{b'_n\}$, is to produce a receiver output that depends on the spreading code design only through sequence $\{x_i\}$ with elements of the form

$$x_i = e^{j\phi} \sum_{n=iD}^{iD+D-1} d_{\lfloor (n+\tau)/D \rfloor} e^{j\omega n} \rho^{b_{n+\tau} - b'_n} . \tag{1}$$

In this model, the sequence $\{d_i\}$ represents the transmitted data sequence, D is the number of spreading-code symbols transmitted per data symbol, and ρ is a root of unity which indicates the spreading codes' PSK modulation factor. If more than one similar transmitter is in operation, then the receiver's output sequence is a weighted sum of sequences of the form (1), each with distinct values for the parameters τ, ω, and τ, and the spreading code $\{b_n\}$.

Different levels of synchronization in the receiver ideally set different parameters to zero in the model (1) of spreading code reception (see Table 1). In a phase-synchronous receiver which has achieved code synchronism with a signal using spreading code $\{b_n\}$, the sequence $\{x_i\}$ determining the receiver's output caused by this signal is

Table 1. Parameters for the model (1) as a function of the level of synchronization in a CDMA receiver

Level	Parameter Values
Frequency sync	$\omega = 0$
Phase sync	$\omega = 0$ and $\phi = 0$
Code sync	$b_n = b'_n \ \forall n$ and $\tau = 0$

$$x_i = D\, d_i \,. \tag{2}$$

On the other hand, another transmitter in the CDMA system creates *multiple-access noise* in a receiver that is not in spreading-code synchronism with its transmission. Assuming that the interfering transmitter is frequency synchronous with the transmission that the receiver is trying to detect, the multiple-access noise caused by the interferer is of the form

$$x_i = e^{j\phi} \sum_{n=iD}^{iD+D-1} d_{\lfloor (n+\tau)/D \rfloor} \rho^{b_{n+\tau} - b'_n} \,. \tag{3}$$

The sequence design objective for CDMA systems then is to minimize the multiple access noise, e.g., $|x_i|$ in (3).

2 Periodic Correlation Designs

In terms of minimizing multiple-access noise, classical sequence set design corresponds to the following conditions:

- Each sequence $\{b_n\}$ has period $N = D$, i.e., one period per data symbol.
- No data modulation, $d_i = 1$ for all i and all transmitters.
- All transmitters frequency synchronous, $\omega = 0$.

In the design of a set of J sequences $\{b_n^{(j)}\}$, $j = 1, \ldots, J$, the design objective is to insure that

$$x_i^{(j,k)} = x_i = e^{j\phi} \sum_{n=iD}^{iD+D-1} \rho^{b^{(j)}_{n+\tau} - b^{(k)}_n} \tag{4}$$

is as small as possible in magnitude for all $j \neq k$. The magnitude is necessary because ϕ is unknown. Since $N = D$ is the period of each spreading code, minimization of either

$$\max_{1 \le j < k \le J} \ \max_{0 \le \tau < D} |P_{j,k}(\tau)| \qquad \text{or} \qquad \max_{1 \le j < k \le J} \sum_{\tau=0}^{D-1} |P_{j,k}(\tau)|^2 \qquad \text{or} \ \ldots ,$$

is a reasonable design procedure, where $P_{j,k}(\tau)$ is the periodic cross-correlation between the PSK modulated spreading codes.

$$P_{j,k}(\tau) = \sum_{n=0}^{N} \rho^{b^{(j)}_{n+\tau} - b^{(k)}_n} \tag{5}$$

For reasons having to do with the acquisition of synchronization, it is also desirable that the autocorrelation $|P_{j,j}(\tau)|$ be as small as possible for $\tau \neq 0$. Inner product bounds (Welch 1974) can be used to lower bound these measures of design quality.

2.1 Constructions Based on the Trace Function

The most intensely studied periodic designs consist of sets of binary sequences with $b_n^{(j)} \in \mathrm{GF}(2^d)$ and $\rho = -1$. The mathematical tool used most often in these designs is the trace function, which linearly maps $\mathrm{GF}(2^b)$ into a subfield $\mathrm{GF}(2^a)$ (i.e., a divides b) according to the relation

$$\mathrm{tr}_a^b(z) = \sum_{i=0}^{\frac{b}{a}-1} z^{2^{ai}} \qquad \forall z \in \mathrm{GF}(2^b)\,. \tag{6}$$

In terms of the trace function, the classic m-sequence design is simply

$$b_n = \mathrm{tr}_1^d(\alpha^n)\,, \tag{7}$$

in which α is a primitive element of $\mathrm{GF}(2^d)$ and the period of $\{b_n\}$ is $N = 2^d - 1$. A list of designs based on properties of the trace function is shown in Table 2.

Table 2. Typical binary designs in terms the trace function. All sequences have period $N = 2^d - 1$ and α is a primitive element of $\mathrm{GF}(2^d)$. The elements α^s are properly chosen primitive elements in the smallest fields containing them, r determines a fixed nonlinearity, and β determines which sequence is selected. In the bent sequences, $\mathbf{M}(\cdot)$ and σ are fixed quantities and z (a parameter of the bent function $f_z(\cdot)$) is the sequence selector. The quantity $P_{\max}$ is the largest absolute value of cross-correlation or out-of-phase autocorrelation.

Family	Conditions	Sequence b_n	Size J	$P_{\max}$
m-sequence	—	$\mathrm{tr}_1^d(\alpha^n)$	1	1
GMW	$m\|d$	$\mathrm{tr}_1^m([\mathrm{tr}_m^d(\alpha^n)]^r)$	1	1
Gold	d odd	$\mathrm{tr}_1^d(\alpha^{sn}) + \mathrm{tr}_1^d(\beta\alpha^n)$	2^d+1	$1+2^{(d+1)/2}$
BK	$d=2r+1$	$\mathrm{tr}_1^d(\beta\alpha^n) + \sum_{l=1}^r \mathrm{tr}_1^d(\alpha^{n(1+2^l)})$	2^d+1	$1+2^{(d+1)/2}$
Kasami	d even	$\mathrm{tr}_1^{d/2}(\mathrm{tr}_{d/2}^d(\alpha^n) + \beta\alpha^{sn})$	$2^{d/2}$	$1+2^{d/2}$
No	d even	$\mathrm{tr}_1^{d/2}([\mathrm{tr}_{d/2}^d(\alpha^n) + \beta\alpha^{sn}]^r)$	$2^{d/2}$	$1+2^{d/2}$
bent	$4\|d$	$f_z(\mathbf{M}(\alpha^n)) + \mathrm{tr}_1^d(\sigma\alpha^n)$	$2^{d/2}$	$1+2^{d/2}$

Notice that the model for Gordon-Mills-Welch (GMW) sequences (Welch and Scholtz 1984) reduces to that for m-sequences when the non-linearity parameter $r = 1$. The effect of the nonlinearity when $r \neq 1$ is to increase the linear span of the sequence and hence potentially increase system security. The same type of nonlinearity can be added (No and Kumar 1987) to the Kasami design (Kasami

1966) to achieve the same effect. The BK design (Boztaş and Kumar (1993)) also can be viewed as a relative of the Gold design (Gold 1967), that possesses larger linear span. The bent sequences (Olsen et al. 1982) also have large linear span, caused by the fact that the bent function $f_z(\cdot)$ is a nonlinear function. Specifically, a function $f(\mathrm{x})$ on binary k-tuples is *bent* (Rothaus 1976) if its Fourier transform

$$F(\lambda) = \sum_{\mathbf{x} \in \mathcal{V}_k} (-1)^{f(\mathbf{x}) + \mathbf{x}^t \lambda} \tag{8}$$

is ± 1 for all transform domain variables. The class of bent functions used in bent-sequence design is defined by

$$f_z(\mathbf{x}) = \mathbf{x}_1^t \mathbf{x}_2 + g(\mathbf{x}_2) + \mathbf{x}^t \mathbf{z}\,, \tag{9}$$

where $\mathbf{x}_1$ and $\mathbf{x}_2$ are the first and second halves of $\mathbf{x}$, the function $g(\cdot)$ is an arbitrary function, and $\mathbf{z}$ serves as the sequence selector variable. The requirement of large linear span, motivated by military applications, may also provide added privacy in commercial applications.

Many of the tools and concepts used in binary designs can be adapted directly to sequences over GF(p), p prime, with ρ a primitive p-th root of unity. Potentially more useful designs for CDMA systems have ρ a primitive fourth root of unity. Recent designs (Solé 1988, Hammons and Kumar 1992) include a code with size and period identical to the Gold codes, but with the maximum cross-correlation $P_{\max}$ lower by a factor of $\sqrt{2}$.

2.2 Fourier Transform Approaches to Periodic Sequence Design

Because correlation calculations can be viewed as inner products and Fourier transforms are inner-product preserving, sequence design can be attempted in the frequency domain. Let $\{b_n^{(j)}\}$ be a sequence of Nth roots of unity with period N and let $\rho = \exp(i2\pi/N)$. Then the discrete Fourier transform $\tilde{\mathbf{b}}^{(j)}$ of one period of the sequence is defined as

$$\tilde{\mathbf{b}}^{(j)} = \mathbb{F}\mathbf{b}^{(j)} = (\tilde{b}_0^{(j)}, \ldots, \tilde{b}_{N-1}^{(j)})^t\,, \quad \text{where} \quad \tilde{b}_m^{(j)} = \frac{1}{\sqrt{N}} \sum_{n=0}^{N-1} b_n^{(j)} \rho^{-mn} \tag{10}$$

$\mathbb{F}$ represents a Fourier operator, and $(\cdot)^t$ denotes the transpose so that $\mathbb{F}$ can also be interpreted as a unitary matrix multiplier. The vector of periodic cross-correlations (see (5)) between two sequences $\{b_n^{(j)}\}$ and $\{b_n^{(k)}\}$ can be written as

$$\mathbf{P}_{j,k} = (P_{j,k}(0), \ldots, P_{j,k}(N-1))^t = \sqrt{N}\mathbb{F}^{-1}[(\mathbb{F}\mathbf{b}^{(j)}) \odot (\mathbb{F}\mathbf{b}^{(k)})]\,, \tag{11}$$

where $\mathbf{x} \odot \mathbf{y}$ is the vector of terms that occur in the inner product of $\mathbf{x}$ and $\mathbf{y}$, i.e., $\mathbf{x} \odot \mathbf{y} = (x_0 y_0^*, \ldots, x_{n-1} y_{N-1}^*)^t$.

The form of (11) suggests the following viewpoint. Let S be a set of sequence periods $\mathbf{b}^{(j)}$, $j = 1, \ldots, J$, with the property that S is *closed* under the operations

$\mathbb{F}$ and $\mathbb{F}^{-1}$, and closed under $\odot$ when distinct sequences are combined. It follows immediately that $\mathbb{F}^{-1}[(\mathbb{F}\mathbf{b}^{(j)}) \odot (\mathbb{F}\mathbf{b}^{(k)})] \in \mathcal{S}$ for $j \neq k$. Let's now assume further that the sequences in $\mathcal{S}$ are composed of elements with unit magnitude, i.e., $|b_n^{(j)}| = 1$ for all n and j. Then

$$|P_{j,k}(\tau)| = \begin{cases} \sqrt{N}, & \text{if } j \neq k, \\ N, & \text{if } j = k \text{ and } \tau = 0, \\ 0, & \text{if } j = k \text{ and } 0 < \tau < N. \end{cases} \tag{12}$$

Sets of sequences with properties quite similar to the set $\mathcal{S}$ do exist.

One possible design (Lerner 1961, Scholtz and Welch 1978) involves the characters of the group $\mathcal{M}(p)$ of integers under addition modulo p, where p is an odd prime. If γ is a generator of this cyclic group, then $\gamma, \gamma^2, \ldots, \gamma^{p-1}$ are the are the distinct elements of the group. Let $l(n)$ be the logarithm to the base γ defined on $\mathcal{M}(p)$, i.e., $n = \gamma^{l(n)}$ for all $n \in \mathcal{M}(p)$. Then let $\mathcal{S}'$ be the set of augmented non-principal characters of the group. That is, $\mathcal{S}' = \{\mathbf{b}^{(j)} : j = 1, \ldots, p-2\}$, where

$$\mathbf{b}^{(j)} = \left(0, \sigma^{jl(1)}, \sigma^{jl(2)}, \ldots, \sigma^{jl(p-1)}\right)^{\mathrm{t}} \quad \text{and} \quad \sigma = \exp(i2\pi/(p-1)). \tag{13}$$

The set $\mathcal{S}'$ of sequence periods is closed under $\odot, \mathbb{F}$, and $\mathbb{F}^{-1}$, and differs from the ideal model of $\mathcal{S}$ only because the first element of each sequence's period is zero. The resulting priodic correlation properties are nearly ideal.

$$|P_{j,k}(\tau)| = \begin{cases} 0, & \text{if } j \neq k \text{ and } \tau = 0, \\ \sqrt{p}, & \text{if } j \neq k \text{ and } 0 < \tau < p, \\ p-1, & \text{if } j = k \text{ and } \tau = 0, \\ -1, & \text{if } j = k \text{ and } 0 < \tau < p. \end{cases} \tag{14}$$

This implementation of this design, as well as some other multiphase designs (e.g., Suehiro and Hatori (1988), Gabidulin (1993)) for large periods is made difficult by the mechanization of the function $l(n)$ and the requirement that the PSK modulator must address $p-1$ different carrier phase shifts relatively accurately.

3 Long Code Designs

Continuing the scenario in which there is no frequency mismatch ($\omega = 0$) and no data modulation ($d_i = 1$ for all i and all transmitters), let's now consider the *long-code design* situation in which the period N of the code is considerably longer than the number D of spreading-code symbols per data symbol. This choice of parameters may be made for transmission security reasons, or for mixing purposes, i.e., so that unfavorable code situations in the asynchronous communications context are not repeated every data symbol.

The design criteria based on signal correlation then change because the correlation value $x_i^{(j,k)}$ ((4) still applies) is not computed over a full period of the spreading code. In this case with $j \neq k$, the multiple-access noise created by one interferer is related to $|x_i^{(j,k)}| = |P_{j,k}(iD, \tau)|$, where $P_{j,k}(m, \tau)$ is the partial period cross-correlation function

$$P_{j,k}(m,\tau) = \sum_{n=m}^{m+D-1} \rho^{b_{n+\tau}^{(j)} - b_n^{(k)}} . \tag{15}$$

If D is relatively prime to N, then for every m there will exist a value of i such that $iD = m \bmod N$, and hence one reasonable design criterion is to design a spreading code set $\mathcal{S}$ that has a small value of the worst partial-period cross-correlation function,

$$\max_{1 \le j < k \le J} \max_{0 \le m < N} \max_{0 \le \tau < N} |P_{j,k}(m,\tau)| .$$

Other criteria that measure the smallness of the partial-period correlation function (15) over the set $\mathcal{S}$ of spreading codes may be equally well suited for particular design philosophies.

3.1 Partial-Period Correlation Analysis – An Illustration

The common practice in communications has been to use sequences with good full-period correlation properties as candidates for long-code designs, and to make a choice based on their partial-period correlation properties. For example, let's consider a CDMA system in which every user employs the same spreading code, but because $N \gg D$ and $N \gg J$, the probability that two transmitters are transmitting their codes in synchronism is negligible. Then an obvious candidate spreading-sequence for this design is a single m-sequence $\{b_n\}$ (see Table 2).

The partial-period correlation calculation for an m-sequence can be reinterpreted in coding theory terms. Let $\mathbf{b}_m$ be a D-tuple from $\{b_n\}$ beginning at position m,

$$\mathbf{b}_m = (b_m, b_{m+1}, \ldots, b_{m+D-1}) . \tag{16}$$

Then for $\tau \neq 0$, the partial-period correlation calculation is given by

$$\begin{aligned} P(m,\tau) &= \sum_{n=m}^{m+D-1} (-1)^{b_{n+\tau} \oplus b_n} = D - 2\mathrm{wt}(\mathbf{b}_{m+\tau} \oplus \mathbf{b}_m) \\ &= D - 2\mathrm{wt}(\mathbf{b}_{m+s(\tau)}) , \end{aligned} \tag{17}$$

where wt(·) is the Hamming weight function. Here we have used the fact that the set $\mathcal{B}(D, \log_2(N+1), -)$,

$$\mathcal{B}(D, \log_2(N+1), -) = \{\mathbf{b}_m : 0 \le m < N\} \cup \{\underline{0}\} , \tag{18}$$

is a linear code with word length D, $\log_2(N+1)$ information bits, and undetermined minimum distance. So, information about the partial-period correlation properties of an m-sequence is equivalent to information about the weight distribution of the binary linear code $\mathcal{B}(D, \log_2(N+1), -)$. In the usual case in which $D > \log_2(N+1)$ as assumed in the specification of (18), this code can be related to other codes as shown in the following diagram.

simplex code		Hamming code
$\mathcal{B}(N, \log_2(N+1), N/2)$	$\longleftarrow$ dual $\longrightarrow$	$\mathcal{B}^{\perp}\ (N, N-\log_2(N+1), 3)$
puncture $\downarrow$ (lose check bits)		shorten $\downarrow$ (lose information bits)
$\mathcal{B}(D, \log_2(N+1), —)$	$\longleftarrow$ dual $\longrightarrow$	$\mathcal{B}^{\perp}\ (D, D-\log_2(N+1), —)$

The code $\mathcal{B}(N, \log_2(N+1), N/2)$ is composed of the all-zeros code word, the full-length m-sequence $(b_0, \ldots, b_{N-1})$, and all its cyclic shifts. This code, often called a simplex code, is then punctured to construct the code whose weight distribution is of interest.

The dual of the simplex code is the Hamming code $\mathcal{B}^{\perp}\ (N, N-\log_2(N+1), 3)$, which in turn must be shortened by surpressing $N-D$ information bits to get the dual of the the code of interest. Because of this latter fact, there are no words of weight 1 or 2 in $\mathcal{B}^{\perp}\ (D, D-\log_2(N+1), —)$, and the number of words of higher weight depends on the choice of m-sequence generator. Typically an additional portion of its weight distribution can be determined by computer. The r-th moment of the weight distribution of $\mathcal{B}(D, D-\log_2(N+1), —)$ can be determined from the first $r+1$ entries of the weight distribution of $\mathcal{B}^{\perp}\ (D, D-\log_2(N+1), —)$ by means of the Pless identities (Pless 1963). These moments can be converted to upper and lower bounds on the desired weight distribution function, by standard techniques (e.g., see Akheizer 1965).

Parts of the structure of partial-period correlation distributions that was described here actually was developed in a series of papers and reports a few decades ago (e.g., see Lindholm 1968, Wainberg and Wolf 1970, Fredricsson 1975). The approach described here was used by Bekir (1978) to investigate both m-sequences and, in a similar fashion, Gold sequence sets. In many cases, the bounds on the weight distribution function appeared to bracket a normal distribution function, supporting the view that multiple-access noise can be modelled as a Gaussian process. The m-sequences whose partial-period correlation distributions most nearly matched Gaussian distributions, were those for which the dual code $\mathcal{B}^{\perp}\ (D, D-\log_2(N+1), —)$ had minimum distance significantly larger than 3.

4 Variations

The prior few sections have illustrated most of the sequence design criteria that have been applied to the design of CDMA sequence sets. Even purely for the purpose of multiple-access noise surpression, these criteria are not optimal. Here are some suggestions for challenging and useful design criteria.

(a) *Consider the effects of data modulation on the interfering signals.* This has been done with some success in the case of single sequence design for good autocorrelation when $N = D$ (Massey and Urhan 1975; see also the survey by Sarwate and Pursley 1980). This author does not know of any work that extends these results to *sets* of data-modulated sequences with good auto- and cross-correlation, or to long-code designs.

(b) *Design sequence sets which are Doppler tolerant.* Specifically, design spreading code sets which produce only a low level of multiple-access noise (see (1)) for all possible choices of ω over a range of possible values. This problem occurs when one receiver attempts to simultaneously receive several transmitters' signals without building separate carrier tracking loops for each one. This may lead to improved performance in these multi-capture receivers.

(c) *Design long codes with locally good correlation properties.* The objective here is to employ long codes which are synchronized to within a few data symbols, i.e., the possible values of τ are controlled by the communication system manager. This means that only certain portions of interfering spreading codes are ever correlated against a particular portion of the correlator's reference code. Hence only these potential crosscorrelations need be minimized as part of the design process. The only work on the *design* of sequences with locally good correlation resulted in relatively short single sequences (Liu 1987).

(d) *Design sequences for new reception algorithms.* Communication theorists are now studying signal reception algorithms which attempt to take advantage of the information that a receiver knows about all of the transmitted signals' code structures (e.g., see the survey by Poor 1992). Typically the optimal structures should nearly eliminate all multiple-access noise effects. In this case, the sequence design criterion may be modified to accomplish other ends, e.g., simplify the new signal reception algorithms.

In summary, (1) bounds on periodic correlation design criteria have have been nearly achieved with 2-phase and 4-phase sequence sets, (2) design criteria that more accurately reflect multiple-access noise levels in systems have not been explored, and (3) new receiver architectures may initiate signal design problems that are not driven by correlation measures.

References

Akheizer, N.I. (1965): The Classical Moment Problem. Hafner, New York.

Bekir, N.E. (1978): Bounds on the distribution of partial correlation for PN and Gold sequences. Ph.D. dissertation, U. of Southern California

Boztaş, S., Kumar, P.V. (1993): Binary sequences with Gold-like correlation but large linear span. IEEE Trans. Inform. Theory (to appear)

Fredricsson, S.A. (1975): Pseudo-randomness properties of binary shift register sequences. IEEE Trans. Inform. Theory **IT-21**(1), 115-120

Gabidulin, E. (1993): Non-binary sequences with the perfect periodic auto-correlation and with optimal periodic cross-correlation. International Symposium on Information Theory, 412

Gold, R. (1967): Optimal binary sequences for spread-spectrum multiplexing. IEEE Trans. Inform. Theory **IT-13**(4), 619-621

Hammons, R., Kumar, P.V. (1992): On a recent 4-phase sequence design. IEEE Second International Symposium on Spread Spectrum Techniques and Applications, Yokohama, Japan, 219-225

Kasami, T. (1966): Weight distribution formula for some classes of cyclic codes. Coordinated Science Lab., Univ. Illinois, Tech. Rep. R-285 (AD 632574)

Lerner, R.M. (1961): Signals having good correlation functions. IEEE WESCON Convention Record

Lindholm, J.H. (1968): An analysis of the pseudo-randomness properties of subsequences of long m-sequences. IEEE Trans. Inform. Theory **IT-14**(4), 569-576

Liu, K.H. (1987): Binary sequences with very small local partial period correlations and local orthogonal sequences. Ph.D. dissertation, U. of Southern California

Massey, J.L., Urhan, J.J., Jr. (1975): Sub-baud Decoding. Proc. 13th Annual Allerton Conf. Circuit and System Theory, 539-547

No, J.S., Kumar, P.V. (1989): A new family of binary pseudorandom sequences having optimal correlation properties and large linear span. IEEE Trans. Inform. Theory **IT-35**(2), 371-379

Olsen, J.D., Scholtz, R.A., Welch, L.R. (1982): Bent-function sequences. IEEE Trans. Inform. Theory **IT-28**(6), 858-864

Pless, V. (1963): Power moment identities on weight distributionsin error correcting codes. Info. and Control **6**, 147-152

Poor, H.V. (1992): Signal processing for wideband communications. IEEE Inform. Theory Newsletter **32**(2), 1-10

Rothaus, O.S. (1976): On "bent" functions. J. Comb. Theory, Series A20, 300-305

Sarwate, D.V., Pursley, M.B. (1980): Crosscorrelation properties of pseudorandom and related sequences. Proc. IEEE **68**(5), 593-619

Scholtz, R.A., Welch, L.R. (1978): Group characters: sequences with good correlation properties. IEEE Trans. Inform. Theory **IT-24**(5), 537-545

Scholtz, R.A., Welch, L.R. (1984): GMW sequences. IEEE Trans. Inform. Theory **IT-30**(3), 548-553

Simon, M.K., Omura, J.K., Scholtz, R.A., Levitt, B.K. (1985): Spread Spectrum Communications. Computer Science Press, Rockville MD

Solé, P. (1988): A Quaternary cyclic code, and a family of quadriphase sequences with low correlation properties. Coding Theory and Applications, Third International Colloquium. (Lecture Notes in Computer Science, 388) Springer, New York Berlin Heidelberg, 193-201

Suehiro, N., Hatori, M. (1988): Modulatable orthogonal sequences and their application to SSMA systems. IEEE Trans. Inform. Theory **IT-34**(1), 93-100

Wainberg, S., Wolf, J.K. (1970): Subsequences of pseudo-random sequences. IEEE Trans. Commun. Tech., **COM-18**(5), 606-612

Welch, L.R. (1974): Lower bounds on the maximum cross correlation of signals. IEEE Trans. Inform. Theory **IT-20**(3), 397-399

Using Groebner Bases to Determine the Algebraic and Transcendental Nature of Field Extensions: return of the killer tag variables

Moss Sweedler
ACSyAM / MSI, 409 College Ave
Cornell University
Ithaca NY 14853

ABSTRACT: Suppose I is a prime ideal in $k[X_1, \cdots, X_n]$ with a given finite generating set and $k(q_1,\ldots,q_m)$ is a finitely generated subfield of the field of fractions Z of $k[X_1, \cdots, X_n]/I$ and c is an element of Z. We present Groebner basis techniques to determine:

* if c is transcendental over $k(q_1,\ldots,q_m)$,
* a minimal polynomial for c if c is algebraic over $k(q_1,\ldots,q_m)$.
* the algebraic or transcendental nature of Z over $k(q_1,\ldots,q_m)$.

The information about c also tells whether c lies in $k(q_1,\ldots,q_m)$, solving the subfield membership problem. Determination of the algebraic or transcendental nature of Z over $k(q_1,\ldots,q_m)$ includes finding the index in case of algebraicity or transcendence degree in case the extension is transcendental. The determination of the nature of Z over $k(q_1,\ldots,q_m)$ is not simply an iterative application of the results for c and only requires computing one Groebner basis.

1 INTRODUCTION:

Certain questions about fields reduce to questions about finitely generated commutative algebras. For example, suppose L is a field containing an element c and a subfield K. One may wish to determine if c is algebraic or transcendental over K and if c is algebraic over K, one may wish to determine a minimal polynomial for c over K. Frequently, in computational algebra, L is the field of rational functions in several variables $X_1,\ldots,X_n$, c is given as a rational function and the subfield K is generated - *as a field* - by a given finite set S of rational functions. In this case, there is a *common denominator* D, where c and S lie in the finitely generated algebra A which is generated by the X_i's and 1/D. The issue of c being transcendental over K reduces to c being transcendental over B, where B is the subalgebra of A which is (*algebra*) generated by S. In section 2 we describe several processes for dealing with this and related situations. Groebner bases are the underlying engine which make both processes run, constructively. *Process one* determines the algebraic or transcendental nature of c over B. The variant - *process two* - determines the algebraic or transcendental nature of A over B in *one shot*, rather than using *process one* repeatedly. Both processes pertain to finitely generated commutative algebras. After the processes are presented, the proofs that the processes perform as promised and application of the processes appear in sections 3 and 4.

Supported in part by NSF & ARO through ACSyAM/MSI at Cornell University, #DAAL03-91-C-0027.

The final application deals with a more general situation than simply subfields of rational function fields. It solves the following problems. Suppose:

I is a prime ideal in $k[X_1, \cdots, X_n]$ with a given finite generating set

$k(q_1,\ldots,q_m)$ is a finitely generated subfield of Z the field of fractions of $k[X_1, \cdots, X_n]/I$

c is an element of Z

Process one may be applied to determine:

* if c is transcendental over $k(q_1,\ldots,q_m)$,
* a minimal polynomial for c if c is algebraic over $k(q_1,\ldots,q_m)$. This also determines if $c \in k(q_1,\ldots,q_m)$.

Process two may be applied to determine the algebraic or transcendental nature of Z over $k(q_1,\ldots,q_m)$. This includes the index in case of algebraicity or transcendence degree in case the extension is transcendental.

CREDITS: In part, this paper deals with transcendence degree. See [Kredel88], [Ollivier89] and [Audoly91] for related work. The source from which this paper springs is [Shannon87]. Dave Shannon is spiritually a joint author of this paper. The subtitle: "Return of the Killer Tag Variables" comes from the fact that [Shannon87], like [Shannon88] and the apocryphal [Spear77], uses tag variables to *knock off* the problem.

2 PROCESSES FOR ALGEBRAS

2.1 THE SETTING: 1. A is a finitely generated commutative algebra over a ground field k and $\{ a_0,\ldots,a_n \}$ is a generating set for A over k. $k[X_0,\ldots,X_n]$ is the polynomial ring in the n variables $\{ X_i \}$ and γ is used to denote the k-algebra map $\gamma: k[X_0,\ldots,X_n] \dashrightarrow A$, $f(X_0,\ldots,X_n) \dashrightarrow f(a_0,\ldots,a_n)$.

2. H_γ is an explicitly given finite subset of $k[X_0,\ldots,X_n]$ which generates the ideal Ker γ.
3. B is a subalgebra of A and B is generated by $\{ b_1,\ldots,b_m \}$.
4. c is an element in A.
5. The b_i's and c are explicitly given as polynomials in the a_i's. I.e. we are given polynomials $B_i(X_0,\ldots,X_n)$ and $C(X_0,\ldots,X_n)$ where $\gamma(B_i) = B_i(a_0,\ldots,a_n) = b_i$ and $\gamma(C) = C(a_0,\ldots,a_n) = c$.

2.2 FIRST CONSTRUCTION: Form the polynomial ring $k[X_0,\ldots,X_n,S,T_1,\ldots,T_m]$. Using any term order where each X_i is greater than any monomial in $k[S,T_1,\ldots,T_m]$ and S is greater than any monomial in $k[T_1,\ldots,T_m]$, construct a Groebner basis G for the ideal generated by:

2.3 $$H_\gamma \cup \{ S - C \} \cup \{ T_i - B_i \}_i$$

Within G let $G_T = G \cap k[T_1,...,T_m]$ and let $G_S = \{ h \in G \cap k[S,T_1,...,T_m] :$ the lead term of h is not divisible by the lead term of any element of $G_T \}$.[1]

2.4 FIRST CONCLUSIONS: a. If $h \in G_S$, then $h \in k[S,T_1,...,T_m]$ and we consider h to be a polynomial of $S,T_1,...,T_m$; i.e. $h = h(S,T_1,...,T_m)$. $h(S,b_1,...,b_m)$ viewed as a polynomial p(S) in B[S] has the property: p(S) is a non-zero polynomial where $p(c) = 0$. Moreover $\text{degree}_S\, h = \text{degree}_S\, p(S)$.

b. If there exists a non-zero polynomial p(S) in B[S] with $p(c) = 0$ then there is a polynomial h in G_S where $\text{degree}_S\, h \le \text{degree}_S\, p(S)$.

c. c is integral over B if and only if G contains an element whose lead term is a pure power of S. In this case, the minimal e where G contains an element with lead term S^e is the same as the minimal degree of integral polynomial satisfied by c over B.

d. Suppose A is an integral domain. Let (B) denote the field of fractions of B. c is algebraic over (B) if and only if G_S is non-empty. In this case, let g be a polynomial in G_S of minimal S-degree. Then $p(S) = g(S,b_1,...,b_m)$ is a minimal degree polynomial in B[S] for c over (B).

Although p(S) is not monic in general, part (**d**) shows that the degree of the field extension is $\text{degree}_S\, g$.

2.5 SECOND CONSTRUCTION: Form the polynomial ring $k[X_0,...,X_n,T_1,...,T_m]$. Using any term order where each X_i is greater than any monomial in $k[X_{i+1},...,X_n,T_1,...,T_m]$,[2] construct a Groebner basis G for the ideal generated by:

2.6
$$H_\gamma \cup \{ T_i - B_i \}_i$$

Within G let $G_T = G \cap k[T_1,...,T_m]$ and let $G_i = \{ h \in G \cap k[X_i,...,X_n,T_1,...,T_m] :$ the lead term of h is not divisible by the lead term of any element of $G \cap k[X_{i+1},...,X_n,T_1,...,T_m] \}$.[3] For $i = 0,...,n$ if G_i is *not empty* define E_i as the minimal positive integer such that there is h_i in G_i with $\text{degree}_{X^i}\, h_i = E_i$.

2.7 SECOND CONCLUSIONS: Suppose A is an integral domain. Let (A) and (B) denote the respective fields of fractions of A and B.

---

1 If G is a reduced Groebner basis then G_S has a simpler description as $(G \cap k[S,T_1,...T_m]) - G_T$. The somewhat more complicated description eliminates the need for a reduced Groebner basis.

2 "$k[X_j,...,X_n,T_1,...,T_m]$" means "$k[T_1,...,T_m]$" for $j \ge n$.

3 If G is a reduced Groebner basis then, G_i has a simpler description as: $(G \cap k[X_i,...,X_n,T_1,...,T_m]) - k[X_{i+1},...,X_n,T_1,...,T_m]$. The somewhat more complicated description eliminates the need for a reduced Groebner basis.

a. If all the G_i's are non-empty then (A) is algebraic over (B) and the index of the extension is $\Pi_i E_i$.

b. If one of the G_i's is empty then (A) is transcendental over (B) and the transcendence degree of the extension is the number of empty G_i's.

3 VERIFICATION: We continue and extend the notation already developed. Γ is used to denote the algebra map $k[X_0,...,X_n,S,T_1,...,T_m] \to A$, $f(X_0,...,X_n,S,T_1,...,T_m) \to f(a_0,...,a_n,c,b_1,...,b_m)$. Γ extends the previous γ. Let H_Γ denote the set $H_\gamma \cup \{ S - C \} \cup \{ T_i - B_i \}_i$ in (2.3). Recall that $\operatorname{Ker}\gamma$ is generated by H_γ

3.1 LEMMA: $\operatorname{Ker}\Gamma$ is generated by H_Γ.

PROOF: Since $c = C(a_0,...,a_n)$ and $b_i = B_i(a_0,...,a_n)$ it follows that H_Γ lies in $\operatorname{Ker}\Gamma$. Now to show that H_Γ actually generates $\operatorname{Ker}\Gamma$. Put any term order on $k[X_0,...,X_n]$ and with respect to this term order let G_γ be a Groebner basis for the ideal generated by H_γ. Since H_γ and G_γ generate the same ideal, it suffices to prove that $\operatorname{Ker}\Gamma$ is generated by $G_\Gamma = G_\gamma \cup \{ S - C \} \cup \{ T_i - B_i \}_i$.

Extend the term order of the previous paragraph to any term order on $k[X_0,...,X_n,S,T_1,...,T_m]$ which has the property that S and all T_i's are greater than any monomial in $k[X_0,...,X_n]$. We show that G_Γ not only generates $\operatorname{Ker}\Gamma$ but is a Groebner basis for $\operatorname{Ker}\Gamma$. We do this by showing that any element of $\operatorname{Ker}\Gamma$ reduces to zero over G_Γ. Since S is the lead term of $S - C$ and T_i is the lead term of $T_i - B_i$, any polynomial in $k[X_0,...,X_n,S,T_1,...,T_m]$ reduces over $\{ S - C \} \cup \{ T_i - B_i \}_i$ to a polynomial in $k[X_0,...,X_n]$. Since Γ extends γ, $\operatorname{Ker}\Gamma \cap k[X_0,...,X_n] = \operatorname{Ker}\gamma$. Hence, reducing any element of $\operatorname{Ker}\Gamma$ over $\{ S - C \} \cup \{ T_i - B_i \}_i$ to an element of $k[X_0,...,X_n]$, yields an element of $\operatorname{Ker}\gamma$. This further reduces to zero over G_γ since G_γ is a Groebner basis for $\operatorname{Ker}\gamma$. **QED**

3.2 PROOF of FIRST CONCLUSIONS (2.4): **a.** Say $h \in G_S$ and form $p(S) = h(S,b_1,...,b_m)$ as in (2.4,a). Then $p(c) = h(c,b_1,...,b_m) = \Gamma(h)$ and this is zero because h lies in $\operatorname{Ker}\Gamma$. To see that $\text{degree}_S\, h = \text{degree}_S\, p(S)$, consider h as a polynomial in $k[T_1,...,T_m][S]$. Write h as:

$$h_0(T_1,...,T_m)S^e + h_1(T_1,...,T_m)S^{e-1} + ... + h_e(T_1,...,T_m)$$

where $h_0(T_1,...,T_m)$ is non-zero. The term order in (2.2) has S larger than any monomial in $k[T_1,...,T_m]$. Hence the lead term of h is the lead term of $h_0(T_1,...,T_m)S^e$ and this is precisely the lead term of h_0 multiplied by S^e. Because h lies in G_S, this lead term is not divisible by any element of G_T. Hence the lead term of h_0 is not divisible by any element of G_T and h_0 does not reduce to zero over G_T.

The choice of term order insures that $G \cap k[T_1,...,T_m] = G_T$ generates $\mathrm{Ker}\,\Gamma \cap k[T_1,...,T_m]$. Hence the fact that h_0 does not reduce to zero over G_T insures that $\Gamma[h_0] = h_0(b_1,...,b_m)$ is non-zero. Thus $p(S) =$

$$h_0(b_1,...,b_m)S^e + h_1(b_1,...,b_m)S^{e-1} + ... + h_e(b_1,...,b_m)$$

is a polynomial of S-degree e in B[S]. This also shows that p(S) is non-zero.

b. Say p(S) is a non-zero polynomial in B[S] with $p(c) = 0$. Write p as $\beta_0 S^e + \beta_1 S^{e-1} + ... + \beta_e$ with β_j's in B and β_0 not equal to zero. Since the b_i's generate B as an algebra, there are polynomials $h_j(T_1,...,T_m)$ where each $\beta_j = h_j(b_1,...,b_m) = \Gamma(h_j)$. Let us assume that $h_0(T_1,...,T_m)$ has been *chosen with minimal lead monomial* among those elements which Γ maps to β_0. Define h in $k[S,T_1,...,T_m]$ by:

$$h = h_0(T_1,...,T_m)S^e + h_1(T_1,...,T_m)S^{e-1} + ... + h_e(T_1,...,T_m)$$

Then $\Gamma(h) = h_0(b_1,...,b_m)c^e + h_1(b_1,...,b_m)c^{e-1} + ... + h_e(b_1,...,b_m) = \beta_0 c^e + \beta_1 c^{e-1} + ... + \beta_e = p(c) = 0$. Hence h lies in $\mathrm{Ker}\,\Gamma$ and so must reduce to zero over the Groebner basis G in (2.2). As in the proof of part **a**, the lead term of h is the lead term of $h_0(T_1,...,T_m)$ times S^e. This lead term must be divisible by (the lead term of) an element g of G since h reduces to zero over G.

We shall show that g lies in G_S. The fact that g divides the lead term of h implies that, $\mathrm{degree}_S\, g \le= \mathrm{degree}_S\, h = e = \mathrm{degree}_S\, p$. Thus g has the claimed "degree_S" property. The way we show that g lies in G_S is to show that h lies in the set **N** defined by:

{ $d \in k[S,T_1,...,T_m]$: the lead term of d is not divisible by the lead term of any element of G_T }

If h lies in **N** and the lead term of g divides the lead term of h, then g lies **N** by transitivity of *divisibility*. This shows that g lies in G_S since $G_S = \mathbf{N} \cap G$.

Can the lead term of h be divisible by the lead term of an element of G_T? Suppose so. Suppose there is f in G_T whose lead term divides the lead term of h. Since f lies in $k[T_1,...,T_m]$, it follows that the lead term of f divides the lead term of $h_0(T_1,...,T_m)$. Thus $h_0(T_1,...,T_m)$ reduces over { f } to an element $e_0(T_1,...,T_m)$ with smaller lead term. Since f lies in G which lies in $\mathrm{Ker}\,\Gamma$, $\Gamma(e_0(T_1,...,T_m)) = \Gamma(h_0(T_1,...,T_m))$ which contradicts the minimality property of h_0. Hence h lies in **N** and g lies in G_S.

c. Suppose c is integral over B. Let p(S) in part (**b**) be an integral polynomial. Thus $\beta_0 = 1$ in the proof of part (**b**) above. Follow that proof using 1 for $h_0(T_1,...,T_m)$

when "pulling back" β_0. It then follows that the g, in the proof, has a lead term which is a pure power of S. And as above: $\text{degree}_S\, g \le e = \text{degree}_S\, p$.

Conversely, say g is an element of G whose lead term is a pure power of S. Since the X_i's are greater than all monomials in $k[S,T_1,...,T_m]$ in the term order (2.2), g must lie in $k[S,T_1,...,T_m]$ and hence in G_S. As in the proof of part (**a**), $g(S,b_1,...,b_m)$ gives a polynomial in $B[S]$ satisfied by v. Moreover, $\text{degree}_S\, g = e = \text{degree}_S\, g(S,b_1,...,b_m)$. Finally, note that $g(S,b_1,...,b_m)$ is a monic polynomial in $B[S]$.

d. If h lies in G_S then part (**a**) shows that $p(S) = h(S,b_1,...,b_m)$ gives a polynomial satisfied by c over B and hence (B). Moreover part (**a**) gives: $\text{degree}_S\, h = \text{degree}_S\, p(S)$. Conversely, assume c is algebraic over (B), let q(S) be a minimal degree polynomial for c over (B). Since the coefficients of q lie in (B) there is a common denominator b in B where bq has coefficients in B. I.e. $p = bq$ is a polynomial in $B[S]$, of same S-degree as q, satisfied by c. By part (**b**), G_S contains an element of this S-degree or less. **QED**

3.3 PROOF of SECOND CONCLUSIONS (2.7): The trick here is to view the information produced by the second construction (2.5) from $n+1$ points of view and each time apply the first conclusion. To be more precise, for $i = 0,...,n$.

VIEW i: $c_i = a_i$, $B_i = k[a_{i+1},...,a_n,b_1,...,b_m]$. In this view X_i plays the role of S in the first process; $X_{i+1},...,X_n,T_1,...,T_m$ plays the role of $T_1,...,T_m$ in the first process and G_i plays the role of G_S in the first process.

The tower of algebras $B_0 \supset B_1 \supset ... \supset B_n \supset B$ gives the tower of fields $(B_0) \supset (B_1) \supset ... \supset (B_n) \supset (B)$. The rest is a straightforward $n+1$ fold application of (2.4) and standard results about algebraicity and transcendentality of towers of fields. **QED**

4 FIELD APPLICATIONS:

4.1 APPLICATION TO FIELDS OF RATIONAL FUNCTIONS AND FINITELY GENERATED SUBFIELDS: The first application is to finitely generated subfields of the field of rational functions: $k(Y_1,...,Y_n)$. Suppose $k(q_1,...,q_m)$ is a finitely generated subfield of $k(Y_1,...,Y_n)$, where $q_i \in k(Y_1,...,Y_n)$. Furthermore, let c be an element of $k(Y_1,...,Y_n)$. Find a common denominator $d \in k[Y_1,...,Y_n]$ where $dq_i \in k[Y_1,...,Y_n]$, $dc \in k[Y_1,...,Y_n]$ and let p_i denote dq_i and let C denote dc. Let A be the subalgebra of $k(Y_1,...,Y_n)$ generated by $\{ 1/d, Y_1,...,Y_n \}$ and let B be the subalgebra of A generated by $\{ p_1/d,...,p_m/d \}$. Note that $c = C/d$ also lies in A. Consider the map $\gamma: k[X_0,X_1,...,X_n] \dashrightarrow A$, $f(X_0,X_1,...,X_n) \dashrightarrow f(1/d,Y_1,...,Y_n)$. It is well known and easy to verify that $\text{Ker}\,\gamma$ is generated by $X_0 d(X_1,...,X_n) - 1$. (Originally d is a polynomial in $k[Y_1,...,Y_n]$

and we are substituting X's for the Y's.) Hence, let $H_\gamma = \{ X_0 - d(X_1,\dots,X_n) \}$. Finally note that $(A) = k(Y_1,\dots,Y_n)$ and $(B) = k(q_1,\dots,q_m)$.

Process one may be applied to determine:

* if c is transcendental over $k(q_1,\dots,q_m)$,
* a minimal polynomial for c if c is algebraic over $k(q_1,\dots,q_m)$. This also determines if $c \in k(q_1,\dots,q_m)$.

Process two may be applied to determine the algebraic or transcendental nature of $k(Y_1,\dots,Y_n)$ over $k(q_1,\dots,q_m)$. This includes the index in case of algebraicity or transcendence degree in case the extension is transcendental.

4.2 COMMENTS: If one just wants to learn the nature of the field extension $k(Y_1,\dots,Y_n)$ over $k(q_1,\dots,q_m)$ then there is no element c and **d** is just the common denominator for $q_1,\dots,q_m$. Also, if $d = 1$, i.e. the q_i's and c - if there is a c - are all polynomials, then drop d and X_0. I.e. let A be the subalgebra $k[Y_1,\dots,Y_n]$ of $k(Y_1,\dots,Y_n)$ and let B be the subalgebra of A generated by $\{ q_1,\dots,q_m \}$. c - if there is a c - lies in A. The map $\gamma{:}k[X_1,\dots,X_n] \to A$ is determined by:

$$f(X_1,\dots,X_n) \to f(Y_1,\dots,Y_n)$$

and $\operatorname{Ker}\gamma = \{ 0 \}$. Let H_γ be the empty set.

4.3 LEMMA: Suppose A is a subalgebra of a larger algebra and d is an element of A which is invertible in the larger algebra. Let $A[1/d]$ denote the subalgebra of the larger algebra which is generated by A and 1/d. Let $\mu{:}A[X] \to A[1/d]$ be the algebra map which sends $f(X) \to f(1/d)$. μ maps $A[X]$ onto $A[1/d]$ and has kernel generated by $dX - 1$.

PROOF: Clearly $\langle dX - 1 \rangle$ lies in the kernel of μ. The opposite inclusion is verified by a little computation. Let $f(X) = a_0X^e + a_1X^{e-1} + \cdots + a_e$ be a polynomial which lies in $\operatorname{Ker}\mu$. If $e = 0$, i.e. f has degree zero, then f must be the zero polynomial which lies in $\langle dX - 1 \rangle$. Hence we may assume that $e \geq 1$. Rewrite f(X) as:

$$(a_0 + a_1d^1 + \cdots + a_ed^e) X^e - (a_1d^1 + \cdots + a_ed^e) X^e + a_1X^{e-1} + \cdots + a_e$$

$0 = f(1/d)$ gives $0 = d^ef(1/d)$ which gives: $0 = a_0 + a_1d^1 + \cdots + a_ed^e$.

Thus the first expression in the rewritten f(X) vanishes, leaving f(X) as:

$$- (a_1d^1 + \cdots + a_ed^e) X^e + a_1X^{e-1} + \cdots + a_e$$

Again, regroup and rewrite, giving f(X) as:

$$a_1 (1 - dX) X^{e-1} + a_2 (1 - (dX)^2) X^{e-2} + \cdots + a_e (1 - (dX)^e)$$

Each $1 - (dX)^i$ equals $1 + dX + \cdots + (dX)^{i-1}$ times $1 - dX$. Hence, f(X) lies in $\langle dX - 1 \rangle$. **QED**

4.4 COROLLARY: Suppose Z is an algebra and the algebra map:

$$\mu{:}k[X_1, \cdots, X_n] \dashrightarrow Z\ ,\ f(X_1, \cdots, X_n) \dashrightarrow f(z_1, \cdots, z_n)$$

has kernel generated by the finite set H_μ. Let $d_1, \cdots, d_m$ be elements in the image of μ which are invertible in Z. Choose polynomials $D_1, \cdots, D_m$ in $k[X_1, \cdots, X_n]$ where $\mu(D_i) = d_i$ and consider the algebra map:

$$\gamma{:}k[W_1, \cdots, W_m, X_1, \cdots, X_n] \dashrightarrow Z$$

$$f(W_1, \cdots, W_m, X_1, \cdots, X_n) \dashrightarrow f(1/d_1, \cdots, 1/d_m, z_1, \cdots, z_n)$$

Then $\operatorname{Ker}\gamma$ is generated by:

$$H_\mu \cup \{\ D_1W_1 - 1\ , \cdots, D_mW_m - 1\ \}$$

PROOF: The map γ factors into the two maps:

$$k[W_1, \cdots, W_m, X_1, \cdots, X_n] \dashrightarrow Z[W_1, \cdots, W_m] \dashrightarrow Z$$

$$f(W_1, \cdots, W_m, X_1, \cdots, X_n) \dashrightarrow f(W_1, \cdots, W_m, z_1, \cdots, z_n)$$

$$\dashrightarrow f(1/d_1, \cdots, 1/d_m, z_1, \cdots, z_n)$$

The kernel of the first map is $\operatorname{Ker}\mu$ extended to $k[W_1, \cdots, W_m, X_1, \cdots, X_n]$. Hence it is generated by H_μ. The first map carries $\{\ D_1W_1 - 1\ , \cdots, D_mW_m - 1\ \}$ to $\{\ d_1W_1 - 1\ , \cdots, d_mW_m - 1\ \}$ which, by the preceding lemma iterated, generates the kernel of the second map. Hence the kernel of the composite - and the composite equals γ - is generated by the given set. **QED**

4.5 APPLICATION TO FINITELY GENERATED FIELDS OVER FINITELY GENERATED SUBFIELDS: Suppose I is a prime ideal in $k[X_1, \cdots, X_n]$ with finite generating set H_I. Let P denote $k[X_1, \cdots, X_n]/I$ and let:

$$\mu{:}k[X_1, \cdots, X_n] \dashrightarrow P$$

be the natural algebra map. Z denotes the field of fractions of P. If $z_i = \mu(X_i)$, then μ may alternatively be described by:

$$f(X_1, \cdots, X_n) \dashrightarrow f(z_1, \cdots, z_n)$$

Suppose we are given a finitely generated subfield $k(q_1, \cdots, q_m)$ of Z and (possibly) an element $c \in Z$. Each q_i and c can be expressed as a quotient $q_i = p_i/d_i$, and $c = p_0/d_0$ with p_i and d_i in P. Let A be the subalgebra of Z generated by P and $\{\ 1/d_i\ \}$. Note that $Z = (A)$. Select D_i in $k[X_1, \cdots, X_n]$ where $\mu(D_i) = d_i$. By the preceding corollary:

$$H_I \cup \{\ D_0W_0 - 1\ , \cdots, D_mW_m - 1\ \}$$

generates the kernel of the algebra map:

$$\gamma{:}k[W_0, \cdots, W_m, X_1, \cdots, X_n] \dashrightarrow A$$

$$f(W_0, \cdots, W_m, X_1, \cdots, X_n) \dashrightarrow f(d_0, \cdots, d_m, z_1, \cdots, z_n)$$

Let H_γ denote $H_I \cup \{ D_0W_0 - 1, \cdots, D_mW_m - 1 \}$ and let B be the subalgebra of A generated by $\{ p_i/d_i \}$ so that $k(q_1, \cdots, q_m) = (B)$. Modulo the renaming $X_0 \cdots, X_n$ to $W_0, \cdots, W_m, X_1, \cdots, X_n$, the techniques of section 2 now apply.

Process one may be applied to determine:

* if c is transcendental over $k(q_1,...,q_m)$,
* a minimal polynomial for c if c is algebraic over $k(q_1,...,q_m)$. This also determines if $c \in k(q_1,...,q_m)$.

Process two may be applied to determine the algebraic or transcendental nature of Z over $k(q_1,...,q_m)$. This includes the index in case of algebraicity or transcendence degree in case the extension is transcendental.

4.6 COMMENTS: If one just wants to learn the nature of the field extension Z over $k(q_1,...,q_m)$ then there is no element c and this permits one to drop W_0. If the q_i's - and c if there is one - actually lie in P, then none of the W_i's are necessary and $A = P$. In (4.1) we found a common denominator and in (4.5) we threw in each denominator separately. Throwing in each denominator separately adds more variables: $W_0, \cdots, W_m$. Finding a common denominator increases the degree of D. One could even do mixed cases, throwing in several partial common denominators. We do not know the best strategy to follow.

REFERENCES:

Audoly,S Bellu,G. Buttu,A and D'Angio',L. (1991). Procedures to investigate injectivity of polynomial maps and to compute the inverse, Lournal Applicable Algebra, 2 91-104

Buchberger,B. (1965). An algorithm for finding a basis for the residue class ring of a zero-dimensional polynomial ideal, Dissertation, Universitaet Innsbruck, Institut fuer Mathematik.

Buchberger,B. (1970). An algorithmic criterion for the solvability of algebraic systems of equations. Aequationes Mathematicae 4/3, 374-383.

Buchberger,B. (1976). A theoretical basis for the reduction of polynomials to canonical forms. ACM Sigsam Bull. 10/3 19-29 1976 & ACM Sigsam Bull. 10/4 19-24.

Buchberger,B. (1979). A criterion for detecting unnecessary reductions in the construction of Groebner bases. Proc. of EUROSAM 79, Lect. Notes in Computer Science 72, Springer 3-21.

Buchberger,B. (1984). A critical-pair/completion algorithm for finitely generated ideals in rings. Decision Problems and Complexity. (Proc of the Symposium "Rekursive Kombinatorik", Muenster, 1983.) E. Boerger, G. Hasenjaeger, D. Roedding, eds. Springer Lecture Notes in Computer Science, 171, page 137.

Buchberger,B. (1985). Groebner bases: an algorithmic method in polynomial ideal theory. Multidimensional Systems Theory. N. K. Boese ed. D. Reidel Pub Co. 184-232.

Kredel,H. and Weispfenning, V. (1988). Computing dimension and independent sets for polynomial ideals. Special Volume of the JSC on the computational aspects of commutative algebra. Vol. 6 1988.

Ollivier,F. (1989). Inversibility of rational mappings and structural identifiabily in automatics. Proc. ISSAC' 89, 43-53, ACM

Shannon,D. and Sweedler,M. (1988). Using Groebner bases to determine algebra membership, split surjective algebra homomorphisms and determine birational equivalence. J. Symbolic Computation, 6, 267-273.

Shannon,D. and Sweedler,M. (1987). Using Groebner bases to determine the algebraic or transcendental nature of field extensions within the field of rational functions. Preprint.

Spear,D. (1977). A constructive approach to commutative ring theory. Proceedings 1977 MACSYMA User's Conference, pp.369-376.

A "Divide and Conquer" Algorithm for Hilbert-Poincaré Series, Multiplicity and Dimension of Monomial Ideals*

Anna Maria Bigatti[1] Pasqualina Conti[2] Lorenzo Robbiano[1] Carlo Traverso[3]

[1] Dipartimento di Matematica, Università di Genova
[2] Istituto di Matematiche Applicate, Università di Pisa
[3] Dipartimento di Matematica, Università di Pisa

1 Introduction

In the computation of the Hilbert-Poincaré series of homogeneous ideals, the known algorithms, [MM], [KP], [BS], [BCR] have a first algebraic step coinciding with the computation of the associated Gröbner basis w.r.t. any ordering and the corresponding initial ideal (the *associated staircase*), and a second combinatorial step that from the staircase computes the Hilbert-Poincaré series.

A well-known classical algorithm computes the Hilbert-Poincaré series from a free resolution, but is practically infeasible because of its complexity. The algorithms of [MM] use an inclusion-exclusion counting technique; the algorithms of [KP] and [BCR] proceed by induction on the dimension; the algorithm of [BS] proceeds by induction on the number of generators of the initial ideal (the cogenerators of the staircase).

Usually, combinatorial algorithms can be speeded up by a "Divide and Conquer" approach: splitting the problem into two smaller problems of approximately the same size. In successful cases this trades a linear step for a logarithmic step, and can reduce from exponential to polynomial complexity.

Our approach explains how to split a staircase through the choice of a monomial (the *pivot*), then we discuss how to design a strategy for the choice of the pivot. The worst case complexity is not improved, since in some extreme cases every splitting is bad, (the computation of Hilbert-Poincaré series is at least as difficult as a NP-complete problem in the number of variables, see [BS]) but in several practical cases the situation is much better; in particular, our algorithm in the best case has a complexity that is a linear factor better than the best case of [BS], and can be specialized, with a choice of the splitting strategy, to the algorithm of [BCR]. In practice, a simple random strategy is quite good, avoids the costly computations involved in choice of an optimal variable of [BCR], and marginally improves the performance even in the optimal Borel-normed case.

* This research was performed with the contribution of C.N.R., M.U.R.S.T, and CEC contract ESPRIT B.R.A. n.6846 POSSO.
E-mail addresses: `bigatti@unimat.to.cnr.it`, `conti@dm.unipi.it`, `robbiano@igecuniv.bitnet`, `traverso@dm.unipi.it`.

The same approach allows to compute the dimension and the multiplicity. If only the dimension is needed, a standard improvement (computing the radical) allows a speeding of the algorithm, and the knowledge of the dimension allows to simplify the finding of the multiplicity.

The algorithms have been implemented, both in CoCoA, [GN] and AlPi, [TD]. Some test cases are given.

2 Staircases

A *staircase* S (also called *Ferrer diagram* or *order ideal of monomials*) is a set of elements of $\mathbf{N}^n$ such that if $(a_1, \ldots, a_n) \in S$ and $b_i \leq a_i$ then $(b_1, \ldots, b_n) \in S$.

On $\mathbf{N}^n$ there is a partial ordering $(a_1, \ldots, a_n) \leq (b_1, \ldots, b_n)$ iff $a_i \leq b_i$ for each i. The corresponding lattice operations $\alpha \wedge \beta$, $\alpha \vee \beta$ are the componentwise min and max.

Elements of $\mathbf{N}^n$ correspond to power products (terms, monic monomials) of $k[x_1, \ldots, x_n]$; the partial ordering corresponds to divisibility ($\alpha \leq \beta \iff X^\alpha \mid X^\beta$). We often identify α with X^α, and use notations referring to both indifferently. In particular, if $\alpha = (a_1, \ldots, a_n) \in \mathbf{N}^n$, $|\alpha| = \sum a_i$ will be called the *degree* of α. The operations $\wedge$ and $\vee$ correspond to GCD and lcm.

We say that an element of $\mathbf{N}^n$ is a *pure power* if it has only one coordinate that is non zero. Otherwise it is called *mixed*.

If $I \subseteq k[x_1, \ldots, x_n]$ is an ideal, and we have an ordering, then the staircase associated to I is the set of exponents $\alpha = (a_1, \ldots, a_n)$ such that no element of I has leading power product equal to X^α. The staircase associated to I can be computed through a Gröbner basis, taking all exponents such that no leading power product of the Gröbner basis divides them, and the X^α constitute a linear basis of $k[X]/I$. Conversely, from the staircase we can recover the leading power products of the reduced Gröbner basis, corresponding to the minimal elements of the complement of the staircase.

The staircase is usually given through these minimal elements, that are called the *minimal cogenerators* of the staircase. Given a set G of elements of $\mathbf{N}^n$, there is a maximal staircase disjoint from G, and it is called the staircase *cogenerated* by G. It is the complementary of the monoideal generated by G.

We denote with $[\alpha_1, \ldots, \alpha_m]$ the staircase cogenerated by $\alpha_1, \ldots, \alpha_m$; if S is a staircase, denote with $[S]$ the minimal set of cogenerators of S. In particular, $[[\alpha_1, \ldots, \alpha_m]]$ is obtained deleting from $\{\alpha_1, \ldots, \alpha_m\}$ all the elements that are multiple of another element (the notation implicitly assumes that no duplications appear in $\{\alpha_1, \ldots, \alpha_m\}$). The algorithm for operating such deletions is an essential tool, and its efficient implementation is very important. The issue is discussed in section 6.

Given two staircases S_1, S_2, we say that S_1 is *strictly smaller* than S_2 if S_1 is a proper subset of S_2 and moreover an injective map ϕ exists from $[S_1]$ to $[S_2]$ such that $\phi(\alpha) \geq \alpha$. Given a staircase, only a finite number of strictly smaller staircases exists.

A *T-staircase* is a translate of a staircase; all that is said for staircases applies, with minor modifications, to T-staircases. Most of what we will prove will be applicable to T-staircases without modifications, and we will not even quote it.

Given a staircase (or a T-staircase) S one defines its Hilbert-Poincaré series being the formal power series in the indeterminate T defined by $\sum_{i=0}^{\infty} d_i T^i$, where d_i is the number of elements of S of degree i, and is denoted by H_S (of course, this being an infinite formula, it is not an algorithm). Clearly, if $S' = \alpha + S$, then $H_{S'} = T^{|\alpha|} H_S$.

If I is a homogeneous ideal of $k[X]$, the Hilbert-Poincaré series of $k[X]/I$ is defined, and coincides with the Hilbert-Poincaré series of the associated staircase. This is the motivation of the interest in computing the Hilbert-Poincaré series of staircases.

Sometimes, one is not interested in the Hilbert-Poincaré series, but only in the *dimension* and the *multiplicity* of the staircase. These are defined as follows. Consider a staircase S, and let $\sum_0^{\infty} d_i T^i$ be its Hilbert-Poincaré series. Define $\phi(r) = \sum_0^r d_i$. Then a well-known theorem states that $\phi(r) = \frac{e}{d!} r^d + O(r^{d-1})$, for suitable integers $e > 0$, $d \geq 0$. Then e is the multiplicity and d is the dimension, and the dimension is smaller than or equal to the number n of variables. The difference $n - d$ is called *codimension*. We denote by e_S, δ_S, c_S the multiplicity, dimension and codimension of S.

Moreover, $H_S = p(T)/(1-T)^d$, where $p(T)$ is a polynomial such that $p(1) = e_S$, $d = \delta_S$. We denote by $\langle S \rangle$ the polynomial $(1 - T)^n H_S$, where n is the number of variables; it has a zero in 1 of order $n - d$. We write $\langle \alpha_1, \ldots, \alpha_m \rangle$ for $\langle [\alpha_1, \ldots, \alpha_m] \rangle$; the aim of the algorithms is to compute $\langle \alpha_1, \ldots, \alpha_m \rangle$ from $(\alpha_1, \ldots, \alpha_m)$.

The description of the algorithm and its correctness do not depend on the unproved assertions stated above, and the algorithm itself is a proof of them.

Everything that will be proved in this paper applies to staircases of modules without changes; a staircase associated to a submodule of a free module is a disjoint union of T-staircases, hence the Hilbert-Poincaré series is the sum of separately computable Hilbert-Poincaré series of T-staircases. Of course, some information can be present relying the different T-staircases, and this could be used to speed up the algorithms, but in general this information is not available. This issue needs further study and experimenting.

3 Splitting a Staircase

We will consider two types of splitting of staircases: as a product of staircases (a *vertical splitting*), and as disjoint union of a staircase and a T-staircase (a *horizontal splitting*).

A vertical splitting is possible if and only if we can identify two disjoint subsets X_1 and X_2 of the variables such that any minimal cogenerator is a power product in either the variables X_1 or X_2. In that case, S is a product of two staircases S_1 and S_2, in X_1 and X_2, each one cogenerated by the corresponding cogenerators of S. We have $H_S = H_{S_1} H_{S_2}$, $\langle S \rangle = \langle S_1 \rangle \langle S_2 \rangle$, since the elements of S of degree d correspond to pairs (α, β), $|\alpha| = d_1$, $|\beta| = d_2$, $d = d_1 + d_2$.

Moreover $e_S = e_{S_1} e_{S_2}$, $\delta_S = \delta_{S_1} + \delta_{S_2}$, $c_S = c_{S_1} + c_{S_2}$ (just apply the definition of multiplicity and dimension).

A horizontal splitting is always possible, unless S is reduced to $\{0\}$: if $\alpha \neq 0$ is an element of the staircase, let $S' = \{\beta \in S \mid \alpha \leq \beta\}$, and $S_1 = S \setminus S'$; then S_1 is a staircase, and S' is a translate of a staircase S_2 by α, $S' = \alpha + S_2$. The cogenerators of S_1 are obtained by deleting from the cogenerators of S all the multiples of α and adding α. The cogenerators of S_2 are obtained as follows: let $\{\beta_i\}$ be a set of cogenerators of S; then a set of cogenerators of S_2 is given by $\beta_i : \alpha$, where the operation : is defined as follows: $(b_1, \ldots, b_m) : (a_1, \ldots, a_m) = (c_1, \ldots, c_m)$, where $c_i = \max(b_i - a_i, 0)$. The operation corresponds to the operation $I : J$ between ideals.

In this case $H_S = H_{S_1} + H_{S'} = H_{S_1} + T^{|\alpha|} H_{S_2}$, by the definition of the Hilbert-Poincaré series; moreover $\langle S \rangle = \langle S_1 \rangle + T^{|\alpha|} \langle S_2 \rangle$, $d_S = \max(d_{S_1}, d_{S_2})$, $c_S = \min(c_{S_1}, c_{S_2})$, $e_S = e_{S_1} + e_{S_2}$ if $d_{S_1} = d_{S_2}$, $e_S = e_{S_1}$ if $d_{S_1} > d_{S_2}$, $e_S = e_{S_2}$ if $d_{S_1} < d_{S_2}$.

The element α is uniquely identified by the splitting, (it is the minimal element of S') and is called the *pivot* of the splitting.

In our applications, we assume that α is smaller than one of the cogenerators of S (in multiplicative notation, properly divides it), and different from 0; this is possible unless the staircase has no cogenerators or all cogenerators are of degree one. In that case, both S_1 and S_2 are strictly smaller than S. This implies that any chain of such splittings must terminate.

Dimension and Multiplicity

A computation of the same type is possible when we do not need the whole Hilbert-Poincaré series, but only the codimension (or the dimension), or the codimension and the multiplicity.

In the case of the codimension, it is sufficient to remark that in a vertical splitting the codimension is the sum of the codimensions of the pieces, in a horizontal splitting the codimension of the staircase is the minimum of the codimensions of the pieces. The algorithm is simplified remarking that when we have found one branch of the computation that gives codimension c, we can abandon all the other branches that give higher codimension, and this can often be checked at an early stage. Moreover, instead of the original staircase, we can take its radical: a staircase that has as generators (non minimal) the same elements in which every positive element of $\mathbf{N}$ is substituted by 1.

The multiplicity in general can be computed together with the codimension remarking that for vertical splittings the multiplicity is the product of the multiplicities, and for horizontal splittings the multiplicity is equal to the sum of the multiplicities if the codimension of the two parts is the same, and it is the multiplicity of the part of lower codimension if the two codimensions are different.

The preliminary computation of the codimension (often simpler, since one can take the radical at once) allows to abandon earlier the branches with higher codimension. This has never been tested, but will be implemented in a short time.

4 Terminating the Algorithm

We can proceed in splitting the staircase until each piece is cogenerated by one element of degree 1, or by no element, but this is impractical. We terminate the splitting when we are reduced to a set of cogenerators consisting of some pure powers and a few elements, pairwise coprime, that are not pure powers.

The following theorem holds:

THEOREM 1. *Assume that a staircase S has a minimal set of cogenerators $\{\pi_1, \ldots, \pi_r, \mu_1, \ldots, \mu_s\}$, such that the π_i are pure powers, and the μ_j are mixed and pairwise coprime. Consider the partition of the variables $\{\Pi_0, \ldots, \Pi_s\}$, where Π_0 is the set of the variables not appearing in the μ_j, and Π_l is the set of the variables appearing in μ_l.*
Then S is split vertically according to $\{\Pi_0, \ldots, \Pi_s\}$ into $s+1$ staircases $S_0, \ldots, S_s$, and every $[S_i]$ contains at most one mixed power.

The proof is immediate. Of course, the degenerate case that Π_0 is empty is possible.

The computation of the Hilbert-Poincaré series of these simple staircases is shown in the two following theorem:

THEOREM 2. *Let S be a staircase in $\mathbf{N}^n$ such that $[S] = \{\pi_1, \ldots, \pi_m\}$, $\pi_i = x_i^{c_i}$.*
Then $H_S = \prod(1 - T^{c_i})/(1 - T)^n$

The proof is obtained through a further vertical splitting in staircases in $\mathbf{N}$, that have either no cogenerators (they coincide with $\mathbf{N}$) or one cogenerator π_i (they coincide with $\{0, 1, \ldots, c_i - 1\}$).

In the first case $H_S = 1 + T + T^2 + \ldots + T^i + \ldots = (1 - T)^{-1}$; in the second case $H_S = 1 + T + \ldots + T^{c_i - 1} = (1 - T)^{-1}(1 - T^{c_i})$.

THEOREM 3. *Let S be a staircase in $\mathbf{N}^{m+r}$, and assume $[S] = \{\pi_1, \ldots, \pi_m, \tau\}$, $\pi_i = x_i^{a_i}$, $\tau = x_1^{b_1} \cdots x_m^{b_m} y_1^{c_1} \cdots y_r^{c_r}$, $a_i > b_i > 0$, $c_j > 0$.*
Then

$$H_S = \left(\prod(1 - T^{a_i}) - T^{|c|}\prod(T^{b_i} - T^{a_i})\right) / (1 - T)^{m+r}$$

where $|c| = \sum c_j$.

The proof is done by considering two special subcases.

If $r = 0$ consider the staircase S' cogenerated by $x_1^{a_1}, \ldots, x_m^{a_m}$ and split it with pivot $\alpha = x_1^{b_1} \cdots x_m^{b_m}$; then S' is disjoint union of S and $\alpha + (S' : \alpha)$, and $S' : \alpha$ is cogenerated by $x_1^{a_1 - b_1}, \ldots, x_n^{a_n - b_n}$; hence H_S is computed by difference.

If $m = 0$ then consider the staircase $S' = \mathbf{N}^r$ and split it with pivot $\beta = y_1^{c_1} \cdots y_r^{c_r}$. Then S' is disjoint union of S and $\beta + S'$, and H_S is computed by difference.

In the general case, split S with pivot $y_1^{c_1} \cdots y_r^{c_r}$; then $S = S_1 \cup (\gamma + S_2)$, $\gamma = y_1^{c_1} \cdots y_r^{c_r}$, S_1 cogenerated by $x_1^{a_1}, \ldots, x_m^{a_m}, y_1^{c_1} \cdots y_r^{c_r}$, S_2 cogenerated by $x_1^{a_1}, \ldots, x_m^{a_m}, x_1^{b_1} \cdots x_m^{b_m}$ and we are reduced to the two previous subcases.

Dimension and Multiplicity

If one is interested only in the dimension or in the multiplicity, these can be computed directly.

In the case of Theorem 2, the dimension is $n-m$ and the multiplicity is $\prod c_i$.

In the case of Theorem 3, we have two subcases:

— if $m = 0$ then the dimension is $r - 1$ and the multiplicity is $|c|$;

— if $m > 0$ then the dimension is r and the multiplicity is $\prod a_i - \prod(a_i - b_i)$.

The proof is immediate by inspection of the formulae given in theorems 2 and 3.

5 The Choice of a Splitting

The vertical splittings appear occasionally, but when they are possible in an early stage of the algorithm their importance is dramatic. They are not so easy to discover, so it is wise to search for them when we have a hint that one might exist. We will see an example in which the algorithm is exponential without vertical splittings, but becomes very simple if we look for them. With this example the other known algorithms perform badly.

5.1 Vertical Splittings

The search for the most general vertical splitting is easy, but often useless; probably it is useful only when the sum of the degrees of the non-pure powers is not much larger than the number of variables.

Here is an algorithm for finding the vertical splittings, S being the set of cogenerators of the staircase.

```
P=empty
REPEAT for T in S
  REPEAT for U in P
    IF GCD(T,U) /= 1 THEN
       T=lcm(T,U)
       delete U from P
  ADD T to P
```

At the end of the algorithm, P is a set of coprime power products; if P contains only one element, then no vertical splitting is possible; otherwise P describes the partition of the variables defining the splitting.

In practice, if in the course of the algorithm T contains all the variables, then we can directly exit the algorithm declaring that no vertical splitting is possible.

Since the set P has never more elements than the set of variables, the complexity cannot exceed $O(n^2m)$, where n is the number of variables and m the number of cogenerators, since the cost of a GCD or a lcm is at most n arithmetic comparisons.

The only type of vertical splitting that is always useful to look for, is when a variable does not appear in the mixed power products. This is recognizable if the lcm of the radical of all the mixed power products does not contain one of the variables, and this only costs nm. These splittings appear frequently, and considerably simplify the algorithm: they do not have influence on the combinatorial part, but the product of a sum of univariate polynomials is simpler to compute than the sum of the products, and the effect of an early splitting is precisely to allow to perform the former instead of the latter.

5.2 Horizontal Splittings: Choice of the Pivot

The horizontal splittings can always be found; an optimal strategy would be to find every time a splitting such that the two pieces have sets of cogenerators that have one half of the cogenerators of the original staircase. This is possible (and easy, see below) when there are two variables, but is impossible in general. A strategy that allows splittings as balanced as possible is very useful. The reason for looking for such a strategy is the following: the algorithm for one splitting is of quadratic complexity (the interreduction of $S : \alpha$); splitting the cost in two at every step is as good as possible.

Consider the case of two variables; then we have a set of cogenerators $\tau_i = x^{n_i}y^{m_i}$, $i = 1, \ldots, d$, and we assume that the n_i are in ascending order (and hence the m_i are in descending order). Then if we take $x^{n_{d/2}}$ as pivot, we split the staircase into two staircases with half as many cogenerators.

There are several possible heuristics for the choice of the pivot. The choice that has appeared to be the best is the following: choose a variable that appears in at least two mixed power products, and some power products in which the variable appears. Then choose as pivot the GCD of these power products. In particular, choosing a random variable among those that appears in most mixed power products, and three random power products (or two if only two exist) among those that contain this variable, the practical performance is often quite satisfying. (In some special cases it seems useful not to choose the three power products at random, but to choose them in a way to have a larger GCD; this heuristics has however not yet been implemented.)

If no variable appears in more than one mixed power product, these are all coprime, and we can terminate the algorithm as described in the previous section.

Probably an uniform strategy like this one is not good in every case, or at every point of the algorithm, and the issue of good heuristics is widely open. However the results even with this rough strategy are quite good.

6 Interreducing a Staircase

The interreduction of a non-minimal set of cogenerators $A = \{\alpha_1, \ldots, \alpha_m\}$ of a staircase S is the costliest part of each step of the algorithm; considerably improved performances can hence be obtained by optimizing this step.

The simplest interreduction algorithm is a double loop:

```
FOR T1 in A DO
  FOR T2 in A-{T1} DO
    IF T1 divides T2 THEN delete T2 from A
```

The cost of this algorithm is m^2 divisibility tests (m being the number of A).

The double loop can be cut in two, if we preliminarly sort A in a way that T_1 can divide T_2 only if T_1 precedes T_2; hence, if for example A is sorted in increasing degree, one half of the comparisons can be avoided. The preliminary sorting is not costly anyway, but we can even arrange the algorithm in such a way that the staircases that are generated are already correctly sorted. In this case the cost is $m^2/2$ divisibility tests.

In the more general situation, this algorithm is probably the best one, but often better algorithms can be obtained using some existing information on the staircase.

If A is obtained from $A' = [S]$ adding one cogenerator T, then one can simply delete from A' all the multiples of T and adding T to the result; one can insert the new element in the correct order, and check only the elements that follow it for divisibility. In this case, the cost is m divisibility tests.

We discuss now the case where a staircase S is obtained from another staircase S' dividing by a power product α. We discuss separately three cases: α is a variable, α is a pure power, α is generic. The more special cases allow more efficient algorithms.

Assume that we have $[S] = \{\alpha_1, \ldots, \alpha_m\}$, (this implying that no α_i divides another α_j) and that the α are already correctly sorted.

In the first case, $\alpha = x_i$, we split A in two, $B = \beta_1, \ldots, \beta_r$, $C = \gamma_1, \ldots, \gamma_s$, the β_j not divisible by x_i, the γ_j divisible by x_i; define $\gamma'_j = \gamma_j/x_i$; then the β_j do not divide any of the γ'_l (they do not divide the $\gamma_l = x_i\gamma_l$), and do not divide each other; the γ'_l may divide some of the β_j but do not divide each other. Moreover, the γ'_l are in correct degree order. Hence it is sufficient to delete from the β_j those elements that are multiple of some γ'_l, and merge the resulting β_j with the γ'_l. If B, C have m_1, m_2 elements respectively, ($m = m_1 + m_2$) the cost is at most $m_1 m_2$ divisibility tests (and the cost of merging is at most m).

In the second case, $\alpha = x_i^r$, divide $\alpha_1, \ldots, \alpha_m$ into $r+1$ subsets $A_0, \ldots, A_r$ according to the divisibility by x_i^j (elements of A_j being divisible by x_i^j). Divide the elements of A_j by x_i^j, obtaining A'_j. Then elements of A'_j are not divisible by elements of A'_k if $k \leq j$. Hence one can obtain the result as follows:

```
FOR i= r down to 1
  delete from A'(i-1) the multiples of elements of A'(i)
  merge in  A'(i-1) the elements of A'(i)
```

and at the end $A'(0)$ is a minimal set of cogenerators.

Let a_i be the cardinality od A_i; then the cost of the algorithm is $\sum_{i=0}^{r-1} a_i \sum_{j=i+1}^{r} a_j$, that is smaller than $a_0(\sum_{j=1}^{s} a_i)^2/2$.

In the third general case, we split $\alpha_1, \ldots, \alpha_m$ in two, $B = \beta_1, \ldots, \beta_r$, $C = \gamma_1, \ldots, \gamma_s$, the β_j coprime with α, the γ_j not coprime with α; define $C' = \{\gamma'_j = \gamma_j : \alpha\}$; then the elements of B do not divide any of the elements of C'. Sort and interreduce C' with the general algorithm, delete from B the elements mutiple of some element of C', then merge B and C': this is the result. If B, C have m_1, m_2 elements respectively, the cost is $m_1 m_2 + m_2^2/2$.

We have never accounted for the cost of merging, or of splitting, that is linear. The overall cost of the algorithm, even in the third case, is very small when the splitting is uneven (many elements coprime with the pivot), and this happens quite frequently in the hardest computations; the fact that the pivots are bad is indeed the reason why the overall algorithm is hard. Hence the improvement is essential for these hard examples. This happens when the cogenerators are easily coprime, for example when the degree of the cogenerators are low compared with the number of variables. This is precisely the case in which the computation of the Hilbert-Poincaré series is especially difficult with all the known algorithms.

7 Comparison with the Other Known Algorithms

The algorithms of [KP] and [Ho] coincide with the present algorithm when the pivot is chosen to be a variable.

The algorithm of [BCR] coincides with the present algorithm, in the following variant: choose one variable x_i, and take as pivot x_i^n, where n is the minimum degree in which x_i appears. The algorithm requires the computation of several splittings for choosing the best variable; the overhead can be frequently reduced by special considerations. When there are several variables and none is especially good, and we have to consider them all, it seems that the cost of computing several splittings is difficult to recover by discovering a relatively better variable.

The algorithm of [BS] is related to ours, with a difference. A staircase S is represented as the difference set $S_1 \setminus S_2$, S_1 obtained removing one of the cogenerators of S, and the $S_2 = S_1 \setminus S$, the T-staircase contained in S_1 composed of the multiples of this cogenerator. Both staircases are simpler (in a different sense than ours) and the termination can be done as in our algorithm. The choice of the cogenerator to remove is guided by heuristics dependent on the ordering.

The "Quick and dirty dimension algorithm" of [BS] coincides with our dimension algorithm if the choice of the pivot is made choosing the first variable of the smallest power product (in a suitable ordering).

These remarks show that our algorithm is "potentially better" than the algorithms of [KP], [Ho] and [BCR] that are a particular case (and indeed the experiments confirm this remark).

For the comparison with the algorithm of [BS], we remark that with our interpretation their dimension algorithm is just a special case of our Hilbert-Poincaré series algorithm. For many examples we remark that their algorithm produces badly balanced splittings, hence it is expected to be worse. The experiments confirm this feeling. The algorithm seems to be superior only when there are few cogenerators of high degree in several variables.

The algorithm of [MM] has mainly theoretical interest and it is known to be practically inefficient.

8 Practical Performance

The algorithm as explained above was implemented in COMMON-LISP and included in AlPi, and in Pascal and included in CoCoA.

The practical comparison of algorithms implemented in a heterogeneous way is hard, since it is difficult to separate the effect of the algorithm and the effect of the clever implementation; moreover some tricks can considerably speed the algorithms, and sometimes a trick can be applied to an algorithm and not to another.

To allow a fair evaluation of the algorithm, in the COMMON-LISP implementation we have included an approximate measure of the complexity. We consider as unit of complexity an operation on power products, such as GCD or lcm.

The interreduction of $[S] \cup \{\tau\}$ requires m operations, if S has m cogenerators (we have to test which power products are multiple of τ and delete them). The interreduction of the staircase $S : \tau$ costs m^2 operations, as we have seen in section 6. The interreduction of pure powers is simpler, and is made separately, so we can take m being the number of mixed cogenerators of S. Hence a rough estimate of the complexity of a single step is m^2.

In our implementation we have put a counter that accumulates m^2, and at the end of the algorithm we can know the number of steps, the maximum recursion depth and the sum of m^2. These data are reported in the tables of the examples.

The algorithm of [BCR] is implemented in CoCoA; an experimental implementation of the present algorithm in CoCoA allows some comparisons. Unfortunately, the limited number of variables allowed in CoCoA does not allow the comparison with the more difficult examples, and moreover the implementation of the algorithm of [BCR] in CoCoA is a highly optimized version, while the implementation of the present algorithm is only experimental. Anyway the behaviour is quite good, and the performance of the new algorithm is worse only in some highly stuctured cases.

A simple modification of the COMMON-LISP implementation could be used to perform the algorithm of [BCR], but the comparison would be unfair since the special structure of the pivots allows many improvements that are impossible in the general algorithm.

A very small modification of the implementation in COMMON-LISP (only a few lines of code) implements the algorithm of [BS]. The performance of this rough implementation is not good, compared with the timings given in [BS], and it is not clear if this is due to an optimization of the implementation or to improvements to the algorithm; one can compare anyway the experimental complexity data (the number of steps, the sum of the m^2) and the algorithm of [BS] appears to be inferior (sometimes dramatically inferior) in all but some special cases with very few cogenerators.

The COMMON-LISP sources are available by anonymous FTP at the address `gauss.dm.unipi.it` (131.114.6.55), in the directory `pub/alpi-cocoa/hilbert`, together with documentation and some files of examples.

We report some data about the computation of some examples. These are some staircases associated to Gröbner bases, and some random examples. With $R_{n,m,d}$ we denote a set of m random power products of degree d in n variables (a random power product being a random product of powers of random degree of a random variable, everything with uniform distribution).

The non-random examples are the Gröbner bases of the following ideals:

1) Gonnet example, [BGK], homogenized, with DegRevLex ordering;

2) Valla example (see [TD]), DegRevLex ordering;

3) [BS], n.4.1

4) [BS], n.4.2

5) [BGK], Butcher, homogenized, DegRevLex

6) [TD], Cohn-2 homogenized, DegRevLex

7) [BGK], Hairer 2, homogenized, DegRevLex, up to degree 10

8) [BGK], Hairer 2, homogenized, DegRevLex, up to degree 11

9) [BGK], Hairer 2, homogenized, DegRevLex, the first 904 elements found (as you understand, we did not find the whole Gröbner basis).

10) to 12) are $R_{50,477,2}$, $R_{100,20,10}$, $R_{10,40,20}$.

The data reported are for our algorithm (BCRT) and for our implementation of [BS] algorithm, and we report the computing time on a SUN Sparc-2, the depth of the iterations, the total number of the iterations, and the sum of m^2 as described above. We report the data for one run of the probabilistic algorithm; the experience is that the deviation is not large.

Our implementation of the [BS] algorithm does not work in AKCL with more than 500 monomials because of bind stack overflow. In this case we have used an implementation of a variant, that is usually better (instead of proceeding by induction on all the cogenerators of the staircase only the mixed cogenerators are considered; moreover induction instead of recursion is used). For this variant, the recursion depth no longer makes sense; we have preceded the timings by a *. The example 10) was too long anyway, and we have interrupted it.

	#	BCRT	depth	iter.	$\sum m^2$	BS	depth	iter.	$\sum m^2$
1	854	23″	17	853	2636718	*4′22″		2371	208144503
2	88	1″31	11	131	30630	9″	88	599	224349
3	1372	1′47″	22	2919	8396358	*13′25″		5274	860272285
4	4785	44′32″	34	100241	208685423	*6h25′		829623	41053923691
5	157	2″15	12	197	64967	17″57	157	341	1307015
6	71	1″	13	121	19220	3″15	71	155	122701
7	437	16″	14	733	517435	2′30″	433	1433	27175765
8	567	24″	14	983	827951	*1′25″		1057	898150
9	904	1′13″	16	1653	2603711	*4′		1279	243426014
10	477	1′15″	41	4701	2345490	> 24h		$> 10^7$	
11	20	2″	17	397	1946	2′	14	16383	267082
12	40	15″	18	2535	78129	26″	40	5025	154333

For the examples 3) and 4) the timings given in [BS] for the new implementation on a SUN Sparc-1 are 3′06″ and 2h53′23″ respectively, and $>$ 4 days and 14h32′24″ for the Macaulay version.

9 Two Elementary Examples

Here we show two simple examples (one from [BCR], the other from [BS]) and our computation is in both cases simpler. We take as pivot the GCD of three power products containing one most frequent variable.

Let $S = [x^4, x^3y^3, x^3y^2z, x^3yz^2, y^2t, yt^5]$. The possible sequences of pivots are either a) (y, y, x^3) or b) (y, x^3), or c) (y^2, y, x^3), with probability $(9/20, 9/20, 1/10)$. The computations run as follows:

a):
$\langle S\rangle = \langle x^4, y\rangle + T\langle x^4, x^3y^2, x^3yz, x^3z^2, yt, t^5\rangle =$
$\langle x^4, y\rangle + T(x^4, y\rangle + T^2\langle x^4, x^3y, x^3z, t\rangle =$
$\langle x^4, y\rangle + T(x^4, y\rangle + T^2\langle x^3, t\rangle + T^5\langle x, y, z, t\rangle.$

b):
$\langle S\rangle = \langle x^4, y\rangle + T\langle x^4, x^3y^2, x^3yz, x^3z^2, yt, t^5\rangle =$
$\langle x^4, y\rangle + T\langle x^3, yt, t^5\rangle + T^4\langle x, y^2, yz, z^2, t\rangle.$

c):
$\langle S\rangle = \langle x^4, y^2, x^3yz^2, yt^5\rangle + T^2\langle x^4, x^3y, x^3z, t\rangle =$
$\langle x^4, y\rangle + T\langle x^4, y, x^3z^2t^5\rangle + T^2\langle x^3, t\rangle + T^5\langle x^4, y, z, t\rangle.$

The [BCR] algorithm corresponds to the pivot sequence (x^3, y, y, y):
$\langle S\rangle = \langle x^3, y^2t, yt^5\rangle + T^3\langle x, y^3, y^2z, yz^2, y^2t, yt^5\rangle =$
$\langle x^3, y\rangle + T\langle x^3, yt, t^5\rangle + T^3\langle x, y\rangle + T^4\langle x, y^2, yz, z^2, yt, t^5\rangle =$
$\langle x^3, y\rangle + T\langle x^3, yt, t^5\rangle + T^3\langle x, y\rangle + T^4\langle x, y, z^2, t^5\rangle + T^5\langle x, y, z, t\rangle.$

The [BS] algorithm computes:
$\langle S\rangle = \langle x^3y^3, x^3y^2z, x^3yz^2, y^2t, yt^5\rangle - T^4\langle y^3. y^2z, yz^2, y^2t, yt^5\rangle =$
$\langle x^3y^2z, x^3yz^2, y^2t, yt^5\rangle - T^6\langle z, t\rangle - T^4\langle y^2z, yz^2, y^2t, yt^5\rangle + T^7\langle z, t\rangle =$
$\langle x^3yz^2, y^2t, yt^5\rangle - T^6\langle z, t\rangle - T^6\langle z, t\rangle - T^4\langle yz^2, y^2t, yt^5\rangle + T^7\langle z, t\rangle + T^7\langle z, t\rangle =$
$\langle y^2t, yt^5\rangle - T^6\langle yt, t^5\rangle - 2T^6\langle z, t\rangle - T^4\langle yz^2, y^2t, yt^5\rangle + T^7\langle yt, t^5\rangle + 2T^7\langle z, t\rangle.$

Here is the second example. Let $S = [ac, ab^2, a^2b, a^3, b^3d]$. Then the pivot is either a or b. In the first case,
$\langle S\rangle = \langle a, b^3d\rangle + T\langle c, b^2, ab, a^2\rangle,$
in the second case,
$\langle S\rangle = \langle b, ac, a^3\rangle + \langle c, b^2, ab, a^2\rangle.$

The [BCR] algorithm has as first pivot either a or c (both variables are excellent, in the terminology of [BCR]), and in the second case:
$\langle S\rangle = \langle c, ab^2, a^2b, a^3, b^3d\rangle + T\langle a, b^3d\rangle =$
$\langle a, c, b^3d\rangle + T\langle c, b^2, ab, a^3\rangle + T\langle a, b^3d\rangle.$

The algorithm of [BS] computes
$\langle S\rangle = \langle ac, ab^2, a^2b, a^3\rangle - T^4\langle a\rangle =$
$\langle ac, ab^2, a^2b\rangle - T^3\langle b\rangle - T^4\langle a\rangle =$
$\langle ac, ab^2\rangle - T^3\langle c, b\rangle - T^3\langle b\rangle - T^4\langle a\rangle =$
$\langle ac\rangle - T^3\langle c\rangle - T^3\langle c, b\rangle - T^3\langle b\rangle - T^4\langle a\rangle$

10 A Simple Bad Example

The computation of the Hilbert-Poincaré series in general, and even computing the dimension, is a problem that is harder than an NP-complete problem (see [BS]), hence bad examples are unavoidable.

Here we study a very simple example that has a very bad behaviour, unless we allow vertical splittings and randomized algorithms: avoiding the general vertical splittings, or taking a "natural" ordering of the variables requires exponential time. The example is the following:

$$I = (x_0x_1, x_1x_2, \ldots, x_{n-1}x_n)$$

A randomized algorithm splits the staircase horizontally in two staircases, one with $n-2$ and one with $n-3$ elements; these can be split vertically, and the expected lengths are in ratio 3 : 1. Hence the expected complexity is polynomial.

If no vertical splittings are allowed, more than $2^{n/3}$ steps are necessary. Moreover, if we always choose as pivot the lowest (or highest) possible variable appearing in more than one monomial, the splittings are always bad, and no vertical splittings are useful.

The algorithm of [BS] in this case has the same type of behaviour; however their heuristics is in this case the worst possible, and even with vertical splittings the algorithm remains exponential.

Indeed, with 42 variables the example can be computed with our implementation in 2″, (with 101 variables it takes 40″); without vertical splittings in 42 variables it takes 6′, and 13h47′ with the algorithm of [BS].

References

[BCR] Bigatti, A.M., Caboara, M., Robbiano, L.,*On the computation of Hilbert-Poincaré series*, AAECC Journal, vol. 2, 1991

[BS] Bayer-Stillman, *Computation of Hilbert functions*, J. of Symbolic Comp. vol. 14 (1992)

[BGK] Boege, W., Gebauer, R., Kredel, H. *Some examples for solving systems of algebraic equations by calculating Gröbner bases*, J. of Symbolic Comp. (1986)

[GN] Giovini, A., Niesi, G. *CoCoA: a user-friendly system for commutative algebra*, DISCO-90 Proceedings, LNCS 429, Springer Verlag (1990)

[Ho] Hollman, J.,*On the computation of the Hilbert series*, Latin 92, Saõ Paulo, LNCS 583, Springer Verlag (1992)

[KP] Kondrat'eva, M.V., Pankrat'ev, E.V.,*A recursive algorithm for the computation of Hilbert polynomial*, EUROCAL 87, LNCS 387, Springer Verlag (1987)

[MM] Möller, M., Mora, T.,*The Computation of the Hilbert Function*, EUROCAL 83, LNCS 162, Springer Verlag (1983)

[TD] Traverso, C., Donati, L.,*Experimenting the Gröbner basis algorithm with the AlPi system*, ISSAC 89 Proceedings, A. C. M. (1989)

An Efficient Algorithm for the Sparse Mixed Resultant

John Canny * and Ioannis Emiris*

Computer Science Division, 571 Evans Hall,
University of California at Berkeley, Berkeley CA 94720.
E-mail: jfc@cs.berkeley.edu and emiris@cs.berkeley.edu.

Abstract. We propose a compact formula for the mixed resultant of a system of $n+1$ sparse Laurent polynomials in n variables. Our approach is conceptually simple and geometric, in that it applies a mixed subdivision to the Minkowski Sum of the input Newton polytopes. It constructs a matrix whose determinant is a non-zero multiple of the resultant so that the latter can be defined as the GCD of $n+1$ such determinants. For any specialization of the coefficients there are two methods which use one extra perturbation variable and return the resultant. Our algorithm is the first to present a determinantal formula for arbitrary systems; moreover, its complexity for unmixed systems is polynomial in the resultant degree. Further empirical results suggest that this is the most efficient method to date for sparse elimination.

1 Introduction

We are given $n+1$ polynomials $f_1, \ldots, f_{n+1} \in \mathbb{C}[x_1, \ldots, x_n]$ and we seek a condition on the coefficients of the f_i that indicates when the system has a solution. Sparsity implies that only certain monomials have non-zero coefficients in the f_i. Such systems may have trivial solutions with some $x_i = 0$ for all coefficient specializations, so we concentrate on solutions $x = \xi$ with $\xi \in (\mathbb{C}^*)^n$, where $\mathbb{C}^* = \mathbb{C} - \{0\}$. Under this assumption, we can deal with the more general case of f_i's which are *Laurent* polynomials in $\mathbb{C}[x_1, x_1^{-1}, \ldots, x_n, x_n^{-1}]$.

We use x^e to denote the monomial $x_1^{e_1} \cdots x_n^{e_n}$, where $e = (e_1, \ldots, e_n) \in \mathbb{Z}^n$ is a multi-exponent. Let $\mathcal{A}_i = \{a_{i1}, \ldots, a_{im_i}\} \subseteq \mathbb{Z}^n$ denote the set of exponents occurring in f_i, then

$$f_i = \sum_{j=1}^{m_i} c_{ij} x^{a_{ij}} , \qquad \text{for } i = 1, \ldots, n+1 , \tag{1}$$

and we suppose $c_{ij} \neq 0$ so that $\mathcal{A}_i$ is uniquely defined given f_i.

Definition 1. The finite set $\mathcal{A}_i \subset \mathbb{Z}^n$ of all monomial exponents appearing in f_i is the *support* of f_i. The *Newton Polytope* of f_i is $Q_i = \text{Conv}(\mathcal{A}_i) \subset \mathbb{R}^n$, the convex hull of $\mathcal{A}_i$.

* Supported by a David and Lucile Packard Foundation Fellowship and by NSF Presidential Young Investigator Grant IRI-8958577.

A polynomial system is *unmixed* if all supports $\mathcal{A}_i$ are the same for $i = 1, \ldots, n+1$, otherwise it is *mixed*.

Definition 2. The *Minkowski Sum* $A + B$ of convex polytopes A and B in $\mathbb{R}^n$ is the set

$$A + B = \{a + b | a \in A, b \in B\} \ .$$

$A + B$ is a convex polytope. Let $\mathrm{Vol}(A)$ denote the usual n-dimensional volume of A.

Definition 3. Given convex polytopes $A_1, \ldots, A_n \subseteq \mathbb{R}^n$, there is a unique real-valued function $MV(A_1, \ldots, A_n)$ called the *Mixed Volume* which is multilinear with respect to Minkowski sum, such that $MV(A_1, \ldots, A_1) = n!\mathrm{Vol}(A_1)$. Equivalently, if $\lambda_1, \ldots, \lambda_n$ are scalars, then $MV(A_1, \ldots, A_n)$ is precisely the coefficient of $\lambda_1 \lambda_2 \cdots \lambda_n$ in $\mathrm{Vol}(\lambda_1 A_1 + \cdots + \lambda_n A_n)$ expanded as a polynomial in $\lambda_1, \ldots, \lambda_n$.

The Newton polytopes offer a convenient model for the sparsity of a polynomial system, in light of the following upper bound on the number of common roots, see [1], [11], [9].

Theorem 4. *[1] Let $f_1, \ldots, f_n \in \mathbb{C}[x_1, x^{-1}, \ldots, x_n, x_n^{-1}]$. The number of common zeros in $(\mathbb{C}^*)^n$ is either infinite, or does not exceed $MV(Q_1, \ldots, Q_n)$. For almost all specialization of the coefficients c_{ij} the number of solutions is exactly $MV(Q_1, \ldots, Q_n)$.*

For systems of $n+1$ polynomials in n unknowns, there are generically no solutions, and the resultant delimits those systems that do have a solution. We adopt the following definition for the *sparse* resultant from [15]; it is identical to the $(\mathcal{A}_1, \ldots, \mathcal{A}_{n+1})$-*resultant* of [4]. Regard a polynomial f_i as a generic point $(c_{i1}, \ldots, c_{im_i}) \in \mathbb{P}^{m_i}$ in the space of all possible polynomials with the given set of exponents $\mathcal{A}_i$, after identifying scalar multiples. Then the input system is a point $c = (c_{11}, \ldots, c_{1m_1}, \ldots, c_{(n+1)1}, \ldots, c_{(n+1)m_{n+1}})$ in $\mathbb{P}^{m_1 - 1} \times \cdots \times \mathbb{P}^{m_{n+1}-1}$. Let $Z_0 = Z_0(\mathcal{A}_1, \ldots, \mathcal{A}_{n+1})$ be the set of all points c such that the system has a solution in $(\mathbb{C}^*)^n$, and let $Z = Z(\mathcal{A}_1, \ldots, \mathcal{A}_{n+1})$ denote the (Zariski) closure of Z_0 in the product of projective spaces. Z is an irreducible algebraic set.

Definition 5. The *sparse resultant* $R(\mathcal{A}_1, \ldots, \mathcal{A}_{n+1})$ of the system (1) is an irreducible polynomial in $\mathbb{Z}[c]$. If $\mathrm{codim}(Z) = 1$ then $R(\mathcal{A}_1, \ldots, \mathcal{A}_{n+1})$ is the defining polynomial of the hypersurface Z. If $\mathrm{codim}(Z) > 1$ then $R(\mathcal{A}_1, \ldots, \mathcal{A}_{n+1}) = 1$.

Throughout this article, it is assumed without loss of generality that the affine lattice generated by $\sum_{i=1}^{n+1} \mathcal{A}_i$ is n-dimensional. Moreover, this lattice is identified with $\mathbb{Z}^n$ after a change of variables, if necessary [21]. Then,

Proposition 6. *[15] The sparse resultant is separately homogeneous in the coefficients $(c_{i1}, \ldots, c_{im_i})$ of each f_i and its degree in these coefficients equals the mixed volume of the other n Newton polytopes $MV(Q_1, \ldots, Q_{i-1}, Q_{i+1}, \ldots, Q_{n+1})$.*

This implies that the total degree $\deg R$ of the resultant equals the sum of all $n+1$ n-fold Mixed Volumes.

The practical significance of this approach relies on the fact that polynomial systems are frequently sparse in several applications such as computer vision, robot kinematics, graphics and geometric modeling. More precise examples include the cyclic n-roots problem, computing the motion from point matches and inverse kinematics. For the later problem, the homogeneous approach leads to an intractable problem, while the custom approach of [14] requires time in the order of milliseconds.

The following section points to previous works on which our approach is based and briefly states our results. Section 3 describes the construction of a matrix M of the correct degree in the coefficients of f_1. Section 4 proves that $\det(M)$ is a multiple of the sparse resultant and is not identically zero. Section 5 shows that the resultant is the Greatest Common Divisor (GCD) of $n+1$ such determinants and sketches two ways to compute it for various specializations. We illustrate the algorithm with an example in Sect. 6 and analyze its complexity in Sect. 7. The article concludes with some open questions.

2 Background and the Present Approach

Our approach consists of regarding the coefficients c_{ij} as indeterminates and expressing the sparse resultant through various determinants in these coefficients. We shall define the resultant as the GCD of $n+1$ such determinants, each of which is a multiple of the resultant and may be thought of as a generalized *inertia form* [23]; Hurwitz showed for the general homogeneous case that the resultant is the GCD of all inertia forms [6]. Alternatively, we may compute the resultant via a series of n divisions of determinants, similarly to Cayley's method [16]. Lastly, our construction is closely related to that of Macauley's [13].

More recently, the sparse unmixed resultant was defined as the Chow form of a projective toric variety in [10], see also [4]. Algorithms for its computation and evaluation were proposed in [20], the most efficient one having complexity higher than polynomial in the degree of the resultant and exponential in n with a quadratic exponent.

For *multigraded* systems, an optimal determinantal formula, called of *Sylvester type*, is given in [22], These systems are unmixed and include polynomials that are homogeneous of degree d_j in each group of variables $\mathbf{x}_j$, where $\mathbf{x}_j$ has l_j+1 variables. The main theorem defines a matrix whose determinant is the resultant for such a system, provided that for each j, $l_j = 1$ or $d_j = 1$.

An explicit formula for the sparse resultant was given in [15] as a *Poisson product* $R' \prod_{\xi \in V(f_1,\ldots,f_n)} f_{n+1}(\xi)$ where R' is a rational function in the coefficients of $f_1, \ldots, f_n$.

Our algorithm requires two randomized steps, the success of which has arbitrarily high probability and can be verified deterministically. The running time for unmixed systems is given in the following restatement of Theorem 24, which makes the algorithm the most efficient to date for this case.

Theorem 7. *Assume that our algorithm executes on an arbitrary unmixed system. Then its asymptotic bit complexity, if we omit logarithmic factors, is polynomial in* $\max_i\{m_i\}$ *and the total degree of the resultant and exponential in* n *with a linear exponent.*

Furthermore, this is the first algorithm that produces a determinantal formula for mixed systems. Although a similar complexity bound as above is not possible in this case, empirical results and a heuristic analysis imply that, for most mixed systems in practice, the algorithm's complexity is given by the above theorem.

3 Matrix Construction

We define and analyze the properties of matrix M associated with the polynomial f_1. Let Q denote the Minkowski Sum of all input Newton polytopes

$$Q = Q_1 + Q_2 + \cdots + Q_{n+1} \subset \mathbb{R}^n \ .$$

If we define an $(n+1)$-argument vector sum

$$\oplus : (\mathbb{R}^n)^{(n+1)} \rightarrow \mathbb{R}^n : (p_1, \ldots, p_{n+1}) \mapsto p_1 + \cdots + p_{n+1} \ ,$$

then Q may be thought of as the image of $Q_1 \times \cdots \times Q_{n+1}$ under $\oplus$. This is clearly a many-to-one mapping; to define a unique inverse $(p_1, \ldots, p_{n+1})$ in $\oplus^{-1}(q) \cap Q_1 \times \cdots \times Q_{n+1}$, for each $q \in Q$, a method from [21] and [2] is employed. Choose $n+1$ sufficiently generic linear forms $l_1, \ldots, l_{n+1} \in \mathbb{Z}[x_1, \ldots, x_n]$ and define, for $1 \leq i \leq n+1$, *lifted* Newton polytopes

$$\hat{Q}_i \triangleq \{(p_i, l_i(p_i)) : p_i \in Q_i\} \subset \mathbb{R}^{n+1} \ .$$

Let the Minkowski Sum of the lifted Newton polytopes be

$$\hat{Q} = \hat{Q}_1 + \cdots + \hat{Q}_{n+1} \subset \mathbb{R}^{n+1} \ .$$

We make use of

Definition 8. Given a convex polytope in $\mathbb{R}^{n+1}$, its *lower envelope* with respect to vector $v \in \mathbb{R}^{n+1}$ is the closure of the subset of all points r on its surface such that, given a point z at infinity in the direction of v, the segment (r, z) intersects the polytope at a point other than r.

Let $\pi : \mathbb{R}^{n+1} \rightarrow \mathbb{R}^n$ denote projection on the first n coordinates, and $h : \mathbb{R}^{n+1} \rightarrow \mathbb{R}$ denote projection on the $(n+1)$-st. Now consider the lower envelope of $\hat{Q}$ with respect to $(0, \ldots, 0, 1)$ and let $s : \mathbb{R}^n \rightarrow \mathbb{R}^{n+1}$ map each point in Q to the point on this envelope that lies in $\pi^{-1}(q)$. Equivalently

$$s(q) \ = \ \hat{q} \in \pi^{-1}(q) \cap \hat{Q} \ , \qquad \text{such that } h(\hat{q}) \text{ is minimized} \ .$$

The lower envelope of $\hat{Q}$ is then $s(Q)$. By construction the l_i's are generic enough so that every point $\hat{q}$ on the lower envelope can be *uniquely* expressed as a sum of points $\hat{q}_1 + \cdots + \hat{q}_{n+1}$ with $\hat{q}_i \in \hat{Q}_i$. This is implemented by picking, for each i, a random integer vector with independent entries whose bit size is $\log c$, for some constant $c > 1$. Then the probability that the genericity condition fails is bounded by $1/c$ [17, Lemma 1].

Let $\hat{\Delta}$ denote the natural (coarsest) polyhedral subdivision of the lower envelope of $\hat{Q}$. Each facet (n-dimensional face) of $\hat{\Delta}$ is a Minkowski sum $\hat{F}_1 + \cdots + \hat{F}_{n+1}$ with $\hat{F}_i$ a face of $\hat{Q}_i$, and since lower envelope points have unique expressions as sums,

$$\sum_{i=1}^{n+1} \dim(\hat{F}_i) = n \ .$$

The image of $\hat{\Delta}$ under π induces a polyhedral subdivision Δ of Q whose cells are of the form $F_1 + \cdots + F_{n+1}$ with the same dimension property, a consequence of which is the following

Remark. For every cell $F_1 + \cdots + F_{n+1}$ in Δ, F_i a face of Q_i, at least one of the F_i is zero-dimensional, i.e. a vertex.

Definition 9. A *mixed cell* of the induced subdivision is a cell which is a sum $F_1 + \cdots + F_{n+1}$ where *exactly one* F_i is a vertex. Thus the remaining F_j for $j \neq i$ are edges.

For selecting the matrix entries in a well-defined manner, we must perturb the Minkowski sum slightly so that each integer lattice point lies in the *interior* of a cell of Δ. Thus we choose a sufficiently small generic vector $\delta \in \mathbb{Q}^n$, and the set of exponents that indexes the rows and columns of M is

$$\mathcal{E} = \mathbb{Z}^n \cap (\delta + Q) \ .$$

If Δ_δ denotes the subdivision obtained by shifting all faces of Δ by δ, the choice of δ is satisfactory if every $p \in \mathcal{E}$ lies in the interior of a cell of Δ_δ. We can now define our selection rule for elements of M based on a function $RC : \mathcal{E} \to \mathbb{Z}^2$, for row content.

Definition 10. (Row content function) Let $p \in \mathcal{E}$ lie in the interior of a cell $\delta + F_1 + \cdots + F_{n+1}$ of Δ_δ. Let i be the largest integer such that F_i is a vertex, so $F_i = a_{ij}$ for some j. Then $RC(p) = (i, j)$.

The row of M indexed by $p \in \mathcal{E}$ contains the coefficients of f_i, and represents a multiple of f_i which is

$$x^{(p - a_{ij})} f_i \tag{2}$$

where $(i, j) = RC(p)$. Let $|\mathcal{E}|$ denote the cardinality of set $\mathcal{E}$; then,

Definition 11. M is an $|\mathcal{E}| \times |\mathcal{E}|$ matrix whose rows and columns are indexed by elements of $\mathcal{E}$, and whose element at row p and column q is as below, for arbitrary $p, q \in \mathcal{E}$ with $RC(p) = (i, j)$:

$$M_{pq} = \begin{cases} c_{ik} & \text{if } q - p + a_{ij} = a_{ik} \text{ for some } k \ , \\ 0 & \text{if } q - p + a_{ij} \notin \mathcal{A}_i \ . \end{cases}$$

Therefore $M_{pp} = c_{ij}$ where $(i, j) = RC(p)$. The matrix is well-defined since it is easily seen that all exponent vectors $p - a_{ij} + a_{ik}$ for $a_{ik} \in \mathcal{A}_i$ lie within $\mathcal{E}$; this is also implied by the discussion in the next section.

4 A Nonzero Multiple of the Resultant

First we prove that the determinant of M is a multiple of the resultant. M represents a linear map $\mathbb{C}^{|\mathcal{E}|} \to \mathbb{C}^{|\mathcal{E}|}$ which we can interpret as the map taking the vector of coefficients of $(g_1, \ldots, g_{n+1})$ to the vector of coefficients of g, where

$$g = g_1 f_1 + \cdots + g_{n+1} f_{n+1} \tag{3}$$

and the support of g is $\mathcal{E}$; in addition, the support of g_i is $\{p - a_{ij} \mid p \in \mathcal{E},\ RC(p) = (i, j)\}$. Thus $|\mathcal{E}|$ is the total number of non-zero coefficients in the g_i's.

Lemma 12. *If there exists $\xi \in (\mathbb{C}^*)^n$ such that $f_1(\xi) = \cdots = f_{n+1}(\xi) = 0$, then* $\det(M) = 0$.

Proof. Assume that M is non-singular. Then the linear map defined by M is surjective and we can choose polynomials $g_1, \ldots, g_{n+1}$ such that g in (3) is a monomial. This monomial must be zero at every solution ξ, which is infeasible for $\xi \in (\mathbb{C}^*)^n$. Hence there can be no solution in $(\mathbb{C}^*)^n$, which is a contradiction. □

Proposition 13. *The sparse resultant divides the determinant of M.*

Proof. The lemma implies that $\det(M) = 0$ on the set Z_0 of specializations of c_{ij} such that the system has a solution in $(\mathbb{C}^*)^n$. Thus it is zero on the closure Z of Z_0, which is exactly the zero set of the resultant $R(\mathcal{A}_1, \ldots, \mathcal{A}_{n+1})$. Since the resultant is irreducible it must divide $\det(M)$. □

To alleviate the possibility that $\det(M)$ is identically zero, we show that under the following specialization of the coefficients c_{ij}, $\det(M) \neq 0$:

$$c_{ij} \mapsto t^{l_i(a_{ij})}$$

so that each c_{ij} becomes an integral power of t where t is a new indeterminate. Observe that the Newton polytope of the specialized f_i as a polynomial in $\mathbb{C}[x_1, \ldots, x_n, t]$ is precisely $\hat{Q}_i$. Let $M(t)$ denote the matrix M under this specialization, and $\det(M)(t)$ denote its determinant, which is a polynomial in t with integer coefficients.

Theorem 14. *The lowest degree term of* $\det(M)(t)$ *is the product of leading diagonal elements of* $M(t)$. *That is, it has coefficient 1 and (integer) exponent*

$$\sum_{p \in \mathcal{E}} l_i(a_{ij})$$

where for each p, $(i,j) = RC(p)$. *Therefore this determinant is non-vanishing.*

This theorem follows from the following series of lemmas.

Lemma 15 (Geometric). *Let* $\hat{p}$ *be a point in the interior of some facet of the subdivision* $\hat{\Delta}$ *of the lower envelope* $s(Q)$. *By construction,* $\hat{p}$ *has a unique expression as a sum of points from* $\hat{Q}_1, \ldots, \hat{Q}_{n+1}$, *and one of these is a vertex* $\hat{a}_{ij} = (a_{ij}, l_i(a_{ij}))$. *Then* $(\hat{p} - \hat{a}_{ij} + \hat{Q}_i) \cap s(Q) = \hat{p}$.

Proof. It suffices to show that every other point $\hat{q} \in \hat{p} - \hat{a}_{ij} + \hat{Q}_i$ lies above the lower envelope. It is easy to see that $\hat{p} - \hat{a}_{ij} + \hat{Q}_i$ is contained in $\hat{Q}$, because it consists of sums of $(n+1)$-tuples of points, one from each polytope. So all points in it are either on or above the lower envelope.

Now displace both $\hat{p}$ and $\hat{q}$ by decreasing their $(n+1)$-st coordinate by the same amount, thus defining points $\hat{p}', \hat{q}' \in \mathbb{R}^{n+1}$. The displacement should be small enough so that the line $(\hat{p}', \hat{q}')$ intersects the lower envelope in the face that contains $\hat{p}$. Let $\hat{p}''$ be this intersection point. $\hat{Q}$ also contains $\hat{p}'' - \hat{a}_{ij} + \hat{Q}_i$.

Clearly (Fig. 1) the vector $\hat{q}' - \hat{p}''$ is smaller than $\hat{q} - \hat{p}$ and in the same direction. Now $\hat{q} - \hat{p}$ is contained in the convex set $\hat{Q}_i - \hat{a}_{ij}$, and it follows that $\hat{q}' - \hat{p}''$ is also contained in $\hat{Q}_i - \hat{a}_{ij}$ (which contains the origin). Thus $\hat{q}' \in \hat{p}'' - \hat{a}_{ij} + \hat{Q}_i \subset \hat{Q}$. So we have demonstrated a point $\hat{q}'$ such that $\pi(\hat{q}) = \pi(\hat{q}')$ but $h(\hat{q}') < h(\hat{q})$. Thus $\hat{q}$ is not on the lower envelope. □

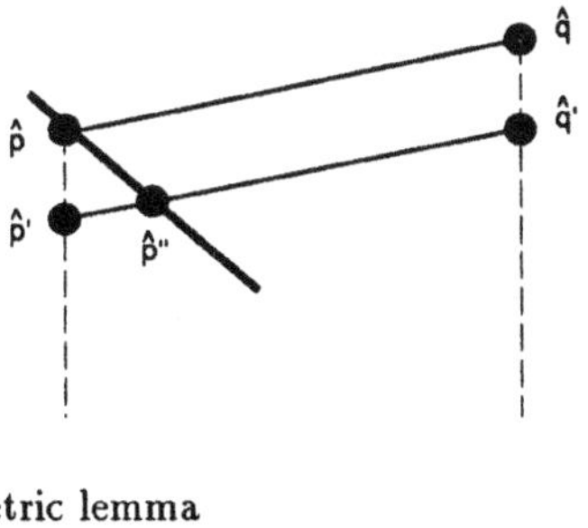

Fig. 1. Proof of the geometric lemma

Define a matrix $M'(t)$ by scaling the rows of $M(t)$:

$$M'_{pq} \stackrel{\Delta}{=} t^{(h(\hat{p}) - l_i(a_{ij}))} M_{pq}$$

for every $q \in \mathcal{E}$, where $(i,j) = RC(p)$ and $\hat{p} = s(p)$. Then the previous lemma leads to an inequality on the degree in t of the M' entries.

Lemma 16. *For all non-zero elements M'_{pq} with $p \neq q$, $\deg(M'_{pq}) > \deg(M'_{qq})$.*

Proof. Let $\hat{p}$ and $\hat{q}_0$ be the points on the lower envelope $s(Q) + \delta$ such that $\pi(\hat{p}) = p$ and $\pi(\hat{q}_0) = q$. Let $(\iota, \gamma) = RC(q)$ and, since $\deg(M_{qq}(t)) = l_\iota(a_{\iota\gamma})$, we have

$$\deg(M'_{qq}(t)) = (h(\hat{q}_0) - l_i(a_{ij})) + l_i(a_{ij}) = h(\hat{q}_0) \ .$$

Note that $\hat{p}$ will lie in the interior of a facet of $\hat{\Delta}_\delta$. Let $\hat{q}$ be the intersection $\pi^{-1}(q) \cap (\hat{p} - \hat{a}_{ij} + \hat{Q}_i)$. The intersection is non-empty because if M_{pq} contains a non-zero coefficient c_{ik}, then $q = p - a_{ij} + a_{ik}$. In fact $\hat{q} = \hat{p} - \hat{a}_{ij} + \hat{a}_{ik}$, hence

$$\deg(M'_{pq}(t)) = h(\hat{p}) - l_i(a_{ij}) + l_i(a_{ik}) = h(\hat{q}) \ .$$

From the previous lemma $\hat{q}$ does not lie on the lower envelope and since $\hat{q}_0$ does lie on the lower envelope, we have $h(\hat{q}) > h(\hat{q}_0)$. □

The previous lemmas are more easily understood by recalling that the $\hat{Q}_i$'s are the Newton polytopes of the specialized system, where t is the $(n+1)$-st variable. More precisely, the Newton polytope of the polynomial in row p is $\hat{Q}_i$ shifted so that its vertex $\hat{a}_{ij}$ lies over p. The row-scaling of M by powers of t corresponds to lifting the Newton polytopes of the rows so that the optimal vertex touches the lower envelope. The rest of the polytope will lie above the lower envelope. Looking down column q of M' corresponds to looking at points in the various Newton polytopes that lie over the lattice point q. There will be a unique point of minimum $(n+1)$-st coordinate on the lower envelope over q corresponding to the leading diagonal element M'_{qq}. All other points will have larger $(n+1)$-st coordinate, therefore the corresponding entries have higher degree in t than that of M'_{qq}.

Proposition 17. *The lowest-degree term of* $\det(M')(t)$ *equals the product of the leading diagonal elements of $M'(t)$, therefore this determinant is non-vanishing.*

Proof. The determinant can be written

$$\det(M') = \sum_{\sigma \in S(\mathcal{E})} (-1)^{\mathrm{sign}(\sigma)} \prod_{q \in \mathcal{E}} M'_{\sigma(q)q}$$

where $S(\mathcal{E})$ is the symmetric group on $\mathcal{E}$. For every σ not equal to the identity, we have $\sigma(q) \neq q$ for some q, so $\deg(M'_{\sigma(q)q}) > \deg(M'_{qq})$ by the previous lemma. Thus

$$\deg(\prod_{q \in \mathcal{E}} M'_{qq}) < \deg(\prod_{q \in \mathcal{E}} M'_{\sigma(q)q})$$

for every permutation σ other than the identity. This implies that the product of leading diagonal entries is a unique lowest power of t and therefore there exists some value $t_0 \neq 0$ of t for which this product is not canceled and $\det(M')(t_0) \neq 0$. □

The main result (Theorem 14) of this section is a straightforward consequence of this proposition by observing that

$$\det(M')(t) = t^\alpha \det(M)(t)$$

where t^α is the product of the scale factors.

5 Computing the Resultant

We show that the degree of $\det(M)$ in the coefficients of the polynomial f_1 equals that of the resultant R. The row content function chooses f_1 if there is no other possibility, which happens precisely at the mixed cells to which Q_1 contributes a vertex. The total volume of these cells equals the mixed volume of the other n Newton polytopes $MV(Q_2, \ldots, Q_{n+1})$. We define an n-dimensional ***half-open integral parallelotope*** HO:

$$HO = \{\sum_{i=1}^{n} r_i \mathrm{e}_i \mid r_i \in [0,1),\ \mathrm{e}_i \in \mathbb{Z}^n\} .$$

Lemma 18. *The number of integer lattice points in a half-open integral parallelotope equals its volume.*

Proof. It follows from [18, Remark, p.335] that the number of these lattice points is $n!\mathrm{Vol}(S)$ where S is the simplex $\mathrm{Conv}(0, \mathrm{e}_1, \ldots, \mathrm{e}_n)$. The volume of the parallelotope HO is also $n!\mathrm{Vol}(S)$. □

Corollary 19. *For any $\delta \in \mathbb{R}^n$, the number of integer lattice points in $HO + \delta$ is* $\mathrm{Vol}(HO)$.

Proof. Imagine that HO is displaced by $t\delta$ as t varies from 0 to 1. Observe that for each facet of HO that is open (or closed) the opposite facet is closed (open), and that the opposite facet is displaced from the first by an integral vector v. Thus as HO moves, whenever a lattice point p enters HO, a corresponding point at $p+v$ exits, and vice versa. Thus the number of lattice points inside HO remains constant. □

A mixed facet of the subdivision $\hat{\Delta}_\delta$ is the Minkowski sum of n edges, hence a parallelotope in $\mathbb{R}^n$. The perturbation by δ guarantees that all lattice points lie in the interior of a facet. So the number of rows containing coefficients of f_1 is precisely $MV(Q_2, \ldots, Q_{n+1})$.

Proposition 20. *The degree of the determinant of M in the coefficients of f_1 equals $MV(Q_2, \ldots, Q_{n+1})$, which equals that of $R(\mathcal{A}_1, \ldots, \mathcal{A}_{n+1})$. Moreover, the degree of* $\det(M)$ *in the coefficients of every other f_j for $j \neq 1$ is at least as large as the respective degree of $R(\mathcal{A}_1, \ldots, \mathcal{A}_{n+1})$.*

For computing R we could use Hurwitz's idea [6] and construct $n+1$ matrices, $M_1, \ldots, M_{n+1}$, where each M_i has the minimum number of rows containing coefficients of f_i. For example, we could modify the row contents function so that it never returns i when there is another choice. Let $D_1, \ldots, D_{n+1}$ be the determinants formed in this way. The GCD of $D_1, \ldots, D_{n+1}$ has the correct degree in all f_i's and, since the GCD is divisible by R, it equals R. Unfortunately, this method does not work when the coefficients of the f_i are specialized. It can be used after a suitable perturbation of the specialized system, but there is a more economical method, essentially the one in [3], with a straightforward adaptation; two variants follow.

5.1 Division Method

Let $g_1, \ldots, g_{n+1}$ be the specialized polynomials. First we choose polynomials $h_1, \ldots, h_{n+1}$ with *random* integer coefficients, such that h_i has support $\mathcal{A}_i$. Then the perturbed system is

$$(f_1, \ldots, f_{n+1}) \mapsto (g_1 + u_1 h_1, \ldots, g_{n+1} + u_{n+1} h_{n+1})$$

where each u_i is a new indeterminate. Define the extraneous factor b_i of each D_i via

$$D_i = b_i R$$

and notice that b_i will be independent of u_i.

Definition 21. Suppose a polynomial $A(u_1, \ldots, u_{n+1})$ has maximum degree d_i in u_i. Then A is said to be *rectangular* if it contains a monomial of the form $u_1^{d_1} u_2^{d_2} \cdots u_{n+1}^{d_{n+1}}$.

Under the specialization above, note that R as well as $D_1, \ldots, D_{n+1}$ will be rectangular, because the coefficient of $u_1^{d_1} u_2^{d_2} \cdots u_{n+1}^{d_{n+1}}$ will be the resultant (or one of the determinants) when each f_i is specialized to h_i.

Define $R^{(j)}(u_1, \ldots, u_j)$ to be the leading coefficient, with respect to total degree, of R considered as a polynomial in $u_{j+1}, \ldots, u_{n+1}$. Define $D_i^{(j)}(u_1, \ldots, u_j)$ and $b_i^{(j)}(u_1, \ldots, u_j)$ analogously and notice that all these polynomials are rectangular. Then $D_i^{(j)} = b_i^{(j)} R^{(j)}$ for all i and j. But notice that since b_i is independent of u_i, $b_i^{(i)} = b_i^{(i-1)}$, which we can use to eliminate b_i:

$$R^{(i)} = \frac{D_i^{(i)}}{D_i^{(i-1)}} R^{(i-1)} \quad . \tag{4}$$

Now notice that $R^{(n+1)}$ is exactly the resultant of the f_i, so setting $u_1 = \cdots = u_{n+1} = 0$ in $R^{(n+1)}$ will give the resultant of the g_i.

The recurrence (4) has initial term $R^{(0)}$ which is some integer that we may set to 1, thus obtaining $R^{(n+1)}$ equal to a scalar multiple of the resultant.

Next observe that the identity (4) is valid for specializations of u_i's so long as no denominator vanishes. So we take $u_1 = u_2 = \cdots = u_{n+1} = u$, so that all the $D_i^{(j)}$'s become univariate polynomials in u. Since they are all rectangular, they have a unique term of highest total degree in the u_i's which cannot cancel, so none of them will vanish under this specialization. Each $D_i^{(j)}(u)$ is easily seen to be the determinant of M under the specialization:

$$(f_1, \ldots, f_{n+1}) \mapsto (g_1 + uh_1, \ldots, g_j + uh_j, h_{j+1}, \ldots, h_{n+1})$$

and the leading coefficient of $D_i^{(j)}$ is once again non-zero for almost all choices of h_i's. Thus we have an almost guaranteed method of constructing the resultant at the cost of adding the single variable u. More precisely, to bound the probability of failure by $1/c$ for some arbitrary $c > 1$ it suffices, by Schwartz's lemma [17,

Lem. 1], to pick the coefficients of each h_i independently, each with $c \log |\mathcal{E}|$ bits. It is possible to detect failure deterministically, in which case new randomized variables must be chosen.

If the g_i's are sufficiently generic, which here means that no $D_i^{(j)}$ vanishes, we may compute $D_i^{(j)}$ as the determinant of M_i under the specialization

$$(f_1, \ldots, f_{n+1}) \mapsto (g_1, \ldots, g_j, h_{j+1}, \ldots, h_{n+1}) \ .$$

5.2 GCD Method

This method requires that the coefficients of the specialized system $g_1, \ldots, g_{n+1}$ be non-zero and chosen from some polynomial ring over $\mathbb{Q}$. Again we choose polynomials h_i with random coefficients, whose size is given by Schwartz's lemma, and specialize

$$(f_1, \ldots, f_{n+1}) \mapsto (g_1 + uh_1, \ldots, g_{n+1} + uh_{n+1}) \ .$$

By Hilbert's irreducibility theorem, R will remain irreducible over $\mathbb{Q}[u]$ after almost all such specializations. Let $D_1(u)$ be the determinant of M_1 with this specialization, and let $b(u)$ be the extraneous factor, $D_1(u) = b(u)R(u)$.

Suppose without loss of generality that M_1 was defined using a linear functional l_1 which is "much larger" than the others. The effect of this is that whenever a vertex a_{1j} of Q_1 contributes to an optimal sum, that vertex will be the one which minimizes l_1. Thus in every row containing coefficients of f_1, the leading diagonal element will be c_{1j}. Now let $D_2(u)$ be the determinant of M under the specialization

$$(f_1, f_2, \ldots, f_{n+1}) \mapsto (x^{a_{1j}}, g_2 + uh_2, \ldots, g_{n+1} + uh_{n+1})$$

with $D_2(u) = b(u)R'(u)$, where $R'(u)$ is the resultant under this new specialization. Therefore

$$R(u) = \frac{D_1(u)}{GCD(D_1(u), D_2(u))}$$

and specializing $u = 0$ gives the resultant of $g_1, \ldots, g_{n+1}$. It is worth remarking that the degree of $b(u)$ is known in advance, namely it is the number of elements of $\mathcal{E}$ that do not lie in mixed facets. Thus the GCD computation is branch-free and reduces to calculation of the appropriate minors of the Sylvester matrix of D_1 and D_2, [12].

Once again if the given g_i's are generic enough, in this case meaning that the specialized resultant under $f_i \mapsto g_i$ is irreducible, and the determinants $D_1(0)$ and $D_2(0)$ are both non-zero, then the resultant can be computed as simply $D_1(0)/GCD(D_1(0), D_2(0))$.

6 An Example

The construction is illustrated for a system of 3 polynomials in 2 unknowns

$$\begin{aligned} f_1 &= c_{11} + c_{12}xy + c_{13}x^2y + c_{14}x \\ f_2 &= c_{21}y + c_{22}x^2y^2 + c_{23}x^2y + c_{24}x \\ f_3 &= c_{31} + c_{32}y + c_{33}xy + c_{34}x \ . \end{aligned}$$

Pick generic functions

$$\begin{aligned} l_1(x,y) &= L^5x + L^4y \\ l_2(x,y) &= L^3x + L^2y \\ l_3(x,y) &= Lx + y \end{aligned}$$

where L is a sufficiently large integer. The input Newton polytopes are shown in Fig. 2 and a subdivision of $Q + \delta$ into 2-dimensional cells is shown in Fig. 3.

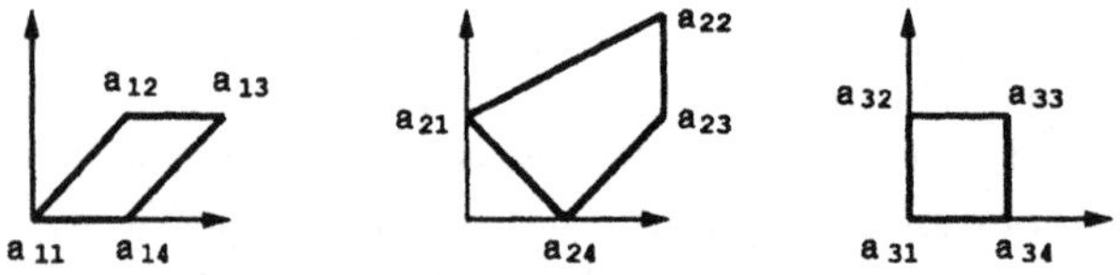

Fig. 2. The Newton polytopes and the exponents a_{ij}

Matrix M_1 appears at (5) with rows and columns indexed by exponent vectors from $\mathcal{E}$. Matrices corresponding to f_2 and f_3 are formed similarly.

$$\begin{array}{c|ccccccccccccccc|}
 & 1,0 & 2,0 & 0,1 & 1,1 & 2,1 & 3,1 & 0,2 & 1,2 & 2,2 & 3,2 & 4,2 & 1,3 & 2,3 & 3,3 & 4,3 \\
1,0 & c_{11} & c_{14} & 0 & 0 & c_{12} & c_{13} & 0 & 0 & 0 & 0 & 0 & 0 & 0 & 0 & 0 \\
2,0 & c_{31} & c_{34} & 0 & c_{32} & c_{33} & 0 & 0 & 0 & 0 & 0 & 0 & 0 & 0 & 0 & 0 \\
0,1 & c_{24} & 0 & c_{21} & 0 & c_{23} & 0 & 0 & 0 & c_{22} & 0 & 0 & 0 & 0 & 0 & 0 \\
1,1 & 0 & 0 & 0 & c_{11} & c_{14} & 0 & 0 & 0 & c_{12} & c_{13} & 0 & 0 & 0 & 0 & 0 \\
2,1 & 0 & 0 & 0 & 0 & c_{11} & c_{14} & 0 & 0 & 0 & c_{12} & c_{13} & 0 & 0 & 0 & 0 \\
3,1 & 0 & c_{24} & 0 & c_{21} & 0 & c_{23} & 0 & 0 & 0 & c_{22} & 0 & 0 & 0 & 0 & 0 \\
0,2 & 0 & 0 & 0 & 0 & 0 & 0 & c_{11} & c_{14} & 0 & 0 & 0 & c_{12} & c_{13} & 0 & 0 \\
1,2 & 0 & 0 & c_{31} & c_{34} & 0 & 0 & c_{32} & c_{33} & 0 & 0 & 0 & 0 & 0 & 0 & 0 \\
2,2 & 0 & 0 & 0 & c_{31} & c_{34} & 0 & 0 & c_{32} & c_{33} & 0 & 0 & 0 & 0 & 0 & 0 \\
3,2 & 0 & 0 & 0 & 0 & c_{31} & c_{34} & 0 & 0 & c_{32} & c_{33} & 0 & 0 & 0 & 0 & 0 \\
4,2 & 0 & 0 & 0 & 0 & 0 & c_{24} & 0 & 0 & c_{21} & 0 & c_{23} & 0 & 0 & 0 & c_{22} \\
1,3 & 0 & 0 & 0 & 0 & 0 & 0 & 0 & c_{31} & c_{34} & 0 & 0 & c_{32} & c_{33} & 0 & 0 \\
2,3 & 0 & 0 & 0 & c_{24} & 0 & 0 & c_{21} & 0 & c_{23} & 0 & 0 & 0 & c_{22} & 0 & 0 \\
3,3 & 0 & 0 & 0 & 0 & 0 & 0 & 0 & 0 & c_{31} & c_{34} & 0 & 0 & c_{32} & c_{33} & 0 \\
4,3 & 0 & 0 & 0 & 0 & 0 & 0 & 0 & 0 & 0 & c_{31} & c_{34} & 0 & 0 & c_{32} & c_{33}
\end{array} \tag{5}$$

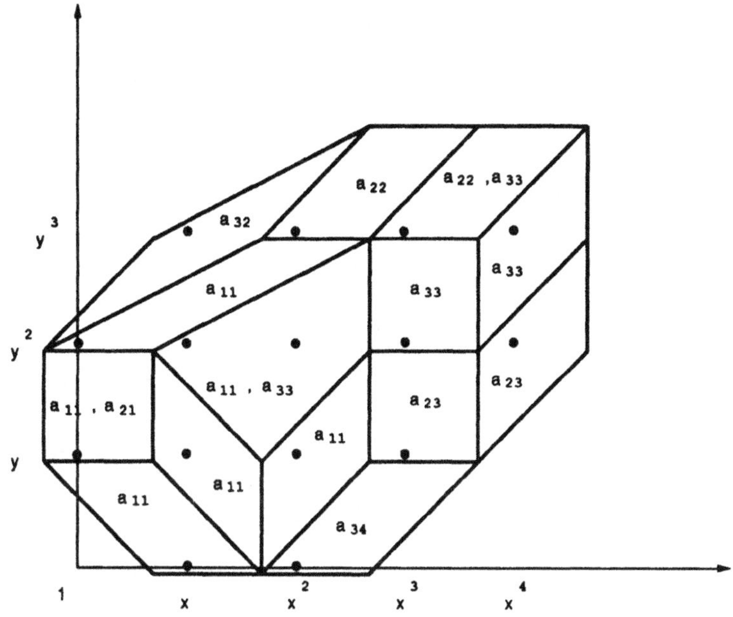

Fig. 3. The induced subdivision Δ_δ of $Q+\delta$; each facet is labeled with the vertices which contribute to optimal sums within that facet

7 Complexity

The change of variables that may be required to ensure that the supports generate the lattice $\mathbb{Z}^n$ involves linear algebra and has complexity which is dominated by that of the later steps.

Identifying the vertices of all Newton polytopes may be reduced to Linear Programming; then we can apply either Khachiyan's Ellipsoid or Karmarkar's algorithm. To bound the bit size of the input exponents a_{ij} we recall that the Newton polytopes have been translated to the origin, thus every exponent is bounded by $|\mathcal{E}|$.

In the case of Karmarkar's algorithm [8] the bit complexity, omitting the logarithmic factors, is $\mathcal{O}(m_i^{5.5}\log^2|\mathcal{E}|)$ for each Q_i and the output is its vertex set, namely $\{a_{i1},\ldots,a_{i\mu_i}\}$ with $\mu_i \leq m_i$, possibly after reindexing. The total complexity for all Newton polytopes is thus $\mathcal{O}(n(\max_i m_i)^{5.5}\log^2|\mathcal{E}|)$.

The most expensive step of the algorithm is to associate an optimal sum of points $p_i \in Q_i$ with every $p \in \mathcal{E}$. To reduce this to Linear Programming we introduce constraints

$$p=\sum_{i=1}^{n+1} p_i=\sum_{i=1}^{n+1}\sum_{j=1}^{\mu_i}\lambda_{ij}a_{ij}$$

where

$$\lambda_{ij} \geq 0 \ , \qquad \text{for } 1 \leq j \leq \mu_i, \qquad\qquad \text{and } \sum_{j=1}^{\mu_i} \lambda_{ij} = 1$$

for each i in $\{1, \ldots, n+1\}$. The objective function forces the lifted point corresponding to p to lie on the lower envelope of $\hat{Q}$ by requiring that

$$\sum_{i=1}^{n+1} \sum_{j=1}^{\mu_i} \lambda_{ij} l_i(a_{ij})$$

is *minimized*, where the l_i's are the generic linear functionals.

Either polynomial-time algorithm may again be used; here we calculate the complexity of Karmarkar's. The bit size of $l_i(a_{ij})$ is constant, once the desired probability of success is fixed. As already seen, each $a_{ij} < |\mathcal{E}|$ so the bit complexity after omitting the logarithmic factors is $\mathcal{O}(n^{5.5}(\max_i \mu_i)^{5.5} \log^2 |\mathcal{E}|)$. Hence, finding the optimal sum for all lattice points $p \in \mathcal{E}$ takes time polynomial in n, $\max_i \mu_i$ and $\mathcal{E}$.

Lastly, we have to extract the resultant from matrix M by one of the described methods. This can be done with linear algebra and the arithmetic complexity is polynomial in the order of M. Since both the matrix order and the input exponents are bounded by $|\mathcal{E}|$, the overall complexity is polynomial in $|\mathcal{E}|$.

This discussion proves

Proposition 22. *For any input system, the bit complexity of our algorithm is polynomial in in n, $\max_i\{m_i\}$ and $|\mathcal{E}|$.*

Now we estimate $|\mathcal{E}|$; unfortunately, only the unmixed case can be treated without requiring additional hypotheses. Consider the unmixed system

$$Q_1 = \cdots = Q_{n+1} \ .$$

Then the total degree of the resultant equals the sum of all $n+1$ n-fold Mixed Volumes, each being equal to $n!\text{Vol}(Q_1)$. Hence

$$\deg R = (n+1)! \ \text{Vol}(Q_1) \ .$$

The Minkowski Sum has volume $\text{Vol}(Q) = n^n \text{Vol}(Q_1)$ and the number of lattice points in it is asymptotically the same [7]. Then $|\mathcal{E}| = \mathcal{O}\left(\frac{n^n \deg R}{(n+1)!}\right)$. Using Sterling's approximation and letting e be the base of natural logarithms, we arrive at

Lemma 23. *For unmixed systems*

$$|\mathcal{E}| = \mathcal{O}(e^n \deg R) \ .$$

Therefore

Theorem 24. *Assume that our algorithm executes on an arbitrary unmixed system. Then its asymptotic bit complexity, if we omit logarithmic factors, is polynomial in* $\max_i\{m_i\}$ *and the total degree of the resultant and exponential in* n *with a linear exponent.*

We cannot obtain the same bounds in general because there exist cases like the following, in which the cardinality $|\mathcal{E}|$ is exponential over the sum of all n-fold mixed volumes. Suppose that all Newton polytopes are hypercubes, with edge length constant for the first n and proportional to n for the last polytope. Then $|\mathcal{E}| > n^n$, while the sum of mixed volumes is $\mathcal{O}(n^2)$, hence the algorithm's complexity is higher than polynomial in $\deg R$.

Nonetheless, our algorithm is roughly as efficient on mixed systems whose Newton polytopes do not differ so drastically as indicated in Theorem 24. Moreover, a greedy version of the algorithm has been implemented on *Maple V* by the first author and P. Pedersen, and preliminary empirical results imply that this approach is efficient for most systems encountered in practice.

8 Open Questions

We are currently looking into ways for decreasing the size of the determinantal formula, the final goal being to obtain Sylvester-type formulas for different systems. Characterizing these systems for which an optimal formula does not exist is another active area [24]. A more theoretical question is on the connection of our technique with Gröbner bases, in light of [19]. Lastly, this approach leads to improved methods for calculating the common roots of sparse polynomial systems [5].

Acknowledgment

We wish to thank the anonymous referee for his comments and Ashu Rege for several discussions.

References

1. Bernstein, D.N.: The number of roots of a system of equations. *Funktsional'nyi Analiz i Ego Prilozheniya*, 9(3):1–4, Jul-Sep 1975.
2. Betke, U.: Mixed volumes of polytopes. *Arch. der Math.*, 58:388–391, 1992.
3. Canny, J.F.: *The Complexity of Robot Motion Planning.* M.I.T. Press, Cambridge, 1988.
4. Gel'fand, I.M., Kapranov, M.M. and Zelevinsky, A.V.: Discriminants of polynomials in several variables and triangulations of Newton polytopes. *Algebra i Analiz*, 2:1–62, 1990.
5. Huber, B. and Sturmfels, B.: Homotopies preserving the Newton polytopes. Manuscript, presented at the "Workshop on Real Algebraic Geometry", August 1992.

6. Hurwitz, A.: Über die Trägheitsformen eines algebraischen Moduls. *Annali di Mat.*, Tomo XX(Ser. III):113–151, 1913.
7. Kantor, J.M.: Sur le polynôme associé à un polytope à sommets entiers. *Comptes rendues de l'Académie des Sciences, Série I*, 314:669–672, 1992.
8. Karmarkar, N.: A new polynomial-time algorithm for linear programming. *Combinatorica*, 4:373–395, 1984.
9. Khovanskii, A.G.: Newton polyhedra and the genus of complete intersections. *Funktsional'nyi Analiz i Ego Prilozheniya*, 12(1):51–61, Jan-Mar 1978.
10. Kushnirenko, A.G.: The Newton polyhedron and the number of solutions of a system of k equations in k unknowns. *Uspekhi Mat. Nauk.*, 30:266–267, 1975.
11. Kushnirenko, A.G.: Newton polytopes and the Bezout theorem. *Funktsional'nyi Analiz i Ego Prilozheniya*, 10(3), Jul-Sep 1976.
12. Loos, R.: Generalized polynomial remainder sequences. In B. Buchberger, G.E. Collins, and R. Loos, editors, *Computer Algebra: Symbolic and Algebraic Computation*, pages 115–137. Springer-Verlag, Wien, 2nd edition, 1982.
13. Macaulay, F.S.: Some formulae in elimination. *Proc. London Math. Soc.*, 1(33):3–27, 1902.
14. Manocha, D. and Canny, J.: Real time inverse kinematics for general 6R manipulators. In *Proc. IEEE Intern. Conf. Robotics and Automation*, Nice, May 1992.
15. Pedersen, P. and Sturmfels, B.: Product formulas for sparse resultants. Manuscript, 1991.
16. Salmon, G.: *Modern Higher Algebra.* G.E. Stechert and Co., New York, 1885. reprinted 1924.
17. Schwartz, J.T.: Fast probabilistic algorithms for verification of polynomial identities. *J. ACM*, 27(4):701–717, 1980.
18. Stanley, R.P.: Decompositions of rational convex polyhedra. In J. Srivastava, editor, *Combinatorial Mathematics, Optimal Designs and Their Applications, Annals of Discrete Math. 6*, pages 333–342. North-Holland, Amsterdam, 1980.
19. Sturmfels, B.: Gröbner bases of toric varieties. *Tôhoku Math. J.*, 43:249–261, 1991.
20. Sturmfels, B.: Sparse elimination theory. In D. Eisenbud and L. Robbiano, editors, *Proc. Computat. Algebraic Geom. and Commut. Algebra*, Cortona, Italy, June 1991. Cambridge Univ. Press. To appear.
21. Sturmfels, B.: Combinatorics of the sparse resultant. Technical Report 020-93, MSRI, Berkeley, November 1992.
22. Sturmfels, B. and Zelevinsky, A.: Multigraded resultants of Sylvester type. *J. of Algebra.* To appear. Also, Manuscript, 1991.
23. van der Waerden, B.L.: *Modern Algebra.* Ungar Publishing Co., New York, 3rd edition, 1950.
24. Weyman, J. and Zelevinsky, A.: Determinantal formulas for multigraded resultants. Manuscript, 1992.

Some Features of Binary Block Codes for Correcting Asymmetric Errors

G. Fang, H. C. A. van Tilborg, F. W. Sun[1] and I. S. Honkala[2]

[1] Department of Mathematics and Computing Science
Eindhoven University of Technology
P.O. Box 513, 5600 MB Eindhoven
The Netherlands
[2] Department of Mathematics
University of Turku
20500 Turku 50, Finland

Abstract. Binary block codes for correcting asymmetric errors are called binary AsEC block codes. With the properties of perfect codes for the binary symmetric channel in mind, natural definitions of perfect, weakly perfect and uniformly weakly perfect binary AsEC block codes are given and their properties are studied.
It is shown that a perfect asymmetric–error–correcting code is trivial or is equal to the repetition code. Also, it is proved that any weakly perfect code which is nontrivial can always be enlarged to a bigger code of the same length and the same distance. As necessary ingredients for the proofs of those two results, several related properties of codes for correcting asymmetric errors are studied as well.

1 Introduction

Perfect codes for correcting symmetric errors have received a lot of attention since the celebrated Hamming codes were discovered for the binary symmetric channels (BSC). An excellent exposition on the existence of nontrivial perfect codes was given by Van Lint [13] (including many references). For additional results on perfect codes, one is also referred to [11] and [14]. For later comparisons, we indicate two main properties of perfect binary codes for the BSC below:

- A binary perfect block code of length n and minimum Hamming distance $2t+1$ (capable of correcting up to t symmetric errors) corresponds to a partition of the binary n–dimensional vector space. This partition consists of a collection of spheres all with the same radius t centered around the codewords. That is to say, all such packing spheres are mutually disjoint and together cover the whole space. So, a perfect code can correct all (symmetric) errors of weight $\leq t$, but none of weight greater than t.
- A binary perfect block code has the highest information rate (or maximum size) among all codes of the same length and error–correcting capability. Therefore, in this sense it can be said that the whole vector space is packed optimally by a perfect code.

The study of perfect codes used for BSC has been generalized in several directions which are mentioned in the comments of Chapter 7 of [14]. In the last two decades, a lot of attention has been paid to the study of codes which are capable of correcting asymmetric errors. Such codes apply to, for instance, some data storage systems and optical communication. However, as far as we know, no literature discusses perfect asymmetric–error–correcting codes, and the definition of such codes has not even been given yet. One can find many results related to bounds on the maximum size of codes and to constructions of codes with high cardinality, given the length and minimum (asymmetric) distance. For this, we refer to the bibliography of [10] and the references listed in Chapter 7 of [12]. More recent papers on this are [3], [4], [2], and [17]. However, most of them have had very little impact on the material presented in this paper.

For the decoding of a code, one should realize that through a binary asymmetric channel, a possibly received word, say $\mathbf{y}$, only comes from the codewords covering it. The strategy of a maximum likelihood decoder is of course to decode the received word $\mathbf{y}$ to one of the codewords of lowest weight covering $\mathbf{y}$. Furthermore, in view of sphere packings, the set to be packed in asymmetric cases is, in general, not the whole vector space anymore as in symmetric cases, but is that consisting of all possibly received words obtained by introducing asymmetric errors from all codewords. In fact, in asymmetric cases the set to be packed is the whole vector space if and only if the all–one vector is a codeword (if only 1–errors are considered). Generally, this set will depend on the specific code. This gives rise to some concept of *perfect packing* and *weakly perfect packing*.

The study of perfect codes and weakly perfect codes capable of correcting asymmetric errors is the main goal of this paper, which warrants additional investigation on AsEC codes at a fundamental level. According to the definitions of such codes, it is conceivable that there are many questions one can address with respect to those codes. However, here we are mainly concerned with the same two properties as mentioned above for perfect codes. To be more precise, we raise the following two questions:

1. Which notion of t–asymmetric–error–correcting perfect codes of length n corresponds to a partition of the binary n–dimensional vector space?
2. Does a nontrivial asymmetric–error–correcting perfect (weakly perfect or uniformly weakly perfect) code exist that reaches the highest information rate among all codes of the same length with the same error–correcting capability?

We restrict ourselves to binary block codes, and to the *binary asymmetric channel* (also called Z–channel [12]). Such a channel is totally error–free for 0's and noisy for 1's, namely the probability of a crossover $0 \rightarrow 1$ is 0 and the probability of a crossover $1 \rightarrow 0$ is p. Using the terminology introduced by Kim and Freiman [7], we refer to the $1 \rightarrow 0$ crossovers as *1–errors*. Such errors are called *asymmetric errors*. Because of this channel it is natural to look at codes that correct only 1–errors. We define perfect, weakly perfect and uniformly weakly perfect binary asymmetric–error–correcting codes in Sect. 2.

Section 3 presents several basic results derived from the definitions which will be needed to prove the main conclusions, which will be presented in Sect. 4. It is shown that any perfect code defined in Sect. 2 has a trivial form that is the repetition code. This answers the first question above. It follows that any asymmetric error–correcting code for which all the Goldbaum inequalities [6] are sharp must have a trivial form too. Further analysis gives a negative answer to the second question above. Hence, any weakly perfect code which is nontrivial can always be enlarged to a bigger code of the same length and the same distance.

2 Definition and Notation

Let $\mathbf{V_n}$ denote the n–dimensional vector space over $GF(2) = \{0, 1\}$. We use the words *vector* or *word* to denote the n–tuples from $\mathbf{V_n}$ (n is called the block length or word length). The all–one vector and the all–zero vector will be abbreviated to $\mathbf{1}$ and $\mathbf{0}$ respectively, and $\mathbf{a}^l$ denotes the all–a vector of length l ($a = 0, 1$). The cardinality of a finite set A is denoted by $|A|$. For any $\mathbf{x} = (x_1, x_2, \cdots, x_n) \in \mathbf{V_n}$ and $\mathbf{y} = (y_1, y_2, \cdots, y_n) \in \mathbf{V_n}$, put $N(\mathbf{x}, \mathbf{y}) = |\{1 \leq i \leq n \mid x_i = 1 \wedge y_i = 0\}|$. If $N(\mathbf{x}, \mathbf{y}) = 0$, i.e. for all i, $x_i = 1$ implies $y_i = 1$, we say that the vector $\mathbf{x}$ *is covered* by the vector $\mathbf{y}$. This can be written as $\mathbf{x} \leq \mathbf{y}$ or $\mathbf{y} \geq \mathbf{x}$. The *Hamming distance* between $\mathbf{x}$ and $\mathbf{y}$ indicates the number of positions in which the two words differ: $d(\mathbf{x}, \mathbf{y}) = N(\mathbf{x}, \mathbf{y}) + N(\mathbf{y}, \mathbf{x})$. The *asymmetric distance* between $\mathbf{x}$ and $\mathbf{y}$ is defined as the maximum of the number of $1 \to 0$ crossovers from $\mathbf{x}$ to $\mathbf{y}$ and from $\mathbf{y}$ to $\mathbf{x}$:

$$\Delta(\mathbf{x}, \mathbf{y}) = max\{N(\mathbf{x}, \mathbf{y}), N(\mathbf{y}, \mathbf{x})\}. \tag{1}$$

An embryonic form of this definition and the statement of the error–correcting capability of codes with this distance function can be found in earlier literature, for instance in [15]. Both the Hamming distance and the asymmetric distance are metrics on $\mathbf{V_n}$, and they are related in the following way:

$$2\Delta(\mathbf{x}, \mathbf{y}) = d(\mathbf{x}, \mathbf{y}) + \mid w(\mathbf{x}) - w(\mathbf{y}) \mid \tag{2}$$

where $w(\mathbf{a}) = |\{1 \leq i \leq n \mid a_i = 1\}|$ denotes the *weight* of the vector $\mathbf{a}$.

A binary *asymmetric error–correcting code* of length n and minimum asymmetric distance Δ, denoted by $C_a(n, \Delta)$, is a non–empty proper subset of $\mathbf{V_n}$ in which any two distinct vectors are at asymmetric distance at least Δ apart and this distance is realized at least once. The vectors in the code are called *codewords*. Therefore, the *minimum asymmetric distance* (often simply called the minimum distance or distance) of a code C is defined as the minimum of the distances between all pairs of codewords: $\Delta = min\{\Delta(\mathbf{x}, \mathbf{y}) | \mathbf{x}, \mathbf{y} \in C \wedge \mathbf{x} \neq \mathbf{y}\}$. The *weight distribution* of C is denoted by $A_0, A_1, \cdots, A_n$. The *weight of* C, indicated by $w(C)$, is defined as the sum of weights of all codewords of C. The following notions will be used throughout the paper.

Definition 1. Let C be a $C_a(n, \Delta)$ code. For any codeword $\mathbf{c} \in C$, $r(\mathbf{c})$ will be used to denote the asymmetric distance to the nearest codeword to $\mathbf{c}$, namely

$$r(\mathbf{c}) = min\{\Delta(\mathbf{x}, \mathbf{c}) | \mathbf{c} \neq \mathbf{x} \wedge \mathbf{x} \in C\} = \Delta(\mathbf{c}, C \backslash \{\mathbf{c}\}). \tag{3}$$

Evidently, $r(\mathbf{c}) \geq \Delta$ for any codeword $\mathbf{c}$ in Definition 1. The idea behind Definition 1 is that for a $C_a(n, \Delta)$ code C, there may exist a codeword $\mathbf{c} \in C$ such that $r(\mathbf{c}) > \Delta$, which means that the word $\mathbf{c}$ can be protected against more errors than other codewords $\mathbf{x}$ with $r(\mathbf{x}) = \Delta$. Codes with this property do exist. Actually in many of linear codes of maximum dimension for correcting asymmetric errors, the weight of any nonzero codeword is greater than the minimum distance of the code (see e.g. [15]). Therefore $r(\mathbf{0})$ is larger than the minimum distance of the code.

The *sphere* with radius t and center $\mathbf{c}$ is defined by:

$$S_a(\mathbf{c}, t) = \{\mathbf{x} \in \mathbf{V_n} \mid \Delta(\mathbf{c}, \mathbf{x}) \leq t \ \wedge \ \mathbf{c} \geq \mathbf{x}\}. \tag{4}$$

In other words, $S_a(\mathbf{c}, t)$ consists of all vectors that can be obtained from $\mathbf{c}$ by introducing up to t 1–errors. Therefore, the cardinality of the sphere $S_a(\mathbf{c}, t)$ equals:

$$| S_a(\mathbf{c}, t) | = \sum_{i=0}^{t} \binom{w(\mathbf{c})}{i}.$$

Unlike in Hamming space, this number relies not only on the radius t but also on the weight of the word $\mathbf{c}$.

Definition 2. Let C be a $C_a(n, \Delta)$ code. Also, let E denote the set of all possibly received words, i.e. $E = \{\mathbf{x} \in \mathbf{V_n} | \exists_{\mathbf{c} \in C} [\mathbf{c} \geq \mathbf{x}]\}$. The code C is called a *weakly perfect* code, for short WP code, if

$$E = \bigcup_{\mathbf{c} \in C} S_a(\mathbf{c}, r(\mathbf{c}) - 1). \tag{5}$$

In particular, if all $r(\mathbf{c})$ are equal to Δ in (5), then C is called a *uniformly weakly perfect* code, for short, UWP code. If $E = \mathbf{V_n}$ in (5), then C is called a *perfect* code.

Obviously, $| C | \leq | E | \leq | \mathbf{V_n} | = 2^n$, and $E = \mathbf{V_n}$ if and only if $\mathbf{1} \in C$. From Definition 2, it follows that any UWP $C_a(n, \Delta)$ code is also a WP $C_a(n, \Delta)$ code. The definition given for perfect $C_a(n, \Delta)$ codes is consistent with that given for the perfect codes used for the binary symmetric channel. Let $A_a(n, \Delta)$ be the maximum number of codewords in a $C_a(n, \Delta)$ code, $W_a(n, \Delta)$ the maximum number of codewords in a WP $C_a(n, \Delta)$ code and $U_a(n, \Delta)$ the maximum number of codewords in a UWP $C_a(n, \Delta)$ code. Of course

$$U_a(n, \Delta) \leq W_a(n, \Delta) \leq A_a(n, \Delta). \tag{6}$$

A $C_a(n, \Delta)$ code is called *trivial* if its cardinality is less than or equal to 2. It is easy to show that a $C_a(n, \Delta)$ code, if $n < 2\Delta$, contains at most two codewords, therefore, it is a trivial code.

To illustrate the existence of WP $C_a(n, \Delta)$ codes, we present some examples.

Example 1. Let C be the repetition code of length n. Then $E = \mathbf{V_n}$, $S_a(\mathbf{0}, n-1) = \{\mathbf{0}\}$ and $S_a(\mathbf{1}, n-1) = \mathbf{V_n}\backslash\{\mathbf{0}\}$. Hence the condition stated in (5) is satisfied, and C is a perfect code.

Example 2. Let C consist of three words: 00000, 11000 and 00111. Then, C is a nontrivial WP $C_a(5,2)$ code and $r(00111) = 3$, whereas $r(\mathbf{0}) = r(11000) = 2$.

Example 3. Let C be the following $C_a(6,2)$ code (all the columns represent codewords):

```
0 0 0 1 0 1 1 0 0 1 1
0 0 0 1 1 0 0 1 0 1 1
0 0 1 0 0 1 0 1 1 0 1
0 0 1 0 1 0 1 0 1 0 1
0 1 0 0 0 0 1 1 1 1 0
0 1 0 0 1 1 0 0 1 1 0
```

It can be readily checked that C is a nontrivial UWP code of size 11 by Definition 2. Also, it is known that $C \cup \{\mathbf{1}\}$ is a $C_a(6,2)$ code of maximum size, i.e., $A_a(6,2) = 12$, and essentially it is unique [3]. So, from (6) and Theorem 16 (see Sect. 4), it follows that $U_a(6,2) = W_a(6,2) = 11$. This code is unique up to permutation.

It is quite obvious that concatenating two $C_a(n, \Delta)$ codes gives a $C_a(2n, 2\Delta)$ code [4]. However, concatenating two weakly perfect codes does not necessarily result in a weakly perfect code. For instance, concatenating the code C in Example 3 with itself does not yield a WP $C_a(12,4)$ code.

If C is a WP $C_a(n, \Delta)$ code, then the code obtained by deleting one of the largest weight codewords of C will be a WP code of length n as well. The proof is easy. Let $\mathbf{x}$ be one of the largest weight codewords of C, and $C' = C\backslash\{\mathbf{x}\}$. Also, let $r(\mathbf{c}) = \Delta(\mathbf{c}, C\backslash\{\mathbf{c}\})$ and $r'(\mathbf{c}) = \Delta(\mathbf{c}, C'\backslash\{\mathbf{c}\})$ for any $\mathbf{c} \in C'$. Then $r(\mathbf{c}) \leq r'(\mathbf{c})$ for any $\mathbf{c} \in C'$. Let E' denote the set consisting of all possibly received words corresponding to C', i.e. $E' = \{\mathbf{y} \in \mathbf{V_n} | \exists_{\mathbf{c} \in C'}[\mathbf{c} \geq \mathbf{y}]\}$. Now we need to show that $E' \subseteq \bigcup_{\mathbf{c} \in C'} S_a(\mathbf{c}, r'(\mathbf{c}) - 1)$. Let $\mathbf{z} \in E'$. Since C is a WP code and $E' \subseteq E$, one has:

$$\mathbf{z} \in \bigcup_{\mathbf{c} \in C'} S_a(\mathbf{c}, r(\mathbf{c}) - 1) \cup S_a(\mathbf{x}, r(\mathbf{x}) - 1).$$

Note that $\Delta(\mathbf{z}, \mathbf{x}) \geq r(\mathbf{x})$. So, $\mathbf{z} \notin S_a(\mathbf{x}, r(\mathbf{x}) - 1)$. Hence

$$\mathbf{z} \in \bigcup_{\mathbf{c} \in C'} S_a(\mathbf{c}, r(\mathbf{c}) - 1) \subseteq \bigcup_{\mathbf{c} \in C'} S_a(\mathbf{c}, r'(\mathbf{c}) - 1).$$

Thus C' is a WP code of length n (but possibly with a larger minimum distance than Δ).

Example 4. Let C be the code defined in Example 3. The set $\{\mathbf{x} \mid (\mathbf{x}, 1) \in C\}$ gives a UWP $C_a(5,2)$ code of size 5 by Definition 2. From (6) and Theorem 16 (see Sect. 4), also from [3], it follows that $U_a(5,2) = W_a(5,2) = 5$.

Example 5. For any integer number w ($0 \leq w < n$), the set of all words of length n and weight at most w together with the all–one vector $\mathbf{1}$ is a perfect $C_a(n,1)$ code.

The *probability of error*, or the *word error rate*, P_{err}, for a particular decoding rule is the probability that the decoder produces a wrong codeword. Assume that $\mathbf{c}_1, \mathbf{c}_2, \cdots, \mathbf{c}_M$ are codewords of a code C which are used with equal probability, then the probability of incorrect decoding of a received word is:

$$P_{err}(C) = \frac{1}{M} \sum_{i=1}^{M} P_i \tag{7}$$

where P_i is the probability of making an incorrect decision given that $\mathbf{c}_i$ is transmitted for $i = 1, 2, \cdots, M$.

For a $C_a(n, \Delta)$ code C, our decoding rules are based on two assumptions. First of all, it is assumed that all codewords are transmitted equally likely during communication. Furthermore, the decoding strategy is that if $\mathbf{x}$ is received, then $\mathbf{x}$ is decoded into a codeword $\mathbf{c}$ where $\mathbf{c} \geq \mathbf{x}$ and $w(\mathbf{c}) = min\{w(\mathbf{y}) | \mathbf{y} \in C \wedge \mathbf{y} \geq \mathbf{x}\}$. This decoding rule results in *maximum–likelihood–decoding*.

To avoid unnecessary complications in the discussions we will make the following conventions on the codes: the minimum asymmetric distance is always assumed to be greater than or equal to 2, and none of the coordinates of codewords is identically zero or identically one.

3 Some results related to $C_a(n, \Delta)$ codes

In this section we shall derive several results regarding $C_a(n, \Delta)$ codes that are interesting in their own right but are also necessary for deriving the results in the next section.

Lemma 3. *Let C be a $C_a(n, \Delta)$ code. Then the following properties hold:*

1. *if C is nontrivial, then $r(\mathbf{c}) \leq n - \Delta$ for any codeword $\mathbf{c} \in C$.*
2. *for any two different codewords $\mathbf{c}_1$ and $\mathbf{c}_2$,*

$$S_a(\mathbf{c}_1, r(\mathbf{c}_1) - 1) \cap S_a(\mathbf{c}_2, r(\mathbf{c}_2) - 1) = \emptyset.$$

Proof. The proof of the first assertion is straightforward and omitted. In order to prove the second assertion, take $r = max\{r(\mathbf{c}_1), r(\mathbf{c}_2)\}$. Since $\Delta(\mathbf{c}_1, \mathbf{c}_2) \geq r$, without loss of generality, we may assume, from (1), that $N(\mathbf{c}_1, \mathbf{c}_2) \geq r$. If there exists a vector $\mathbf{x}$ such that $\mathbf{x} \in S_a(\mathbf{c}_1, r(\mathbf{c}_1) - 1) \cap S_a(\mathbf{c}_2, r(\mathbf{c}_2) - 1)$, then from (4), one has that $\mathbf{x} \leq \mathbf{c}_1$, $\mathbf{x} \leq \mathbf{c}_2$ and $N(\mathbf{c}_1, \mathbf{x}) \leq r(\mathbf{c}_1) - 1$. This implies that

$$r - 1 \geq r(\mathbf{c}_1) - 1 \geq N(\mathbf{c}_1, \mathbf{x}) \geq N(\mathbf{c}_1, \mathbf{c}_2) \geq r,$$

which is a contradiction. □

It is well known that $C_a(n, \Delta)$ codes can correct up to $\Delta - 1$ asymmetric errors (see e.g. [12]), and hence they are sometimes referred to as $(\Delta - 1)$-AsEC codes. However, possibly there exists a codeword $\mathbf{c}$ such that $r(\mathbf{c}) > \Delta$. It follows that the minimum distance criterion of judging capabilities of correcting errors of codes can be generalized. This motivates us to introduce the *average error-correcting capability* of a code. The *average error-correcting capability* of a $C_a(n, \Delta)$ code C is defined:

$$\overline{r}(C) = \frac{1}{|C|} \sum_{\mathbf{c} \in C} (r(\mathbf{c}) - 1). \tag{8}$$

For the average error-correcting capability of the code C, one remark needs to be given here, namely if $\mathbf{c}$ is a codeword of C of weight less than Δ, then the error-correcting capability of $\mathbf{c}$ may be referred to as any number which is greater than $w(\mathbf{c})$. Hence in the sense of error-correcting capability, $r(\mathbf{c})$ does not give an appropriate measure, nor does $\overline{r}(C)$. However, this will not give much additional scope in the error-correcting capability of the code C, so we still adopt the definition in (8) as the average error-correcting capability of C. Another thing which should be noticed is that sometimes a codeword $\mathbf{c}$ may correct more than $r(\mathbf{c}) - 1$ errors. The following Theorem 4 shows that if a $C_a(n, \Delta)$ code does not contain the all-one vector $\mathbf{1}$, then changing a maximal weight codeword into $\mathbf{1}$ may increase the average error-correcting capability of the code. The proof, being very simple, is omttied.

Theorem 4. *Let C be a $C_a(n, \Delta)$ code with average error-correcting capability $\overline{r}(C)$, and $\mathbf{x}$ be a maximum weight codeword of C. If $\mathbf{x} \neq \mathbf{1}$, then the code $C' = (C \backslash \{\mathbf{x}\}) \cup \{\mathbf{1}\}$ has average error-correcting capability $\overline{r}(C') \geq \overline{r}(C)$.*

Lemma 5. *If C is a $C_a(n, \Delta)$ code with the property: $w(C) \leq w(C')$ for any $C_a(n, \Delta)$ code C' with $|C'| = |C|$, then $r(\mathbf{c}) = \Delta$ for any $\mathbf{c} \in C$.*

Proof. Without loss of generality, we may assume that C consists of the M codewords: $\mathbf{c}_1, \cdots, \mathbf{c}_M$ with nondecreasing weight order, i.e., $w(\mathbf{c}_i) \leq w(\mathbf{c}_j)$ for $1 \leq i < j \leq M$. Since C is of minimum weight, the codeword $\mathbf{c}_1$ must be the all-zero vector $\mathbf{0}$. If $r(\mathbf{c}_1) \geq \Delta + 1$, then any $1 \rightarrow 0$ change in $\mathbf{c}_2$ can be introduced such that the resulting code is still a $C_a(n, \Delta)$ code, which contradicts the assumption on the weight of C. Similarly, if $r(\mathbf{c}) \geq \Delta + 1$ for $\mathbf{c} \neq \mathbf{c}_1$, then introducing any $1 \rightarrow 0$ change in $\mathbf{c}$ will give the same contradiction. □

Theorem 6. *A $C_a(n, \Delta)$ code C can always be transformed into another $C_a(n, \Delta)$ code C' so that $r(\mathbf{c}) = \Delta$ for any codeword $\mathbf{c}$ of C'.*

Proof. Without loss of generality, we may assume that $\mathbf{0} \in C$. The same argument as used in the proof of Lemma 5 can be applied here. Note that the weight of C is finite. Therefore, the process of making $1 \rightarrow 0$ changes must terminate in a finite number of steps. □

Theorem 6 admits the following heuristic interpretation. Given a $C_a(n,\Delta)$ code C. From Lemma 3, it follows that all the packing spheres centered at the codewords of C are disjoint in the space $\mathbf{V_n}$. If there is $r(\mathbf{c})$ which is greater than Δ, then, the corresponding center $\mathbf{c}$ can be replaced by a word $\mathbf{y}$ so that $S_a(\mathbf{y},\Delta-1)$ is a proper subset of $S_a(\mathbf{c},r(\mathbf{c})-1)$, keeping it disjoint from the other packing spheres. Thus, by this technique, all the packing spheres with larger radii ($\geq \Delta$) can be reduced to smaller packing spheres that are still disjoint.

Theorem 6 also shows that any $C_a(n,\Delta)$ code of maximum size can be assumed to satisfy that for any codeword, it has distance Δ to the set of all other codewords. An interesting question is whether this condition is necessary for any nontrivial $C_a(n,\Delta)$ code of maximum size. From Theorem 6, it follows that the answer is positive for codes which are unique up to permutation. But, in general, it is not true. Two counterexamples are shown below.

Example 6. The following four words: 0000000, 1110000, 0001110 and 1111111 form a $C_a(7,3)$ code of maximum size. However, the minimum distance from the last codeword to the other three codewords is 4 which is greater than the minimum distance of the code.

For Example 6, it is easy to show that $A_a(6,3) = A_a(7,3) = 4$. In the following, we shall give another example which shows that $C_a(n,\Delta)$ codes of size $A_a(n,\Delta)$ exist so that at least one codeword is of minimum distance greater than Δ to all other codewords, even though the relation $A_a(n-1,\Delta) < A_a(n,\Delta)$ holds.

Example 7. The $C_a(17,6)$ code consisting of the following eight words:

00000000000000000
11111100000000000
00000011111100000
11000011000011110
00111000111011100
10011110010110011
01100101101101011
11111111111111111

is of maximum size. From [4], it follows that $A_a(16,6) = 7 < A_a(17,6) = 8$. However, the minimum asymmetric distance from the last codeword $\mathbf{1}$ to the other codewords is 7.

Lemma 7. *Let C be a $C_a(n,\Delta)$ code. If $w(\mathbf{c}) < r(\mathbf{c})$ for a certain $\mathbf{c} \in C$, then $\mathbf{c}$ is of minimum weight in C, i.e, $w(\mathbf{c}) < min\{w(\mathbf{x})|\mathbf{x} \in C \wedge \mathbf{x} \neq \mathbf{c}\}$.*

Proof. Suppose that there is a codeword $\mathbf{x}$ in C such that $w(\mathbf{x}) \leq w(\mathbf{c})$. Then $\Delta(\mathbf{x},\mathbf{c}) \leq w(\mathbf{c}) < r(\mathbf{c})$, which contradicts the definition of $r(\mathbf{c})$. □

Lemma 8. *For a maximum size $C_a(n,\Delta)$ code C, $r(\mathbf{c}) \leq 2\Delta - 1$ for any codeword $\mathbf{c} \in C$.*

Proof. Suppose that there exists a codeword $\mathbf{c}$ such that $r(\mathbf{c}) \geq 2\Delta$. Let $\mathbf{y}$ be any word with distance Δ from $\mathbf{c}$. Since C is of maximum size, $\mathbf{y}$ cannot be added to C as a codeword without affecting the minimum distance Δ. Hence, $\mathbf{y}$ has distance $\leq \Delta - 1$ to a codeword in C, say $\mathbf{x}$. From the triangle inequality, we get the following contradiction:

$$2\Delta \leq \Delta(\mathbf{x}, \mathbf{c}) \leq \Delta(\mathbf{x}, \mathbf{y}) + \Delta(\mathbf{y}, \mathbf{c}) \leq \Delta - 1 + \Delta = 2\Delta - 1.$$

So the lemma is proved. □

From Examples 6 and 7, it follows that for a maximum size $C_a(n, \Delta)$ code, the bound $2\Delta - 1$ shown in Lemma 8 certainly is not tight but also cannot be replaced by Δ. The following theorem strengthens this bound.

Theorem 9. *Let C be a $C_a(n, \Delta)$ code with $A_a(n, \Delta)$ codewords. If $n \leq 2\Delta$, then $r(c) = \Delta$ for any codeword $\mathbf{c}$ of C. If $n > 2\Delta$ and $\mathbf{c} \in C$, then*

$$r(\mathbf{c}) < \begin{cases} 2\Delta - w(\mathbf{c}), & \text{for } 0 \leq w(\mathbf{c}) < \Delta; \\ 3\Delta/2, & \text{for } \Delta \leq w(\mathbf{c}) \leq n - \Delta; \\ 2\Delta - n + w(\mathbf{c}), & \text{for } n - \Delta < w(\mathbf{c}) \leq n. \end{cases}$$

Proof. When $n < 2\Delta$, C is the repetition code. So, the assertion holds. If $n = 2\Delta$, the following four words: $\mathbf{0}^{\Delta}\mathbf{0}^{\Delta}$, $\mathbf{1}^{\Delta}\mathbf{0}^{\Delta}$, $\mathbf{0}^{\Delta}\mathbf{1}^{\Delta}$ and $\mathbf{1}^{\Delta}\mathbf{1}^{\Delta}$ form a $C_a(2\Delta, \Delta)$ code of maximum size, which is unique up to permutation. Hence, $r(\mathbf{c}) = \Delta$ for any $\mathbf{c} \in C$. Let $n > 2\Delta$ and $\mathbf{c} \in C$. Without loss of generality, we may assume that $\mathbf{c} = \mathbf{1}^w\mathbf{0}^{n-w}$ where w is the weight of the codeword $\mathbf{c}$. Since C is of maximum size, it must contain a unique codeword which has weight less than Δ. Suppose that $\mathbf{c}$ is such a codeword. Because $n - w \geq \Delta$ and $r(\mathbf{c}) < 2\Delta$ (using Lemma 8), we can put:

$$\mathbf{a} = \mathbf{1}^w\mathbf{1}^{r(\mathbf{c})-\Delta}\mathbf{0}^{n-w-(r(\mathbf{c})-\Delta)}.$$

From the triangle inequality, the asymmetric distance from $\mathbf{a}$ to any codeword other than $\mathbf{c}$ must be at least Δ. Moreover, the weight of the word $\mathbf{a}$ must be less than Δ, otherwise a $C_a(n, \Delta)$ code with larger size can be obtained by replacing $\mathbf{c}$ with $\mathbf{a}$ and $\mathbf{0}$ (if $\mathbf{c} = \mathbf{0}$, then one can simply add $\mathbf{a}$ into C), which contradicts the assumption on the size of C. Therefore

$$w(\mathbf{a}) = w + r(\mathbf{c}) - \Delta < \Delta.$$

This means that $r(\mathbf{c}) < 2\Delta - w(\mathbf{c})$. Similarly, one can prove that for the unique highest weight codeword $\mathbf{c}$ ($n - \Delta < w(\mathbf{c}) \leq n$), $r(\mathbf{c}) < 2\Delta - n + w(\mathbf{c})$. For any other codeword $\mathbf{c}$, by using Lemma 7, $w(\mathbf{c}) \geq r(\mathbf{c})$. Put

$$\begin{aligned} \mathbf{a} &= \mathbf{1}^w\mathbf{1}^{r(\mathbf{c})-\Delta}\mathbf{0}^{n-w-(r(\mathbf{c})-\Delta)} \\ \mathbf{b} &= \mathbf{0}^{r(\mathbf{c})-\Delta}\mathbf{1}^{w-(r(\mathbf{c})-\Delta)}\mathbf{0}^{n-w}. \end{aligned}$$

Hence from the triangle inequality, it follows that for any $\mathbf{c}' \in C$ and $\mathbf{c}' \neq \mathbf{c}$,

$$\Delta(\mathbf{a}, \mathbf{c}') \geq \Delta(\mathbf{c}', \mathbf{c}) - \Delta(\mathbf{c}, \mathbf{a}) \geq r(\mathbf{c}) - (r(\mathbf{c}) - \Delta) = \Delta.$$

Similarly, $\Delta(\mathbf{b}, \mathbf{c}') \geq \Delta$. Thus, the asymmetric distance between $\mathbf{a}$ and $\mathbf{b}$ must be less than Δ. Indeed, if this is not the case, replacing $\mathbf{c}$ by $\mathbf{a}$ and $\mathbf{b}$ in C will result in a $C_a(n, \Delta)$ code of larger size, which is not possible. Therefore

$$\Delta(\mathbf{a}, \mathbf{b}) = 2(r(\mathbf{c}) - \Delta) < \Delta.$$

This leads to $r(\mathbf{c}) < 3\Delta/2$. □

Theorem 9 tells us that in a $C_a(n, \Delta)$ code of maximum size there are at most two codewords at asymmetric distance greater than or equal to $3\Delta/2$ to all other codewords. Specifically, one has that $r(\mathbf{0}) < 2\Delta$ and $r(\mathbf{1}) < 2\Delta$ if $\mathbf{0}$ and $\mathbf{1}$ are these codewords. Furthermore, applying Theorem 9 to the case $\Delta = 2$ presents some interesting results. In fact, if C is a $C_a(n, 2)$ of maximum cardinality with $1 \leq w(\mathbf{c}) \leq n-1$ for all $\mathbf{c} \in C$, then from Theorem 9, it follows that $r(\mathbf{c}) = 2$ for all $\mathbf{c} \in C$. Though, in general all $r(\mathbf{c})$ are not necessary equal to Δ for a nontrivial $C_a(n, \Delta)$ code of maximum size, it is at least necessary for many $C_a(n, \Delta)$ codes with maximum size $A_a(n, \Delta)$, as shown in the following theorem.

Theorem 10. *Let C be a $C_a(n, 2)$ code with $|C| = A_a(n, 2)$. If $n \not\equiv 1$ or 3 (mod 6), then $r(\mathbf{c}) = 2$ for all $\mathbf{c} \in C$.*

Proof. Without loss of generality, the all-zero vector $\mathbf{0}$ may be assumed to be a codeword. By Theorem 9 and for reasons of symmetry, we only need to prove that $r(\mathbf{0}) = 2$. Assume $r(\mathbf{0}) > 2$. Then from Theorem 9 it follows that $r(\mathbf{0}) = 3$. This means that apart from the all-zero vector $\mathbf{0}$, all the other codewords of C have weight at least three. If the length n is not congruent to 1 or 3 modulo 6, then the set of codewords of C of weight 3 cannot form a Steiner triple system [8]. Hence, there exists at least one word of weight 2 which has asymmetric distance greater than or equal to 2 to all codewords of weight 3 and trivially is at distance greater than or equal to 2 to all codewords of weight greater than 3 as well as to $\mathbf{0}$. So this word of weight 2 can be added to C without decreasing the minimum distance, which contradicts the assumption that the code C is of maximum size. □

The question whether any $C_a(n, 2)$ code C of maximum size must satisfy that $r(\mathbf{c}) = 2$ for any $\mathbf{c} \in C$ is still open. From the proof of Theorem 10, one only needs to prove that no $C_a(n, 2)$ code of maximum size exists that contains $\mathbf{0}$, and in which the codewords of weight 3 form a Steiner triple system $S(2, 3, n)$. At the present, we could only check this for $n = 7$ and one case for $n = 9$ (based on the results in [3]). Therefore, we are left with:

Conjecture. Any $C_a(n, 2)$ code C of size $A_a(n, 2)$ satisfies that $r(\mathbf{c}) = 2$ for all $\mathbf{c} \in C$.

We shall now investigate the significance of taking the all-one vector $\mathbf{1}$ as a codeword in a $C_a(n, \Delta)$ code (this will come back when WP $C_a(n, \Delta)$ codes will be discussed in the next section). It has been observed by Kløve [9] that the all-one vector $\mathbf{1}$ and the all-zero vector $\mathbf{0}$ may always be assumed to be codewords in a $C_a(n,\Delta)$ code. This observation is useful in constructing codes [17], [2] and [4]. However, for a fixed length n and distance Δ, two or more $C_a(n,\Delta)$ codes may exist, all of maximum size but not equivalent to each other. In this case, their performance cannot be judged by simply comparing their average error–correcting capabilities and their information rates, since both parameters may be equal. For example, let C_1 and C_2 be defined as follows:

$$C_1 = \begin{matrix} 00000 \\ 11000 \\ 00110 \\ 01101 \\ 10011 \\ 11111 \end{matrix} \text{ and } C_2 = \begin{matrix} 00000 \\ 11000 \\ 00110 \\ 01101 \\ 10011 \\ 11110 \end{matrix} \tag{9}$$

then $|C_1| = |C_2| = A_a(5,2) = 6$. So, they have the same information rate. Also, for both codes the minimum distance from a certain codeword to all other codewords is 2. Hence they have the same average error–correcting capability. To distinguish their performance we shall study their respective probabilities of erroneous decoding.

When error probability is taken into account, one arrives at the following question: let C be a $C_a(n,\Delta)$ code of maximum size without the all–one vector and let $C' = (C\backslash\{\mathbf{f}\}) \cup \{\mathbf{1}\}$ where $\mathbf{f}$ is one of the maximal weight codewords of C, does the inequality $P_{err}(C) \geq P_{err}(C')$ hold ? This question was essentially solved by Weber [16] (unpublished), which shows that if $\mathbf{f}$ is a codeword of C of weight $w(\mathbf{f}) = n - j$ $(1 \leq j \leq \Delta - 1)$, then $P_{err}(C') \leq P_{err}(C)$. Therefore, for the above two codes C_1 and C_2, we have that $P_{err}(C_1) \leq P_{err}(C_2)$. In fact, one can show that $P_{err}(C_1) < P_{err}(C_2)$. Thus, C_1 can be said to be better than C_2 in the sense that it has a lower error probability. Below, a slightly more general result than that obtained by Weber will be presented in Theorem 12 for which the proof is essentially due to Weber.

For the decoding of a $C_a(n,\Delta)$ code C, one should realize that a received word $\mathbf{y}$ only comes from the codewords covering it. In other words, the received set E depends on the specific code C, which is not the case for the BSC where any word in $\mathbf{V_n}$ can be a possibly received word. The strategy of a maximum likelihood decoder is of course to decode a received word $\mathbf{y}$ to a codeword of lowest weight covering $\mathbf{y}$. This fact is stated in Lemma 11. We denote the probability of receiving $\mathbf{y}$ given that $\mathbf{x}$ is transmitted as $P(\mathbf{y}|\mathbf{x})$.

Lemma 11. *Let C be a $C_a(n,\Delta)$ code and let $\mathbf{y}$ be a received word. If $\mathbf{x}_1, \mathbf{x}_2 \in C$ with $\mathbf{x}_i \geq \mathbf{y}$ $(i = 1, 2)$ and $w(\mathbf{x}_2) \geq w(\mathbf{x}_1)$, then $P(\mathbf{y}|\mathbf{x}_1) \geq P(\mathbf{y}|\mathbf{x}_2)$.*

Theorem 12. *Let C be a nontrivial $C_a(n,\Delta)$ code and let $\mathbf{f}$ be a codeword of C of maximum weight. Then, $P_{err}(C') \leq P_{err}(C)$ where $C' = (C\backslash\{\mathbf{f}\}) \cup \{\mathbf{g}\}$ and $\mathbf{g} \geq \mathbf{f}$.*

Proof. Obviously, C' is still an AsEC code of length n and minimum distance at least Δ. Without loss of generality, it is sufficient to consider the case: $\mathbf{f} = \mathbf{0}^j\mathbf{1}^{n-j}$ and $\mathbf{g} = \mathbf{0}^{j-1}\mathbf{1}^{n-j+1}$ for a certain $j \geq 1$. Define:

$$A = \{\mathbf{a} \in \mathbf{V_n} \mid \mathbf{f} \geq \mathbf{a} \wedge \forall_{\mathbf{c} \in C \setminus \{\mathbf{f}\}} [\mathbf{c} \not\geq \mathbf{a}]\}.$$

If $\mathbf{a} \in A$, we denote $\mathbf{a}'$ as the word obtained by changing the jth coordinate of $\mathbf{a}$ (which is 0 because $\mathbf{f} \geq \mathbf{a}$) to 1, and further define the set consisting of all such words $\mathbf{a}'$ by A'. Without loss of generality, we may assume that only the words of A are decoded into the codeword $\mathbf{f}$ (clearly, if there is more than one codeword of the same maximum weight, then according to the decoding rule, some words other than the words of A might be decoded into $\mathbf{f}$ as well. But the assumption does not change the overall error probability). Let $P\{correct|\mathbf{x}\}$ denote the probability of making the correct decoding decision given that $\mathbf{x}$ is transmitted. According to (7) and Lemma 11, we only need to consider two probabilities: $P\{correct|\mathbf{f}\}$ and $P\{correct|\mathbf{g}\}$. Note that $A \cap A' = \emptyset$. One obtains:

$$\begin{aligned} P(correct|\mathbf{g}) &\geq \textstyle\sum_{\mathbf{a} \in A} P(\mathbf{a}|\mathbf{g}) + \sum_{\mathbf{a}' \in A'} P(\mathbf{a}'|\mathbf{g}) \\ &= \textstyle\sum_{\mathbf{a} \in A} p^{n-j+1-w(\mathbf{a})}(1-p)^{w(\mathbf{a})} + \\ &\quad \textstyle\sum_{\mathbf{a}' \in A'} p^{n-j+1-w(\mathbf{a}')}(1-p)^{w(\mathbf{a}')} \\ &= (p + (1-p)) \textstyle\sum_{\mathbf{a} \in A} p^{n-j-w(\mathbf{a})}(1-p)^{w(\mathbf{a})} \\ &= \textstyle\sum_{\mathbf{a} \in A} P(\mathbf{a}|\mathbf{f}) = P(correct|\mathbf{f}). \end{aligned}$$

Thus, $P_{err}(C') \leq P_{err}(C)$ which completes the proof. □

Weber's original result is a particular case of Theorem 12 by taking $\mathbf{g} = \mathbf{1}$ and $\mathbf{f}$ a codeword of weight greater than $n - \Delta$. From Theorems 4 and 12 together with the observation by Kløve, it follows that the all–one vector $\mathbf{1}$ should always be included in a $C_a(n, \Delta)$ code. However, in the next section it will be shown that the all–one vector $\mathbf{1}$ cannot be a codeword in any nontrivial WP code for correcting asymmetric errors.

4 On the rate of weakly perfect codes

In this section, we will present the answers to the two questions stated in Sect. 1. We shall show that any perfect $C_a(n, \Delta)$ code must have a trivial form. This implies that for the binary asymmetric channel only the repetition code gives rise to a partition of $\mathbf{V_n}$. Furthermore, it will be shown that any nontrivial WP $C_a(n, \Delta)$ code of maximum size can always be enlarged to a $C_a(n, \Delta)$ code of the same length and distance. This means that $W_a(n, \Delta) < A_a(n, \Delta)$ for nontrivial cases for any n and Δ.

Lemma 13. *Let C be a WP $C_a(n, \Delta)$ code containing the all–one vector $\mathbf{1}$. Then C must be the repetition code.*

Proof. Since $\mathbf{1} \in C$, the received set E corresponding to C equals the whole space $\mathbf{V_n}$. From (5) in Definition 2, it follows that

$$\mathbf{V_n} = \bigcup_{\mathbf{c} \in C} S_a(\mathbf{c}, r(\mathbf{c}) - 1). \tag{10}$$

Let $r = r(\mathbf{1})$. Obviously, $r \geq \Delta \geq 2$ and the weight distribution of C satisfies $A_{n-1} = A_{n-2} = \cdots = A_{n-r+1} = 0$. Furthermore, we have

$$A_{n-r} = \binom{n}{n-r}. \tag{11}$$

Otherwise, there would exist at least one word of weight $n - r$ which does not belong to the right-hand side of (10), which is not possible. If $n - r \geq 1$ in (11), there must be two different codewords of weight $n - r$ such that they are at asymmetric distance 1 apart, which disagrees with $\Delta \geq 2$. So, $n - r = 0$, that is, $n = r$. This shows that the code C is the repetition code which is trivial. □

Theorem 14. *A $C_a(n, \Delta)$ code C is perfect if and only if C is the repetition code.*

Proof. Because any perfect $C_a(n, \Delta)$ code contains the all–one vector $\mathbf{1}$, the assertion directly follows from Lemma 13 and Example 1. □

Instead of applying the sphere packing concept to the whole space $\mathbf{V_n}$, Goldbaum [6] derives an upper bound on the maximum size of an asymmetric error–correcting code which is only based on the constraints on the vectors of length n having a certain weight i. He shows that the weight distribution $A_0, A_1, \ldots, A_n$ of a $C_a(n, \Delta)$ code satisfies:

$$\sum_{j=0}^{\Delta-1} \binom{i+j}{j} A_{i+j} \leq \binom{n}{i} \tag{12}$$

for $i = 0, 1, \ldots, n$. According to Lemma 13, we can conclude that:

Theorem 15. *The only $C_a(n, \Delta)$ code ($n \geq \Delta \geq 2$) for which all the Goldbaum inequalities (12) are sharp is the the repetition code.*

Proof. By taking $i = n$ and $i = n - \Delta$ respectively in (12), one will be in the same situation as described in the proof of Lemma 13. So, the result is the same as there. □

Theorem 14 shows that no nontrivial $C_a(n, \Delta)$ code exists such that the union of all packing spheres centered at the codewords covers the whole space $\mathbf{V_n}$. However, in practice, we are only interested in those words which are in the received set E corresponding to C instead of the whole space $\mathbf{V_n}$. In general, the cardinality of E is greater than that of the union S of all packing spheres centered at the codewords. But for WP codes, both sets are equal, i.e., $E = S$, by Definition 2. Therefore, in this case any possibly received word can always be decoded to a unique codeword (maybe incorrectly). Specifically, a *UWP* $C_a(n, \Delta)$ code can correct all errors of weight $\leq \Delta - 1$, and none of weight greater than or equal to Δ. Of course, we expect that the cardinality (or information rate) of a *WP* code is as large as possible. Thus, one question arises, namely, whether a nontrivial *WP* $C_a(n, \Delta)$ code exists such that it contains $A_a(n, \Delta)$ codewords. In the following, we shall show that the answer to this question is negative.

Theorem 16. *If* $n \geq 2\Delta$*, then* $W_a(n, \Delta) < A_a(n, \Delta)$.

Proof. Let C be a *WP* $C_a(n, \Delta)$ code with cardinality $W_a(n, \Delta)$ and with $n \geq 2\Delta$. Let $A_0, \cdots, A_n$ be the weight distribution of C. Note that $\sum_{i=n-\Delta+1}^{n} A_i \leq 1$. If $\sum_{i=n-\Delta+1}^{n} A_i = 0$, then the all–one vector $\mathbf{1}$ can be always added to C such that the enlarged code is still a $C_a(n, \Delta)$ code. So the claim is ture in this case. If $A_n = 1$, then from Lemma 13 it follows that C is the repetition code. Because $A_a(n, \Delta) \geq 4$ when $n \geq 2\Delta$, the assertion holds for this trivial case too. Now, suppose that

$$\sum_{i=n-\Delta+1}^{n-1} A_i = 1. \qquad (\text{hence } A_n = 0)$$

Then, there will be an index j $(1 \leq j \leq \Delta - 1)$ such that $A_{n-j} = 1$. Let $\mathbf{a}$ represent the codeword of weight $n - j$, and without loss of generality, the word $\mathbf{a}$ may be assumed that $\mathbf{a} = \mathbf{0}^j\mathbf{1}^{n-j}$. We want to prove that $|C| < A_a(n, \Delta)$. Suppose that the contrary holds, namely $|C| = A_a(n, \Delta)$. From Theorem 9, it follows that $w(\mathbf{a}) - r(\mathbf{a}) \geq n - 2\Delta + 1$. Since $n \geq 2\Delta$ and $w(\mathbf{a}) = n - j$, one has that $n - j - r(\mathbf{a}) \geq 1$. Define the following three words of length n (note that $2 \leq \Delta \leq r(\mathbf{a})$):

$$\begin{aligned}
\mathbf{x}_1 &= \mathbf{0}^j\ \mathbf{1}^{n-j-r(\mathbf{a})-1}\ 100\ \mathbf{0}^{r(\mathbf{a})-2},\\
\mathbf{x}_2 &= \mathbf{0}^j\ \mathbf{1}^{n-j-r(\mathbf{a})-1}\ 010\ \mathbf{0}^{r(\mathbf{a})-2},\\
\mathbf{x}_3 &= \mathbf{0}^j\ \mathbf{1}^{n-j-r(\mathbf{a})-1}\ 001\ \mathbf{0}^{r(\mathbf{a})-2}.
\end{aligned}$$

Obviously, the above three words are all of weight $n - j - r(\mathbf{a})$. Therefore, none of them can be contained in the sphere $S_a(\mathbf{a}, r(\mathbf{a}) - 1)$. Since C is weakly perfect, every word of weight $n - j - r(\mathbf{a})$ covered by $\mathbf{a}$ must be covered by one of the codewords of C of weight i where $n - j - r(\mathbf{a}) \leq i \leq n - r(\mathbf{a})$. Note that the minimum asymmetric distance from the codeword $\mathbf{a}$ to the other codewords is $r(\mathbf{a})$ and $\mathbf{a}$ is of the highest weight. With this in mind, one will arrive at the fact that the three words $\mathbf{x}_1$, $\mathbf{x}_2$ and $\mathbf{x}_3$ must be covered uniquely by three different other codewords of C respectively, and these three codewords are necessarily of the following forms respectively:

$$\begin{aligned}
\mathbf{a}_1 &= \mathbf{b}_1\ \mathbf{1}^{n-j-r(\mathbf{a})-1}\ 100\ \mathbf{0}^{r(\mathbf{a})-2},\\
\mathbf{a}_2 &= \mathbf{b}_2\ \mathbf{1}^{n-j-r(\mathbf{a})-1}\ 010\ \mathbf{0}^{r(\mathbf{a})-2},\\
\mathbf{a}_3 &= \mathbf{b}_3\ \mathbf{1}^{n-j-r(\mathbf{a})-1}\ 001\ \mathbf{0}^{r(\mathbf{a})-2}.
\end{aligned}$$

where $\mathbf{b}_1$, $\mathbf{b}_2$ and $\mathbf{b}_3$ are of length j. However, since $\Delta(\mathbf{a}_1, \mathbf{a}_2) \geq \Delta$ and the length of $\mathbf{b}_1$ and $\mathbf{b}_2$ is $j \leq \Delta - 1$, one must have $w(\mathbf{b}_1) = j = \Delta - 1$ and $w(\mathbf{b}_2) = 0$ (or vice versa). Then $w(\mathbf{b}_3)$ must equal zero by $\Delta(\mathbf{a}_1, \mathbf{a}_3) \geq \Delta$. This results in $\Delta(\mathbf{a}_2, \mathbf{a}_3) = 1$ which is a contradiction. Hence, $|C| < A_a(n, \Delta)$ if $n \geq 2\Delta$. □

It is known that for a $C_a(n, \Delta)$ code, the inequality $n < 2\Delta$ implies that the code is trivial. So, $n < 2\Delta$ gives rise to $U_a(n, \Delta) = W_a(n, \Delta) = A_a(n, \Delta) = 2$. Hence, Theorem 16 states that the rate of any nontrivial $C_a(n, \Delta)$ code of maximum size is always greater than that of WP $C_a(n, \Delta)$ codes. Two direct corollaries to Theorem 16 are:

Corollary 17. *If $n \geq 2\Delta$, then no nontrivial WP $C_a(n, \Delta)$ code can contain a codeword of weight greater than $n - \Delta$. Therefore, any WP $C_a(n, \Delta)$ code with $n \geq 2\Delta$ can always be enlarged with the all-one vector* $\mathbf{1}$ *to a bigger $C_a(n, \Delta)$ code.*

Proof. From Lemma 13, we only need to show the assertion in the corollary is true for $A_n = 0$ and $n \geq 2\Delta$. Let C be a nontrivial WP $C_a(n, \Delta)$ code containing the codeword $\mathbf{a}$ shown in the proof of Theorem 16. From Lemma 7, it follows that $w(\mathbf{a}) \geq r(\mathbf{a})$. If $w(\mathbf{a}) \geq r(\mathbf{a}) + 1$, the same contradiction stated in the proof of Theorem 16 will be obtained. On the other hand, one can readily verify that the relation $w(\mathbf{a}) = r(\mathbf{a})$ will lead to $|C| \leq 2$ which is trivial, which contradicts the assumption that C is nontrivial. □

Corollary 18. *$U_a(n, \Delta) < A_a(n, \Delta)$ for $n \geq 2\Delta$.*

5 Conclusion

For the asymmetric distance metric, the notion of the minimum distance $r(\mathbf{c})$ from a certain codeword $\mathbf{c}$ to all other codewords is introduced. It is shown that for any maximum size AsEC code C of length n and minimum distance Δ, $r(\mathbf{c})$ is less than $3\Delta/2$ if $\mathbf{c}$ is not of the lowest weight or the highest weight. For the trivial case $n \leq 2\Delta$, $r(\mathbf{c}) = \Delta$ for any codeword $\mathbf{c}$. Specifically, all $r(\mathbf{c})$ ($\mathbf{c} \in C$) are equal to Δ when C is unique up to permutation. This also holds for all 1–AsEC codes of length n and maximum size when n is not congruent to 1 or 3 modulo 6.

The only AsEC codes of length n and minimum distance Δ (≥ 2) for which the spheres (in the sense of (4)) with radius $r(\mathbf{c}) - 1$ around codeword $\mathbf{c}$ where $\mathbf{c}$ runs over all codewords form a partition of the whole vector space $\mathbf{V_n}$, are the repetition code.

An important other result is that for given length n and minimum distance Δ with $n \geq 2\Delta$, $W_a(n, \Delta) < A_a(n, \Delta)$, in words: the maximum size of a nontrivial $C_a(n, \Delta)$ code is always greater than the size of any weakly perfect $C_a(n, \Delta)$ code. It would be interesting to find $C_a(n, \Delta)$ codes of cardinality $A_a(n, \Delta)$ which become weakly perfect upon removal of the codeword of maximum weight.

Acknowledgment. The authors acknowledge with appreciation a number of comments by J. H. Weber.

References

1. Constantin, S. D., Rao, T.R.M.: On the theory of binary asymmetric error correcting codes. Information and Contr. **40**, 20–36, 1979
2. Etzion, T.: New lower bounds for asymmetric and unidirectional codes. IEEE Trans. Inform. Theory, IT–**37**, 1696–1704, Nov. 1991
3. Fang, G., van Tilborg, H. C. A.: Bounds and Constructions of Asymmetric or Unidirectional Error-Correcting Codes. Applicable Algebra in Engineering, Communication and Computing, **3**(4), 269–300, Dec. 1992
4. —: New tables of AsEC and UEC codes. Report 91–WSK–02, Eindhoven University of Technology, The Netherlands, August 1991
5. Fang, G., van Tilborg, H. C. A., Sun, F. W.: Weakly perfect binary block codes for correcting asymmetric errors. Proc. of the International Symposium on Communications, Taiwan, Tainan, 57–60, Dec. 1991
6. Goldbaum, I. Y.: Estimate for the number of signals in codes correcting nonsymmetric errors (in Russian). Automat. Telemekh., **32**, 94–97, 1971 (English translation: Automat. Rem. Control, **32**, 1783–1785, 1971)
7. Kim, W. H., Freiman, C. V.: Single error correcting codes for asymmetric channels. IRE Trans. Information Theory, IT–**5**, 62–66, June 1959
8. Kirkman, T. P.: On a problem in combinations. Cambridge and Dublin Math. J., **2**, 191–204, 1847
9. Kløve, T.: Upper bounds on codes correcting asymmetric errors. IEEE Trans. Inform. Theory, IT–**27**, 128–131, Jan. 1981
10. —: Error correcting codes for the asymmetric channel. Rep. 18–09–07–81, Dept. Mathematics, University of Bergen, Norway, July 1981
11. Macwilliams, F. J., Sloane, N. J. A.: The Theory of Error-Correcting Codes. Amsterdam: North-Holland, 1979
12. Rao, T. R. N., Fujiwara, E.: Error-control coding for computer systems. Prentice Hall Series in Computer Engineering, Prentice Hall, 1989
13. Van Lint, J. H.: A Survey of Perfect Codes. Rocky Mountain J. Math., **5**, 199–224, 1975
14. —: Introduction to Coding Theory. Graduate Texts in Mathematics, **86**, New York: Springer-verlag, 1982
15. Varshamov, R. R.: Some features of linear codes that correct asymmetric errors (in Russian). Doklady Akad. Nauk. SSSR **157**(3), 546–548, 1964 (English translation: Soviet Physics-Doklady **9**, 538–540, Jan. 1965)
16. Weber, J. H.: private communication
17. Weber, J. H., de Vroedt, C., Boekee, D. E.: Bounds and constructions for binary codes of length less than 24 and asymmetric distance less than 6. IEEE Trans. Inform. Theory. IT–**34**, 1321–1331, Sept. 1988

Fixed-Parameter Complexity and Cryptography

Michael R. Fellows[1] and Neal Koblitz[2]

[1] Department of Computer Science
University of Victoria
Victoria, B.C. V8W 3P6, Canada
mfellows@csr.uvic.ca

[2] Department of Mathematics GN-50
University of Washington
Seattle, Washington 98195, U.S.A
koblitz@math.washington.edu

Abstract. We discuss the issue of the parameterized computational complexity of a number of problems of interest in cryptography. We show that the problem of determining whether an n-digit number has a prime divisor less than or equal to n^k can be solved in expected time $f(k)n^3$ by a randomized algorithm that employs elliptic curve factorization techniques (this result depends on an unproved but plausible number-theoretic conjecture). An analogous computational problem concerning discrete logarithms is directly relevant to some proposed cryptosystem implementations. Our result suggests caution about implementations which fix a parameter such as the size or Hamming weight of keys. We show that several parameterized problems of relevance to cryptography, including k-Subset Sum, k-Perfect Code, and k-Subset Product are likely to be intractable with respect to fixed-parameter complexity. In particular, we show that they cannot be solved in time $f(k)n^\alpha$, where α is independent of k, unless a similar result holds for the well-studied and apparently resistant k-Clique problem.

1 Introduction

In this paper we introduce and study the *parameterized* computational complexity of a number of problems relevant to cryptography. One of the main reasons for considering parameterized complexity is that for many presumably hard computational problems, only a small range of parameter values may actually be used in practice. For example, it has been suggested that exponents of small Hamming weight might be used to speed up implementations of cryptosystems based on discrete logarithms [1].

It should be observed that, when considering computational problems of a certain complexity that are defined for two or more inputs, fixing one of the input parameters may result in fixed-parameter versions of the problems having widely varying complexity behavior. For example, both of the following graph-theoretic problems are NP-complete in their general (unparameterized) form [11].

Vertex Cover
Input: A graph $G = (V, E)$ and a positive integer parameter k.
Question: Is there a set of vertices $V' \subseteq V$ of cardinality at most k such that for every edge $uv \in E$ either $u \in V'$ or $v \in V'$?

Clique
Input: A graph $G = (V, E)$ and a positive integer parameter k.
Question: Is there a set of vertices $V' \subseteq V$ of cardinality k such that for every pair of vertices $u, v \in V'$ we have $uv \in E$?

But we have the following contrasting results concerning the computational complexity of the parameterized versions of these problems: (1) for each fixed value of the parameter k, Vertex Cover is solvable in linear time [2], (2) for each fixed value of the parameter k, the best known algorithm for Clique runs in time $O(n^{2.376k/3})$ [18].

In this paper, by a *parameterized problem* we mean a computational problem whose input includes a positive integer parameter that we will invariably denote by k. A parameterized problem is said to be *fixed-parameter tractable* if it is solvable in time $f(k)n^{\alpha}$ for each fixed parameter value k, where α is a constant independent of k. In [5], [6] a completeness theory for parameterized tractability is introduced, and a number of well-known problems are shown to be complete for various levels of a hierarchy of complexity classes of parameterized problems. In particular, *Clique* is shown to be complete for the bottom level of the hierarchy, denoted $W[1]$. The reason for studying the issue of fixed-parameter tractability by means of a completeness program is that while, for example, we strongly conjecture that Clique is not fixed-parameter tractable, a direct demonstration of this would imply $P \neq NP$. Thus, our focus is on the relative hardness of problems with respect to fixed-parameter tractability.

In this paper we report a number of both "positive" and "negative" results concerning the parameterized complexity of problems of relevance to cryptography, and we point out some potentially important problems for which the question of fixed-parameter tractability is so far unresolved.

The main positive result, presented in the next section, is that whether an n-digit integer has a "small" prime factor (less than n^k, where k is the parameter) is randomized fixed-parameter tractable, subject to a plausible number-theoretic conjecture. (By *randomized fixed-parameter tractable* we mean that there is a randomized algorithm to solve the problem in expected running time on arbitrary input $f(k)n^{\alpha}$, where α is a constant independent of k.) The argument uses elliptic curve factorization methods.

In section 3 we obtain hardness results for certain parameterized problems of interest in cryptography. We show that the following problems are hard for $W[1]$, and thus cannot be fixed-parameter tractable without a breakthrough for the well-known Clique problem.

Subset Sum
Input: A set of integers $X = \{x_1, \ldots, x_n\}$, an integer s and the parameter k.
Question: Is there a subset $X' \subseteq X$ of cardinality k such that the sum of the integers in X' equals s?

Subset Product
Input: A set of integers $X = \{x_1, \ldots, x_n\}$, integers a and m, and the parameter k.
Question: Is there a set of integers $X' \subseteq X$ of cardinality k such that the product of the integers in X' is congruent to a mod m?

In section 4 we describe some parameterized problems of direct relevance to proposed implementations of cryptosystems, for which the question of fixed-parameter tractability (or apparent intractability) remains open.

2 Finding Small Factors is Randomized Fixed-Parameter Tractable

In this section we describe a randomized algorithm for finding a "small" prime factor of a positive integer N that runs in expected time $f(k)\log^3 N$, where k is a parameter describing the size of the factor. More precisely, we consider the complexity of the following parameterized search problem.

Bounded Factor Factorization
Input: An n-bit positive integer N that has a prime factor $p < n^k$.
Output: A prime factor $p < n^k$.

Theorem 1. *Under Conjecture A below concerning the distribution of "smooth" numbers, the problem Bounded Factor Factorization is randomized fixed-parameter tractable.*

Proof. This theorem under a somewhat stronger number-theoretic conjecture is a direct corollary of Lenstra's result (2.10) of [15]. Namely, when specialized to the situation $p < n^k$, Lenstra's running time takes the form

$$n^2 \exp\left(\sqrt{(2+o(1))k \log n \log(k \log n)}\right).$$

For fixed k we set $g_k(n) = 1.5\sqrt{k \log n \log(k \log n)} - \log n$. Since $g_k(n)$ is bounded from above as a function of n, the function $G(k) = \sup_{n \geq 2} g_k(n)$ is well defined. Then the above bound running time is $O(e^{G(k)} n^3)$, as desired.

However, in our special situation the smoothness conjecture in [15] can be weakened. In order to describe the weakened conjecture and also make our discussion more self-contained, we return to the elliptic curve factorization algorithm in [15]. We shall make a different optimal choice of parameters, so that the function bounding the smoothness probability depends only on k.

Let $p < n^k$ be a prime factor of the n-bit integer N. Let $n' = n^k + 1 + 2n^{k/2}$; this is an upper bound on the number of points on an elliptic curve mod p. Define m to be

$$m = \prod_{\text{primes } l \leq n} l^{\alpha_l}, \qquad \text{where} \qquad \alpha_l = \left[\frac{\log n'}{\log l}\right],$$

i.e., m is the product of all primes $l \leq n$, each raised to the largest power that is $\leq n'$. Note that m can be computed in time $O(n\sqrt{n}\log^2 n)$ (to make a list of primes $l \leq n$ using trial division to test primality) plus $O(k^2 n \log n)$ (to raise all of the l to the α_l) plus $O(k^2 \log^4 n)$ (to compute the product), i.e., in time $O(k^2 n^2)$.

Following [15], we choose a random pair (E, P) consisting of an elliptic curve E modulo N and a point P on it. Then we try to compute the point mP on E mod N. This takes time

$$O(\log m \log^2 N) = O(\frac{n}{\log n} \cdot k \log n \cdot n^2) = O(kn^3).$$

We are almost certain to find the factor p as soon as our random choice of (E, P) has the property that $\#(E \bmod p)$ is n-smooth ("n-smooth" means that it is not divisible by any prime $> n$), because if that is the case, then the order of P on E mod p divides m. (For more details, see [15].) As explained in [15], as the pair (E, P) varies the number $\#(E \bmod p)$ is essentially uniformly distributed in the interval $(p + 1 - \sqrt{p}, p + 1 + \sqrt{p})$. Thus, to prove the theorem it suffices to show that the probability that an integer in that interval is n-smooth is bounded from below by a nonzero function of k (independent of n).

We now use the Canfield–Erdös–Pomerance lower bound for the number $\Phi(x, x^{1/k})$ of positive integers $\leq x$ that have no prime factors greater than $x^{1/k}$. According to Theorem 3.1 of [4], if $x \geq 1$ and $k \geq 3$, then

$$\Phi(x, x^{1/k}) \geq x\, \rho(k),$$

where

$$\rho(k) = \exp\left(-k\left(\log k + \log\log k - 1 + \frac{\log\log k - 1}{\log k} + C\frac{(\log\log k)^2}{\log^2 k}\right)\right),$$

in which C is an absolute constant. In other words,

Corollary 2 of Theorem 3.1 of [4]. *The function $\rho(k)$ is $\leq$ the probability that an integer in the interval $(0, x)$ is $x^{1/k}$-smooth.*

This is not quite what is needed to prove the theorem. As in [15], one needs a conjecture that says that the Canfield–Erdös–Pomerance lower bound — perhaps with a weaker choice of function $\rho(k)$ — holds on a smaller interval. Namely, one needs the following

Conjecture A. *There exists a positive function $\tilde{\rho}(k)$ such that the above corollary holds with the interval $(0, x)$ replaced by the interval $(x+1-\sqrt{x}, x+1+\sqrt{x})$ and with $\rho(k)$ replaced by $\tilde{\rho}(k)$.*

Under Conjecture A, there is at least a $\tilde{\rho}(k)$ probability that a random integer in the interval $(p+1-\sqrt{p}, p+1+\sqrt{p})$ is $p^{1/k}$-smooth, and *a fortiori* n-smooth. Thus, in the above algorithm the expected time to find p is

$$O\left(\frac{kn^3}{\tilde{\rho}(k)}\right).$$

The elliptic curve method is the only known factorization algorithm that can be used to prove the above result, because, as observed in [15], it is the only subexponential general purpose factorization method that tends to find small factors more quickly than large factors. We have the following Corollary (see also [16] §7 and [20]).

Corollary 3. *Under Conjecture A, the following decision problems are randomized fixed-parameter tractable.*

Small Prime Divisor
Input: An n-bit integer N and a positive integer parameter k.
Question: Does N have a nontrivial divisor less than n^k?

Polynomially Smooth Number
Input: An n-bit integer N and a positive integer parameter k.
Question: Is N n^k-smooth, i.e., is every prime divisor of N bounded by n^k?

Remark. n^k-smoothness of n-digit numbers is a natural number-theoretic property that arises in the study of polynomial-time complexity. For example, the concept plays a central role in the demonstration that primality belongs to the complexity class UP (*Unique* P) [9].

3 Intractable Parameterized Problems

The starting point for this section is the complexity theory for parameterized problems introduced in [5] and [6]. A *parameterized problem* is defined formally to be a subset $L \subseteq \Sigma^* \times \Sigma^*$. A parameterized problem L is *fixed-parameter tractable* if membership in L can be decided for $(x, k) \in \Sigma^* \times \Sigma^*$ in time $f(k)|x|^\alpha$, where α is a constant independent of k. In [5], [6], [7], [8] a hierarchy of complexity classes of parameterized problems is defined, and a wide variety of familiar parameterized computational problems are shown to be complete for these classes.

The complexity classes of parameterized problems in the W hierarchy intuitively correspond to the depth of circuits used to check $k \log n$ nondeterministic bits of information in a potential solution. The reader is best refered to [5–8] for the details. But just as NP-hardness can be concretely discussed in terms of polynomial-time reductions from Satisfiability to the problem in question, we can discuss hardness for the lowest level of the W hierarchy in terms in reductions from Clique. Since Clique is a well-known problem and apparently difficult with respect to fixed-parameter tractability, this seems reasonable for our present purposes.

The following parameterized problem provides the basis for our hardness results. For a graph $G = (V, E)$ we use the notation $N[v]$ to denote the *closed neighborhood* of a vertex v: the set of vertices consisting of v together with all vertices adjacent to v in G. A *perfect code* is a set of vertices $V' \subseteq V$ with the property that for each vertex $v \in V$ there is precisely one vertex in $N[v] \cap V'$.

Perfect Code
Input: A graph $G = (V, E)$ and a positive integer parameter k.
Question: Does G have a k-element perfect code?

In [8] it is shown that the Perfect Code problem belongs to the complexity class $W[2]$ and is hard for the class $W[1]$. This tells us concretely that demonstrating fixed-parameter tractability for Perfect Code would be at least as difficult as demonstrating fixed-parameter tractability for the well-known and apparently resistant Clique problem, and no harder than demonstrating fixed-parameter tractability for the Minimum Dominating Set problem (see [11] for the definition).

The Appendix contains a description of the reduction of the Clique problem to the Perfect Code problem. As an aside, we note that the Perfect Code problem is directly relevant to the rudimentary public-key cryptosystem described in [10]. The Subset Sum and Subset Product problems (defined in §1) are also of some interest in cryptography. We show that they are hard for $W[1]$, or equivalently and concretely: they are at least as hard as the Clique problem with respect to fixed-parameter tractability.

Theorem 4. *Subset Sum is hard for* $W[1]$.

Proof. We reduce from Perfect Code. Let $G = (V, E)$ be a graph for which we wish to determine whether G has a perfect code of size k. Suppose for convenience that the vertex set of the graph V is $\{0, ..., n-1\}$. We can easily compute the set of positive integers $X = \{x[i,j] : 1 \leq i \leq k,\ 0 \leq j \leq n-1\})$ and the positive integer s defined by

$$x[i,j] = (k+1)^{n+k-i} + \sum_{u \in N[j]} (k+1)^u$$

$$s = \sum_{t=0}^{n+k-1} (k+1)^t$$

Then X has a subset of cardinality k summing to s if and only if G has a k-element perfect code. The correctness of this transformation is easily observed if the numbers of X are represented in base $k+1$, and it is noted that there can be no carries in a sum of k integers from X expressed in this way.

Theorem 5. *Subset Product is hard for* $W[1]$.

Proof. We again reduce from Perfect Code. Let $G = (V, E)$ be a graph for which we wish to determine whether G has a perfect code of size k, and suppose the vertex set of G is $V = \{0, \ldots, n-1\}$. We may assume $k \leq n$.

We first produce n positive integers m_i, $i = 0, \ldots, n-1$, satisfying:
(1) for $i = 0, \ldots, n-1$, we have $2^k < m_i$, and
(2) the m_i are pairwise relatively prime.

This can be accomplished in time polynomial in n by finding the first n primes $p_0, \ldots, p_{n-1}$ larger than k. Trial division is sufficient for this, and requires time easily bounded by n^2 by elementary considerations. Let $m_i = 2^{p_i} - 1$. By Theorem 3.13 of [21], the m_i are pairwise relatively prime.

The m_i are to be used in conjunction with the Chinese Remainder Theorem. Let

$$m = \prod_{i=0}^{n-1} m_i \quad ,$$

We produce the following input for the Subset Product problem.

For each vertex $i \in V$, let x_i be the unique positive integer in the range $0 \leq x_i \leq m-1$ for which $x_i \equiv 2 \pmod{m_j}$ if $j \in N[i]$, and $x_i \equiv 1 \pmod{m_j}$ if $j \notin N[i]$. From the proof of the Chinese Remainder Theorem, which uses the extended Euclidean algorithm, we see that computing all of the x_i can be accomplished in time $O(n^5)$. For our input to the Subset Product problem the set X consists of the x_i, and we set a equal to 2.

Because of properties (1) and (2) in our construction of the m_i, the only way that a product of k of the x_i can equal 2 (by the Chinese Remainder Theorem) is for exactly one of the chosen x_i to be congruent to $2 \pmod{m_j}$ for each j, $0 \leq j \leq n-1$. By the definition of the x_i, the corresponding vertices of G are a perfect code. The converse is equally direct.

4 Open Problems

Suppose we are given a generator g of the multiplicative group of the field F_p of p elements. The discrete logarithm problem in F_p^* is, given $a \in F_p^*$, to find an integer x such that $a = g^x$.

Generally, there is thought to be a close relationship between the complexity of factoring an n-bit integer and the complexity of solving the discrete log problem in F_p^* for an n-bit prime p. See [19] and [22] for discussions of this question. Until recently, the best algorithms for both problems had time estimates of the form $L(n, 1/2)$, where for $0 \leq \gamma \leq 1$ one defines $L(n,\gamma) = \exp\Big(O(n^\gamma \log^{1-\gamma} n)\Big)$. Then with the invention of the number field sieve for factoring [3], the time estimate for factoring was brought down to $L(n, 1/3)$. Soon after, the number field sieve was also applied to the discrete log problem [12], bringing the time estimate for discrete log down to $L(n, 1/3)$ as well.

If we suppose that factorization and discrete log have roughly the same complexity, then it is natural to ask whether or not the same is true of parameterized versions of these problems. That is, we ask whether there is a parameterized version of the discrete logarithm problem that shares the same complexity behavior as that which was established for factorization in Theorem 2.1. Here are two possible discrete log analogues of Bounded Factor Factorization.

Bounded Size Discrete Logarithm
Input: An n-bit prime p, a generator g of F_p^*, an element $a \in F_p^*$ and a positive integer parameter k.
Question: Is there a positive integer $x < n^k$ such that $a = g^x$?

Bounded Hamming Weight Discrete Logarithm
Input: An n-bit prime p, a generator g of F_p^*, an element $a \in F_p^*$ and a positive integer parameter k.
Question: Is there a positive integer x whose binary representation has at most k 1's (that is, x has Hamming weight at most k), such that $a = g^x$?

These two problems are natural analogues of Bounded Factor Factorization. Note that in Bounded Factor Factorization and in both of the corresponding discrete log search problems, the bound on the length of the desired output is the same: $k \log n$. (The output is, respectively: the prime factor $p < n^k$, the exponent $x < n^k$, the locations of the $\leq k$ ones in the binary representation of x.)

Remark. Note that Bounded Hamming Weight Discrete Logarithm is a subproblem of the parameterized Subset Product problem. Namely, in the latter problem set $m = p$ and take the set X to be the residues mod p of g^{2^i}, $0 \leq i \leq p-2$.

The following questions are of interest.

Question 1. Is Bounded Size Discrete Logarithm randomized fixed-parameter tractable?

Question 2. Is Bounded Hamming Weight Discrete Logarithm randomized fixed-parameter tractable?

Question 2 is of practical significance, because the use of exponents of fairly small Hamming weight has been suggested in order to speed up cryptosystems based on discrete log. See, for example, [1]. If Question 2 has an affirmative answer, then limiting the Hamming weight of the exponent may compromise the security of the system.

Finally, the above open questions concerning F_p^* have analogues in other groups. For example:

Bounded Hamming Weight Elliptic Curve Discrete Logarithm
Input: An n-bit prime power q, an elliptic curve E defined overF_q, a point $P \in E$, a point $Q \in E$, and a positive integer parameter k.
Question: Is there a positive integer x of Hamming weight at most k such that $Q = xP$?

This problem is also of practical interest, because proposals for using elliptic curve cryptosystems in smart cards have included the idea of placing an upper bound on the Hamming weight of the integer multiples of points (see [17]). A somewhat different version of this problem has a bearing on the security of the type of implementation of elliptic curve cryptosystems proposed in [14]:

Bounded Hamming Weight Elliptic Curve Discrete Logarithm — Version 2
Input: A (small) prime power q, an elliptic curve E defined over F_q, an F_{q^n}-point P of E, an F_{q^n}-point Q, and a positive integer parameter k.
Question: Are there two disjoint sets $S_1, S_2 \subset \{0, 1, \ldots, n-1\}$ of total cardinality at most k such that $\sum_{j \in S_1} \varphi^j(P) - \sum_{j \in S_2} \varphi^j(P) = Q$, where φ denotes the q-th power Frobenius map?

Remark. Recall that discrete logarithm problems are *self-reducible*; this means that tractability of the decision problem implies tractability of the corresponding search problem. For example, if we have an algorithm for the decision problem Bounded Hamming Weight Discrete Logarithm, and if a certain instance of this problem has a positive answer, then we can *find* the discrete logarithm by $\leq n$ repetitions of the decision problem algorithm (with decreasing values of k). This seems to be a fundamental difference between discrete logarithm and factoring.

5 Appendix: On the Parameterized Complexity of the Perfect Code Problem

Theorem 6. *Perfect Code is hard for* $W[1]$.

Proof. We reduce from the Independent Set problem [11]. (A set of k vertices is an *independent set* in G if and only if it is a k-clique in the complement of G. Thus the Clique and Independent Set problems are essentially identical.) Let $G = (V, E)$ be a graph. We show how to produce a graph $H = (V', E')$ that has a perfect code of size $k' = \binom{k}{2} + k + 1$ if and only if G has a k-element independent set. The vertex set V' of H is the union of the sets of vertices:
$V_1 = \{a[s] : 0 \leq s \leq 2\}$
$V_2 = \{b[i] : 1 \leq i \leq k\}$
$V_3 = \{c[i] : 1 \leq i \leq k\}$
$V_4 = \{d[i, u] : 1 \leq i \leq k, u \in V\}$
$V_5 = \{e[i, j, u] : 1 \leq i < j \leq k, u \in V\}$
$V_6 = \{f[i, j, u, v] : 1 \leq i < j \leq k,\ u, v \in V\}$
The edge set E' of H is the union of the sets of edges:
$E_1 = \{a[0]a[i] : i = 1, 2\}$

$E_2 = \{a[0]b[i] : 1 \le i \le k\}$
$E_3 = \{b[i]c[i] : 1 \le i \le k\}$
$E_4 = \{c[i]d[i,u] : 1 \le i \le k, u \in V\}$
$E_5 = \{d[i,u]d[i,v] : 1 \le i \le k,\ u,v \in V\}$
$E_6 = \{d[i,u]e[i,j,u] : 1 \le i < j \le k, u \in V\}$
$E_7 = \{d[j,v]e[i,j,u] : 1 \le i < j \le k, v \in N[u]\}$
$E_8 = \{e[i,j,x]f[i,j,u,v] : 1 \le i < j \le k, x \ne u, x \notin N[v]\}$
$E_9 = \{f[i,j,u,v]f[i,j,x,y] : 1 \le i < j \le k, u \ne x \text{ or } v \ne y\}$

Suppose C is a perfect code of size k' in H. Since $a[1]$ and $a[2]$ are pendant vertices attached to $a[0]$, neither vertex belongs to C because both cannot belong to C, and if only one belongs to C, then C fails to be a dominating set. It follows that $a[0] \in C$. This implies that none of the vertices in V_2 and V_3 belong to C, and it implies also that exactly one vertex in each of the cliques formed by the edges of E_5 belongs to C. Note that each of these k cliques has n vertices indexed by V, the vertex set of G (this is the *selection* gadget). Let I be the set of vertices of G corresponding to the elements of C in these cliques. We argue that I is an independent set of order k in G.

Suppose $u, v \in I$ and that $uv \in E$. Then there are indices $i < j$ between 1 and k such that (without loss of generality) $d[i,u] \in C$ and $d[j,v] \in C$. By the definition of E_6 and E_7 each of these vertices is adjacent to $e[i,j,u]$, which contradicts that C is a perfect code in H. Thus I is an independent set in G.

Conversely, we argue that if $J = \{u_1, ..., u_k\}$ is a k-element independent set in G, then H has a perfect code C_J of size k'. We may take C_J to be the following set of vertices:

$$C_J = \{a[0]\} \cup \{d[i,u_i] : 1 \le i \le k\} \cup \{f[i,j,u_i,u_j] : 1 \le i < j \le k\}$$

That C_J is a perfect code can be verified directly from the definition of H.

References

1. G. B. Agnew, R. C. Mullin, I. M. Onyszchuk, and S. A. Vanstone: An implementation for a fast public-key cryptosystem, *J. Cryptology*, vol. 3 (1991), pp. 63–79.
2. J. Buss and J. Goldsmith: Nondeterminism within *P*, *SIAM J. Computing*, to appear.
3. J. P. Buhler, H. W. Lenstra, Jr., and C. Pomerance: Factoring integers with the number field sieve, to appear.
4. E. R. Canfield, P. Erdős, and C. Pomerance: On a problem of Oppenheim concerning "Factorisatio Numerorum," *J. Number Theory*, vol. 17 (1983), pp. 1–28.
5. R. G. Downey and M. R. Fellows: Fixed-parameter tractability and completeness, *Congressus Numerantium*, vol. 87 (1992), pp. 161–178.
6. R. G. Downey and M. R. Fellows: Fixed-parameter intractability, *Proceedings of the Seventh Annual IEEE Conference on Structure in Complexity Theory*, 1992, pp. 36–49.
7. R. G. Downey and M. R. Fellows: Fixed-parameter tractability and completeness I: basic results, to appear.

8. R. G. Downey and M. R. Fellows: Fixed-parameter tractability and completeness II: on completeness for $W[1]$, to appear.
9. M. R. Fellows and N. Koblitz: Self-witnessing polynomial-time complexity and prime factorization, *Proceedings of the Seventh Annual IEEE Conference on Structure in Complexity Theory*, 1992, pp. 107–110.
10. M. R. Fellows and N. Koblitz: Kid krypto, *Advances in Cryptology — Crypto '92*, Springer-Verlag, to appear.
11. M. Garey and D. S. Johnson: *Computers and Intractability: A Guide to the Theory of NP-Completeness*, W. H. Freeman, 1979.
12. D. Gordon: Discrete logarithms in $GF(p)$ using the number field sieve, *SIAM J. Discrete Math.*, to appear.
13. D. Gordon: Discrete logarithms in $GF(p^n)$ using the number field sieve, Preprint.
14. N. Koblitz: CM-curves with good cryptographic properties, *Advances in Cryptology — Crypto '91*, Springer-Verlag, 1992, pp. 279–287.
15. H. W. Lenstra, Jr.: Factoring integers with elliptic curves, *Annals Math.*, vol. 126 (1987), pp. 649–673.
16. H. W. Lenstra, Jr. and C. Pomerance: A rigorous time bound for factoring integers, *J. Amer. Math. Soc.*, vol. 5 (1992), pp. 483–516.
17. A. Menezes and S. A. Vanstone: The implementation of elliptic curve cryptosystems, *Advances in Cryptology — Auscrypt '90*, Springer-Verlag, 1990, pp. 2–13.
18. J. Nesetřil and S. Poljak: On the complexity of the subgraph problem, *Commen. Math. Univ. Carol.*, vol. 26 (1985), pp. 415–419.
19. A. Odlyzko: Discrete logarithms and their cryptographic significance, *Advances in Cryptology — Eurocrypt '84*, Springer–Verlag, 1985, pp. 224–314.
20. C. Pomerance: Fast, rigorous factorization and discrete logarithm algorithms, in D. S. Johnson, T. Nishizeki, A. Nozaki, H. S. Wilf, eds., *Discrete Algorithms and Complexity*, Academic Press, 1987, pp. 119–143.
21. K. Rosen: *Elementary Number Theory and Its Applications*, 3rd ed., Addison-Wesley, 1993.
22. P. van Oorschot: A comparison of practical public-key cryptosystems based on integer factorization and discrete logarithms, in G. Simmons, ed., *Contemporary Cryptology: The Science of Information Integrity*, IEEE Press, 1992, pp. 289–322.

A Class of Algebraic Geometric Codes from Curves in High-Dimensional Projective Spaces

G. L. Feng and T. R. N. Rao Fellow IEEE
the Center for Advanced Computer Studies,
University of Southwestern Louisiana, Lafayette, LA. 70504, USA

Abstract

Most of the research work in the area of algebraic geometric (AG) codes deals with the construction of AG codes from plane algebraic geometric curves. But, some work pertains to the construction of the AG codes from non-planar algebraic geometric curves. However, longer AG codes must have relatively larger genus and should only be the codes constructed from non-planar curves. In this paper, we present a new construction of a class of AG codes from curves in high-dimensional projective spaces. For this construction, it is easy to determine the designed minimum distance and find the parity check matrix, and the decoding up to the designed minimum distance is fast. Furthermore, this approach can be easily understood by most engineers.

I. Introduction

The most important development in the theory of error-correcting codes in recent years is the introduction of methods from algebraic geometry to construct linear codes [1-3]. These so-called *algebraic geometric codes (AG codes)* were introduced by Goppa in 1980. In 1982, Tsfasman, Vladut and Zink [4] obtained an extremely exciting result: the existence of a sequence of AG codes which exceeds the Gilbert-Varshamov bound [5]. For this paper they received the IEEE Information Theory Group Paper Award for 1983. Since then, many papers dealing with algebraic geometric codes have followed [6-16]. However, most of these papers deal only with the AG codes obtained from plane curves.

The greatest advantage of AG codes is that they offer more flexibility in the choice of code parameters. Most importantly, for a fixed finite field $\mathbf{F}_q$, there are AG codes having any length. It is known that the coordinates of AG codes are the rational points of algebraic geometric curves. Thus, greater the number of rational points, longer the length of the AG code. Over $\mathbf{F}_q$, the number of rational points of any plane algebraic geometric curve is obviously less than $q^2 + q + 1$. Thus, actually useful AG codes should be the AG codes constructed from curves in high-dimensional projective spaces (HDAG codes). In [17], Pellikaan et. al constructed a large class of codes from non-planar curves. Justesen et. al in [6], first gave a description of algebraic-geometric codes defined only by monomials. Following this description of AG codes, in [18], a simple approach for the construction of AG codes from affine plane algebraic geometric curves has been proposed. This new approach can be easily understood by most engineers. In this paper, we generalize the results in [18] to the case of curves in high-dimensional projective spaces and further present a construction of a class of HDAG codes. The codes considered here are essentially the algebraic-geometric Reed-Muller codes [19].

The paper is organized as follows. In the next section, for easy reference, we review and improve some preliminary results in [18] for HDAG codes. In Section III, an approach for the construction of HDAG codes is presented. For a large class of curves in high-dimensional projective spaces the corresponding HDAG codes can be easily constructed and the designed minimum distances of constructed codes can also be determined by the method in Section II. In Section IV, we show that the fast decoding procedure [8,15,16] is very efficient for these AG codes constructed by the approach in Section III from curves in high-dimensional projective spaces. Finally, some problems are proposed for further research in Section V.

II. New Construction of AG codes from Curves in High-dimensional Space

In this section we introduce a new approach for the construction of linear codes, which include some current AG codes. Then we present an essential method, which plays a key role in determining the minimum distance bound for the linear codes constructed by the approach.

Let $H \triangleq \{ \mathbf{p}_1, \mathbf{p}_2, \ldots , \mathbf{p}_r, \ldots , \mathbf{p}_v \}$ be a sequence of F_q^n vectors, where $\mathbf{p}_r \triangleq (p_r(Q_1), p_r(Q_2), \ldots , p_r(Q_n))$, and $Q_i = (x_{1,i},\cdots,x_{m,i})$ for $i = 1, 2, \cdots , n$, are rational points of an algebraic geometric curve in a m-dimensional space, and p_r is a polynomial with m-variables. Hereinafter, the polynomial p_r sometimes denotes a vector $\mathbf{p}_r$. Let S(r) be the linear space over F_q spanned by the first r vectors of H. Now let us define the *order* of a polynomial $f(x_1,\cdots,x_m)$. We associate with each variable x_i a positive integer $\underline{x_i}$, which is called the *order* of x_i. We define the order of $x_1^{a_1} x_2^{a_2} \cdots x_m^{a_m}$ as $a_1\underline{x_1} + \cdots + a_m\underline{x_m}$. The order of a polynomial $f(x_1,\cdots,x_m)$ is defined as the maximum of the orders of all $x_1^{a_1} x_2^{a_2} \cdots x_m^{a_m}$ having non-zero coefficients in f and is denoted as $\underline{f}$. Then the polynomial order satisfies the following conditions:

$$\underline{f+g} = \underline{f}, \quad \text{if } \underline{f} > \underline{g}, \tag{2.1}$$

$$\underline{f \cdot g} = \underline{f} + \underline{g}. \tag{2.2}$$

Let the order of a constant be 0.

In order to construct AG codes from algebraic geometric curves in high-dimensional projective spaces, we are interested in such H:

$$p_1(x_1,\cdots,x_m) = 1, \quad \underline{p_i(x_1,\cdots,x_m)} < \underline{p_{i+1}(x_1,\cdots,x_m)}, \tag{2.3}$$

and

$$p_i \cdot p_j \in S(r) \quad \text{and} \quad p_i \cdot p_j \notin S(r-1), \tag{2.4}$$

where $\underline{p_i} + \underline{p_j} = \underline{p_r}$, $r = \phi(i,j)$, and $\phi(i,j)$ is a function of i and j.

Remark: The polynomials sometimes denote vectors. Usually, when $i + j \geq g^*$, $\phi(i,j) = i + j - g^* + 1$, where g^* is a constant related to H. Later, we will define this constant g^*.

Let d_v be an integer such that

$$\text{any } d_v - 1 \text{ or fewer columns of } \mathbf{H}_v \text{ are linearly independent },\tag{2.5}$$

where

$$\mathbf{H}_r \triangleq \begin{bmatrix} \mathbf{p}_1 \\ \mathbf{p}_2 \\ | \\ \mathbf{p}_r \end{bmatrix} .\tag{2.6}$$

In practice, d_v should be easily estimated.

Now let us consider the basic properties of H. Let I be the set of orders of polynomials in H, that is, $I \triangleq \{ p_i \mid i = 1, 2, ..., v \}$. If an integer $p \notin I$ and $0 \le p \le p_v$, p is called a *gap* of I. Let g^* denote the number of all gaps of I. We call g^* the *genus* of H. Later we will see that the action of g^* is similar to that of g for the current AG codes.

Hereinafter, we always assume that H satisfies (2.3), (2.4), and that d_v can be estimated. From (2.4) and (2.1)-(2.2), we know that H satisfies the following condition:

$$\textit{If } s, t \in I \text{ and } s + t \le p_v(x_1, \cdots, x_m) , \quad \textit{then } s + t \in I .\tag{2.7}$$

Lemma 2.1: If $2g^* \le s \le p_v$, then $s \in I$, that is, from $2g^*$, the orders of polynomials in H are consecutive.

Proof: Assume $2g^* \le s \le p_v$ and $s \notin I$, then we have
$s = 1 + s - 1 = 2 + s - 2 = \cdots = g^* + s - g^*$,
where $1, s-1, 2, s-2, \cdots, g^*-1, s-g^*+1$ are distinct from each other as well as from $g^*, s-g^*$, for $s > 2g^*$. From (2.7), it is known that there is at least one gap among 1 and $s-1$, among 2 and $s-2$, $\cdots$, among g^* and $s-g^*$. Thus, there are at least g^* gaps in $[0, s-1]$. On the other hand, s is a gap. Therefore, there are at least g^*+1 gaps in I. This is in contradiction with the assumption that there are g^* gaps in I. □

Now we describe a new approach for the construction of AG codes from algebraic geometric curves in high-dimensional projective spaces and present a method for determining the minimum distance bound of such AG codes without directly using the Riemann-Roch theorem.

Let C_r be a linear code over F_q defined by $\mathbf{H}_r$ as a parity check matrix. Obviously, the dimension of the linear code C_r is $n - r$. Such linear codes C_r, are the AG codes from curves in high-dimensional projective spaces and are the subject of this paper. In the following, we will show that its minimum distance lower bound can be easily determined.

Let $\mathbf{r}$ be a received vector, $\mathbf{c}$ be a codeword of C_r , and $\mathbf{e}$ be an error vector. We have $\mathbf{r} = \mathbf{c} + \mathbf{e}$. Define

$$s_i = \mathbf{p}_i \cdot \mathbf{r}^T ,\tag{2.8}$$

and

$$S_{ij} = \mathbf{p}_{i,j} \cdot \mathbf{r}^T , \tag{2.9}$$

where $\mathbf{p}_{i,j} \triangleq (p_i(Q_1)\cdot p_j(Q_1), p_i(Q_2)\cdot p_j(Q_2), \cdots, p_i(Q_n)\cdot p_j(Q_n))$.

From (2.3), (2.4) and the definition of C_r, we have

$$s_i = \mathbf{p}_i \cdot \mathbf{e}^T \quad \text{for } 1 \le i \le r , \tag{2.10}$$

and

$$S_{ij} = \mathbf{p}_{i,j} \cdot \mathbf{e}^T , \quad \text{if } \mathbf{p}_{i,j} \in S(r) . \tag{2.11}$$

We call these syndromes *known*. When $\mathbf{e} = \vec{0}$, these *known* syndromes are all equal to zero. Other syndromes are termed *unknown*.

Suppose $\mathbf{p}' \notin S(r)$ and $\mathbf{p}' \in S(r+1)$, then we say that s_{r+1} and $s'_{r+1} \triangleq \mathbf{p}'\cdot\mathbf{r}^T$ are consistent and s'_{r+1} is a consistent element of s_{r+1}. Obviously, s'_{r+1} can be expressed as $\sum_{i=1}^{r+1} a_i s_i$ and the coefficient a_{r+1} is not zero. There may be many such $\mathbf{p}'$. We use s'_{r+1} to express any syndrome defined by some $\mathbf{p}'$ different from $\mathbf{p}_{r+1}$. Thus, $S_{i,j} = s_{r+1}$ or s'_{r+1}, if $\phi(i,j) = r+1$. We have this lemma:

Lemma 2.2: If $s_i = 0$ for $1 \le i \le r+1$, then $s'_{r+1} = 0$. If $s_i = 0$ for $1 \le i \le r$, and $s_{r+1} \ne 0$, then $s'_{r+1} \ne 0$.

For matrix $\mathbf{S} = [\, S_{ij} \,]_{1 \le i,j \le n}$, we denote $S_{\le u, \le v} \triangleq [\, S_{ij} \,]_{1 \le i \le u, 1 \le j \le v}$. If some elements of matrix $\mathbf{S}$ are *unknown*, we call it an *incomplete* matrix, and denote it as $\overline{\mathbf{S}}$. We are only interested in *well-behaving* matrices, in which whenever S_{uv} is *known*, all of $S_{\le u, \le v}$ are *known*.

We define

$$\overline{\mathbf{S}^{(r+1)}} \triangleq [\, S_{ij} \,]_{1 \le i,\, j \le r+1} . \tag{2.12}$$

Lemma 2.3: $\overline{\mathbf{S}^{(r+1)}}$ is a *well-behaving* matrix.

Proof: If S_{hk} is *known*, then $\phi(h,k) \le r$. From (2.4), for any $1 \le i \le h$ and $1 \le j \le k$, $\phi(i,j) \le r$. Hence, S_{ij} is *known* too. □

Lemma 2.4: Suppose $\mathbf{r}$ is a codeword $\mathbf{c} \triangleq (c_1, c_2, \ldots, c_n)$. If there are d nonzero s_{r+1} and its consistent elements s'_{r+1} in $\overline{\mathbf{S}^{(r+1)}}$, then the weight of the codeword $\mathbf{c}$ is at least d.

Proof: The matrix $\overline{\mathbf{S}^{(r+1)}}$ can be decomposed into $\mathbf{XYX}^T$, where

$$\mathbf{X} = \begin{bmatrix} \mathbf{H}_r \\ \mathbf{h}_{r+1} \end{bmatrix} ,$$

$$\mathbf{Y} = \begin{bmatrix} c_1 & 0 & \cdots & 0 \\ 0 & c_2 & \cdots & 0 \\ | & | & | & | \\ 0 & 0 & \cdots & c_n \end{bmatrix}.$$

Since $\mathbf{H}_r \cdot \mathbf{c}^T = \vec{0}$, i.e., all syndromes $s_i = 0$, for $1 \le i \le r$. However, there are d nonzero s_{r+1} and its consistent elements. From Lemma 2.2 and Lemma 2.3, $\overline{\mathbf{S}^{(r+1)}}$ is a well-behaving matrix and the *known* S_{ij} are all zero. Thus, rank $(\overline{\mathbf{S}^{(r+1)}}) \ge d$. Therefore, rank $(\mathbf{Y}) \ge d$. Hence, the weight of $\mathbf{c}$ is at least d. □

Theorem 2.1: Suppose $\mathbf{H}_r$ is a parity check matrix of a linear code C_r. If the number of s_{w+1} and its consistent elements in $\overline{\mathbf{S}^{(w+1)}}$ is at least d_r^*, for $r \le w < v$ and $d_r^* \le d_v$, then the minimum distance of the code C_r is at least d_r^*.

Proof: For any nonzero codeword $\mathbf{c}$ of C_r, if the value of s_{r+1} is not zero, from Lemma 2.4, the weight of $\mathbf{c} \ge d_r^*$. If the value of s_{r+1} is zero, then from (2.5) and $d_r^* \le d_v$, we know that there is the smallest $p < v$ such that $s_h = 0$ for $1 \le h \le p$, and $s_{p+1} \ne 0$. Thus, $\mathbf{c}$ is also a codeword of codes C_{r+1}, ..., C_p. Since $s_{p+1} \ne 0$ and the number of s_{p+1} and s'_{p+1} in $\overline{\mathbf{S}^{(p+1)}} \ge d_r^*$. Regarding $\mathbf{c}$ as a codeword of C_p, from Lemma 2.4, it is known that the weight of $\mathbf{c} \ge d_r^*$. Therefore, the minimum distance of C_r is at least d_r^*.
□

In order to easily construct such H, we are interested in the case that $p_i(x_1,\cdots,x_m)$ is a simple polynomial. A *monomial* is a polynomial with exactly one nonzero term, i.e. it takes the form $x_1^{a_1}\cdots x_m^{a_m}$, $a_1, a_2, \cdots, a_m \ge 0$ in the current context. For more convenience, in the following we also restrict $p_i(x_1,\cdots,x_m)$ to be monomials, that is, $H \triangleq \{ x_1^{a_1^{(i)}},\cdots,x_m^{a_m^{(i)}} \mid i = 1, 2, \ldots, v \}$. We have

$$\underline{x_1^{a_1^{(i)}},\cdots,x_m^{a_m^{(i)}}} \triangleq a_1^{(i)} \cdot \underline{x_1} + \cdots + a_m^{(i)} \cdot \underline{x_m}.$$

From (2.3) we have $x_1^{a_1^{(1)}},\cdots,x_m^{a_m^{(1)}} = 1$, that is, $a_s^{(1)} = 0$ for $1 \le s \le m$. Let α_s be the value such that there are at most α_s rational points having the same x_t for $1 \le t \le m$ and $t \ne s$. We have the following useful lemma:

Lemma 2.5: If

$$\{ x_1^{c_1},\cdots,x_m^{c_m} \mid 0 \le c_t \le \alpha_t - 1,\ t \ne s \text{ and } 1 \le t \le m;\ 0 \le c_s \le q-2 \} \subseteq H_v, \quad (2.13)$$

then the columns of $\mathbf{H}_v$ are linearly independent, that is,

$$d_v = v + 1. \quad (2.14)$$

Proof: Without loss of the generality, let $s = m$ and $0 \le t \le m-1$. Assume that (2.13) is satisfied and that (2.14) is not satisfied. Suppose the $(l_{\mu_1}^{(k_1)}\cdots l_{\mu_m}^{(k_m)})$-th monomials for $\lambda = 1, 2, \cdots, m$, $k_\lambda = 1, \cdots, w_\lambda$, and $\mu_\lambda = 1, \cdots, w_{k_\lambda}$ are linearly dependent, where $x_{\lambda, l_{\mu\lambda}^{(k\lambda)}}$ have same value x_{λ,k_λ}. From (2.13), we have

$$\sum_{\mu_m=1}^{w_{k_m}} \cdots \sum_{\mu_1=1}^{w_{k_1}} C_{l_{\mu_1}^{(k_1)} \cdots l_{\mu_m}^{(k_m)}} \cdot x_{1,l_{\mu_1}^{(k_1)}}^{i_1} \cdots x_{m,l_{\mu_m}^{(k_m)}}^{i_m} = 0 ,$$

for $t = 1, 2, ..., m-1$; $i_t = 0, 1, \cdots, \alpha_t - 1$; $i_m = 0, 1, ..., q-2$,

where $C_{l_{\mu_1}^{(k_1)} \cdots l_{\mu_m}^{(k_m)}} \neq 0$. Since $x_{m,l_{\mu_m}^{(k_m)}} = x_{m,k_m}$, we have

$$\sum_{\mu_m=1}^{w_{k_m}} \left(\sum_{\mu_{m-1}=1}^{w_{k_{m-1}}} \cdots \sum_{\mu_1=1}^{w_{k_1}} C_{l_{\mu_1}^{(k_1)} \cdots l_{\mu_m}^{(k_m)}} \cdot x_{1,l_{\mu_1}^{(k_1)}}^{i_1} \cdots x_{m-1,l_{\mu_{m-1}}^{(k_{m-1})}}^{i_{m-1}} \right) \cdot x_{m,l_{\mu_m}^{(k_m)}}^{i_m} = 0,$$

for $i_m = 0, 1, \cdots, q-2$.

Since the number of distinct nonzero x_{m,k_m} is at most $q-1$, from the above equations, we have

$$\sum_{\mu_{m-1}=1}^{w_{k_{m-1}}} \cdots \sum_{\mu_1=1}^{w_{k_1}} C_{l_{\mu_1}^{(k_1)} \cdots l_{\mu_m}^{(k_m)}} \cdot x_{1,l_{\mu_1}^{(k_1)}}^{i_1} \cdots x_{m-1,l_{\mu_{m-1}}^{(k_{m-1})}}^{i_{m-1}} = 0,$$

for $t = 1, 2, ..., m-1$; $i_t = 0, 1, \cdots, \alpha_t - 1$.

On the other hand, since the number of distinct $x_{m-1,l_{\mu_{m-1}}^{(k_{m-1})}}$ is at most α_{m-1} and they are distinct from each other (because $x_{m,l_{\mu_m}^{(k_m)}}$ are same), we have,

$$\sum_{\mu_{m-2}=1}^{w_{k_{m-2}}} \cdots \sum_{\mu_1=1}^{w_{k_1}} C_{l_{\mu_1}^{(k_1)} \cdots l_{\mu_m}^{(k_m)}} \cdot x_{1,l_{\mu_1}^{(k_1)}}^{i_1} \cdots x_{m-2,l_{\mu_{m-2}}^{(k_{m-2})}}^{i_{m-2}} = 0,$$

for $t = 1, 2, ..., m-2$; $i_t = 0, 1, \cdots, \alpha_t - 1$.

Repeatedly using this procedure, we have:

$$C_{l_{\mu_1}^{(k_1)} \cdots l_{\mu_m}^{(k_m)}} = 0 .$$

This is in contradiction with the assumption that $C_{l_{\mu_1}^{(k_1)} \cdots l_{\mu_m}^{(k_m)}} \neq 0$. □

Lemma 2.6: If $2g^* \leq s \leq \underline{x_1^{a_1^{(v)}}, \cdots, x_m^{a_m^{(v)}}}$, then $s \in I$, that is, from $2g^*$, the orders of polynomials in H are consecutive.

Proof: Assume $2g^* \leq s \leq x_1^{a_1^{(v)}}, \cdots, x_m^{a_m^{(v)}}$ and $s \notin I$, then we have

$$s = 1 + \overline{s-1 = 2} + s - 2 = \cdots = g^* + s - g^*,$$

where 1, $s-1$, 2, $s-2, \cdots$, g^*-1, $s-g^*+1$ are distinct from each other as well as from g^*, $s-g^*$, for $s > 2g^*$. It is known that there exists at least one gap among 1 and $s-1$, among 2 and $s-2$, $\cdots$, among g^* and $s-g^*$. Thus, there are at least g^* gaps in $[0, s-1]$. On the other hand, we know that s is a gap. Therefore, there are at least g^*+1 gaps in I. This is in contradiction with the assumption that there are g^* gaps in I. □

We have the following main theorem regarding the designed minimum distance:

Theorem 2.2: If $r > g^*$ and $r - g^* + 1 \leq d_v$, the minimum distance of a $(n, n-r)$ linear code C_r defined by $\mathbf{H}_r$, is at least r-g^*+1.

Proof: Let us consider $\overline{\mathbf{S}^{(r+1)}}$ associated with C_r. From the definition, for $i, j \leq r+1$, $S_{ij} = \sum_{k=1}^{n} r_k x_{1,k}^{a_1^{(i)}}, \cdots, x_{m,k}^{a_m^{(i)}} x_{1,k}^{a_1^{(j)}}, \cdots, x_{m,k}^{a_m^{(j)}}$. We construct a new integer matrix $\mathbf{P}^{(r+1)} = [\, p_{ij} \,]_{(r+1)\times(r+1)}$, where $p_{ij} = x_1^{a_1^{(i)}}, \cdots, x_m^{a_m^{(i)}} + x_1^{a_1^{(j)}}, \cdots, x_m^{a_m^{(j)}}$. From (2.4) and the definition of s_{r+1} and s'_{r+1}, it is known that the number of s_{r+1} and s'_{r+1} in $\overline{\mathbf{S}^{(r+1)}}$, is equal to the number of $x_1^{a_1^{(r+1)}}, \cdots, x_m^{a_m^{(r+1)}}$ in $\mathbf{P}^{(r+1)}$.

On the other hand, let $\mathbf{A}$ be this matrix $[\, A_{ij} \,]_{(p_{r+1}+1)\times(p_{r+1}+1)}$ and $A_{ij} = i + j - 2$, where $p_{r+1} \triangleq x_1^{a_1^{(r+1)}}, \cdots, x_m^{a_m^{(r+1)}}$. Obviously, $\mathbf{P}^{(r+1)}$ can be obtained by deleting all rows and columns associated with the gaps in I from $\mathbf{A}$. Thus, $\mathbf{P}^{(r+1)}$ can be obtained by deleting g^* rows and g^* columns from $\mathbf{A}$. Since the original number of p_{r+1} in $\mathbf{A}$ is p_{r+1}+1, the number of p_{r+1} in $\mathbf{P}^{(r+1)}$ should be p_{r+1}+1-2g^*.

From the definition of g^*, the condition of $r > g^*$, and Lemma 2.6, we have that $p_{r+1}+1-2g^* = r - g^* + 1$. Obviously, this value increases with r. From Theorem 2.1, the minimum distance of C_r is at least $r - g^* + 1$. □

The value of r-g^*+1 is called the designed minimum distance. In the proof of Theorem 2.2, when $r < 2g^*$, there are some rows and columns which are associated with the gaps and in which there are common p_{r+1}. Thus, deletion of some p_{r+1}'s may be counted twice. Therefore, when $r \leq 2g^*$, this theorem can more precisely be stated as follows:

Theorem 2.3: Let d_r be the number of $x_1^{a_1^{(r+1)}}, \cdots, x_m^{a_m^{(r+1)}}$ in $\mathbf{P}^{(r+1)}$, $r > g^*$, and $d_r \leq d_v$. Then the minimum distance of code C_r defined by $\mathbf{H}_r$, is at least d_r.

Thus, when $g^* < r \leq 2g^*$, we can determine a more precise minimum distance bound using Theorem 2.3.

From the above discussion, it is clear that the smaller g^* is, the better C_r is. In the following, we will give an approach to find H for a given affine curve in a high-dimensional projective space such that

(1*) (2.3), (2.4), and d_v can be estimated, and
(2*) g^* is as small as possible.

III. A Class of AG Codes from Curves in High-dimensional Projective Spaces

In this section, we introduce a class of AG codes from the following type of curves in high-dimensional projective spaces:

$$f_s(x_1, x_2, \cdots, x_{s+1}) = 0 \qquad \text{for} \quad s = 1, 2, \ldots, m-1, \tag{3.1}$$

namely,

$$\left\{ \begin{array}{l} f_1(x_1, x_2) = 0, \\ f_2(x_1, x_2, x_3) = 0, \\ \cdots \\ f_{m-1}(x_1, x_2, \cdots, x_m) = 0, \end{array} \right.$$

where

$$f_s(x_1, x_2, \cdots, x_{s+1}) = x_s^{a_s} + x_{s+1}^{b_s} + g_s(x_1, x_2, \cdots, x_{s+1}), \tag{3.2}$$

$$gcd(\ a_s, b_s\ ) = 1 \quad \text{and} \quad deg\ g_s(x_1, x_2, \cdots, x_{s+1}) < \min\{\ a_s, b_s\ \}. \tag{3.3}$$

Let us show some examples of such curves in high-dimensional projective spaces.

Example 3.1 A Hermitian-like curve in 3-dimensional space over GF(2^4) is defined by the following equation:

$$\left\{ \begin{array}{l} x_1^5 + x_2^4 + x_2 = 0\ , \\ x_2^5 + x_3^4 + x_3 = 0\ . \end{array} \right. \tag{3.4}$$

Example 3.2 A Hermitian-like curve in m-dimensional space over GF(q^2), where $q = p^r$, is defined by the following equation:

$$x_s^{p^r+1} + x_{s+1}^{p^r} + x_{s+1} = 0, \qquad \text{for } s = 1, 2, \cdots, m-1\ . \tag{3.5}$$

Example 3.3: A curve in 3-dimensional space over GF(2^4) is defined by the following equation:

$$\left\{ \begin{array}{l} x_1^5 + x_2^4 + x_2 = 0\ , \\ x_2^5 + x_3^3 + x_3 = 0\ . \end{array} \right. \tag{3.6}$$

Now let us introduce a recursive procedure to construct AG codes from the curve in a high-dimensional projective space defined by (3.1). From the previous section we know that the key is to find H. For the curve defined by (3.1), we have the following theorem.

Theorem 3.1: For the curve in a high-dimensional projective space defined by (3.1), let H_m, $x_s^{(m)}$, g_m^* be determined by the following recursive procedure, then H_m satisfies (1*) and (2*).

(1) $H_m = \{\ x_1^{i_1} \cdots x_{m-1}^{i_{m-1}} x_m^{i_m} \mid x_1^{i_1} \cdots x_{m-1}^{i_{m-1}} \in H_{m-1} \text{ and } 0 \le i_m \le b_{m-1} - 1\ \}$.

(2) $\underline{x_s^{(m)}} = \underline{x_s^{(m-1)}} b_{m-1}$ for $s = 1, 2, \cdots, m-1$ and $\underline{x_m^{(m)}} = \prod_{s=1}^{m-1} a_s \triangleq \Pi_{m-1}$.

(3) $2g_m^* = b_{m-1}(2g_{m-1}^* - 1) + \Pi_{m-1}(b_{m-1} - 1) + 1$.

where $H_2 = \{\ x_1^{i_1} x_2^{i_2} \mid 0 \le i_1 \le q-2,\ 0 \le i_2 \le b_1 - 1\ \}$, $\underline{x_1^{(2)}} = b_1$, $\underline{x_2^{(2)}} = a_1$, and $2g_2^* = (a_1-1)(b_1-1)$.

Proof: From Theorem 3.3 in [18], we know that Theorem 3.1 is true for the case of $m = 2$. Now we suppose that Theorem 3.1 is true for the case of $m-1$, and attempt to prove that it is true for the case of m.

Let us prove that H_m satisfies (1*).

Let $(i_1, \dots, i_m) \neq (i'_1, \dots, i'_m)$ and $x_1^{i_1} \cdots x_m^{i_m}, x_1^{i'_1} \cdots x_m^{i'_m} \in H_m$, now we prove that

$$\underline{x_1^{i_1} \cdots x_m^{i_m}} \neq \underline{x_1^{i'_1} \cdots x_m^{i'_m}} . \tag{3.7}$$

From (2.1) and (2) we have,

$$\underline{x_1^{i_1} \cdots x_m^{i_m}} = b_{m-1} \cdot \underline{x_1^{i_1} \cdots x_{m-1}^{i_{m-1}}} + \Pi_{m-1} \cdot i_m ,$$

and

$$\underline{x_1^{i'_1} \cdots x_m^{i'_m}} = b_{m-1} \cdot \underline{x_1^{i'_1} \cdots x_{m-1}^{i'_{m-1}}} + \Pi_{m-1} \cdot i'_m .$$

If $i_m \neq i'_m$, then from $\gcd(a_{m-1}, b_{m-1}) = 1$, we know that (3.7) is true. If $i_m = i'_m$, then $(i_1, \dots, i_{m-1}) \neq (i'_1, \dots, i'_{m-1})$. Since H_{m-1} satisfies (1*) and (2*), we have $\underline{x_1^{i_1} \cdots x_{m-1}^{i_{m-1}}} \neq \underline{x_1^{i'_1} \cdots x_{m-1}^{i'_{m-1}}}$. Thus, (3.7) is also true.

On the other hand, $1 \in H_m$. Therefore, (2.3) is satisfied. For convenience, let the elements of H be arranged as $x_1^{i_1^{(\lambda)}} \cdots x_m^{i_m^{(\lambda)}}$, for $\lambda = 1, 2, \dots, v$ and let $\underline{x_1^{i_1^{(\lambda)}} \cdots x_m^{i_m^{(\lambda)}}} < \underline{x_1^{i_1^{(\lambda+1)}} \cdots x_m^{i_m^{(\lambda+1)}}}$.

For any s and t, $(x_1^{i_1^{(s)}} \cdots x_m^{i_m^{(s)}}) \cdot (x_1^{i_1^{(t)}} \cdots x_m^{i_m^{(t)}}) = x_1^{i_1^{(s)}+i_1^{(t)}} \cdots x_m^{i_m^{(s)}+i_m^{(t)}}$. Since H_{m-1} satisfies (1*) and (2*), $x_1^{i_1^{(s)}+i_1^{(t)}} \cdots x_{m-1}^{i_{m-1}^{(s)}+i_{m-1}^{(t)}} = x_1^{i_1^{(\mu)}} \cdots x_{m-1}^{i_{m-1}^{(\mu)}} \in H_{m-1}$, for some μ. Now we consider two cases:

(a) Case 1: $i_m^{(s)} + i_m^{(t)} \leq b_{m-1}$. Obviously, from (1), i.e. the definition of H_m, $x_1^{i_1^{(s)}+i_1^{(t)}} \cdots x_m^{i_m^{(s)}+i_m^{(t)}} \in H_m$ and $x_1^{i_1^{(s)}+i_1^{(t)}} \cdots x_m^{i_m^{(s)}+i_m^{(t)}} = x_1^{i_1^{(\mu')}} \cdots x_{m-1}^{i_{m-1}^{(\mu')}} x_m^{i_m^{(\mu')}}$, for some μ' which is related to μ.

(b) Case 2: $i_m^{(s)} + i_m^{(t)} > b_{m-1}$. By repeatedly using (3.1), we see that $x_1^{i_1^{(s)}+i_1^{(t)}} \cdots x_m^{i_m^{(s)}+i_m^{(t)}}$ can be written as a linear combination of the elements of H_m, and that each element has order less than or equal to $\underline{x_1^{i_1^{(s)}} \cdots x_m^{i_m^{(s)}}} + \underline{x_1^{i_1^{(t)}} \cdots x_m^{i_m^{(t)}}}$. Thus, we have the same result as in the case (a).

On the other hand, since H_{m-1} satisfies (2.4), that is, $x_1^{i_1^{(s)}+i_1^{(t)}} \cdots x_{m-1}^{i_{m-1}^{(s)}+i_{m-1}^{(t)}} = x_1^{i_1^{(\mu)}} \cdots x_{m-1}^{i_{m-1}^{(\mu)}} \notin S^{(m-1)}(\mu\text{-}1)$, which is a linear space spanned by the first μ monomials of H_{m-1}, we have $x_1^{i_1^{(s)}+i_1^{(t)}} \cdots x_m^{i_m^{(s)}+i_m^{(t)}} = x_1^{i_1^{(\mu')}} \cdots x_{m-1}^{i_{m-1}^{(\mu')}} x_m^{i_m^{(\mu')}} \notin S^{(m)}(\mu'\text{-}1)$. Therefore, H_m also satisfies (2.4).

From (3.1), it is easily seen that $\alpha_t = b_{t-1}$ for $t = 2, 3, \dots, m$ and $\alpha_1 = a_1$. From Lemma 3.2, $d_v = v + 1$, where $v = (q-1) \cdot \prod_{s=1}^{m-1} b_s$. Thus, (1*) is satisfied.

Now we are going to prove that H_m satisfies (2*). First we prove the following:

$$\textit{If } w \geq b_{m-1}(2g^*_{m-1} - 1) + \Pi_{m-1}(b_{m-1} - 1) + 1, \textit{ then } w \textit{ can be } \sum_{s=1}^{m} i_s \underline{x_s^{(m)}}, \tag{3.8}$$

where $0 \leq i_1 \leq q-2$ and $0 \leq i_{t+1} \leq b_t - 1$ for $1 \leq t \leq m-1$.

Suppose that $w \bmod b_{m-1} \equiv i$, where $0 \le i \le b_{m-1}-1$. Let $i_m = i$, then $(w - i_m x_m) \bmod b_{m-1} = (w - i\Pi_{m-1}) \bmod b_{m-1} = 0$. Hence, $w - i_m x_m$ can be written as $b_{m-1}\overline{w'}$. Since $w \ge 2[b_{m-1}(2g_{m-1}\text{-}1) + \Pi_{m-1}(b_{m-1}-1) + 1]$, we obtain $w' \ge 2g_{m-1}$. From the definition of g_{m-1}, we know that w' can be written as

$$w' = \sum_{s=1}^{m-1} i_s \cdot \underline{x_s^{(m-1)}} \quad \text{for} \quad 0 \le i_s \le b_s \ \text{ for } \ s \ne 1 \ \text{ and } \ 0 \le i_1 \le q-2 ,$$

namely,

$$b_{m-1} w' = \sum_{s=1}^{m-1} i_s \cdot \underline{x_s^{(m)}} \quad \text{for} \quad 0 \le i_s \le b_s \ \text{ for } \ s \ne 1 \ \text{ and } \ 0 \le i_1 \le q-2 .$$

Thus, (3.8) is true. Now we prove that

$$2g_m^* - 1 = b_{m-1}(2g_{m-1}-1) + \Pi_{m-1} b_{m-1} \ \textit{can not be written as} \ \sum_{s=1}^{m} i_s \cdot \underline{x_s^{(m)}}, \tag{3.9}$$

where $0 \le i_1 \le q-2$ and $0 \le i_{t+1} \le b_t - 1$, for $1 \le t \le m-1$.

From the definition of g_{m-1}, we know that

$$2g_{m-1} - 1 \ne \sum_{s=1}^{m-1} i'_s \cdot \underline{x_s^{(m-1)}} , \quad \text{for} \ \ 0 \le i'_s \le a_s \ , \ \ s \ne 1 \ \text{ and } \ 0 \le i'_1 \le q-2 .$$

Assume that $b_{m-1}(2g_{m-1}\text{-}1) + \Pi_{m-1}(b_{m-1}-1) = \sum_{s=1}^{m-1} i_s \underline{x_s^{(m)}} + i_m \underline{x_m^{(m)}}$, for $0 \le i_s \le b_s - 1$ and $0 \le i_1 \le q-2$, then

$$b_{m-1} \cdot (2g_{m-1} - 1) + \Pi_{m-1}(b_{m-1}-1) = b_{m-1} \cdot \sum_{s=1}^{m-1} i_s \cdot \underline{x_s^{(m-1)}} + i_m \cdot \Pi_{m-1},$$

namely,

$$b_{m-1} \cdot \Big(2g_{m-1} - 1 - \sum_{s=1}^{m-1} i_s \cdot \underline{x_s^{(m-1)}} \Big) = \Pi_{m-1} \cdot ((i_m - (b_{m-1}-1))).$$

Since $\gcd(a_{m-1}, b_{m-1}) = 1$ and $0 \le i_m \le b_{m-1}-1$, the two sides of the above equation must be equal to zero, i.e., $2g_{m-1} - 1 - \sum_{s=1}^{m-1} i_s \cdot \underline{x_s^{(m-1)}} = 0$, for $0 \le i_s \le b_s$ and $i_m - b_{m-1} - 1 = 0$. This is in contradiction with that $2g_{m-1} - 1 - \sum_{s=1}^{m-1} i_s \cdot \underline{x_s^{(m-1)}} \ne 0$. Thus, (3.9) is true.

From (3.8), (3.9), and Lemma 2.1, we know that g_m^* is the genus of H_m. Thus, H_m satisfies (2*).

Combining the above proofs, the proof of Theorem 3.1 is completed. □

Theorem 3.1 can be written as follows:

Theorem 3.1': For the curve in a high-dimensional projective space defined by (3.1), let H_m, $x_s^{(m)}$ and g_m^* be determined by the following recursive procedure. Then H_m satisfies (1*) and (2*).

(1) $H_m = \{ x_1^{i_1} \cdots x_{m-1}^{i_{m-1}} x_m^{i_m} \mid 0 \le i_1 \le q-2 \text{ and } 0 \le i_s \le b_{s-1}-1 \text{ for } 2 \le s \le m \}$.

(2) $\underline{x_s} = \prod_{i=1}^{m-s} b_i \prod_{j=m-s+1}^{m-1} a_j$.

(3) $g_m^* = \frac{1}{2} \sum_{i=1}^{m-1} [(\prod_{s=1}^{i} a_s - 1) \cdot (\prod_{t=i+1}^{m-1} b_t) \cdot (b_i - 1)]$.

Now let us demonstrate the application of Theorem 3.1' to the above three examples and also show how to apply Theorem 2.2 to the above two examples.

Example 3.1': Let us consider the 3-dimensional Hermitian-like curve in Example 3.1, where $m = 3$. From the first equation of Example 3.1 and Theorem 3.3 in [18], we have

$H_2 = \{ x_1^{i_1} \cdot x_2^{i_2} \mid 0 \le i_1 \le 14 \text{ and } 0 \le i_2 \le 3 \}$,

$x_1^{(2)} = 4,\ x_2^{(2)} = 5,\ g_2^* = 6$.

Using Theorem 3.1', we have

$H_3 = \{ x_1^{i_1} \cdot x_2^{i_2} \cdot x_3^{i_3} \mid 0 \le i_1 \le 14 \text{ and } 0 \le i_2,\ i_3 \le 3 \}$,

$\underline{x_1} = 16,\ \underline{x_2} = 20,\ \underline{x_3} = 25,\ g_3^* = (4\cdot(2\cdot 6 - 1) + 5^2\cdot(4 - 1) + 1)/2 = 60$.

For example, let $r = 128$, i.e., $\mathbf{H}_{128} = \{ x_1^{i_1} x_2^{i_2} x_3^{i_3} \mid 0 \le i_1 \le 8, 0 \le i_2 \le 3, 0 \le i_3 \le 3 \}$. From Theorem 2.2, the designed minimum distance of C_{128} is $128 - 60 + 1 = 69$.

Example 3.2': Now we consider the m-dimensional Hermitian-like curve over $GF(q^2)$ in Example 3.2. From the first equation of Example 3.2 and Theorem 3.3 in [18], we have

$H_2 = \{ x_1^{i_1} \cdot x_2^{i_2} \mid 0 \le i_1 \le p^{2r} - 2 \text{ and } 0 \le i_2 \le p^r - 1 \}$,

$x_1^{(2)} = p^r,\ x_2^{(2)} = p^r + 1,\ g_2^* = p^r p^r - 1/2$.

Using Theorem 3.1', we have

$H_m = \{ x_1^{i_1} \cdots x_m^{i_m} \mid 0 \le i_1 \le p^{2r} - 2 \text{ and } 0 \le i_s \le p^r - 1 \text{ for } s \ne 1 \}$,

$x_s^{(m)} = p^{(m-s)r} \cdot (p^r + 1)^{(s-1)r}$,

$\overline{g_m} = [(p^r - 1)(p^r + 1)^{m+1} - (p^r - 1)(p^r + 1)^3 p^{r(m-2)} + (p^r - 1) p^{r(m-1)} - p^{r(m-2)} + 1]$.

Example 3.3': Let us consider the 3-dimensional curve defined by (3.6). From the first equation of Example 3.1 and Theorem 3.3 in [18], we have

$H_2 = \{ x_1^{i_1} \cdot x_2^{i_2} \mid 0 \le i_1 \le 14 \text{ and } 0 \le i_2 \le 3 \}$,

$x_1^{(2)} = 4,\ x_2^{(2)} = 5,\ g_2^* = 6$.

Using Theorem 3.1', we have

$H_3 = \{ x_1^{i_1} \cdot x_2^{i_2} \cdot x_3^{i_3} \mid 0 \le i_1 \le 14 \text{ and } 0 \le i_2 \le 3, 0 \le i_3 \le 2 \}$,

$\underline{x_1} = 12,\ \underline{x_2} = 15,\ \underline{x_3} = 25,\ g_3^* = (3\cdot(2\cdot 6 - 1) + 5^2\cdot(3 - 1) + 1)/2 = 42$.

For example, let $r = 96$, i.e., $\mathbf{H}_{96} = \{ x_1^{i_1} x_2^{i_2} x_3^{i_3} \mid 0 \le i_1 \le 8, 0 \le i_2 \le 3, 0 \le i_3 \le 2 \}$. From Theorem 2.2, the designed minimum distance of C_{96} is $96 - 42 + 1 = 55$.

IV. A Fast Decoding Procedure

In this section, we prove that the AG codes derived by the approach in the previous section can be fast decoded up to the designed minimum distance using the decoding procedure in [15,16].

Let Q = $\{ Q_1, Q_2, \cdots, Q_n \}$ be a set of n rational points on the curve X in a high-dimensional projective space over F_q. We consider the AG codes C_r defined by a parity check matrix H_r, obtained by the construction in the previous section from X over F_q.

Now we introduce some important concepts. Let $A=[A_{i,j}]_{0\le i,j<n}$ be a Hankel matrix, i.e. $A_{i,j}=a_{i+j}$. A (p,w)*-block Hankel matrix* is a $pw \times pw$ matrix of the form $A=[A_{i,j}]_{0\le i,j<p}$, where each $A_{i,j} = A_{i+j}$ is an $w\times w$ matrix. In [16], a modified Gauss elimination (*algorithm BH*) on A, which can be accomplished in $O(w^3 p^2)$ accumulations, is presented. Let $A' = [A_{i,j}]_{i\in I', j\in J'}$, we say that $A = [A_{i,j}]_{i\in I, j\in J}$ can be *embedded* in A′, if $I\subset I'$ and $J\subset J'$. If A′ is a Hankel (resp. block Hankel) matrix, then A is called a *embedded* Hankel (resp. block Hankel) matrix.

If a matrix A can be embedded in a Hankel, or a block Hankel matrix, then many matrix problems for A can be solved by applying efficient algorithms on the embedding host. The complexity is often no more than that for the embedding host. This is a popular technique in numerical linear algebra. Algorithm BH can be easily adopted for all embedded matrices. We refer to this modification as Algorithm BH*. Algorithm BH* has been discussed in detail in [16].

Let **r**, **e** and **c** be a received vector, an error vector and a code vector, respectively. Then we define the syndromes as in (2.8) and (2.9) and the syndrome matrix $\overline{S^{(r+1)}}$ as in (2.12). From Theorem 3.1, we have the following theorem.

Theorem 4.1: $\overline{S^{(r+1)}}$ is an embedded block Hankel matrix.

Proof: Let $w \triangleq \prod_{s=1}^{m-1} a_s$. Now we consider a submatrix $(S_{i,j})_{\xi\le i\le \xi+w-1, \lambda\le j\le\lambda+w-1}$ of $\overline{S^{(r+1)}}$, where $x_1^{a_1^{(\xi)}},\cdots,x_m^{a_m^{(\xi)}}$ and $x_1^{a_1^{(\lambda)}},\cdots,x_m^{a_m^{(\lambda)}}$ have the condition that $a_s^{(\xi)} = a_s^{(\lambda)} = 0$ for all $2\le s\le m$.

From the definitions (2.8)-(2.11), for i, $j \ge w(w+1)/2$, we have

$$\begin{aligned} S_{i,j} &= \mathbf{p}_{i,j}\cdot \mathbf{r}^T \\ &= \sum_{\mu=1}^{n} x_{1,v}^{a_1^{(i)}}\cdots x_{m,v}^{a_m^{(i)}}\cdot x_{1,v}^{a_1^{(j)}}\cdots x_{m,v}^{a_m^{(j)}}\cdot r_\mu \\ &= \sum_{\mu=1}^{n} x_{1,v}^{a_1^{(i+w)}}\cdots x_{m,v}^{a_m^{(i+w)}}\cdot x_{1,v}^{a_1^{(j-w)}}\cdots x_{m,v}^{a_m^{(j-w)}}\cdot r_\mu \\ &= \mathbf{p}_{i+w,j-w}\cdot \mathbf{r}^T \\ &= S_{i+w,j-w} \end{aligned}$$

Thus, $\overline{S^{(r+1)}}$ is an embedded block Hankel matrix. The size of each block is w. □

From the decoding procedure in [8] and [15], that is, majority decoding procedure, any $\lfloor (d-1)/2 \rfloor$ or fewer errors can be corrected by performing Gauss elimination on the syndrome matrix $\overline{S^{(r+1)}}$, where d is the number of s_{r+1} and s'_{r+1}. From Theorem 2.2, d is the designed minimum distance. On the other hand, the derived BH* algorithm in [16] has the complexity $O(w^3 p^2)$, where $n = w \cdot p$. We follow the model in [7], if $a_1 \sim a_2 \sim \cdots \sim a_{m-1} \sim q^{1/2}$ and $n \sim q^{(m+1)/2}$, then $w = \prod_{s=1}^{m-1} a_s \sim q^{(m-1)/2}$, i.e. $w \sim n^{\frac{m-1}{m+1}}$ and $p \sim n^{\frac{2}{m+1}}$. Hence, $O(w^3 p^2) = O(n^{3-\frac{2}{m+1}})$. Thus, using the majority decoding procedure given in [8] and [15] and Algorithm BH*, the decoding of C_r up to the designed minimum distance with the complexity $O(n^{3-\frac{2}{m+1}})$, can be realized. This result first was mentioned by Professor Høholdt and Professor Jensen in a conversation during revision of this paper.

V. Conclusions

In this paper, we have introduced a new construction of linear codes and have also presented a very general method to determine the minimum distance lower bounds. Based on this new construction, we have shown a simple construction (their basis formed only by monomials) for AG codes from curves in high-dimensional projective spaces without directly using the theory about algebraic geometry curves, and have further indicated that their designed minimum distance can be easily determined.

However, this is a prime work on deriving general AG codes without directly using deep knowledge of algebraic geometric curves. For more general work on AG codes or for all curves in high-dimensional projective spaces, the construction of AG codes using only a part of monomials as the basis, is still an interesting and open research problem.

Acknowledgements

The authors are deeply grateful to Mr. M. S. Kolluru for his useful discussions and valuable comments during the preparation of this paper. The authors would like to thank the referees, Professors O. Moreno, H. Elbrond Jensen, and T. Høholdt for their many valuable suggestions and comments on the presentation of this paper. During revision of this paper, Professor Hoholdt and Professor Jensen told one of the authors that they have got a decoding procedure with the complexity $O(n^{3-\frac{2}{m+1}})$.

References

[1] V. D. Goppa, "Codes associated with divisors." *Problemy Peredachi Informatsii* **13** (1977) 33-39.

[2] V. D. Goppa, "Codes on algebraic curves." *Soviet math. Dokl.* **24** (1981) 75-91.

[3] V. D. Goppa, "Algebraic-geometric codes." *Math. USSR Izvestiya3 21* (1983) 75-91.

[4] M. A. Tsfasman, S. G. Vladut, and T. Zink, "Modular curves, Shimura curves and Goppa codes, better than Varshamov-Gilbert bound." *Math. Nachr.* **104** (1982) 13-28.

[5] F. J. MacWilliams and N. J. A. Sloane, *The Theory of Error-Correcting Codes.* Amsterdam: North-Holland, 1977.

[6] J. Justesen, K. J. Larsen, A. Havemose, H. E. Jensen, and T. Høholdt, "Construction and decoding of a class of algebraic geometric codes." *IEEE Trans. on Information Theory* vol. IT-35, pp. 811-821, July, 1989.

[7] J. Justesen, K. J. Larsen, H. E. Jensen, and T. Høholdt, "Fast decoding of codes from algebraic plane curves." *IEEE Trans. on Information Theory* vol. IT-38, pp. 111-119, Jan., 1992.

[8] G. L. Feng and T. R. N. Rao, "Decoding algebraic-geometric codes up to the designed minimum distance." presented at *AAECC-9* New Orleans, USA, Oct 7-12, 1991, *IEEE Trans. on Information Theory* Vol IT-39, pp. 37-45, Jan., 1993.

[9] R. Pellikaan, "On a decoding algorithm for codes on maximal curves." *IEEE Trans. on Information Theory* vol. IT-35, pp. 1228-1232, Nov., 1989.

[10] S. C. Porter, B.-Z. Shen, and R. Pellikaan, "Decoding geometric Goppa codes using an extra place." *IEEE Trans. on Information Theory,* Vol IT-38, pp. 1663-1676, Nov., 1992.

[11] H. Stichtenoth, "A note on Hermitian codes over GF(q^2)." *IEEE Trans. on Information Theory* vol. IT-34, pp. 1345-1348, Sep., 1988.

[12] A. N. Skorobogatov and S. G. Vladut, "On the decoding of algebraic-geometric codes." *IEEE Trans. on Information Theory* vol. IT-36, pp. 1051-1061, Sep., 1990.

[13] S. G. Vladut, "Decoding algebraic-geometric codes over F_q for $q \geq 16$." *IEEE Trans. Information Theory* vol. IT-36, pp. 1461-1463, Nov., 1990.

[14] J. H. van Lint, "Algebraic geometric codes." *Coding Theory and Design Theory, IMA Volume in Mathematics and its Applications* **20** pp. 137-162, 1988.

[15] G. L. Feng, V. K. Wei, T. R. Rao, and K. K. Tzeng, "True Designed-Distance Decoding of a Class of Algebraic-Geometric Codes, Part I: A New Theory without Riemann-Roch Theorem," to appear in *IEEE Trans. on Information Theory.*

[16] G. L. Feng, V. K. Wei, T. R. Rao, and K. K. Tzeng, "True Designed-Distance Decoding of a Class of Algebraic-Geometric Codes, Part II: Fast Algorithms and Toeplitz-Block Toeplitz Matrices," to appear in *IEEE Trans. on Information Theory.*

[17] R. Pellikaan, B. -Z. Shen, and G. J. M. van Wee, "Which Linear Codes are Algebraic-Geometric ?", *IEEE Trans. on Information Theory*, vol.37, No.3. pp.383-602, May, 1991.

[18] G. L. Feng and T. R. N. Rao, "A Novel Approach for Construction of Algebraic Geometric Codes from Affine Plane Curves," submitted to *IEEE Trans. on Information Theory*.

[19] A. B. Sorensen, "Weighted Reed-Muller Codes and Algebraic-Geometric Codes," *IEEE Trans. on Information Theory*, vol.38, No.6. pp.1821-1826, Nov., 1992.

A NEW CLASS OF SEQUENCES: MAPPING SEQUENCES

Guang Gong

Fondazione Ugo Bordoni

Via B. Castiglione, 59, I-00142, Roma, Italy

Abstract: A Galois field $GF(q^m)$ can be regarded as a linear vector space over $GF(q)$. The elements of $GF(q^m)$ can be represented by m-tuples of elements belonging to $GF(q)$. In this way, a sequence over $GF(q^m)$ becomes a sequence over $GF(q)$. We call the sequence over $GF(q)$ *a mapping sequence*. A change of basis of $GF(q^m)$ over $GF(q)$ can change the period, the linear span and the autocorrelation function of a mapping sequence. The aim of this work is to investigate the above properties of mapping sequences. It is shown that the sufficient and necessary conditions which guarantee the periods and the linear spans of mapping sequences reach the maximum values. The special kind of bases of $GF(q^m)$ over $GF(q)$ found is one in which the autocorrelation functions of such mapping sequences are 3-valued. We point out that mapping sequences are of considerable theoretical importance. The result of autocorrelation functions can be used to solve the minimum distance of one kind of burst error correcting codes. The practical importance relies on the fact that the set of generalized mapping sequences contains a lot of well-known sequences (i. e., multiplexed sequences, clock controlled sequences, GMW sequences or generalized GMW seqences and No sequences).

Work carried out in the framework of the agreement between the Italian PT Administration and the Fondazione "Ugo Bordoni".

I . Definition and Algebraic Structure of Mapping Sequences

Let $a = \{a(k)\}_{k\geq 0}$ be a periodic sequence over $GF(q^m)$, $\{\beta_0, \cdots, \beta_{m-1}\}$ be a basis of $GF(q^m)$ over $GF(q)$. Any element of a can be represented as follows:

(1) $$a(k) = a_0(k)\beta_0 + \cdots + a_{m-1}(k)\beta_{m-1} \quad a_j(k) \in GF(q).$$

We map $a(k)$ into $a_0(k), \cdots, a_{m-1}(k)$. This map generates the following sequence:

(2) $$u = a_0(0), \cdots, a_{m-1}(0), a_0(1), \cdots a_{m-1}(1), \cdots.$$

We write $u = \{u(k)\}_{k\geq 0}$. Notice that u is a periodic sequence over $GF(q)$.

Definition 1: The *q-ary* sequence u, defined by (2), is called the *mapping sequence* generated by the sequence a under the basis $\{\beta_0, \cdots, \beta_{m-1}\}$ of $GF(q^m)$ over $GF(q)$. (we call u the *mapping sequence* for a short).

Let $a_j = \{a_j(k)\}_{k\geq 0} (j = 0, \cdots, m-1)$, we call a_j *the j-th component sequence* of u (or a) over $GF(q)$.

From this point on, we suppose that

- $f(x)$ is the minimal polynomial of the sequence a over $GF(q^m)$ (the polynomial of the lowest degree over $GF(q^m)$ which generates the sequence a);
- $h(x)$ is the minimal polynomial of the mapping sequences u over $GF(q)$;
- $f_j(x)$ is the minimal polynomial of the component sequences a_j of u over $GF(q)$ $(j=0, \cdots, m-1)$.

In this section, we will show the relationships among $f(x)$, $h(x)$, and $f_j(x) (j= 0, \cdots, m-1)$ and the representations of the elements of the mapping sequence u.

Now we list some notations which are used throughout this paper.

- $s^{(r)}$ represent the sequence $s(0), s(r), s(2r), \cdots$, recalling that $s^{(r)}$ is called a r-decimation of the sequences s;
- L represent the left shift operator, i. e., $L^k(s) = s(k), s(k+1), \cdots$; let $g(x) = \sum_{k=0}^{T} c_k x^k$ be a polynomial over $GF(q^m)$, then $g(L)(s) = \sum_{k=0}^{T} c_k L^k(s)$;
- $Tr_{GF(t^n)/GF(t)}(x) = Tr_1^n(x) = x + x^t + \cdots + x^{t^{n-1}}$, $x \in GF(t^n)$, where t is a power of a prime;

- $\sigma_i: \eta \to \eta^{q^i}$ ($\eta \in GF(q^m)$, $i = 0,\cdots,m-1$), $\{\sigma_i | i = 0,\cdots,m-1\}$ is the automorphism group of $GF(q^m)$ over $GF(q)$; as for the polynomial $g(x)$ over $GF(q^m)$, $\sigma_i(g(x)) = \sum_{k=0}^{T} c_k^{q^i} x^k$ where the c_ks are defined by $g(x) = \sum_{k=0}^{T} c_k x^k (i = 0,\cdots,m-1)$, which are called the *conjugates of* $g(x)$ *with respect to* $GF(q)$;
- $LS(.)$ represents the linear span of a sequence.

Remark: For theory of finite fields and a survey of linear recurring sequences, the reader is referred to [1]. Basic symbols of theory of finite fields used in this paper are the same as those used in [1].

From (1), we have

$$(3) \qquad u(im + j) = a_j(i), \ (i = 0,1,\cdots; j = 0,\cdots,m-1).$$

Thus a_j is a m-decimation of the sequence $L^j(u)$. Therefore

$$f_j(L^m)(L^j u) = f_j(L)(a_j) = 0 \quad (j = 0,\cdots,m-1).$$

Hence, we have proved the following lemma.

Lemma 1. Let $a_j, u, f_j(x)$ and $h(x)$ be defined as above, then

(i) $a_j = [L^j(u)]^{(m)} \quad (j = 0,\cdots,m-1)$.

(ii) $h(x)$ is a factor of the least common multiple of $f_0(x^m),\cdots,f_{m-1}(x^m)$.

In order to establish Lemma 3, we need the following lemma.

Lemma 2. Suppose $g(x)$ is an irreducible polynomial over $GF(q^m)$ of degree T and $\sigma_i(g(x)) (i = 0, 1, \cdots, m-1)$ are distinct, where σ_i are the automorphisms of $GF(q^m)$ over $GF(q)$, as defined above. Let η be a root of $g(x)$ is the extension $GF(q^{mT})$ of $GF(q^m)$ and $G(x) = \prod_{i=0}^{m-1} \sigma_i(g(x))$. Then $G(x)$ is the minimal polynomial of η over $GF(q)$ so that $G(x)$ is irreducible over $GF(q)$.

Proof.

Since $\{\sigma_i | i = 0,\cdots,m-1\}$ is the automorphism group of $GF(q^m)$ over $GF(q)$, we have

$$\sigma_j[\prod_{i=0}^{m-1}\sigma_i(g(x))] = \prod_{i=0}^{m-1}(\sigma_j\sigma_i)(g(x)) = \prod_{i=0}^{m-1}\sigma_i(g(x)),$$

whence

$$G(x) = \prod_{i=0}^{m-1}\sigma_i(g(x)) \in GF(q)[x].$$

From $g(\eta)=0$, we have $G(\eta)=0$.

Now we show the conjugates $\eta^{q^t}(t=0,1,\cdots,mT-1)$ of η with respect to GF(q) are distinct.

Recalling $g(x) = \sum_{j=0}^{T} c_j x^j$ and $\sigma_i(g(x)) = \sum_{j=0}^{T} c_j^{q^i} x^j$, then

$$\sigma_i(g(\eta^{q^i})) = \sum c_j^{q^i}\eta^{q^i j} = (\sum c_j\eta^j)^{q^i} = 0,$$

i.e. η^{q^i} is a root of $\sigma_i(g(x))$.

Notice that $\sigma_i(g_1(x)g_2(x)) = \sigma_i(g_1(x))\sigma_i(g_2(x))$, where $g_j(x) \in$ GF$(q^m)[x]$, $j = 1,2$. Hence we have $\sigma_i(g(x))(i=1,\cdots,m-1)$ are irreducible over GF(q^m) of degree T. Therefore $(\eta^{q^i})^{q^{mt}}(t = 0,1,\cdots,T-1)$ are distinct. Let

$$R(i) = \{(\eta^{q^i})^{q^{mt}} \mid t = 0,1,\cdots,T-1\}.$$

According to the condition $\sigma_i(g(x)) \neq \sigma_j(g(x))$ for $i\neq j$, we have

$$R(i) \cap R(j) = \phi.$$

Consequently, we obtain

$$\bigcup_{i=0}^{m-1} R(i) = \{\eta^{q^t} \mid t = 0,1,\cdots,mT-1\}$$

where $\eta^{q^t}(t = 0,1,\cdots,mT-1)$ are distinct. Thus $G(x)$ is the minimal polynomial of η over GF(q) and $G(x)$ is irreducible over GF(q) of degree mT.

Lemma 3. Suppose $f(x) = \prod_{s=1}^{r} g_s(x)$ where $g_s(x)$ $(s=1,\cdots,r)$ are distinct irreducible polynomials over GF(q^m) of degree n, and $\sigma_i(g_s(x))(i=0,\cdots,m-1)$ are distinct for each $s \in \{1,\cdots,r\}$. Let $F(x)$ represent the product of $G_s(x) = \prod_{i=0}^{m-1}\sigma_i(g_s(x))$ where $\{g_s(x) \mid s=i_1,\cdots,i_t, t\leqslant r\}$ is maximal such that the g_s are not conjugates of each other with respect to GF(q). Then

(i) $f_j(x) = F(x)$ $\quad (j = 0,\cdots,m\text{-}1)$.

(ii) $h(x) \mid F(x^m)$ so that $LS(u) \leqslant m\, LS(a_j)$.

Proof. Let α_s be a fixed root of $g_s(x)$ in the extention $F_s = GF(q^{mn_s})$ of $GF(q^m)$. Form the reference [1], the elements of a can be represented as follows:

(4) $$a(k) = Tr_1^{n_1}(\theta_1\alpha_1^k) + \cdots + Tr_1^{n_r}(\theta_r\alpha_r^k), \qquad \text{for } k = 0, 1, \cdots.$$

where $\theta_s \in F_s (s = 1, \cdots, r)$.

Let $\{\gamma_0, \cdots, \gamma_{m-1}\}$ be the dual basis of the basis $\{\beta_0, \cdots, \beta_{m-1}\}$ of $GF(q^m)$ over $GF(q)$, then

(5) $$Tr_1^m(\gamma_i\beta_j) = \begin{cases} 0 & i \neq j \\ 1 & i = j. \end{cases}$$

Now by multipling (1) by γ_j and applying the trace function $Tr_1^m(.)$, using (5), we obtain

(6) $$a_j(k) = Tr_1^m(\gamma_j a(k)).$$

Then by substituting (4) into the above identity, we have

(7) $$a_j(k) = Tr_1^{mn_1}(\gamma_j\theta_1 a_1^k) + \cdots + Tr_1^{mn_r}(\gamma_j\theta_r a_r^k).$$

From Lemma 2, we have $G_s(x)$ is the minimal polynomial of α_s over GF(q). Notice that if α_s is a conjugate of α_i with respect to $GF(q)$, then they have the same minimal polynomial over $GF(q)$. Therefore, (7) shows that $f_j(x)$ is equal to the product of $G_s(x) (s = i_1, \cdots, i_t, t \leqslant r)$. Hence, (i) is true.

From (ii) of Lemma 1, we have $h(x) \mid F(x^m)$. Therefore, $LS(u) \leqslant$ degree of $F(x^m) = m$(degree of $F(x)$) $= mLS(a_j)$.

Q. E. D.

From (2) and (7) in the proof of Lemma 3, we have the following corollary.

Corollary 1. Let $f(x)$ be defined as in Lemma 3. Then the elements of u can be represented by:

(8) $$u(im + j) = Tr_1^{mn_1}(\gamma_j\theta_1\alpha_1^i) + \cdots + Tr_1^{mn_r}(\gamma_j\theta_r\alpha_r^i)$$
$$(i = 0, 1, \cdots; j = 0, \cdots, m\text{-}1)$$

where $\theta_s \in F_s (s = 1, \cdots, r)$. Moreover, if $f(x)$ is an irreducible polynomial over $GF(q^m)$ of degree n, α is a root of $f(x)$ in the extension $GF(q^{nm})$ of $GF(q^m)$, then the elements of u can be represented by

(9) $$u(im + j) = Tr_1^{mn}(\theta\gamma_j\alpha^i), \; \theta \in GF(q^{nm})$$

and $f_j(x) = F(x) = \prod_{i=0}^{m-1} \sigma_i(f(x))$.

Ⅱ . Period and Linear Span of a Mapping Sequence

Let $u, a, a_j, h(x), f(x), f_j(x), \{\gamma_0, \cdots, \gamma_{m-1}\}$ and $\{\beta_0, \cdots, \beta_{m-1}\}$ have the same meaning as in Section Ⅰ .

In this section, we suppose that the polynomial $f(x)$ is irreducible over $GF(q^m)$. Thus the elements of the mapping sequence u can be represented by (9) of Corollary 1 in Section Ⅰ .

In order to establish the main results of this section, we need the following lemma whose proof follows directly from the definitions of trace functions and bases of finite fileds.

Lemma 4. If $Tr_1^h(\eta\zeta_i) = 0$ $(i = 0, \cdots, h-1)$, where $\eta, \zeta_i \in GF(q^h)$ and $\{\zeta_0, \cdots, \zeta_{h-1}\}$ is a basis of $GF(q^h)$ over $GF(q)$, then $\eta=0$.

Now we state the following theorems.

Theorem 1. Let $f(x)$ be an irreducible polynomial over $GF(q^m)$ of degree n and $\sigma_i(f(x))$ $(i=0, \cdots, m-1)$ be distinct, a be a root of $f(x)$ in the extension $GF(q^{nm})$ of $GF(q^m)$.

Case 1. $n>1$.

Let the symbol $per(.)$ represent the period of a sequence, then

$$per(u) = m\,per(a)$$

for an arbitrary basis of $GF(q^m)$ over $GF(q)$.

Case 2. $n=1$.

Let $T_0 = \{\gamma_j\gamma_{j+s}^{-1} \mid j = 0, 1, \cdots, m-s-1\}$ and $T_1 = \{\gamma_j\gamma_{j+s-m}^{-1} \mid j = m-s, \cdots, m-1\}$, then $per(u) = m\,per(a)$ if and only if the dual basis $\{\gamma_0, \cdots, \gamma_{m-1}\}$ satisfies one of the following conditions:

(i) $|T_0|>1$ or $|T_1|>1$ for $s=1, \cdots, m-1$.

(ii) If $|T_0|=1$ and $|T_1|=1$, then $\alpha^s \in T_1 \Rightarrow \alpha^{s+1} \notin T_2$ for $s=1,\cdots,m-1$.

Proof. From the definition of mapping sequences, we have that $per(u)$ is a factor of $mper(a)$. Thus we can write

$$per(u) = rm + s \quad (0 \leqslant s < m, 0 \leqslant r \leqslant per(a)).$$

Let $k=im+j$ $(i=0,1,\cdots; j=0,\cdots,m-1)$, then we have

$$u(k) = u(im + j) = u((i + r)m + (j + s)) = u(k + per(u)) \text{ for } k = 0,1\cdots.$$

Using (9) of Corollary 1 in Sec. I, we obtain that

$$\mathrm{Tr}_1^{nm}(\theta\gamma_j\alpha^i) = \mathrm{Tr}_1^{nm}(\theta\gamma_{j+s}\alpha^{i+r}) \qquad \text{for } j + s < m$$

$$\mathrm{Tr}_1^{nm}(\theta\gamma_j\alpha^i) = \mathrm{Tr}_1^{nm}(\theta\gamma_{j+s-m}\alpha^{i+r+1}) \qquad \text{for } j + s \geqslant m,$$

Where $i=0,1,\cdots$. According to Lemma 2 of Sec. I, the minimal polynomial of α over $GF(q)$ has degree nm, thus $\{1,\alpha,\cdots,\alpha^{nm-1}\}$ is a basis of $GF(q^{nm})$ over $GF(q)$. From Lemma 4, the above two identities yield

(10) $$\gamma_j = \gamma_{j+s}\alpha^r \qquad \text{for } j + s < m$$

(11) $$\gamma_j = \gamma_{j+s-m}\alpha^{r+1} \qquad \text{for } j + s \geqslant m.$$

Case 1. n>1. Notice that $\alpha \notin GF(q^m)$ for $n>1$ and the γ_i are pairwise different, so (10) and (11) show that $r=per(a)$ and $s=0$. Thus $per(u)=mper(a)$.

Case 2. n=1. Similarly, from (10) and (11), we obtain that $r=per(a)$ and $s=0$ if and only if $\{\gamma_0,\cdots,\gamma_{m-1}\}$ satisfies one of (i) and (ii). Thus the result is true.

Q. E. D.

Theorem 2. Let $u, a, a_j, h(x), f(x), \{\gamma_0,\cdots,\gamma_{m-1}\}$ and α be defined as in Theorem 1. Recalling that LS (.) represent the linear span of a sequence. Then $LS(u)=mLS(a_0)$ if and only if

(12) $$\det \mathbf{A} = \det \begin{bmatrix} \gamma_0 & \gamma_1 & \cdots & \gamma_{m-1} \\ \gamma_1 & \gamma_2 & \cdots & \alpha\gamma_0 \\ \cdots & & & \\ \gamma_{m-1} & \alpha\gamma_0 & & \alpha\gamma_{m-2} \end{bmatrix} \neq 0.$$

Moreover, if $m \leqslant n$, then $LS(u)=mLS(a_0)$ for any basis of $GF(q^m)$ over $GF(q)$.

Proof. We use (9) of Corollary 1 in Sec. I to represent the elements of u. Then

$$u = \mathrm{Tr}_1^{nm}(\theta\gamma_0), \cdots, \mathrm{Tr}_1^{nm}(\theta\gamma_{m-1}), \mathrm{Tr}_1^{nm}(\theta\gamma_0\alpha), \cdots, \mathrm{Tr}_1^{nm}(\theta\gamma_{m-1}\alpha), \cdots.$$

Notice that $LS(a_0) = nm$. Let

$$\mathbf{u}_{im+j} = (\mathrm{Tr}_1^{nm}(\theta\gamma_j\alpha^i), \cdots, \mathrm{Tr}_1^{nm}(\theta\gamma_{m-1}\alpha^i), \mathrm{Tr}_1^{nm}(\theta\gamma_0\alpha^{i+1}), \cdots, \mathrm{Tr}_1^{nm}(\theta\gamma_{j-1}\alpha^{i+nm-1}))$$

and

$$\mathrm{W} = \{\mathbf{u}_{im+j} \mid i = 0, \cdots, nm-1; j = 0, \cdots, m-1\}.$$

Since $\mathbf{u}_{im+j}$ is a state vector of u, $LS(u) = mLS(a_0) = nm^2$ if and only if W is linearly independent over GF (q). Now we exhibit the condition where W is linearly independent over GF(q).

Suppose there are $c_{ij} \in \mathrm{GF}(q)$ such that

$$\sum_{i=0}^{nm-1}\sum_{j=0}^{m-1} c_{ij}\mathbf{u}_{im+j} = 0. \tag{13}$$

We write

$$x_j = \sum_{i=0}^{nm-1} c_{ij}\alpha^i \quad (j = 0, \cdots, m-1); \tag{14}$$

$$B_j = x_0\gamma_j + \cdots + x_{m-j-1}\gamma_{m-1} + x_{m-j}\alpha\gamma_0 + \cdots + x_{m-1}\alpha\gamma_{j-1} (j = 0, \cdots, m-1). \tag{15}$$

Now (14) yields

$$\mathrm{Tr}_1^{nm}(B_j\alpha^i) = 0 \qquad (i = 0, \cdots, nm-1; j = 0, \cdots, m-1). \tag{16}$$

Since $\{1, \alpha, \cdots, \alpha^{nm-1}\}$ is a basis GF(q^{nm}) over GF(q), from Lemma 4, the above identities show that

$$B_j = 0 \quad (j = 0, \cdots, m-1). \tag{17}$$

From (15) and (17), we obtain

$$\begin{cases} \gamma_0 x_0 + \gamma_1 x_1 + \cdots + \gamma_{m-1}x_{m-1} = 0 \\ \gamma_1 x_0 + \gamma_2 x_1 + \cdots + \alpha\gamma_0 x_{m-1} = 0 \\ \cdots \\ \gamma_{m-1}x_0 + \alpha\gamma_0 x_1 + \cdots + \alpha\gamma_{m-2}x_{m-1} = 0 \end{cases} \tag{18}$$

which is a homogeneous system of m linear equations in the same number m of unknowns x_j with coefficients over GF (q^{nm}). The matrix of coefficients of this homogeneous system is A, defined by (12). The homogeneous system (18) has the solution zero if and only if det A$\neq$0.

Thus if det A$\neq$0, then (19) has the solution zero, i.e., we have $x_j = 0$ $(j=0, \cdots, m-1)$. Recalling that $\{1, \alpha, \cdots, \alpha^{nm-1}\}$ is a basis of GF(q^{nm}) over GF(q), from (14), we obtain $c_{ij} = 0$ for $i = 0, \cdots, nm-1, j = 0, \cdots, m-1$. Therefore, W is linearly independent over GF(q).

If det A$=0$, according to the meaning of x_j defined by (14), it is easy to see that W is linearly dependent over GF(q). The proof of the first result is completed.

We can write det A as follows,

$$\det A = \theta_{m-1}\alpha^{m-1} + \cdots + \theta_1\alpha + \theta_0 \quad (\theta_j \in GF(q^m)) \tag{19}$$

where $\theta_{m-1} = \gamma_0^m \neq 0$. Then det A is a polynomial of degree $(m-1)$ over GF(q^m). Since the minimal polynomial $f(x)$ of α over GF(q^m) is an irreducible polynomial of degree n, then det A$\neq 0$ for $m-1<n$. Therefore, we obtain that $LS(u)=nm^2=mLS(a_0)$ for all bases of GF(q^m) over GF(q) for $m \leqslant n$.

Q. E. D.

We have the following corollary whose proof follows directly from the above Theorem 2 and Lemma 3 in Sec. I .

Corollary 2. Suppose $m \leqslant n$. Then $h(x)=F(x^m)$ for an arbitrary basis of GF(q^m) over GF(q), where $F(x)=\prod_{i=0}^{m-1}\sigma_i(f(x))$.

Corollary 3. Let $n>1$, then there exist some bases of GF(q^m) over GF(q) such that $LS(u)=mLS(a_0)$. The number $BL_q(m)$ of such bases of GF(q^m) over GF(q) is bounded by

$$BL_q(m) \geqslant N_q(m)/m$$

where $N_q(m)$ is the number of monic irreducible polynomials over GF(q) of degree m.

Proof. We choose

$$\{\gamma_0, \cdots, \gamma_{m-1}\} = \{1, \gamma, \cdots, \gamma^{m-1}\}, \tag{20}$$

where the minimal polynomial of γ is a monic irreducible polynomial over GF(q) of degree m. Then

$$\det A = -(\alpha-\gamma^m)^{m-1}.$$

Since $n>1$, $\alpha \notin GF(q^m)$, so that det A$\neq 0$. From Theorem 12, we obtain $LS(u)= mLS(a_0)$. According to (20), the second result is reached immediately (See Theorem

3. 25 of [1] as for the number $N_q(m)$).

Q. E. D.

Remark. There is another situation which shows that $LS(u) = mLS(a_0)$ for any basis of GF (q^m) over GF (q). This is obtained by forcing that $F(x^m)$ remains irreducible over GF(q). That is, from Lemma 2 in Sec. Ⅱ and Theorem 3. 35 of [1], we can establish the following statement.

If all prime factors of m are the factors of the period of a_0, then $h(x) = F(x^m)$ so that $LS(u) = mLS(a_0)$ for any basis of GF(q^m) over GF(q).

Ⅲ. Autocorrelation Function of a Mapping Sequence

Hereafter, we suppose that $f(x)$ is primitive over GF(q^m), so α is a primitive element in GF(q^{nm}). From Corollary 1 in Sec. Ⅰ, we have the component sequences a_j $(j = 0, \cdots, m-1)$ of u (*or* a) are m-sequences over GF (q) generated by $F(x) = \prod_{i=0}^{m-1} \sigma_i(f(x))$. Let $\gamma = \alpha^d$, where $d = \frac{q^{nm}-1}{q^m-1}$, then γ is a primitive element in GF(q^m).

We write

$$(21) \qquad \gamma_j = \gamma^{e_j} = \alpha^{de_j}, e_j \in \{0, 1, \cdots, q^m-2\} \quad (j = 0, \cdots, m-1).$$

From (9) of Corollary 1 in Sec. Ⅰ, we can represent the elements of u as follows

$$(22) \qquad u(im + j) = \mathrm{Tr}_1^{mn}(\theta\alpha^{i+de_j}) \quad (i = 0, 1, \cdots; j = 0, \cdots, m-1)$$

Let $q = p^r$, p is a prime number, then the canonical additive character [16] of GF(q) is defined by

$$(23) \qquad \chi(x) = e^{2\pi i \mathrm{Tr}_1^k(x)/p}, \ x \in GF(q)$$

where $\mathrm{Tr}_1^k(x) = \mathrm{Tr}_{GF(p^k)/GF(p)}(x) = x + x^p + \cdots + x^{p^{k-1}}$. In particular, we have:

$$(24) \qquad \chi(x)\overline{\chi(y)} = e^{2\pi i \mathrm{Tr}_1^k(x)/p}\,\overline{e^{2\pi i \mathrm{Tr}_1^k(y)/p}} = e^{2\pi i[\mathrm{Tr}_1^k(x) - \mathrm{Tr}_1^k(y)]/p} = \chi(x-y).$$

In order to estabish the main results of this section, we need the following lemma whose proof follows directly from the definition of the character χ (23) and the trace function $\mathrm{Tr}_1^{nm}(.)$ on GF(q^{nm}).

Lemma 5.

$$\sum_{x \in GF(q^{nm})} \chi(\mathrm{Tr}_1^{nm}(\zeta x)) = \begin{cases} 0 & \Leftrightarrow \quad \zeta \neq 0 \\ q^{nm} & \Leftrightarrow \quad \zeta = 0. \end{cases}$$

The (periodic) autocorrelation function of u is defined by

$$C(\tau) = \sum_{k=0}^{m(q^{nm}-2)} \chi(u(k))\,\overline{\chi(u(k+\tau))}. \tag{25}$$

Now we state the main results of this section.

Let

$$S_0 = \{e_j\text{-}e_{j+s} \mid j = 0, \cdots, m\text{-}s\text{-}1\} \text{ and } S_1 = \{e_j\text{-}e_{j+s-m} \mid j = m\text{-}s, \cdots, m\text{-}1\}. \tag{26}$$

Theorem 3. The autocorrelation function $C(\tau)$ of the mapping sequence u, defined by (25), is $(m+1)$-valued function and satisfies the following equalities.

Case 1. $n>1$

$$C(\tau) = \begin{cases} m(q^{nm}\text{-}1 & \tau \equiv 0 (\mathrm{mod}\ m(q^{nm}\text{-}1)) \\ -m + Nq^{nm} & \tau \equiv rm + s (\mathrm{mod}\ m(q^{nm}\text{-}1)) (1 \leqslant s \leqslant m\text{-}1),\ \frac{r}{d} \text{ occurs } N \\ & \text{times in } S_0 \text{ or } \frac{r+1}{d} \text{ occurs } N \text{ time in } S_1 \\ -m & \text{otherwise} \end{cases}$$

Case 2. $n=1$.

$$C(\tau) = \begin{cases} m(q^m\text{-}1) & \tau \equiv 0 (\mathrm{mod}\ m(q^m\text{-}1)) \\ -m + (N_0 + N_1)q^m & \tau \equiv rm + s (\mathrm{mod}\ m(q^m\text{-}1)\ (1 \leqslant s \leqslant m\text{-}1), \\ & r \text{ occurs } N_0 \text{ times } in\ S_0 \textbf{ and} \\ & r+1 \text{ occurs } N_1 \text{ times in } S_1 \\ -m & \text{otherwise} \end{cases}$$

Proof.

Let $k=im+j, \tau=rm+s, 0 \leqslant j, s \leqslant m\text{-}1$. From (22)-(25), we have

$$C(\tau) = \sum_{k=0}^{m(q^{nm}-2)} \chi(u(k))\,\overline{\chi(u(k+\tau))}$$

$$= \sum_{j=0}^{m-2} \sum_{i=0}^{q^{nm}-2} \chi[\mathrm{Tr}_1^{nm}(\theta\alpha^{i+de_j}) \text{-} \mathrm{Tr}_1^{nm}(\theta\alpha^{i+r+\delta+de_{j+s-\delta m}})]. \tag{27}$$

where

(28) $$\delta = \begin{cases} 0 & j \in \{0, \cdots, m-s-1\} \\ 1 & j \in \{m-s, \cdots, m-1\}. \end{cases}$$

Let us denote

(29) $$\Gamma(j,\delta) = \alpha^{de_j} - \alpha^{r+\delta+de_{j+s-\delta m}}$$

(30) $$N_0 = |\{j \mid \Gamma(j,0) = 0, j = 0, \cdots, m-s-1\}|$$

(31) $$N_1 = |\{j \mid \Gamma(j,1) = 0, j = m-s, \cdots, m-1\}|.$$

According to Lemma 5, we obtain that

(32) $$C(\tau) = \sum_{j=0}^{m-1} \sum_{x \in GF(q^{nm})} \chi[\mathrm{Tr}_1^{nm}(\theta_x \Gamma(j,\delta))] - m.$$

$$= -m + (N_0 + N_1) q^{nm}.$$

(i) Let $s=0$. Notice that $\alpha^r \neq 1$ for $r \not\equiv 0 (\mathrm{mod}(q^{nm}-1))$. Then we obtain that $C(0) = m(q^{nm}-1)$. and $C(rm) = -m$, $r \not\equiv 0 (\mathrm{mod}(q^{nm}-1))$.

(ii) Let $s \neq 0$. If there exists $j \in \{0, \cdots, m-s-1\}$ such that $\Gamma(j,0)=0$. Since α is a primitive element in $GF(q^{nm})$, $\Gamma(j,0)=0 \Leftrightarrow \alpha^{de_j} - \alpha^{r+de_{j+s}} = 0 \Leftrightarrow de_j - r - de_{j+s} \equiv 0 (\mathrm{mod}$ $(q^{nm}-1)) \Leftrightarrow r \equiv (e_j - e_{j+s})d$ $(\mathrm{mod}\ (q^{nm}-1))$, i. e.,

(33) $$\Gamma(j,0) = 0 \Leftrightarrow r \equiv (e_j - e_{j+s})d \quad (\mathrm{mod}(q^{nm}-1)), (j = 0, \cdots, m-s-1).$$

Similarly, we obtain that

(34) $$\Gamma(j,1) = 0 \Leftrightarrow r + 1 \equiv (e_j - e_{j+s-m})d \quad (\mathrm{mod}\ (q^{nm}-1)), (j = m-s, \cdots, m-1).$$

According to $d>1$ and $d=1$ for $n>1$ and $n = 1$, respectively, (33) and (34) show that the other part of the results is true.

Q. E. D.

Corollary 4. Let m be a prime number. Then there exist some bases of $GF(q^m)$ over $GF(q)$ such that

(37) $$C(\tau) \in \{m(q^{nm} - 1), -m + q^{nm}, -m\}$$

for both of $n>1$ and $n=1$. The number $BC_q(m)$ of such bases of is bounded by the number $NB_q(m)$ of different normal bases of $GF(q^m)$ over $GF(q)$, i,e.,

$$BC_q(m) \geqslant NB_q(m).$$

Proof. Let $\{\gamma, \cdots, \gamma^{q^{m-1}}\}$ be a normal basis of $GF(q^m)$ over $GF(q)$. We choose that

$\{\gamma^{e_0},\cdots,\gamma^{e_{m-1}}\}=\{\gamma,\cdots,\gamma^{q^{m-1}}\}$. Then

$$S_0 \cup S_1 = \{1-q^s, q(1-q^s), \cdots, q^{m-1}(1-q^s)\},$$

where S_0 and S_1 are defined by (26). The right-hand side of the above identity is the cyclotomic coset *mod* q^m-1 containing $(1-q^s)$ [2]. Since m is prime, then

$$|\{1-q^s, q(1-q^s), \cdots, q^{m-1}(1-q^s)\}| = m \qquad \text{for } s = 1, \cdots, m-1.$$

Therefore, we obtain that $|S_0|=m-s$, $|S_1|=s$ and $r\in S_0 \Rightarrow r+1\notin S_1$. From Theorem 3, the autocorrelation function $C(\tau)$ of u under the basis $\{\gamma,\cdots,\gamma^{q^{m-1}}\}$ satisfies (37) for both of $n>1$ and $n=1$. The second result is immediat (See Theorem 3.73 of [1] as for the number $NB_q(m)$).

Q. E. D.

Remark. The results of this section can be used to solve the minimum distance of one kind of burst error correction codes [2].

IV. Generalized Mapping Sequences

Definition 2. Let $v(x)$ be an irreducible polynomial over GF(q) of degree m, α be a root of $v(x)$ in the extension GF(q^m) of GF(q), let $\{\gamma_0,\cdots,\gamma_{T-1}\}$ be *a subset* of GF(q^m). Let $k=iT+j, i=0,1,\cdots, 0\leqslant j\leqslant T-1$,

$$u(k) = u(iT + j) = Tr_1^m(\gamma_j\alpha^i), k = 0, 1, \cdots$$

then $u=\{u(k)\}_{k\geqslant 0}$ is called *the generalized mapping sequence* over GF(q).

Notice that the proof of Theorem 1-3 in Section II - III is not based on the supposition that $\{\gamma_0, \cdots, \gamma_{m-1}\}$ is a basis of GF(q^m) over GF(q). Therefore the generalized mapping sequences have the same results of Theorem 1-2 for $n=1$ in Section II where the basis $\{\gamma_0,\cdots,\gamma_{m-1}\}$ is replaced by a subset $\{\gamma_0,\cdots,\gamma_{T-1}\}$ of GF(q^m). As far as Theorem 3 in Section III is concerned, there need a slight change, since the elements of the set $\{\gamma_0,\cdots,\gamma_{m-1}\}$ can be equal to zero elements.

We write $e_j=\infty$ if $\gamma_j=0$. Suppose that the e_j and e_{j+s} in the sets S_0 and S_1, defined by (26) in Section III, satisfy $e_j\neq\infty$ and $e_{j+s}\neq\infty$, and we denote the cardinal number of the set $\{j\,|\,e_j=\infty \text{ and } e_{j+s}=\infty, j=0,\cdots,m-1\}$ by N_∞, where

$j+s$ is computed by mod m. Then the auto correlation function of the generalized mapping sequence u satisfies the following identity,

$$C(\tau)=\begin{cases} T(q^m\text{-}1) & \tau\equiv 0 \quad (\text{mod } T(q^m\text{-}1)) \\ -T+(N_0+N_1+N_\infty)q^m & \tau\equiv rm+s \quad (\text{mod } T(q^m\text{-}1)) \\ & \quad (1\leqslant s\leqslant T\text{-}1), \\ & r \text{ roccures } N_0 \text{ times in } S_0 \text{ and} \\ & r+1 \text{ occures } N_1 \text{ times in } S_1 \\ -T+N_\infty q^m & \text{otherwise} \end{cases}$$

Remark. According to the above definition of generalized mapping sequences, if a sequence over GF(q) can be arranged into an array which has m columns in which every column is a sequence generated by the same irreducible polynomial over GF(q) or a zero sequence, then this sequence is a generalized mapping sequence. So, the generalized mapping sequences contain a lot of well-known sequences. For example, the multiplexed sequences [3] and the clock-controlled sequences [4, 5] can be obtained by taking $q=2$, $T=q^m\text{-}1$. The No sequences [7] can be obtained by taking $q=2$, $T=q^m+1$. Thus we have the upper bound of the linear span of No sequences is $m(q^m+1)$, this result was not found in [7]. GMW sequences [6] and generized GMW sequences [8] can be obtained by taking $T=\frac{q^N-1}{q^m-1}$, m is a factor of the positive integer N, where $q=2$ and q is a prime number, respectively.

The applications of generalized mapping sequences can be found in the reference [9].

References

[1] R. Lidl, H. Niederreiter: *Finite Fields*, Encyclopedia of Mathematics and its Applications, Volume 20, Addison-Wesley, 1983.

[2] F. J. MacWilliams, N. J. A. Sloane: *The Theory of Error-Correcting Codes*, North-Holland, New York, 1977.

[3] S. M. Jennings: "Multiplexed Sequences: Some Properties of the Minimum Polynomial", Proceedings of Euro-crypt' 82, Lecture Notes in Computer Science, Vol. 149, Springer-Verlag, 1983, pp. 189-206.

[4] T. Beth, F. Piper: "The Stop-and-Go Generators", Advances in Crytology, Proc. of Eurocrypt' 84, Springer-Verlag, 1985, pp. 88-92.

[5] D. Gollmann, W. G. Chambers: " Clock-controlled Shift Register: A Review", IEEE J. on Selected Areas in Comm. Vol. 7, No. 4, May 1989, pp. 525-533.

[6] R. A. Scholtz, L. R. Welch: GMW Sequences", IEEE Trans. Inform. Theory, vol. IT-30, no. 3, May 1984, pp. 548-553.

[7] Jong-Seon No, P. V. Kumar: " A New Family of Binary Pseudorandom Sequences Having Optimal Periodic Correlation Properties and Large Linear Span", IEEE Trans. on Inform. Theory, Vol. 35, No. 2, March 1989, pp. 371-379.

[8] M. Antweiler, L. Bomer: "Complex Sequences over GF(p^M) with a Two-level Autocorrelation Function and a Large Linear Span", IEEE Trans. on Inf. Theory, vol. 38, No. 1, January 1992, pp. 120-130.

[9] Guang Gong, " On the q-ary Sequences Generated by Mapping GF(q^m) sequences into GF(q)", FUB report: 3B00692, January 1992, Rome, Italy.

A Zero-Test and an Interpolation Algorithm for the Shifted Sparse Polynomials

Dima Grigoriev*[1] and Marek Karpinski **[2]

[1] Dept. of Computer Science, The Pennsylvania State University, University Park, PA 16802

[2] Dept. of Computer Science, University of Bonn, 5300 Bonn 1, and International Computer Science Institute, Berkeley, California

Abstract. Recall that a polynomial $f \in F[X_1, \ldots, X_n]$ is t -sparse, if $f = \sum \alpha_I X^I$ contains at most t terms. In [BT 88], [GKS 90] (see also [GK 87] and [Ka 89]) the problem of interpolation of t-sparse polynomial given by a black-box for its evaluation has been solved. In this paper we shall assume that F is a field of characteristic zero. One can consider a t-sparse polynomial as a polynomial represented by a straight-line program or an arithmetic circuit of the depth 2 where on the first level there are multiplications with unbounded fan-in and on the second level there is an addition with fan-in t.

In the present paper we consider a generalization of the notion of sparsity, namely we say that a polynomial $g(X_1, \ldots, X_n) \in F[X_1, \ldots, X_n]$ is *shifted t-sparse* if for a suitable nonsingular $n \times n$ matrix A and a vector B the polynomial $g(A(X_1, \ldots, X_n)^T + B)$ is t-sparse. One could consider g as being represented by a straight-line program of the depth 3 where on the first level (with the fan-in $n+1$) a linear transformation $A(X_1, \ldots, X_n)^T + B$ is computed. One could also consider a shifted t-sparse polynomial as t-sparse with respect to other coordinates $(Y_1, \ldots, Y_n)^T = A(X_1, \ldots, X_n)^T + B$.

We assume that a shifted t-sparse polynomial g is given by a black-box and the problem we consider is to construct a transformation $A(X_1, \ldots, X_n)^T + B$. As the complexity of the designed below algorithm (see the Theorem in which we describe the variety of all possible A, B and the corresponding t-sparse representations of $g(A(X_1, \ldots, X_n)^T + B)$) depends on d^{n^4} where d is the degree of g, we could first interpolate g within time $d^{O(n)}$ and suppose that g is given explicitly. It would be interesting to get rid of d in the complexity bounds as it is usually done in the interpolation of sparse polynomials ([BT 88], [GKS 90], [Ka 89]). The main technical tool we rely on is the criterium of t-sparsity based on

---

* Work partially done while visiting the Dept. of Computer Science, University of Bonn. On leave from the Steklov Mathematical Institute, Fontanka 27, St. Petersburg, 191011 Russia

** Supported in part by Leibniz Center for Research in Computer Science, by the DFG Grant KA 673/4-1 and by the SERC Grant GR-E 68297

Wronskian ([GKS 91], [GKS 92]), the latter criterium has a *parametrical* nature (so we can select t-sparse polynomials from a given *parametrical* family of polynomials) unlike the approach in [BT 88] using BCH-codes. We could directly consider (see the Theorem) the multivariate polynomials (section 3), but to make the exposition clearer before that we first study (see the proposition) the one-variable case (section 2). First at all we recall (section 1) the criterium of t-sparsity and based on it interpolation method for t-sparse multivariable polynomials.
In the last section 4 we design a zero-test algorithm for shifted t-sparse polynomials with the complexity independent on d.

1 A Criterium of t-sparsity and the Interpolation

Let $p_1, \ldots, p_n$ be pairwise distinct primes and denote by D a linear operator mapping $D : X_1 \to p_1X_1, \ldots, D : X_n \to p_nX_n$. We recall a criterium of t-sparsity (cf. also [BT 88]).
Lemma 1. ([GKS 91], [GKS 92]) *A polynomial $f \in F[X_1, \ldots, X_n]$ is t-sparse if and only if the Wronskian*

$$W_f(X_1, \ldots, X_n) = \det \begin{pmatrix} f & Df & \ldots D^t f \\ Df & D^2 f & \ldots D^{t+1} f \\ \vdots & \vdots & \vdots \\ D^t f & D^{t+1} f & \ldots D^{2t} f \end{pmatrix} \in F[X_1, \ldots, X_n]$$

vanishes identically.

An interpolation method from [BT 88] (see also [KY 88]) actually considers the Wronskian $W_f(1, \ldots, 1)$ at the point $(1, \ldots, 1)$ and is based on the following
Lemma 2. ([BT 88]) *If f is exactly t-sparse (i.e., f contains exactly t terms), then the reduced Wronskian does not vanish*

$$\bar{W}_f(1, \ldots, 1) = \det \begin{pmatrix} f(1, \ldots, 1) & (Df)(1, \ldots, 1) & \ldots (D^{t-1} f)(1, \ldots, 1) \\ \vdots & \vdots & \vdots \\ (D^{t-1} f)(1, \ldots, 1) & (D^t f)(1, \ldots, 1) & \ldots (D^{2t-2} f)(1, \ldots, 1) \end{pmatrix} \neq 0$$

at the point $(1, \ldots, 1)$.

Thus, if $f = \sum \alpha_I X^I$ is exactly t-sparse and if a (characteristic) polynomial $\chi(Z) = \sum_{0 \le j \le t} \gamma_j Z^j \in \mathbf{Z}[Z]$ has as its t roots p^I for all exponent vectors I occuring in f (where for $I = (i_1, \ldots, i_n)$ we denote $p^I = p_1^{i_1} \cdots p_n^{i_n}$), then $\sum_{0 \le j \le t} \gamma_j D^j f = 0$ and hence

$$\begin{pmatrix} f & Df & \ldots D^t f \\ \vdots & \vdots & \vdots \\ D^t f & D^{t+1} f & \ldots D^{2t} f \end{pmatrix} (\gamma_0, \ldots, \gamma_t)^T = 0\,.$$

Therefore, a linear system

$$\begin{pmatrix} f(1,\ldots,1) & (Df)(1,\ldots,1) & \ldots (D^t f)(1,\ldots,1) \\ \vdots & \vdots & \vdots \\ (D^t f)(1,\ldots,1) & (D^{t+1} f)(1,\ldots,1) & \ldots (D^{2t} f)(1,\ldots,1) \end{pmatrix} (Y_0,\ldots,Y_t)^T = o$$

has (up to a constant multiple) a unique (by lemma 2) solution $(Y_0,\ldots,Y_t) = (\gamma_0,\ldots,\gamma_t)$ which gives the coefficients of χ, thereby its roots p^I and finally I.

2 One-variable Shifted Sparse Polynomials

A polynomial $g \in F[X]$ is called *shifted t-sparse* if for an appropriate b a polynomial $g(X-b)$ is t-sparse (so the origin is shifted from 0 to b). If t is the least possible, we say that g is *minimally shifted t-sparse*, this notion relates also to the multivariable case. Let $F = \mathbf{Q}$. Usually we take b from the algebraic closure $\bar{\mathbf{Q}}$ (we could also consider b from $\mathbf{R}$). Assume that the bit-size of the (rational) coefficients of g does not exceed M.

Consider a new variable Y and an $\mathbf{Q}(Y)$-linear transformation of the ring $\mathbf{Q}(Y)[X]$ mapping $D_1 : X \to p_1 X + (p_1 - 1)Y$. Denote

$$\mathcal{W}_g(X,Y) = \det \begin{pmatrix} g & D_1 g & \ldots D_1^t g \\ \vdots & \vdots & \vdots \\ D_1^t g & D_1^{t+1} g & \ldots D_1^{2t} g \end{pmatrix} \in \mathbf{Q}[X,Y]$$

Lemma 3. *g is shifted t-sparse if and only if for some $Y = b$ a polynomial $\mathcal{W}_g(X,b)$ vanishes identically. Moreover in this case a polynomial $g(X-b)$ is t-sparse.*

Proof. If $g(X-b)$ is t-sparse, then the expansion $g = \sum_j \beta_j (X+b)^j$ into the powers of $(X+b)$ contains at most t terms. Lemma 1 implies that $\mathcal{W}_g(X,b)$ vanishes identically. The other direction follows also from lemma 1 which completes the proof.

Observe that for almost every b the polynomial $g(X-b)$ has exactly $(d+1)$ terms, where $d = \deg(g)$, since in the polynomial $g(X-Y) \in \mathbf{Q}[X,Y]$ the coefficient in the power X^S is a polynomial in Y of degree exactly $d-S$, $0 \le S \le d$.

Lemma 3 provides an algorithm for finding t such that g is minimal shifted t-sparse which runs in time $d^{O(1)}$ (trying successively $t = 1, 2, \ldots$), moreover this algorithm finds all $Y = Y_0$ such that $g(X - Y_0)$ is t-sparse. Namely, one writes down a polynomial system in Y equating to zero all the coefficients in the powers of X, thus the system contains $d^{O(1)}$ equations of degrees at most $d^{O(1)}$. So, one can prove the following proposition.

Proposition. *There is an algorithm which for one-variable polynomial g finds the minimal t and all Y_0 for which $g(X-Y_0)$ is t-sparse in time $(Md)^{O(1)}$. The number of such Y_0 does not exceed $d^{O(1)}$.*

One of the purposes of the sparse analysis is to get rid of d in the complexity bounds. We can write down a system in b with a less (for small t) number of equations, when b is supposed to belong to $\mathbf{R}$. So, assume that the expansion $g = \sum_j \beta_j (X+b)^j$ contains at most t terms for some $b \in \mathbf{R}$. Then for any fixed $Y = Y_0 \in \mathbf{R}$ a polynomial $(D_1^K g)(X, Y_0) = \sum_j \beta_j (p_1^K (X+Y_0) - Y_0 + b)^j$ for $K \geq 0$. Therefore the polynomial $\mathcal{W}_g(X, Y_0)$ has at most $2^{O(t^4)}$ real roots because of [Kh 91] since one can consider $(2t+1)t$ powers of linear polynomials $(p_1^K(X+Y_0)-Y_0+b)^j, \quad 0 \leq K \leq 2t$ as the elements of a Pfaffian chain [Kh 91].

Thus Y satisfies the conditions of lemma 3 if and only if it satisfies the following system of polynomial equations (cf. lemma 5 below)

$$\mathcal{W}_g(0, Y) = \mathcal{W}_g(1, Y) = \ldots = \mathcal{W}_g(2^{O(t^4)}, Y) = 0 .$$

Each of the polynomials from the latter system can be represented by a black-box for its evaluation. As each of these polynomials $\mathcal{W}_g(s, Y)$ contains $(2t+1)t$ powers $(p_1^K(s+Y) - Y + b)^j, \quad 0 \leq K \leq 2t$ the system has at most $2^{O(t^4)}$ real solutions (by the same argument relying on [Kh 91] as above), thus the number of such $Y = Y_0$ that $g(X - Y_0)$ is t-sparse is less than $2^{O(t^4)}$.

3 Multivariate Shifted Sparse Polynomials

Consider now n^2+n new variables $Z_{i,j}, Y_i, \quad 1 \leq i, j \leq n$ and a $\mathbb{Q}(\{Z_{ij}, Y_i\}_{1 \leq i,j \leq n})$-linear transformation D_n of the ring $\mathbb{Q}(\{Z_{ij}, Y_i\}_{1 \leq i,j \leq n})[X_1, \ldots, X_n]$ mapping

$$D_n X = ZPZ^{-1}(X - Y) + Y$$

where vectors $X = (X_1, \ldots, X_n)^T, Y = (Y_1, \ldots, Y_n)^T$, matrices $Z = (Z_{ij}),\ P = \begin{pmatrix} p_1 & & 0 \\ & \ddots & \\ 0 & & p_n \end{pmatrix}$. Similarly, as above denote

$$\mathcal{W}_g(X, Y, Z) = \det \begin{pmatrix} g & D_n g & \ldots D_n^t g \\ \vdots & \vdots & \vdots \\ D_n^t g & D_n^{t+1} g & \ldots D_n^{2t} g \end{pmatrix} \in \mathbb{Q}(Z)[X, Y] .$$

Lemma 4. *g is shifted t-sparse if and only if for some Z_0, Y_0 such that $\det Z_0 \neq 0$, the polynomial $\mathcal{W}_g(X, Y_0, Z_0)$ vanishes identically. Moreover, in this case a polynomial $g(Z_0 X + Y_0)$ is t-sparse.*

The proof is similar to the proof of lemma 3 taking into account that

$$(D_n g)(ZX + Y) = g(ZPZ^{-1}(ZX + Y - Y) + Y) = g(ZPX + Y) .$$

As in section 2 lemma 4 provides a test for minimal shifted t-sparsity trying successively $t = 1, 2, \ldots$ running in time $d^{O(n^4)}$ (see [CG 83] for solving system of polynomial equations and inequalities). Moreover, the algorithm finds algebraic conditions (equations and inequality $\det Z \neq 0$) on all Z, Y for which $g(ZX+Y)$ is t-sparse.

So, these Z, Y form a constructive set $U \subset \bar{\mathbf{Q}}^{n^2+n}$ given by a system $h_1 = \ldots = h_k = 0$, $\det Z \neq 0$ where $h_1, \ldots, h_k \in \mathbf{Q}[\{Z_{ij}, Y_i\}_{1\leq i,j\leq n}]$, then $\deg(h_1), \ldots, \deg(h_k) \leq d^{O(1)}$, $k \leq d^{O(1)}$. Applying the algorithm from [CG 83] one can find the irreducible over $\mathbf{Q}$ components $\bar{U} = \bigcup_l U^{(l)}$ of the closure (in the Zariski topology) $\bar{U}$. For each component $U^{(l)}$ the algorithm from [CG 83] produces firstly, some polynomials $h_1^{(l)}, \ldots, h_{N(l)}^{(l)} \in \mathbf{Q}[\{Z_{ij}, Y_i\}]$ such that $U^{(l)} = \{h_1^{(l)} = \ldots = h_{N(l)}^{(l)} = 0\}$ and secondly, a general point of $U^{(l)}$, namely the following fields isomorphism

$$\mathbf{Q}(U^{(l)}) \simeq \mathbf{Q}(T_1, \ldots, T_m)[\theta]$$

where $\mathbf{Q}(U^{(l)})$ is the field of rational functions on $U^{(l)}$, $m = \dim(U^{(l)})$, linear forms $T_1, \ldots, T_m$ in variables $\{Z_{ij}, Y_i\}_{1\leq i,j\leq n}$ constitute a transcendental basis of $\mathbf{Q}(U^{(l)})$ and θ is algebraic over $\mathbf{Q}(T_1, \ldots, T_m)$. The algorithm produces a minimal polynomial $\phi(Z) \in \mathbf{Q}(T_1, \ldots, T_m)[Z]$ of θ, the linear forms $T_S(\{Z_{ij}, Y_i\})$, $1 \leq S \leq m$, a linear form $\theta(\{Z_{ij}, Y_i\})$, and the expressions for the coordinate functions $Z_{i,j}(T_1, \ldots, T_m, \theta), Y_i(T_1, \ldots, T_m, \theta)$ as rational functions in $T_1, \ldots, T_m, \theta$. The degrees of the polynomials $h_1^{(l)}, \ldots, h_{N(l)}^{(l)}$ do not exceed $d^{O(n^2)}$, the bit-size of any of the (rational) coefficients occuring in these polynomials can be bounded by $M^{O(1)}d^{O(n^2)}$ and the algorithm runs in time $M^{O(1)}d^{O(n^4)}$.

Denote $\tilde{U}^{(l)} = U^{(l)} \setminus \{\det Z = 0\}$ (some of $\tilde{U}^{(l)}$ can be empty), remark that $U = \bigcup_l \tilde{U}^{(l)}$.

For any point $(Z_0, Y_0) \in \tilde{U}^{(l)}$ the polynomial $g(Z_0X + Y_0)$ is exactly t-sparse, therefore by lemma 2 the following linear system

$$\begin{pmatrix} g(X_0, Y_0, Z_0) & D_n g(X_o, Y_0, Z_0) & \ldots D_n^t g(X_0, Y_0, Z_0) \\ \vdots & \vdots & \vdots \\ D_n^t g(X_0, Y_0, Z_0) & D_n^{t+1} g(X_0, Y_0, Z_0) & \ldots D_n^{2t} g(X_0, Y_0, Z_0) \end{pmatrix} (\gamma_0, \ldots, \gamma_{t-1}, 1) = 0$$

has a unique solution, where the vector $X_0 = Z_0^{-1}((1, \ldots, 1)^T - Y_0)$. As $\gamma_0, \ldots, \gamma_{t-1} \in \mathbf{Z}$ (see section 1) and $\gamma_0, \ldots, \gamma_{t-1}$ can be represented as the rational functions in $(Z, Y) \in \tilde{U}^{(l)}$, we conclude taking into account the irreducibility of $U^{(l)}$ that $\gamma_0, \ldots, \gamma_{t-1}$ are constants on $\tilde{U}^{(l)}$. Thus, the exponent vectors I (see section 1) are the same for all the points $(Z, Y) \in \tilde{U}^{(l)}$.

So, for $(Z, Y) \in \tilde{U}^{(l)}$ one can write t-sparse representation of the polynomial

$$g = \sum_I C_I(Z, Y)(Z^{-1}(X - Y))^I \tag{1}$$

where the coefficients $C_I(Z, Y)$ depend on Z, Y. The equality (1) is equivalent to a system of equalities

$$g(ZX^{(0)} + Y) = \sum_I C_I(Z, Y)(Z^{-1}(X^{(0)} - Y))^I$$

where $X^{(0)}$ runs over all the vectors from $\{0,\ldots,d\}^n$. Adding to the latter system the system $\det Z \neq 0$, $h_1^{(l)} = \ldots = h_{N(l)}^{(l)} = 0$ determining $\tilde{U}^{(l)}$ we come to a parametrical (with the parameters $\{Z_{ij}, Y_i\}$) linear in C_I system which one can solve invoking the algorithm from [H 83] (see also [CG 84]) in time $M^{O(1)}d^{O(n^4)}$. This algorithm yields some disjoint decomposition of $\tilde{U}^{(l)} = \bigcup_S U_S^{(l)}$ where each $U_S^{(l)}$ is a constructive set and also yields the rational functions $\bar{C}_{I,S}^{(l)}(\{Z_{ij}, Y_i\}) \in \mathbf{Q}(\{Z_{ij}, Y_i\})$ such that $C_I = \bar{C}_{I,S}^{(l)}(\{Z_{ij}, Y_i\})$ for every point $\{Z_{ij}, Y_i\} \in U_S^{(l)}$ (thus each C_I is a piecewise-rational function on $\tilde{U}^{(l)}$).

The algorithm yields also polynomials $h_{S,0}^{(l)}, \ldots, h_{S,N_S^{(l)}}^{(l)} \in \mathbf{Q}[\{Z_{ij}, Y_i\}]$ such that $U_S^{(l)} = \{h_{S,0}^{(l)} \neq 0, h_{S,1}^{(l)} = \ldots = h_{S,N_S^{(l)}}^{(l)} = 0\}$. From [H 83] (see also [CG 84]) we get the bounds on the degrees $\deg(h_{S,q}^{(l)}), \deg(\bar{C}_{I,S}^{(l)}) \leq d^{O(n^2)}$ and the bound $M^{O(1)}d^{O(n^2)}$ for the bit-size of every (rational) coefficients of all the yielded rational functions.

Thus, we have proved the following theorem (cf. proposition above).

Theorem. *There is an algorithm which finds a minimal t and produces a constructive set $U \subset \bar{\mathbf{Q}}^{n^2+n}$ of all $\{Z_{ij}, Y_i\}_{1\leq i,j\leq n}$ such that $g(ZX+Y)$ is t-sparse, in the form $U = \bigcup_l \mathcal{U}^{(l)}$ and for each constructive set $\mathcal{U}^{(l)}$ the algorithm produces polynomials $\mathcal{H}_0^{(l)}, \ldots, \mathcal{H}_{\mathcal{N}^{(l)}}^{(l)} \in \mathbf{Q}[\{Z_{ij}, Y_i\}]$ such that $\mathcal{U}^{(l)} = \{\mathcal{H}_0^{(l)} \neq 0, \mathcal{H}_1^{(l)} = \ldots = \mathcal{H}_{\mathcal{N}^{(l)}}^{(l)} = 0\}$. Also the algorithm produces t exponent vectors and for each exponent vector I a rational function $C_I^{(l)}(\{Z_{ij}, Y_i\}) \in \mathbf{Q}(\{Z_{ij}, Y_i\})$ which provide t-sparse representations of*

$$g = \sum_I C_I^{(l)}(\{Z_{ij}, Y_i\})(Z^{-1}(X-Y))^I$$

which is valid for every point $(\{Z_{ij}, Y_i\}) \in \mathcal{U}^{(l)}$. The degrees of all produced rational functions $\mathcal{H}_S^{(l)}$, $C_I^{(l)}$ do not exceed $d^{O(n^2)}$, the bit-size of the coefficients of these rational functions can be bounded by $(Md^{n^2})^{O(1)}$ and the running time of the algorithm is at most $(Md^{n^4})^{O(1)}$.

Again when Z_{ij}, Y_i belong to $\mathbf{R}$ we could write down a polynomial system on Z, Y with a less number of equations. For this purpose we need the following

Lemma 5. *If g is a shifted t-sparse polynomial, then for any Z_0, Y_0 such that $\det Z_0 \neq 0$ for at least one of $X_1^{(0)} = 1, \ldots, n^{O(n)}2^{O(t^4)}$, a polynomial $\mathcal{W}_g(X_1^{(0)}, X_2, \ldots, X_n, Y_0, Z_0) \in \mathbf{R}[X_2, \ldots, X_n]$ does not vanish identically, provided that $\mathcal{W}_g(X, Y_0, Z_0) \in \mathbf{R}[X]$ does not vanish identically.*

Proof. Let for some $Z^{(0)}, Y^{(0)}$ a polynomial $g(Z^{(0)}X + Y^{(0)})$ be t-sparse, i.e.

$$g = \sum_J \beta_J \prod_{1\leq i\leq n} ((Z^{(0)})^{-1}(X - Y^{(0)}))_i^{j_i}$$

where $J = (j_1, \ldots, j_n)$ and the sum has at most t items (by $((Z^{(0)})^{-1}(X-Y^{(0)}))_i$ we denote i-th coordinate of the vector $(Z^{(0)})^{-1}(X - Y^{(0)})$). Then

$$(D_n^K g)(X, Y_0, Z_0) = \sum_J \beta_J \prod_{1\leq i\leq n} ((Z^{(0)})^{-1}((Z_0 P^K Z_0^{-1}(X - Y_0) + Y_0) - Y^{(0)}))_i^{j_i}$$
$$\text{for } 0 \leq K \leq 2t\,.$$

Thus $\mathcal{W}_g(X, Y_0, Z_0)$ is a polynomial in $(2t + 1)t$ products of the form like in the latter expression and these products can be considered as the elements of a Pfaffian chain. [Kh 91] entails (cf. also [GKS 93]) that the sum of Betti numbers of the variety $\{\mathcal{W}_g(X, Y_0, Z_0) = 0\} \subset \mathbf{R}^n$ is less than $n^{O(n)}2^{O(t^4)}$. As in particular $(n-1)$-th Betti number $b^{n-1} < n^{O(n)}2^{O(t^4)}$ we conclude the statement of the lemma (cf. [GKS 93]).

Thus, Y, Z satisfy the conditions of lemma 4 if and only if $\det Z \neq 0$ and they satisfy the following $n^{O(n^2)}2^{O(nt^4)}$ equations.

$$\mathcal{W}_g(X_1^{(0)}, \ldots, X_n^{(0)}, Y, Z) = 0, \qquad X_1^{(o)}, \ldots, X_n^{(0)} \in \{1, \ldots, n^{O(n)}2^{O(t^4)}\}$$

4 Zero-test for shifted sparse polynomials

Let g be shifted t-sparse polynomial. Then (see lemma 5) for at least one of $X_1^{(0)} = 1, \ldots, n^{O(n)}2^{(t^2)}$ a polynomial $g(X_1^{(0)}, X_2, \ldots, X_n) \in \mathbb{Q}[X_2, \ldots, X_n]$ does not vanish identically. Thus for zero-test one can compute $g(X_1^{(0)}, \ldots, X_n^{(0)})$ for $n^{O(n^2)}2^{O(nt^2)}$ points $(X_1^{(0)}, \ldots, X_n^{(0)}) \in \{1, \ldots, n^{O(n)}2^{O(t^2)}\}^n$. Then g vanishes identically if and only if all the results of computation vanish. Thus, the complexity of zero-test does not depend on d.

Acknowledgement. The authors would like to thank C. Schnorr for initiating the question about the shifted sparse polynomials.

References

[BT 88] Ben-Or, M. & Tiwari, P., *A deterministic algorithm for sparse multivariate polynomial interpolation*, Proc. 20 STOC ACM, 1988, pp. 301-309.

[CG 83] Chistov, A. & Grigoriev, D., *Subexponential-time solving systems of algebraic equations*, Preprints LOMI E-9-83, E-10-83, Leningrad, 1983.

[CG 84] Chistov, A. & Grigoriev, D., *Complexity of quantifier elimination in the theory of algebraically closed fields*, Lect. Notes Comp. Sci. **176**, 1984, pp. 17-31.

[GK 87] Grigoriev, D. & Karpinski, M., *The matching problem for bipartite graphs with polynomially bounded permanents is in NC*, Proc. 28 FOCS IEEE, 1987, pp. 166-172.

[GKS 90] Grigoriev, D., Karpinski, M. & Singer, M., *Fast parallel algorithms for sparse multivariate polynimial interpolation over finite fields*, SIAM J. Comput. **19**, N 6, 1990, pp. 1059-1063.

[GKS 91] Grigoriev, D., Karpinski, M. & Singer, M., *The interpolation problem for k-sparse sums of eigenfunctions of operators*, Adv. Appl. Math. **12**, 1991, pp. 76-81.

[GKS 92] Grigoriev, D., Karpinski, M. & Singer, M., *Computational complexity of sparse rational interpolation* , to appear in SIAM J. Comput.

[GKS 93] Grigoriev, D., Karpinski, M. & Singer, M., *Computational complexity of sparse real algebraic function interpolation*, to appear in Proc. Int. Conf. Eff. Meth. Alg. Geom., Nice, April 1992 (Progr. in Math. Birkhäuser).

[H 83] Heintz, J., *Definability and fast quantifier elimination in algebraically closed fields*, Theor. Comp. Sci. **24**, 1983, pp. 239-278.

[Ka 89] Karpinski, M., *Boolean Circuit Complexity of Algebraic Interpolation Problems*, Technical Report TR-89-027, International Computer Science Institute, Berkeley, 1989; in Proc. CSL'88, Lecture Notes in Computer Science **385**, 1989, pp. 138-147.

[Kh 91] Khovanski, A., *Fewnomials*, Transl. Math. Monogr., AMS 88, 1991.

[KY 88] Kaltofen, E. & Yagati, L., *Improved sparse multivariate interpolation*, Report 88-17, Dept. Comput. Sci., Rensselaer Polytechnic Institute, 1988.

Parallelization of Quantifier Elimination on a Workstation Network *

Hoon Hong

Research Institute for Symbolic Computation
Johannes Kepler University
A-4040 Linz, Austria
e-mail: hhong@risc.uni-linz.ac.at

Abstract. This paper reports our effort to parallelize on a network of workstations the partial cylindrical algebraic decomposition based quantifier elimination algorithm over the reals, which was devised by Collins and improved by the author. We have parallelized the lifting phase of the algorithm, so that cylinders are constructed in parallel. An interesting feature is that the algorithm sometimes appears to produce super-linear speedups, due to speculative parallelism. Thus it suggests a possible further improvement of the sequential algorithm via simulating parallelism.

1 Introduction

This paper reports our effort to parallelize on a network of workstations a quantifier elimination algorithm for the first order theory of real closed fields. Since Tarski's first algorithm [24], various improvements and new algorithms have been devised and analyzed [5, 1, 21, 3, 9, 4, 11, 8, 22, 14, 15, 6, 19, 16, 18]. In spite of these significant improvements, however, the practical applicability of the methods is still limited due to the enormous computational time required. Thus the parallelization seems to be a natural step to take in bringing down the computation time.

We chose to parallelize the partial cylindrical algebraic decomposition based method [15], mainly because it is the only algorithm which has been fully implemented and in use, as far as we are aware. The algorithm is an improvement of Collins' method [5]. The improved algorithm has the same asymptotic time bound for the worst case (doubly exponential in the number of variables), but it is dramatically faster than the original for small inputs, which can be tackled using currently available machines.

There exist algorithms with better asymptotic complexity (both in sequential and parallel) [9, 22, 8, 11, 12, 13, 10]. However, these methods are not fully implemented yet, and also the analysis of [16] suggests that their computational

* This research was carried out in the framework of the Austrian science foundation (FWF) project S5302-PHY (Parallel Algebraic Computation) and the European project (ESPRIT II) POSSO (Polynomial Systems Solving).

requirement for small inputs might be quite big. As soon as the sequential implementation of these methods are available, it will be interesting to investigate the behavior of these algorithms on small inputs and also parallelize them.

In [23] Saunders, Lee, and Abdali report their work on parallelizing Collins' original algorithm [5] on a shared memory machine. Their experiment show about 50% efficiency, due to the following two main bottlenecks: (1) A few tasks took very long, causing several other processors to be idle. (2) The lock operations on the shared heap caused heavy overhead. We tackled these two problems as follows:

- As mentioned earlier, we parallelized the author's *improved sequential algorithm* [15] which, unlike the original algorithm, allows the cylinders constructed in almost arbitrary order, thus provides possibility for achieving better load balancing.
- We used a *network of workstations* where each processor accesses its own local memory and carries out garbage collection also independently of each other. Thus the overhead of lock operation does not exist. The communication overhead was made negligible by ensuring coarse granularity of tasks.

The improved sequential algorithm [15] is, in one respect, a "search" algorithm. Parallelization of such algorithms sometimes produces an interesting experimental experience of "super"-linear speedups, due to "speculative" parallelism. This phenomenon can be understood by considering the following simple scenario. Suppose that I am searching for a treasure, and I come to the beginning of two different paths. At the ends of both paths, a treasure lies, but I do not know the fact. Assume that it takes 10 seconds to follow one path and 1 second for the other. Now if I carry out a "sequential" search by checking one path after the other, the average search time is 5.5 seconds. But if I and my friend carry out a "parallel" search by checking both paths at once, then the average (also maximum) search time is 1 second. Thus I and my friend will experience a "super"-linear speedup of 5.5.

It is important to note that this advantage could also be had in an improved sequential version, one which simulates the parallel search through a kind of timesharing.

2 Overview of Sequential Algorithm

In this section, we give a high level description of the partial cylindrical algebraic decomposition based quantifier elimination algorithm [15]. Let $F^* = (Q_{f+1}x_{f+1})\cdots(Q_r x_r)F(x_1,\ldots,x_r)$ be a prenex formula of the first order theory of real closed fields where Q_i's are quantifiers and F is a quantifier free matrix. The algorithm, given such a formula, produces an equivalent quantifier-free formula $F'(x_1,\ldots,x_f)$. The algorithm proceeds in three steps: projection, truth invariant decomposition, and solution. We have parallelized only the second step, and thus we will go into some details of this step, while giving only brief descriptions of the other steps.

Let P_r be a set of all the polynomials occurring in the formula F^*. The algorithm begins by "projecting" the r-variate polynomials in P_r into a set P_{r-1} of $(r-1)$-variate polynomials. Then P_{r-1} is projected again into P_{r-2}, and so forth, until P_1 is obtained. This constitutes the first step (projection step) of the algorithm. For the exact definition of "projection", see [15, 14].

From the projection polynomials P and the input formula F^*, the next step of the algorithm decomposes $\mathbb{R}^f$ into a finite disjoint connected subsets (called cells) such that the truth of F^* is constant throughout each cell. More concretely, this step computes a set of points in $\mathbb{R}^f$ where each point belongs to one and only one cell. These points are called "sample points". As mentioned earlier, this is the step we parallelized, and thus we will give a sufficiently detailed description of how this step is done, but after the brief description of the final step (solution step).

The final step of the algorithm constructs a quantifier-free formula F' which is equivalent to F^*. This is done by "reasoning" with the signs of the projection polynomials on each cells of the F^*-truth invariant decomposition, and also by applying three-valued logic minimization for simplification. Again see [15, 18] for how it is done.

Now we describe how the second step is done. First we introduce some definitions in order to facilitate the subsequent discussions.

Definition 1.

- A *cell tree* is a tree where a node on level k correspond to a cell in $\mathbb{R}^k$, and thus to its sample point. Each node is labeled with one of the three truth values: `true`, `false`, and `undetermined`. We will denote the truth value of a node c by $v(c)$. From now on, we will simply say "tree" instead of "cell tree".
- A *candidate node* is a leaf whose truth value is `undetermined`.
- The *initial tree* is a tree consisting of only the root node labeled `undetermined`.
- A *final tree* is a tree without any candidates nodes. The set of the leaves of a final tree corresponds to a F^*-truth invariant decomposition.
- An *expansion* of a leaf is an operation of constructing its (immediate) children nodes. This corresponds to "stack" construction in the terminology of [2]. This operation involves various computation with algebraic numbers, and can be very expensive. □

Roughly the algorithm works as follows: it begins with the initial tree, and repeatedly extends the tree by expanding a candidate node until a final tree is obtained. However the actual algorithm is a bit more complicated since sometimes certain portion of the tree might be pruned. Thus during the execution of the algorithm, the tree sometimes grows and other times shrinks. The algorithm is as follows:

Algorithm 1 (Truth Invariant Decomposition)
In: F^* is a quantified formula. P is the projection polynomials for F^*.
Out: D is a F^* truth-invariant decomposition of $\mathbb{R}^f$ where f is the number of free variables in F^*.
(1) [Initialize.] Set D to be the initial tree.

(2) [Choose.] Choose a candidate node c from the current tree D.

(3) [Expand.] Expand the chosen node c, and attempt to determine the truth values of some of its children nodes.

(4) [Prune.] Determine the truth values of as many ancestors of c as possible, referring to the quantifiers, removing from the tree the subtree of each node whose truth value is thus determined.

(5) [Done?] If there is a candidate node in the current tree, then loop to Step (2). Otherwise we are done. □

Step (4) of the algorithm is based on the following theorem proved in [5]:

Theorem 2. *Let $(Q_{f+1}x_{f+1})\cdots(Q_r x_r)F(x_1,\ldots,x_n)$ be an input quantified formula, where each Q_i is a quantifier. Let c be a node of level k, and let $c_1,\ldots,c_n$ be its children nodes. If $Q_{k+1}=\exists$, then $v(c)=\bigvee_{i=1}^{n} v(c_i)$. If $Q_{k+1}=\forall$, then $v(c)=\bigwedge_{i=1}^{n} v(c_i)$.* □

This theorem provides a way to determine the truth value of a node from the truth values of its children. Step (4) applies this theorem repeatedly, determining the truth value of a parent node from the truth values of its children, until it cannot be applied any more.

Note that Step (2) of the algorithm is *non-deterministic* in the sense that it does not specify which candidate node to be chosen in case there is more than one in the current tree. It can be shown that the output of this algorithm does not depend on the choices made. But the computing time does depend on the choices. Since it is difficult (impossible) to determine the optimal choices, usually various strategies are used instead. For the experiments reported in this paper, we used a strategy which seems to work well for most inputs: roughly it chooses a node whose expansion is guessed to be cheapest in time. See [15] for the details.

The asymptotic complexity of this algorithm is the same as that of Collins' algorithm. Let r be the number of variables in the input sentence, m the number of polynomials, d the total degree, and L the bit length bound for the coefficients. The worst case computing time of both algorithm is bounded above by $L^3(md)^{2^{O(r)}}$. Our objective in implementing a parallel version of such a doubly exponential algorithm on a network of workstations is clearly not in changing the asymptotic complexity, but in reducing the constants (which do not show up in the asymptotic analysis).

3 Parallelization

In this section we modify the sequential algorithm for coarse-grain parallelism on bounded number of workstations. The most time consuming part of the sequential algorithm is Step (3) where a chosen node is expanded. Thus we should try to parallelize the executions of this step.

We begin by noting that in Step (2) of the sequential algorithm there can be more than one candidate node. So the natural idea is to expand several candidate nodes in parallel. A naive approach may be to carry out in parallel Step (3) for all candidate nodes, and wait until all are finished, then carry out Step (4) all together. In other words the parallel processes are synchronized at the beginning of Step (4). However, this approach has several problems:

- The computation times for Step (3) can be very *heterogeneous*. For example, it might take only a few milli-seconds for one candidate node, but a few minutes for another candidate node. Thus the load might not be balanced, making some processors idle.
- It might turn out that during Step (4) a sub-tree built by a processor is pruned out. But this fact will be discovered *only after* that the sub–tree has been already completely built.
- The *communication overhead* becomes dominant, erasing the gain obtained by parallelism. This is so because for some candidate nodes, the computation time taken by expansion is so small (a few milliseconds) that the communication overhead cannot be ignored. Further, an input formula of a moderate size often requires enormously many expansions (for example, several hundreds of thousands). Since each of these expansions require at least one message passing, even when the overhead of each message passing is relatively small, the accumulated overhead can be huge.

A better way is to use the model of a *manager and workers*, in which the manager repeatedly assigns a task to the workers and collects the results from the workers. The manager might *abort* the computation of a worker, if the results from other workers indicate that the outcome of the computation carried out by that worker is no longer needed.

The manager keeps track of the current tree. In particular it records whether a node has been assigned to a worker or not. This is done by attaching to each node one more boolean label. The manager also keeps track of the status of each worker: working or idle.

In this way, a better load balance can be obtained, since there is no explicit synchronization any more. While a worker is working on a big task, other workers do not have to wait at the synchronization barrier, but instead can continue with other candidate nodes. It also detects useless computations early and aborts them.

The communication overhead can be minimized by letting a worker construct a larger subtree over a candidate node, not just the subtree consisting of its children. The manager controls the size of subtrees by setting two limits:

N_{max}: The maximum number of expansions carried out by the worker.

T_{max}: Roughly the maximum computing time that the worker can spend on expansions. More precisely, before the worker carries out each expansion, it checks whether it has already used up the allotted time, and if so, it reports to the manger the subtree it has built thus far.

The following algorithm incorporates these ideas.

Algorithm 2 (Manager)

(1) [Initialize.] Set the current tree to be the initial tree.

(2) [Assign tasks.] Simultaneously assign an unassigned candidate node to each idle worker, also setting the two limits: N_{max} and T_{max}.
(3) [Wait and Attach.] Wait for a result from a worker. (A result from a worker is a *subtree* over the candidate node c assigned to the worker.) Attach the subtree to the current tree.
(4) [Prune.] Determine the truth values of as many ancestors of c as possible, referring to the quantifiers, removing from the tree the subtree of each node whose truth value is thus determined.
(5) [Abort.] Abort the computation of each worker whose result is not needed any more.
(6) [Done?] If there is no candidate node in the current tree, we are done. Otherwise go to Step (2). □

Algorithm 3 (Worker)
(1) [Wait and Initialize.] Wait for a candidate node and N_{max} and T_{max} from the manager. Let c_0 be the candidate node it received from the manager. Set the current tree to be the single node c_0. Set $i = N_{max}$. Set $T = T_{max}$.
(2) [Choose.] Choose a candidate node c from the current tree.
(3) [Expand.] Expand the chosen node c, and attempt to determine the truth values of some of its children.
(4) [Prune.] Determine the truth values of as many ancestors of c as possible, referring to the quantifiers, removing from the tree the subtree of each node whose truth value is thus determined.
(5) [Done?] Set $i \leftarrow i - 1$ and set $T \leftarrow T$ minus the time consumed by Steps (2), (3) and (4). If one of the following holds, report the current tree to the manager: (a) there is no candidate node in the current tree, or (b) $i = 0$, or (c) $T \leq 0$. Otherwise goto Step (2). □

When the manager aborts the computation of a worker, the worker immediately jumps to Step (1). Note that Step (2) of the manager algorithm does not specify which candidate node to be assigned to which worker. Likewise Step (2) of the worker algorithm does not specify which candidate node to be chosen in case there is more than one candidate node.

Therefore, before executing this algorithm, one needs to decide the following (which we will call a *strategy*): N_{max}, T_{max}, an ordering for manager node choice (scheduling), and an ordering for worker node choice. In devising suitable strategies, one should consider the following conflicting factors which influence the efficiency of the algorithm:

- *Load balancing*: If N_{max} or T_{max} are chosen too big, then the load might not be balanced well.
- *Communication overhead*: If N_{max} or T_{max} are too small, then the communication overhead might become dominant.
- *Speculativeness*: Depending on the chosen strategy the parallel algorithm might carry out fewer (or cheaper) expansions than the sequential algorithm, or vice versa.

Currently we do not know any general strategy which works well for all inputs, thus in this preliminary investigation we used the following (naive) strategy.

Strategy 1 (Used in the experiments)

- $N_{max} = \begin{cases} 1 & \text{if the level of the chosen node} \leq 2 \\ \infty & \text{otherwise} \end{cases}$
  The reason for $N_{max} = 1$ for the lower levels is to produce sufficiently many candidate node at the beginning in order to help load balancing. The reason for $N_{max} = \infty$ for the higher levels is to reduce communication overhead by making task sizes sufficiently large.
- T_{max} = 10 seconds. This choice was made because we thought that 10 seconds of idling of workers is tolerable.
- *Manager node choice*: the depth first order. This choice was made because it can be efficiently implemented.
- *Worker node choice*: the same strategy as the one used by the sequential algorithm. Roughly it chooses a node whose expansion is guessed to be cheapest. For the details see [15]. □

4 Experiments

We have implemented both the sequential and parallel algorithms in C language on the computer algebra library SAC-2 [7]. Originally SAC-2 was written in the programming language ALDES [20], but we translated the whole library into C. The current implementation runs on DEC stations.

We carried out experiments on both the sequential and the parallel algorithms on various quantifier elimination problems arising from diverse application areas. Due to page limit, we show only three representative problems among them.

Quartic: $(\forall x)x^4 + ax^3 + bx^2 + cx + d > 0.$

Quadratic:
$$(\exists x)(\exists y)(\exists z)$$
$$[[ay^2 + by + c = 0 \wedge dz + e \neq 0 \wedge fz + g \neq 0]$$
$$\Longrightarrow$$
$$[ax^2 + bx + c = 0 \wedge dx + e \neq 0 \wedge fx + g \neq 0]].$$

Torus: $(\exists b)(\exists c)\ (b^2 + c^2 = 9/4\ \wedge\ x^2 + (y-b)^2 + (z-c)^2 = 1/4).$

Figure 1 reports experimental data for the example inputs. The horizontal axes stand for the number of workstations. The vertical axes stand for:

- Speedup: the elapsed (wall-clock) time taken by the sequential program / that taken by the parallel program.
- Utilization: the average running time (cpu-time) of the parallel program / its elapsed time.
- Communication: the communication time / the elapsed time.
- Work ratio: the total running time of the parallel program / the running time of the sequential program.

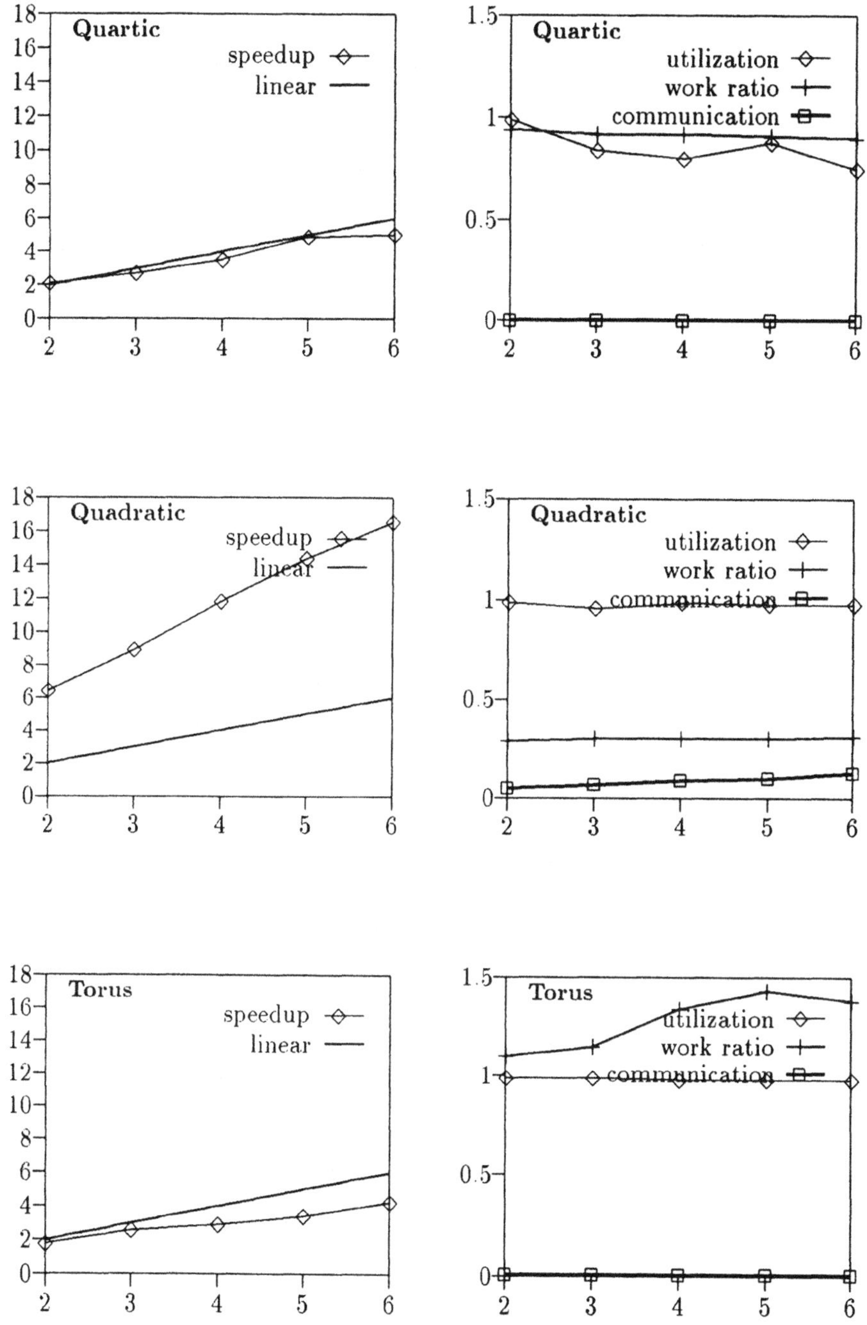

Fig. 1. Experimental Results for Three Example Inputs

For the quartic problem, the speedup is almost linear until 5 workers are used, but then it drops when 6 workers are used. It is mainly because the utilization drops to about 75%. More detailed data (not included here) show that there is exactly one worker with 100% utilization. We conjecture that this happens because while one worker is finishing up the last task, the other workers are idle. Note that the communication overheads are negligible.

The quadratic problem shows super–linear speedups. This is mainly due to the small work ratios (about 30%), which means that the parallel program carried out far fewer or cheaper expansions than the sequential program. Note that there are some communication overheads (in the range of 3–15%), but their effects on the speedup is mitigated by the huge drop in the work ratios.

The torus problem shows the opposite behavior to the quadratic problem: in spite of very high utilization and negligible communication overhead, it shows relatively low speedup. It is mainly due to the large work ratios, which means that the parallel program carries out more (expensive) expansions than the sequential program.

5 Conclusion

In this paper we reported our experience in parallelizing on a network of workstations the partial cylindrical algebraic decomposition based quantifier elimination algorithm. An interesting experience is that the parallel algorithm sometimes "appears" to produce super-linear speedups, which is due to searching several parts of the cell tree at once, resulting in more effective pruning. Thus it suggests a further improvement of the sequential algorithm by simulating the parallel search (for example, through time-sharing). In fact, in order to measure the effect of (pure) parallelism, one will have to compare the parallel algorithm with such a simulating sequential algorithm.

I thank for the anonymous referees for their many valuable suggestions for improving the paper. I also would like to thank Dan Teodosiu for implementing the underlying communication primitives.

References

1. D. S. Arnon. Algorithms for the geometry of semi-algebraic sets. Technical Report 436, Computer Sciences Dept, Univ. of Wisconsin-Madison, 1981. Ph.D. Thesis.
2. D. S. Arnon, G. E. Collins, and S. McCallum. Cylindrical algebraic decomposition I: The basic algorithm. *SIAM J. Comp.*, 13:865–877, 1984.
3. M. Ben-Or, D. Kozen, and J. H. Reif. The complexity of elementary algebra and geometry. *J. Comput. System Sci.*, 32(2):251–264, 1986.
4. J. Canny. Some algebraic and geometric computations in PSPACE. In *Proceedings of the 20th annual ACM symposium on the theory of computing*, pages 460–467, 1988.
5. G. E. Collins. Quantifier elimination for the elementary theory of real closed fields by cylindrical algebraic decomposition. In *Lecture Notes In Computer Science*, pages 134–183. Springer-Verlag, Berlin, 1975. Vol. 33.

6. G. E. Collins and H. Hong. Partial cylindrical algebraic decomposition for quantifier elimination. *Journal of Symbolic Computation*, 12(3):299–328, September 1991.
7. G. E. Collins and R. Loos. *The SAC-2 Computer Algebra System.* Research Institute for Symbolic Computation, Johannes Kepler University, Linz, Austria A-4040.
8. N. Fitchas, A. Galligo, and J. Morgenstern. Precise sequential and parallel complexity bounds for quantifier elimination over algebraically closed fields. *Journal of Pure and Applied Algebra*, (67):1–14, 1990.
9. D. Yu. Grigor'ev. The complexity of deciding Tarski algebra. *Journal of Symbolic Computation*, 5(1,2):65–108, 1988.
10. J. Heintz, M-F. Roy, and T. Recio. Algorithms in real algebraic geometry and applications to computational geometry. *DIMACS*, 6, 1991.
11. J. Heintz, M-F. Roy, and P. Solernó. On the complexity of semialgebraic sets. In *Proc. IFIP*, pages 293–298, 1989.
12. J. Heintz, M-F. Roy, and P. Solernó. Single exponential path finding in semialgebraic sets I: The case of smooth compact hypersurface. In *Proceedings of AAECC-8*, 1990.
13. J. Heintz, M-F. Roy, and P. Solernó. Single exponential path finding in semialgebraic sets II: The general case. In *Abhyankar's conference proceedings*, 1990.
14. H. Hong. An improvement of the projection operator in cylindrical algebraic decomposition. In *International Symposium of Symbolic and Algebraic Computation ISSAC-90*, pages 261–264, 1990.
15. H. Hong. *Improvements in CAD-based Quantifier Elimination.* PhD thesis, The Ohio State University, 1990.
16. H. Hong. Comparison of several decision algorithms for the existential theory of the reals. Technical Report 91-41.0, Research Institute for Symbolic Computation, Johannes Kepler University A-4040 Linz, Austria, 1991.
17. H. Hong. Parallelization of quantifier elimination on workstation network. Technical Report 91-55.0, Research Institute for Symbolic Computation, Johannes Kepler University A-4040 Linz, Austria, 1991.
18. H. Hong. Simple solution formula construction in cylindrical algebraic decomposition based quantifier elimination. In *International Conference on Symbolic and Algebraic Computation ISSAC-92*, pages 177–188, 1992.
19. L. Langemyr. The cylindrical algebraic decomposition algorithm and multiple algebraic extensions. In *Proc. 9th IMA Conference on the Mathematics of Surfaces*, September 1990.
20. R. G. K. Loos. The algorithm description language ALDES (Report). *ACM SIGSAM Bull.*, 10(1):15–39, 1976.
21. S. McCallum. *An Improved Projection Operator for Cylindrical Algebraic Decomposition.* PhD thesis, University of Wisconsin-Madison, 1984.
22. J. Renegar. On the computational complexity and geometry of the first-order theory of the reals (part III). *Journal of Symbolic Computation*, 13(3):329–352, 1992.
23. B. D. Saunders, H. R. Lee, and S. K. Abdali. A parallel implementation of the cylindrical algebraic decomposition algorithm. In *International Symposium of Symbolic and Algebraic Computation*, pages 298–307, 1990.
24. A. Tarski. *A Decision Method for Elementary Algebra and Geometry.* Univ. of California Press, Berkeley, second edition, 1951.

Hyperplane Sections of Fermat Varieties in P^3 in Char. 2 and Some Applications to Cyclic Codes

H. Janwa[1]* and R. M. Wilson[2]†

[1] Centre of Advanced Study in Mathematics, University of Bombay, Bombay 400098
[2] Department of Mathematics, California Institute of Technology, Pasadena, CA 91125

Abstract. We consider the cyclic codes $C_s^{(t)}$ of length $2^s - 1$ generated by $m_1(X)m_t(X)$ where $m_i(X)$ is the minimal polynomial of a primitive element of $GF(2^s)$, and ask when these codes have minimum distance ≥ 5. Words of weight ≤ 4 in these codes are directly related to rational points in $GF(2^s)$ on the curves corresponding to the polynomials $X^t + Y^t + Z^t + (X+Y+Z)^t$ over the algebraic closure of $GF(2)$. Study of the singularities and absolutely irreducible components of these polynomials leads to results on the minimum distance of the codes.

1 Introduction

A fundamental problem in coding theory is to determine the minimum distance of a cyclic code given its generating polynomial. In this article, we study a very special case. Given integers $s \geq 3$ and $t \geq 3$ with t odd, let ω be a primitive element in the finite field $GF(2^s)$ and let $m_i(X)$ denote the minimal polynomial of ω^i over $GF(2)$. Let $C_s^{(t)}$ be the binary cyclic code of length $n = 2^s - 1$ generated by the product $g(X) = m_1(X)m_t(X)$.

For example, $C_s^{(3)}$ is the 2-error-correcting BCH code of length $n = 2^s - 1$; it has minimum distance ≥ 5 (and equal to 5 except when $s = 3$ in which case the code has length and minimum distance both equal to 7). The problem we address is the question of when the codes $C_s^{(t)}$ are 2-error-correcting, i.e. have minimum distance ≥ 5. For t fixed, the degree of $m_t(X)$ is exactly s for all s sufficiently large with respect to t and then $C_s^{(t)}$ has dimension $n - 2s$. For such values of s the sphere packing bound then shows that the minimum distance of $C_s^{(t)}$ is at most 6; but Theorem 12 of [8] shows that the minimum distance is at most 5 whenever $s \geq 4$.

For some values of t the codes $C_s^{(t)}$ or equivalent codes have arisen in discussion of the Preparata codes and their generalizations; see [1]. The following

---

* Much of the work on this article was done while the author was a Bateman Research Instructor at Caltech from 1987–1989.
† This work was supported in part by NSF Grant DMS-8703898-02.

theorem is proved in [8] (the first part is Theorem 14 and the second follows from Theorem 17 of that reference by changing primitive elements).

Theorem 1. *If $t = 2^i + 1$, $i \geq 1$, then the code $C_s^{(t)}$ is 2-error-correcting whenever $(s,i) = 1$. If $t = 2^{2i} - 2^i + 1$, $i \geq 1$, then the code $C_s^{(t)}$ is 2-error-correcting whenever s is odd and $(s,i) = 1$.*

We shall give another proof of Theorem 1 in Sections 5 and 6 by very different methods from those of [1] or [8]. In fact, we generalize the second part in that *we do not require the hypothesis that s is odd.*

By Theorem 1, when $t = 3, 5, 9, 13, 17, 33, 57, 65, \ldots$, there are infinitely many values of s for which the codes $C_s^{(t)}$ are 2-error-correcting. This is not true for $t = 7$. In [9], it was shown that for $t = 7$, $C_s^{(7)}$ is not 2-error-correcting for $s > 5$ except possibly for $s = 11$, 15, or 17. In [5] the present authors have confirmed that $C_s^{(7)}$ is also not 2-error-correcting for $s = 11, 15, 17$ and have given a formula for the number of words of weight 4 in $C_s^{(7)}$. For $s = 5$, the code $C_5^{(7)}$ is 2-error-correcting "by accident"; it is equivalent to $C_5^{(5)}$, which is 2-error-correcting by Theorem 1, since if we consider a new primitive element $\alpha = \omega^7$, then the minimal polynomial of α is our original $m_7(x)$ and the minimal polynomial of α^5 is $m_1(x)$.

We will give other values of t for which $C_s^{(t)}$ fails to be 2-error-correcting for more than a "few" small values of s. It may be that given t, the code $C_s^{(t)}$ is 2-error-correcting for infinitely many values of s only when t has one of the forms $2^i + 1$ or $2^{2i} - 2^i + 1$ appearing in Theorem 1.

The method of [8] was to relate the words of weight four in $C_s^{(7)}$ to the zeros in $GF(2^s)$ of a binary polynomial $g_4(X,Y)$ of degree 4 in two variables. The polynomial $g_4(X,Y)$ is absolutely irreducible, i.e. irreducible over $\overline{GF(2)}$, and a form of Weil's theorem shows that it has many zeros, and hence words of weight four, exist for sufficiently large s. In this paper we study the analogous polynomials for general t.

We introduce the polynomial

$$f_t(X,Y,Z) := X^t + Y^t + Z^t + (X+Y+Z)^t$$

over $GF(2)$. This is a "hyperplane section" of $X^t + Y^t + Z^t + W^t$. We will be interested in zeros of f_t, but only those with distinct coordinates, and so we also study

$$g_t(X,Y,Z) := \frac{X^t + Y^t + Z^t + (X+Y+Z)^t}{(X+Y)(Y+Z)(Z+X)}$$

of degree $t-3$ over $GF(2)$. (However it may still be the case that $g_t(X,Y,Z)$ has zeros (x,y,z) with some two coordinates equal.) As noted in [9] (or see Section 2), there will be words of weight 3 or 4 in $C_s^{(t)}$ if and only if $g_t(X,Y,Z)$ has zeros (x,y,z) in $GF(2^s)$ with x, y, and z distinct.

It will be convenient to use the same notation for the affine parts of the homogeneous polynomials above. That is, we write

$$f_t(X,Y) := X^t + Y^t + 1 + (X+Y+1)^t,$$

$$g_t(X,Y) := \frac{X^t + Y^t + 1 + (X+Y+1)^t}{(X+Y)(Y+1)(X+1)};$$

the context should make clear whether we are thinking projectively in three variables, or affinely in two.

In Section 5, we note that if $t = 2^i + 1$, then g_t factors into $2^i - 2$ linear factors over $GF(2^i)$. In Section 6, we prove that if $t = 2^{2i} - 2^i + 1$, then g_t factors into $2^i - 2$ factors of degree $2^i + 1$ over $GF(2^i)$. We derive Theorem 1 from these factorizations. We are unable to factor g_t for any other values of t and are tempted to conjecture that g_t is absolutely irreducible if t does not have one of these two forms.

Values of t for which the absolutely irreducibility of g_t can be proved are given in Sections 4 and 7. For these values of t, as we point out in Section 2, there are only finitely many 2-error-correcting codes $C_s^{(t)}$.

2 The code $C_s^{(t)}$ and the curve $g_t(X,Y,Z)$

The check matrix of $C_s^{(t)}$ is

$$\begin{pmatrix} 1 & \omega & \omega^2 & \dots & \omega^{n-1} \\ 1 & \omega^t & \omega^{2t} & \dots & \omega^{(n-1)t} \end{pmatrix}.$$

We coordinatize $C_s^{(t)}$ by $0, 1, \dots n-1$. If $C_s^{(t)}$ had a codeword of weight 4 in the position i, j, k, l, then we would have

$$\omega^i + \omega^j + \omega^k + \omega^l = 0$$

and

$$(\omega^i)^t + (\omega^j)^t + (\omega^k)^t + (\omega^l)^t = 0.$$

That is, we should have a solution (x, y, z, w) to the following algebraic set in the projective three-space P^3 over $GF(2^s)$.

$$X + Y + Z + W = 0$$
$$X^t + Y^t + Z^t + W^t = 0,$$

with distinct nonzero elements x, y, z, w in $GF(2^s)$. Conversely, such a solution would provide a solution to the given system and lead to a vector of weight 4 in $C_s^{(t)}$. A solution with distinct elements x, y, z, w one of which is 0 corresponds to a vector of weight 3 in $C_s^{(t)}$.

Finding a solution to the above algebraic set in P^3 is equivalent to finding a solution to the following projective algebraic set in the projective two-space P^2 over $GF(2^s)$:

$$X^t + Y^t + Z^t + (X+Y+Z)^t = 0.$$

So it is easy to see that we have:

Proposition 1. *The following are equivalent:*

(i) *the curve $g_t(X,Y,Z)$ has no rational points (x,y,z) over $GF(2^s)$ with distinct coordinates;*
(ii) *the curve $g_t(X,Y)$ has no rational points (x,y) over $GF(2^s)$ with $x \neq y$, $x \neq 1$, and $y \neq 1$;*
(iii) *the code $C_s^{(t)}$ is 2-error-correcting.*

Corollary. *If $g_t(X,Y,Z)$ is absolutely irreducible, then the code $C_s^{(t)}$ is 2-error-correcting for only a finite number of values of s.*

Proof: The corollary follows from well known results from algebraic geometry. If $g_t(X,Y,Z)$ is absolutely irreducible of degree $d = t-3$ over $GF(2)$, a form of Weil's theorem due to Schmidt [11] shows that the number N_s of (affine) zeros of $g_t(X,Y)$ in $GF(2^s)$ satisfies

$$|N_s - 2^s| < (t-4)(t-5)2^{s/2} + (t-3)^2$$

for all s. The number of zeros (x,y) of $g_t(X,Y)$ with $x=1$, $y=1$, or $x=y$ is bounded by $3t$ (cf. Theorem 2), so it is clear that $g_t(X,Y)$ has other zeros in $GF(2^s)$ provided that s is large compared with t. □

3 Some calculations

We used a computer to calculate the number of rational points on $g_t(X,Y,Z)$ for small values of s and t. The following observation was helpful in reducing the time required for these computations.

Theorem 2. *Fix s and t, t odd. The number of projective points (x,y,z) with distinct coordinates over $GF(2^s)$ on $g_t(X,Y,Z)$ is*

$$\sum_{c \in GF(2^s)} (\mathrm{freq}(c))^2 - 2^{s+1}$$

where freq(c) *is the number of solutions in $GF(2^s)$ of*

$$X^t + (X+1)^t = c.$$

The total number of rational points over $GF(2^s)$ on the curve $g_t(X,Y,Z)$ is

$$\sum_{c \in GF(2^s)} (\mathrm{freq}(c))^2 - 2^{s+1} + 3M + \delta$$

where M is the number of solutions in $GF(2^s)$ of

$$X^{t-1} + (X+1)^{t-1} = 0$$

and where δ is 1 if $t \equiv 1 \pmod 4$ and 0 if $t \equiv 3 \pmod 4$.

Proof: Let

$$g_t'(X,Y,U) := g_t(X,Y,X+U) = \frac{X^t + Y^t + (X+U)^t + (Y+U)^t}{(X+Y)U(X+Y+U)}.$$

The number of projective points (x,y,z) on $g_t(X,Y,Z)$ is of course the same as the number of projective points (x,y,u) on $g_t'(X,Y,U)$.

The number of projective points (x,y,z) on $g_t(X,Y,Z)$ with distinct coordinates is the number of projective points on $g_t'(X,Y,U)$ with the properties $x \neq y$, $u \neq 0$, $x+y+u \neq 0$, which is the number of affine points on

$$g_t'(X,Y) := g_t'(X,Y,1) = \frac{X^t + Y^t + (X+1)^t + (Y+1)^t}{(X+Y)(X+Y+1)}$$

with $x \neq y$ and $x+y \neq 1$. Now the number of solutions (x,y) of

$$X^t + (X+1)^t + Y^t + (Y+1)^t = 0$$

is obviously the sum $\sum_c (\text{freq}(c))^2$. From this we must subtract the 2^{s+1} pairs (a,a) and $(a,a+1)$, $a \in GF(2^s)$.

The number of projective points (x,y,z) on $g_t(X,Y,Z)$ with $x = z$ is equal to the number of projective points on the part at infinity of $g_t'(X,Y,U)$, i.e. $g_t'(X,Y,0)$. By differentiating $g_t'(X,Y,U)$ w.r.t. U and then assigning $U \leftarrow 0$, we have

$$g_t'(X,Y,0) = \frac{(X^{t-1} + Y^{t-1})}{(X+Y)^2}.$$

Replace Y by $X+V$ and substitute $V \leftarrow 1$ and we see that there are M such points with $x = z \neq y$. Similarly, there are M points with $x = y \neq z$ and M points with $y = z \neq x$. Finally, it is not hard to check that $(1,1,1)$ is a zero of $g_t(X,Y,Z)$ if and only if $t \equiv 1 \pmod 4$; see e.g. the proof of Theorem 3. □

The following table lists the number of projective zeros of $g_t(X,Y,Z)$ with *distinct coordinates* over $GF(2^s)$ for $t = 3,5,7,\ldots,65$ and $s = 2,3,\ldots,16$. The total computation time was about 12 minutes (in 8086 assembly language on a 25MHz 80386 PC).

A zero in row t and column s means that $C_s^{(t)}$ is 2-error-correcting. If $g_t(X,Y,Z)$ were absolutely irreducible, one would expect instead to find a number "close to" 2^s in that position.

$t\backslash s$	2	3	4	5	6	7	8	9	10	11	12	13	14	15	16
3	0	0	0	0	0	0	0	0	0	0	0	0	0	0	0
5	8	0	32	0	128	0	512	0	2048	0	8192	0	32768	0	131072
7	8	24	8	0	32	168	344	528	848	1848	4232	8736	16640	31944	64184
9	0	48	0	0	384	0	0	3072	0	0	24576	0	0	196608	0
11	8	48	8	0	128	0	488	480	1568	2112	3944	8736	15968	33408	58856
13	8	0	8	0	128	0	488	0	2048	0	8168	0	32768	0	131048
15	0	48	168	0	48	0	168	552	840	1848	3912	8736	19152	30288	66696
17	8	0	224	0	128	0	3584	0	2048	0	57344	0	32768	0	917504
19	8	0	224	0	128	168	416	0	1088	3168	4280	8736	20672	30480	63104
21	0	24	0	0	960	168	192	528	480	1848	3072	8736	16800	31944	74688
23	8	48	224	0	128	0	464	480	848	3168	5864	6552	17648	30168	64496
25	8	48	32	0	128	168	416	480	2048	2112	4472	9984	14624	32688	73088
27	0	0	0	0	480	0	1488	0	1560	3960	4752	9048	13776	32760	70320
29	8	48	8	0	128	0	488	408	488	1320	6848	7800	15464	35928	75848
31	8	0	224	840	8	168	224	0	848	1848	3920	8736	16640	32760	66752
33	0	0	0	960	0	0	0	0	30720	0	0	0	0	983040	0
35	8	24	32	960	32	0	416	528	2048	0	6032	6240	16640	39504	70784
37	8	48	8	0	128	168	344	1488	2048	2112	5864	9360	16640	30648	61016
39	0	48	0	960	48	0	0	264	1800	3432	16104	12168	15456	38688	74400
41	8	0	8	0	128	168	344	1008	2048	3168	7016	9360	20672	32760	77144
43	8	48	8	0	128	0	2600	480	848	0	3944	8736	81152	33408	71912
45	0	0	168	0	480	0	168	288	1320	3168	8712	8424	19656	30600	71688
47	8	0	224	960	8	168	416	0	1928	2112	4544	9048	20840	35760	63008
49	8	24	224	0	32	0	416	528	2048	2112	5840	11232	15968	38064	60032
51	0	48	0	0	48	0	6240	552	1440	1848	4512	13416	20328	37728	69216
53	8	48	224	0	128	0	704	408	3728	1848	5720	7176	15296	31488	77024
55	8	0	32	0	8	168	512	1224	2288	1848	5576	9984	18488	38760	66752
57	0	48	0	0	48	0	0	3072	0	0	24240	0	0	196608	0
59	8	0	8	0	8	168	488	0	1088	4224	6032	9048	17816	33480	80072
61	8	0	224	0	8	168	464	792	848	2640	4856	7488	21512	35880	65360
63	0	24	0	960	3720	0	336	528	960	0	3720	8736	19152	32904	64176
65	8	48	32	0	3968	0	512	3072	2048	0	253952	0	32768	196608	131072

4 Singularities of the curve defined by $g_t(X, Y, Z)$

Let us fix t and write simply f for $f_t(X, Y, Z)$ in this section. We have

$$f_X = X^{t-1} + (X + Y + Z)^{t-1},$$
$$f_Y = Y^{t-1} + (X + Y + Z)^{t-1},$$
$$f_Z = Z^{t-1} + (X + Y + Z)^{t-1}.$$

It is clear that no point $P = (x, y, 0)$ at ∞ is a singular point of f, because the simultaneous vanishing of the partial derivatives at P would imply that

$x = y = 0$. Therefore the only singularities of f are projective points of the type $P = (\alpha, \beta, 1)$ for some α, β in $GF(2)$. Let $\lambda = \alpha + \beta + 1$. Then $f_{Z|P} = 0$ implies that λ is a $(t-1)$-st root of unity. Therefore $f_{X|P} = 0$ and $f_{Y|P} = 0$ imply that α and β are also $(t-1)$-st root of unity. For such α, β (so that $\alpha + \beta + 1$ is also a $(t-1)$-st root of unity), clearly $f(P) = 0$.

Let l denote the largest odd integer dividing $t - 1$. Then α, β and $\alpha + \beta + 1$ are l-th roots of unity. Let ψ be a primitive l-th root of unity in some extension of $GF(2)$ and consider the binary cyclic code B_l of length l whose generating polynomial is the minimal polynomial of ψ over $GF(2)$. (This is a so-called *maximal* cyclic code.) The code B_l has a check matrix

$$H = [1, \psi, \psi^2, \ldots, \psi^{(l-1)}].$$

Let $\alpha = \psi^i$, $\beta = \psi^j$, and let $\lambda = \psi^k$ where i, j, k are non-zero and distinct modulo l. Then $1 + \alpha + \beta + \lambda = 1 + \psi^i + \psi^j + \psi^k = 0$. By looking at the check matrix H, it is clear that the cyclic code B_l has a codeword of weight 4, namely the binary vector with a 1 in the positions $0, i, j, k$. Conversely, it is clear that every codeword of weight 4 in the cyclic code B_l yields a singular point of f with distinct coordinates and hence a singular point of $g_t(X, Y, Z)$.

For example, if $t = 2^m - 1 \geq 15$, then $l = 2^{m-1} - 1$ and B_l is a Hamming code which has many codewords of weight 4, so the curve defined by $g_t(X, Y, Z)$ is a singular curve.

Let m_P denote the multiplicity of a projective point $P = (\alpha, \beta, 1)$ on the curve defined by f. Let us write

$$\begin{aligned} f(X+\alpha, Y+\beta) &= 1 + \sum_{j=0}^{t} \binom{t}{j} \left[\alpha^{t-j} X^j + \beta^{t-j} Y^j + (1+\alpha+\beta)^{t-j}(X+Y)^j\right] \\ &= h_0 + h_1 + h_2 + \cdots \end{aligned} \tag{1}$$

where each $h_i(X, Y)$ is a homogeneous form of degree i (or 0). By definition, m_P is the integer m so that $h_i = 0$ for $i < m$ but $h_m \neq 0$.

Theorem 3. *Suppose $t \equiv 3 \pmod 4$, say $t = 2l + 1$ with l an odd integer. If the maximal cyclic code B_l of length l has no codewords of weight 4, in particular if the minimum distance of B_l is at least 5, then the curve defined by $g_t(X, Y, Z)$ is a nonsingular curve. Hence $g_t(X, Y, Z)$ is absolutely irreducible.*

Proof: As we have seen, the fact that B_l has no codewords of weight 4 means that $g_t(X, Y, Z)$ has no singular points $(\alpha, \beta, 1)$ with $\alpha, \beta, 1$ distinct. We now show that $t \equiv 3 \pmod 4$ implies that there are no singular points $P = (\alpha, \alpha, 1)$. Recall that $\alpha^{t-1} = 1$ when such a point P is singular.

Note that $\binom{t}{2}$ is odd, so the value of h_2 in (1) is $(1 + \alpha^{t-2})(X^2 + Y^2)$ and this is not the zero polynomial if $\alpha \neq 1$. That is, the multiplicity of P on f is 2 when $\alpha \neq 1$. The multiplicity of such a point P on $(X+Y)(Y+Z)(Z+X)$ is 1, so the multiplicity of P on $g_t(X, Y, Z)$ is 1 and P is not a singular point.

Finally, consider $P = (1,1,1)$. The congruence $t \equiv 3 \pmod 4$ also implies that $\binom{t}{3}$ is odd, so the value of h_2 in (1) is $XY^2 + YX^2 \neq 0$. Thus P has multiplicity 3 on f_t, as it does on $(X+Y)(Y+Z)(Z+X)$, and so $(1,1,1)$ is not on the curve $g_t(X,Y,Z)$ at all. □

Corollary 1. *The curve $g_t(X,Y,Z)$ is nonsingular for those values of $t = 2l+1$ where l is an odd integer such that $2^r \equiv -1 \pmod l$ for some r.*

Proof: By Theorem 3 we need only show that B_l has no words of weight 4.

As above, let ψ be a primitive l-th root of unity in some extension of $GF(2)$. Any words of weight 4 belong to the even weight subcode of B_l, which is generated by the product of $x+1$ and the minimal polynomial of ψ over $GF(2)$. Since -1 is a power of 2 modulo l, we find

$$\psi^{-2}, \psi^{-1}, \psi^{0}, \psi^{1}, \psi^{2}$$

among the roots of this binary generating polynomial. By the BCH bound, the even weight subcode has minimum distance ≥ 6. □

We remark that the condition of Corollary 1 holds whenever l is a prime $\equiv \pm 3 \pmod 8$, since 2 is then a quadratic nonresidue and Euler's criterion gives $2^{(l-1)/2} \equiv -1 \pmod l$. The values of $l < 100$ that satisfy the condition of the Corollary are: 3, 5, 9, 11, 13, 17, 19, 25, 27, 29, 33, 37, 41, 43, 53, 57, 59, 61, 65, 67, 81, 83, 97, 99.

Corollary 2. *The curve $g_t(X,Y,Z)$ is nonsingular for those values of $t = 2l+1$ where l is a prime ≥ 17 such that the order of 2 modulo l is $(l-1)/2$.*

Proof: By Theorem 3 we need only show that B_l has no words of weight 4.

In this case B_l is a quadratic residue code of length l and has minimum distance at least $\lceil\sqrt{l}\rceil \geq 5$ from the square root bound. □

The values of $l < 100$ satisfying the condition of Corollary 2 are: 17, 23, 41, 47, 71, 79, 97.

For the first several odd values 7, 15, 21, 31 of l that are not covered by either corollary above, the code B_l does have words of weight 4. B_{21} has 84 codewords of weight 4; for the other three values of l, B_l is a Hamming code.

5 The case $t = 2^i + 1$

Let $t = 2^i + 1$. Then

$$\begin{aligned} f_t(X+1, Y+1) &= (X+1)^t + (Y+1)^t + 1 + (X+Y+1)^t = YX^{2^i} + XY^{2^i} \\ &= Y \prod_{\alpha \in GF(2^i)} (X + \alpha Y). \end{aligned}$$

If we replace X by $X+1$ and Y by $Y+1$ above, and then divide by $(X+1)(Y+1)(X+Y)$, we have the following factorization:

Theorem 4. *When $t = 2^i + 1$,*

$$g_t(X,Y) = \prod_{\alpha \in GF(2^i)\setminus\{0,1\}} (X + \alpha Y + 1 + \alpha).$$

Corollary. *Let $t = 2^i + 1$. If $(i,s) = 1$, then the cyclic code $C_s^{(t)}$ of length $2^s - 1$ with generator $m_1(X)m_t(X)$ has minimum distance 5.*

Proof: Suppose that $g_t(a,b) = 0$ for $a, b \in GF(2^s)$ with $a \neq 1$, $b \neq 1$, and $a \neq b$. From the factorization of $g_t(X,Y)$ we would have $a + \alpha b + (1+\alpha) = 0$ for some $\alpha \in GF(2^i)\setminus\{0,1\}$. But then we have $\alpha = (a+1)/(b+1) \in GF(2^s) \cap GF(2^i) = \{0,1\}$, which is a contradiction. □

For completeness, we note that when $t = 2^i + 1$ and $(i,s) > 1$, the code $C_s^{(t)}$ has words of weight 3 and so is certainly not 2-error-correcting. To see this, we need a non-trivial solution for $X^t + Y^t + (X+Y)^t = 0$ in $GF(2^s)$. But (x,y) is a solution for every $x, y \in GF(2^s) \cap GF(2^i)$.

6 The case $t = 2^{2i} - 2^i + 1$

Let $t = 2^{2i} - 2^i + 1$. We will describe the factorization of $g_t(X,Y)$ into absolutely irreducible components in Theorem 5 below.

We will use the following proposition (it is, for example, Problem 5-25 in [2]).

Proposition A. *Let F be a projective plane curve of degree n over an algebraically closed field with no multiple components and N simple components. Then we have*

$$\sum \frac{m_P(m_P - 1)}{2} \leq \frac{(n-1)(n-2)}{2} + N - 1,$$

where the sum is taken over all the multiple points P of F.

We also require the following proposition, a form of Hensel's lemma. We say that a polynomial $p(X,Y)$ is *regular* in X when the degree of $p(X,0)$ as a polynomial in X is equal to the total degree of $p(X,Y)$. We remark that all factors $q(X,Y)$ of such a polynomial $p(X,Y)$ must also be regular in X. The polynomials $g_t(X,Y)$ are regular in X for all t.

Proposition B. *Let $C(X,Y)$ be a polynomial over a field $\mathbf{F}$ that is regular in X. If*

$$C(X,0) = a_0(X)b_0(X)$$

where $a_0(X)$ and $b_0(X)$ are relatively prime polynomials in $\mathbf{F}[X]$, then there is at most one pair $A(X,Y), B(X,Y)$ of polynomials over any extension of $\mathbf{F}$ with the properties that

$$C(X,Y) = A(X,Y)B(X,Y), \quad A(X,0) = a_0(X), \quad \text{and} \quad B(X,0) = b_0(X).$$

If such polynomials $A(X,Y)$ and $B(X,Y)$ exist, then all their coefficients lie in $\mathbf{F}$.

Proof: Let $C(X,Y)$ have total degree d and write $C(X,Y) = \sum_{i=0}^{d} c_i(X)Y^i$. Of course $\deg(c_i) \leq d-i$ for $i = 0,1,\ldots,d$. Assume that $a_0(X)$ and $b_0(X)$ are relatively prime of degrees s and t, respectively. Suppose that $A(X,Y)$ and $B(X,Y)$ have the above properties and write

$$A(X,Y) = \sum_{i=0}^{s} a_i(X)Y^i, \qquad B(X,Y) = \sum_{i=0}^{t} b_i(X)Y^i.$$

The total degrees of $A(x,y)$ and $B(x,y)$ are also s and t, respectively.

Since $a_0(X)$ and $b_0(X)$ are relatively prime, the equation $a_0(X)g(X) + h(X)b_0(X) = f(X)$ will have a unique solution pair $g(X), h(X)$ with $\deg g(X) < t$ and $\deg h(X) < s$ whenever $f(X)$ is given of degree $< d$. And $g(X), h(X) \in \mathbf{F}[X]$ whenever $g(X), h(X) \in \mathbf{F}[X]$. But we have, of course,

$$a_0(X)b_k(X) + a_k(X)b_0(X) = c_k(X) - \sum_{i=1}^{k-1} a_i(X)b_{k-i}(X),$$

and so we see, inductively, that if $a_i(X)$ and $b_i(X)$ are known for $i = 0,1,\ldots,k-1$, then $a_k(X)$ and $b_k(X)$ are uniquely determined as elements of $\mathbf{F}[X]$ by this equation and the constraints on their degrees. □

We remark that it is not difficult to implement the recursive calculation of $a_k(X)$ and $b_k(X)$ described above, for example, in Mathematica. Consider $g_{13}(X,Y)$, which has 42 terms when expanded. We have $g_{13}(X,0) = (X^2+X+1)^5$. We tried $a_0(X) = (X+a)^5$ and $b_0(X) = (X+a+1)^5$ where a satisfies $a^2+a+1=0$ in $GF(4)$. We found two polynomials

$$\begin{aligned} A(X,Y) &= (1+a+aX+aX^4+X^5) \\ &\quad + (1+X+aX+aX^2+X^3+X^4+aX^4)Y \\ &\quad + (X+X^2+aX^2)Y^2 + aXY^3 + aY^5 \end{aligned}$$

and (the conjugate of the first)

$$\begin{aligned} B(X,Y) &= (a+X+aX+X^4+aX^4+X^5) \\ &\quad + (1+aX+X^2+aX^2+X^3+aX^4)Y \\ &\quad + (X+aX^2)Y^2 + (X+aX)Y^3 + (1+a)Y^5 \end{aligned}$$

whose product may be checked to be $g_{13}(X,Y)$.

Theorem 5. *When* $t = 2^{2i} - 2^i + 1$,

$$g_t(X,Y) = \prod_{\alpha \in GF(2^i)\setminus\{0,1\}} p_\alpha(X,Y)$$

where for each $\alpha \in GF(2^i)\setminus\{0,1\}$, $p_\alpha(X,Y)$ is an absolutely irreducible polynomial of degree 2^i+1 over $GF(2^i)$ such that

$$p_\alpha(X,0) = (X-\alpha)^{2^i+1}.$$

Proof: We first check that $f_t(X,Y)$ does not have multiple components. Suppose $f_t(X,Y)$ had a nonconstant multiple component in an extension of $GF(2)$. Then the partial derivatives $f_t(X,Y)_X$ and $f_t(X,Y)_Y$ would have a nonconstant common factor. Since $t = 2^{2i} - 2^i + 1$, we have

$$\begin{aligned} f_t(X,Y)_X =& tX^{t-1} + t(X+Y+1)^{t-1} = \left(X^{2^i-1} + (X+Y+1)^{2^i-1}\right)^{2^i} \\ =& \Bigg(\prod_{\alpha\in GF(2^i)\setminus\{0\}} (Y+(1+\alpha)X+1)\Bigg)^{2^i} \end{aligned}$$

and $f_t(X,Y)_Y$ has an analogous factorization. But no factor $Y+\beta X+1$, $\beta \neq 1$, of $f_t(X,Y)_X$ is equal to a scalar multiple of any factor $\gamma Y + X + 1$, $\gamma \neq 1$, of $f_t(X,Y)_Y$.

We now consider the singularities of the projective curve associated to $f_t(X,Y,Z)$ and their multiplicities. Here $t-1 = 2^i l$ where $l = 2^i - 1$ is odd. From our discussion in Section 4, the singularities of $f_t(X,Y,Z)$ are the projective points $P = (\alpha,\beta,1)$ where α, β, and $\alpha+\beta+1$ are l-th roots of unity, i.e. nonzero elements of $GF(2^i)$. There are $2^{2i} - 3\cdot 2^i + 3$ such singular points.

With the homogeneous polynomials h_i as defined in Equation (1), we have $h_0 = 1 + \alpha^t + \beta^t + (1+\alpha+\beta)^t = 0$, and $h_1 = \alpha^{t-1}X + \beta^{t-1}Y + (1+\alpha+\beta)^{t-1}(X+Y) = 0$. Now note that for $j \geq 2$,

$$\binom{t}{j} = \frac{(2^{2i}-2^i+1)(2^{2i}-2^i)}{j(j-1)}\binom{t-2}{j-2}.$$

So $\binom{t}{j}$ is clearly even for $2 \leq j < 2^i$, implying that $h_j = 0$ for these values of j. When $j = 2^i$ or 2^i+1, $\binom{t}{j}$ is odd. For $j = 2^i$, we have $\alpha^{t-j} = \beta^{t-j} = (1+\alpha+\beta)^{t-j} = 1$. Therefore, $h_j(X,Y) = X^j + Y^j + (X+Y)^j = 0$. However, for $j = 2^i+1$, we have $x^{t-j} = x$ for $x = \alpha$, β, and $1+\alpha+\beta$, and therefore, $h_j(X,Y) = \alpha X^j + \beta Y^j + (1+\alpha+\beta)(X+Y)^j \neq 0$. Therefore, $m_P = 2^i+1$ for all singular points P of $f_t(X,Y)$.

Let N denote the number of components of $f_t(X,Y)$. We have from Proposition A

$$(2^{2i} - 3\cdot 2^i + 3)(2^i+1)(2^i) \leq (2^{2i}-2^i)(2^{2i}-2^i-1) + 2(N-1)$$

which implies $N \geq 2^i + 1$. In summary, we see that $g_t(X,Y)$ has at least $2^i - 2$ absolutely irreducible components.

Write $g_t(X,Y) = \sum_{i=0}^{t-3} c_i(X)Y^i$. We have

$$c_0(X) = g_t(X,0) = \frac{X^t + (X+1)^t}{X(X+1)}$$
$$= \left(\frac{X^{2^i} - X}{X(X-1)}\right)^{2^i+1} = \prod_{\alpha \in GF(2^i)\setminus\{0,1\}} (X-\alpha)^{2^i+1}. \tag{2}$$

From the definition of $g_t(X,Y)$ and elementary manipulations,

$$(X^2+1)\,c_0(X) + (X^2+X)\,c_1(X) = (X+1)^{t-1}.$$

It is clear in view of this and the factorization in (2) that $c_0(X)$ and $c_1(X)$ are relatively prime polynomials in X. This implies that if $g_t(X,Y)$ factors as $A(X,Y)B(X,Y)$ in any extension of $GF(2)$, we have, in addition to $g_t(X,0) = A(X,0)B(X,0)$, that $A(X,0)$ and $B(X,0)$ are relatively prime polynomials in X. In view of (2), $A(X,0)$ and $B(X,0)$ are, up to a scalars, the (2^i+1)-st powers of complementary factors of $(X^{2^i} - X)/(X(X-1))$. If $A(X,Y)$ is not a scalar, then $A(X,0)$ is also nontrivial by the regularity in X.

The results of the previous two paragraphs imply that $g_t(X,Y)$ has exactly $2^i - 2$ components, exactly one of which, call it $p_\alpha(X,Y)$, has the property that $p_\alpha(X,0) = (X-\alpha)^{2^i+1}$ for each $\alpha \in GF(2^i) \setminus \{0,1\}$.

Finally, Proposition B implies that each $p_\alpha(X,Y)$ has coefficients in $GF(2^i)$.
□

Corollary. *Let* $t = 2^{2i} - 2^i + 1$. *If* $(i,s) = 1$, *then the cyclic code* $C_s^{(t)}$ *of length* $2^s - 1$ *with generator* $m_1(X)m_t(X)$ *has minimum distance 5.*

Proof: Suppose that $g_t(x,y) = 0$ for some $x,y \in GF(2^s)$. Then, with the notation of Theorem 5, $p_\alpha(x,y) = 0$ for some $\alpha \in GF(2^i)\setminus\{0,1\}$. Since $(i,s) = 1$, there is an automorphism of $GF(2)$ fixing $GF(2^s)$ pointwise but taking α to α^2, namely $\sigma : z \mapsto z^{2^m}$ where $m \equiv 0 \pmod{s}$ and $m \equiv 1 \pmod{i}$. Applying this automorphism to $p_\alpha(X,Y)$ produces a different absolutely irreducible factor of $g_t(X,Y)$, namely $p_{\alpha^2}(X,Y)$. Since σ fixes x and y, we have $p_{\alpha^2}(x,y) = 0$ also. This means that (x,y) is a singular point of $g_t(X,Y)$. Therefore x,y are nonzero elements of $GF(2^i)$ as well as elements of $GF(2^s)$. We conclude that $x = y = 1$.
□

For completeness, we note that when $t = 2^{2i} - 2^i + 1$ and $(i,s) > 1$, the code $C_s^{(t)}$ has words of weight 3 and so is certainly not 2-error-correcting. To see this, we need a non-trivial solution for $X^t + Y^t + (X+Y)^t = 0$ in $GF(2^s)$. But (x,y) is a solution for every $x,y \in GF(2^s) \cap GF(2^i)$.

7 Small values of t

Consider the odd values of $t \leq 65$. We have factored $g_t(X,Y)$ for $t = 3,5,9,13,17,33,57,65$. By the Corollaries to Theorem 3 and the remarks in that section, we know for $t = 7,11,19,23,27,35,39,47,51,55,59$ that $g_t(X,Y)$ is nonsingular and hence absolutely irreducible. We show here that $g_t(X,Y)$ is absolutely irreducible for $t = 15,21,25$. Using slightly different but related methods and with more extensive computer aid, G. McGuire [10] has shown that $g_t(X,Y)$ is absolutely irreducible for the remaining cases $t = 29,31,37,41,43,45,49,53,61,63$.

We begin with several remarks.

Since $g_t(X,Y)$ is regular in both X and Y, any factor of $g_t(X,Y)$ has the same degree in X as it does in Y, equal to the total degree of the factor.

Note that an irreducible binary polynomial $p(X)$ of degree d remains irreducible over $GF(2^s)$ when $(s,d) = 1$. If $a(X)$ of degree $e > 0$ is irreducible over $GF(2^s)$ and divides $p(X)$, then adjoining a root α of $a(X)$ to $GF(2^s)$ gives us an extension $GF(2^{se})$ containing the root α of $p(X)$, whence $GF(2^d) \subseteq GF(2^{se})$. Then d divides se; so $e = d$.

If $A(X,Y)$ is irreducible of degree n over $GF(2)$, then for some r dividing n, $A(X,Y)$ factors as the product of r absolutely irreducible components of degree n/r over $GF(2^r)$. To see this, let $P(X,Y)$ be an absolutely irreducible factor of $A(X,Y)$ of degree d. We can assume that $P(X,Y)$ has at least one coefficient equal to 1. Let r be the number of distinct conjugates of $P(X,Y)$, including $P(X,Y)$ itself, under powers of the Frobenius automorphism. In particular, the coefficients of $P(X,Y)$ lie in $GF(2^r)$. Since some coefficient is 1, no conjugate is a scalar multiple of another unless they are equal. The product of these distinct conjugates is a binary polynomial dividing $A(X,Y)$; hence it is $A(X,Y)$. That is, $rd = n$.

Theorem 6. *The polynomial $g_t(X,Y)$ is absolutely irreducible for* $t = 15,21,25$.

Proof: We experimented with factoring $g_t(X,q(X))$ over $GF(2)$ for various small binary polynomials $q(X)$. We found

(i) $g_{15}(X, X^7 + X^6 + X^5) = (3)(3)^2(4)(71)$,

(ii) $g_{21}(X, X^6 + X^5 + X^3 + 1) = (6)(25)(77)$, and

(iii) $g_{25}(X, X^6 + X^4) = (2)(3)(4)(7)(8)(11)(97)$.

The notation means, for example, that the polynomial in case (i) is the product of a binary irreducible polynomial of degree 3, the square of another binary irreducible polynomial of degree 3, and binary irreducible polynomials of degrees 4 and 71. We tried many other substitutions, but these were among the most useful. The reader can use a symbolic manipulation program like Mathematica or Maple, as we did, to factor polynomials in a single variable modulo 2 and check our claims.

Assume for the moment that $g_t(X,Y)$ is irreducible over $GF(2)$ for these values $t = 15,21,25$. If not absolutely irreducible, $g_t(X,Y)$ is the product of polynomials of degrees $\leq (t-3)/2$ over $GF(2^{t-3})$ or a subfield. Then

$g_{15}(X, X^7 + X^6 + X^5)$ is the product of polynomials in one variable of degrees ≤ 42, $g_{21}(X, X^6 + X^5 + X^3 + 1)$ is the product of polynomials of degrees ≤ 54, and $g_{25}(X, X^6 + X^4)$ is the product of polynomials of degrees ≤ 66 over $GF(2^{12})$, $GF(2^{18})$, and $GF(2^{22})$, respectively. But the factors of respective degrees 71, 77, and 91 remain irreducible over $GF(2^{12})$, $GF(2^{18})$, and $GF(2^{22})$, respectively, and we have contradictions.

It remains to show that $g_t(X, Y)$ is irreducible over $GF(2)$ for $t = 15, 21, 25$.

Now $g_{15}(X, 0) = (1 + X + X^3)^2(1 + X^2 + X^3)^2$ so any proper binary factor of $g_{15}(X, Y)$ has degree 3, 6, or 9. If we make the substitution $Y \leftarrow X^7 + X^6 + X^5$, then a proper binary factorization of $g_{15}(X, Y)$ will give a factorization of $g_{15}(X, X^7 + X^6 + X^5)$ into binary factors of degree $\leq 9 \times 7 = 63$ in X, a contradiction.

We checked that $g_{21}(X, 0)$ is the product of three distinct irreducible binary polynomials of degree 6. Thus if not irreducible over $GF(2)$, $g_{21}(X, Y)$ has a binary factor of degree 6 and then $g_{21}(X, X^6 + X^5 + X^3 + 1)$, would have a binary factor of degree 36. But it doesn't.

Finally, consider $g_{25}(X, Y)$. If $g_{25}(X, Y)$ is not irreducible over $GF(2)$, then it has a proper binary factor $A(X, Y)$ of degree d with $1 \leq d \leq 11$. Then $A(X, X^6 + X^4)$ would be a binary factor of degree $6d$ of $g_{25}(X, X^6 + X^4) = (2)(3)(4)(7)(8)(11)(97)$, so $d \leq 5$. Also, $A(X, X^7 + X^3)$ would be a binary factor of degree $7d$ of $g_{25}(X, X^7 + X^3) = (2)(3)(17)(31)(44)(57)$, but there are no such factors for $1 \leq d \leq 5$. □

Acknowledgment

The authors wish to thank Gary McGuire for discussions leading to improvements in the exposition.

References

1. R. Baker, J. H. van Lint and R. M. Wilson, "On the Preparata and Goethals codes," *IEEE Trans. Inform. Theory*, vol. IT-29 (1983), pp. 342–345.
2. W. Fulton, *Algebraic Curves*, Benjamin, New York, 1969.
3. J. W. P. Hirschfeld, *Finite Projective Spaces of Three Dimensions*, Clarendon Press, Oxford, 1985.
4. R. Hartshorne, *Algebraic Geometry*, Springer Graduate Texts in Math., vol. 52, Berlin, Germany: Springer-Verlag, 1976.
5. H. Janwa and R. M. Wilson, "Rational points on the Klein quartic and the binary cyclic codes $\langle m_1 m_7 \rangle$", manuscript.
6. E. Kunz, *Introduction to Commutative Algebra and Algebraic Geometry*, Birkhäuser, Boston, 1985.
7. A. K. Lenstra, "Factoring multivariate polynomials over finite fields," *Journal of Computer and System Sciences*, vol. 30–31, 235–284, 1985.
8. J. H. van Lint and R. M. Wilson, "On the minimum distance of cyclic codes," *IEEE Trans. Inform. Theory*, vol. IT-32, no. 1, pp. 23–40, Jan. 1986.
9. J. H. van Lint and R. M. Wilson, "Binary cyclic codes generated by $m_1 m_7$," *IEEE Trans. Inform. Theory*, vol. IT-32 (1986), p. 283.
10. G. McGuire, personal communication.

11. W. M. Schmidt, "A lower bound for the number of solutions of equations over finite fields," *Journal of Number Theory*, vol. 6 (1974), pp. 448–480.
12. W. M. Schmidt, *Equations over Finite Fields*, Springer Lecture in Math., vol. 536, p. 210. Berlin, Germany: Springer-Verlag, 1976.
13. J.-P. Serre, "Rational points on curves over finite fields; q large: Parts I and II," *Lectures given at Harvard University*, September–December 1985. Notes by Fernando Gouvêa.
14. K. S. Williams, "Eisenstein's criteria for absolute irreducibility over a finite field," *Canad. Math. Bull.*, vol. 9, 1966, 575–580.

Analysis of Coppersmith's Block Wiedemann Algorithm for the Parallel Solution of Sparse Linear Systems*

Erich Kaltofen

Department of Computer Science, Rensselaer Polytechnic Institute
Troy, NY 12189-3590, USA
Inter-net: kaltofen@cs.rpi.edu

1 Introduction

Douglas Wiedemann's (1986) algorithm for computing the N-dimensional solution vector of a system of N linear equations over a finite field $\mathbf{K}$ is efficient for sparse unstructured inputs because its running time is bounded by $3N$ multiplications of the coefficient matrix B by vectors and $O(N^2 \log N)$ additional arithmetic operations in the coefficient field. It only needs $O(N)$ additional storage for field elements. Wiedemann's algorithm can be generalized to arbitrary fields, p-adic lifting, and to the problem of computing the rank of a sparse matrix (Kaltofen and Saunders 1991). The method is randomized and computes first the sequence of field elements

$$a^{(i)} = u^{\mathrm{tr}} B^i v \in \mathbf{K} \quad \text{for } 0 \leq i \leq 2N-1,$$

where u and possibly v are vectors with random entries from $\mathbf{K}$. The key property is that this sequence is generated by a linear recursion that with high probability corresponds to the minimum polynomial of B.

There are several implementations of Wiedemann's algorithm. We have observed that for large systems such as the ones arising in the sieve-based integer factoring algorithms (A. K. Lenstra et al. 1990), where N can be as large as 100,000 and B can have as many as 5 million non-zero entries, the running time is dominated by the $3N$ multiplications of B by vectors. In order to speed this bottleneck by use of parallelism, Don Coppersmith (1991) has proposed to use simultaneously m random vectors for u and n random vectors for v. The sequence now is a sequence of $m \times n$ matrices

$$a^{(i)} = x^{\mathrm{tr}} B^i y \in \mathbf{K}^{m \times n} \quad \text{where } x^{\mathrm{tr}} \in \mathbf{K}^{m \times N} \text{ and } y \in \mathbf{K}^{N \times n}.$$

Clearly, the individual entries in $a^{(i)}$ can be computed independently and in parallel. Coppersmith then has cleverly generalized the Berlekamp/Massey algorithm needed to compute a linear recurrence that generates this sequence and

*This material is based on work supported in part by the National Science Foundation under Grant No. CCR-90-06077 and under Grant No. CDA-88-05910.

observed experimentally that over the Galois field with 2 elements $\mathsf{K} = \mathsf{F}_2$ that linear recurrence is determined by the first

$$\frac{N}{m} + \frac{N}{n} + O(1)$$

matrices $a^{(i)}$. Thus the algorithm, when executed in a parallel/distributed setting performs much faster.

Unfortunately, the blocking of the projections introduces substantial difficulties in the analysis of the method. Aside from the experimental evidence, Coppersmith gives a heuristic mathematical argument on the expected running time of the algorithm. It is these difficulties that this paper further clarifies. We can prove that a certain variant of the block algorithm with high probability runs as conjectured provided that the finite field K is of sufficient cardinality.

Our results also impact the sequential complexity of sparse linear system solving. Suppose, e.g., that B is non-singular, and that $\epsilon > 0$. Using blocking, one can find a solution vector $x = B^{-1}b$, where $b \in \mathsf{K}^N$, by

$$(1+\epsilon)N + O(1) \text{ multiplications of } B \text{ times vectors,}$$

and $O(N^2 \log N \operatorname{loglog} N)$ arithmetic operations in K, needing $O(N)$ additional storage for field elements. The algorithm chooses $O(N)$ random field elements and is successful with probability $1 - (N+2)^2/\operatorname{card}(\mathsf{K})$. Here the constants implied by the big-O notation grow with $1/\epsilon$.

Our analysis is based on the observation that generically, i.e., when the projection blocks x and y are symbolic, the block method can be specialized to the original Wiedemann algorithm. From that specialization one then can prove that certain necessary rank conditions must hold generically. By the commonly used Zippel/Schwarz lemma those rank conditions will thus hold with high probability for random blocks. A further problem is the generalization of the Berlekamp/Massey algorithm to sequences of matrices. Instead, we use the equivalent problem of solving a block Toeplitz homogeneous linear system, which we can accomplish efficiently by the standard Wiedemann method.

Notation: We write K^N for the set of column vectors over K, and 0^N for the N-dimensional zero vector; $0^{N\times M}$ is the $N \times M$ zero matrix. Vector and matrix transposition is indicated by superscript ^{tr}. We indicate a block matrix whose entries are matrices or vectors by vertical and horizontal strokes, such as $\left[\begin{array}{c|c} a & a'' \\ \hline a' & a''' \end{array}\right]$. Note that lower case symbols, such as x and y, may also denote matrices.

2 Linearly generated sequences

We now discuss some basic facts about linearly generated sequences of elements in a vector space V over the field K. The sequence

$$\{a_i\}_{i=0}^{\infty}, \quad \text{where } a_i \in \mathsf{V},$$

is *linearly generated* over $\mathbf{K}$ if there exist $c_0, c_1, \ldots, c_N \in \mathbf{K}$, $N \geq 0$, $c_L \neq 0$ for some L with $0 \leq L \leq N$, such that

$$\forall j \geq 0: c_0 a_j + \cdots + c_N a_{j+N} = 0.$$

The polynomial $c_0 + c_1\lambda + \cdots + c_N\lambda^N$ is called a *generating polynomial* for $\{a_i\}_{i=0}^{\infty}$. The set of all generating polynomials for $\{a_i\}_{i=0}^{\infty}$ together with the zero polynomial forms an ideal in $\mathbf{K}[\lambda]$. The unique polynomial generating that ideal, normalized to have leading coefficient 1, is called the *minimum polynomial* of a linearly generated sequence $\{a_i\}_{i=0}^{\infty}$. Every generating polynomial is a multiple of the minimum polynomial.

Let W be also a vector space over $\mathbf{K}$, and let $L: \mathsf{V} \longrightarrow \mathsf{W}$ be a linear map from V to W. Then the sequence $\{L(a_i)\}_{i=0}^{\infty}$ is also linearly generated by a minimum polynomial that divides the minimum generating polynomial of $\{a_i\}_{i=0}^{\infty}$. Let $B \in \mathbf{K}^{N \times N}$ be a square matrix over a field. The sequence of $N \times N$ matrices $\{B^i\}_{i=0}^{\infty}$ is linearly generated, and its minimum polynomial is the minimum polynomial of B, which will be denoted by f^B. For any column vector $b \in \mathbf{K}^N$ the sequence $\{B^i b\}_{i=0}^{\infty}$, where $B^i b \in \mathbf{K}^N$, is also linearly generated by f^B. However, its minimum polynomial, denoted by $f^{B,b}$, can be a proper divisor of f^B. For any row vector $u^{\mathrm{tr}} \in \mathbf{K}^{1 \times N}$ the sequence $\{u^{\mathrm{tr}} B^i b\}_{i=0}^{\infty}$, where $u^{\mathrm{tr}} B^i b \in \mathbf{K}$, is linearly generated as well, and its minimum polynomial, denoted by $f_u^{B,b}$, is again a divisor of $f^{B,b}$. Wiedemann's method is based on the fact that for random vectors u and b with high probability $f_u^{B,b} = f^B$ (c.f. Proposition 1 in §4).

The minimum generator for a sequence $\{a_i\}_{i=0}^{\infty}$ of field elements $a_i \in \mathbf{K}$ can be computed by the Berlekamp/Massey algorithm (Massey 1969). This algorithm will determine the minimum polynomial $f^{(\min)}$ of such a sequence from the first $2M$ elements, where $M = \deg(f^{(\min)})$. If more elements are given, the computed minimum polynomial cannot change. Therefore, we have the following lemma.

Lemma 1. *Suppose* $\{a_i\}_{i=0}^{\infty}$, *where* $a_i \in \mathbf{K}$, *is linearly generated by the minimum polynomial* $f^{(\min)}$. *Let* $M = \deg(f^{(\min)})$ *and let* $M' \geq M$. *Suppose a polynomial* g *with* $\deg(g) \leq M'$ *linearly generates a sequence*

$$\{a_0, a_1, \ldots, a_{2M'-1}, a'_{2M'}, a'_{2M'+1}, \ldots\}$$

whose first $2M'$ *elements agree with* $\{a_i\}_{i=0}^{\infty}$. *Then* $a'_i = a_i$ *for all* $i \geq 2M'$ *and* g *is a polynomial multiple of* $f^{(\min)}$.

3 Coppersmith's block Wiedemann algorithm

In order to prepare for later discussion, we first give a particular variant of Wiedemann's (1986, §III, first paragraph) coordinate recurrence method for solving a homogeneous linear system. This variant already accounts for some changes necessitated by the later block version. Let $B \in \mathbf{K}^{N \times N}$ be a singular matrix, where $\mathbf{K}$ is a finite field; we seek a non-zero vector $w \in \mathbf{K}^N$ such that $Bw = 0$.

Step W1: Pick random vectors $u^{\text{tr}} \in \mathbf{K}^N$ and $v \in \mathbf{K}^N$. For any integers $M' \geq M \geq N$, compute

$$b = Bv, \quad a^{(i)} = u^{\text{tr}} B^i b, \quad 0 \leq i \leq M + M' - 1.$$

(The letters u and b now agree with the ones in Wiedemann's paper.) This requires at least $2M + 1$ multiplications of B by vectors.

Step W2: Compute a non-zero solution to the linear homogeneous $M' \times (M+1)$ Toeplitz system

$$\begin{bmatrix} a^{(M)} & \cdots & a^{(1)} & a^{(0)} \\ a^{(M+1)} & a^{(M)} & a^{(2)} & a^{(1)} \\ \vdots & \ddots & & \vdots \\ a^{(M+M'-1)} & \cdots & & a^{(M'-1)} \end{bmatrix} \begin{bmatrix} c^{(M)} \\ c^{(M-1)} \\ \vdots \\ c^{(0)} \end{bmatrix} = 0^{M'},$$

Define the generating polynomial

$$f(\lambda) = c^{(L)}\lambda^L + c^{(L+1)}\lambda^{L+1} + \cdots + c^{(M)}\lambda^{(M)}, \quad c^{(\ell)} = 0 \text{ for } 0 \leq \ell < L,\ c^{(L)} \neq 0.$$

Such a polynomial can be determined, e.g., by the Berlekamp/Massey algorithm, which then requires, for $M' = M = N$, $O(N^2)$ arithmetic operations in $\mathbf{K}$. Here we introduce unnecessary generality for the later analysis of the block Wiedemann method. Note that

$$u^{\text{tr}} B^j f(B) b = 0 \quad \text{for all } 0 \leq j \leq M - 1,$$

which implies that $f(\lambda)$ is a polynomial multiple of $f_u^{B,b}(\lambda)$. With probability no less than $1 - N/\text{card}(\mathbf{K})$, $f(\lambda)$ is a polynomial multiple of the polynomial $f^{B,b}(\lambda)$, i.e.,

$$c^{(L)} B^L b + c^{(L+1)} B^{L+1} b + \cdots + c^{(M)} B^M b = 0. \tag{1}$$

Step W3: Compute

$$\widehat{w} = c^{(L)} v + c^{(L+1)} Bv + \cdots + c^{(M)} B^{M-L} v.$$

This requires at most $M - L$ additional multiplications of B times a vector. One may argue as follows that $\widehat{w} \neq 0^N$ with probability at least $1 - 1/\text{card}(\mathbf{K})$ (Coppersmith 1992): for $v' = v + w_0$, where $w_0 \in \text{kernel}(B)$, the vector $b = Bv'$ and hence the sequence $a^{(i)}$ does not change. However,

$$\begin{aligned} \widehat{w}' &= c^{(L)} v' + c^{(L+1)} Bv' + \cdots + c^{(M)} B^{M-L} v' \\ &= \widehat{w} + c^{(L)} w_0. \end{aligned}$$

Therefore in the set of vectors $v + \text{kernel}(B)$, at most one vector can produce $\widehat{w}' = 0$. Note that the solution $c^{(0)}, \ldots, c^{(M)}$ is computed without any information on w_0.

Suppose now that $\widehat{w} \neq 0^N$. Finally, determine the first integer i such that $B^i\widehat{w} = 0^N$ and return $w = B^{i-1}\widehat{w}$. By (1), this should happen, with high probability, for an integer $i \leq L+1$. At most $L+1$ more multiplications of B by a vector are required.

Let $m, n < N$. Coppersmith's (1991) block version essentially uses

$$\begin{array}{lll} x^{\mathrm{tr}} \in \mathbf{K}^{m\times N} & \text{in place of} & u^{\mathrm{tr}}, \\ z \in \mathbf{K}^{N\times n} & \text{in place of} & v, \quad \text{and} \\ y = Bz \in \mathbf{K}^{N\times n} & \text{in place of} & b = Bv. \end{array}$$

(The letters B, x, y, and z, agree with the ones in Coppersmith's paper.) Thus the sequence is one of the $m \times n$ matrices

$$a^{(i)} = x^{\mathrm{tr}} B^i y \in \mathbf{K}^{m\times n}, \quad 0 \leq i.$$

(Coppersmith further transposes these matrices.) The main point is that a non-trivial linear dependence of the type (1) can be found from roughly $N/m + N/n$ sequence elements $a^{(i)}$. A brief description of a variant of the block Wiedemann algorithm follows:

Step C1: Pick random vectors $x_1, \ldots, x_m$, $z_1, \ldots, z_n \in \mathbf{K}^N$. Let

$$x^{\mathrm{tr}} = \begin{bmatrix} x_1^{\mathrm{tr}} \\ \hline \vdots \\ \hline x_m^{\mathrm{tr}} \end{bmatrix}, \quad y = B \cdot [\, z_1 \mid \ldots \mid z_n \,].$$

Compute

$$a^{(i)} = x^{\mathrm{tr}} B^i y, \quad \text{for all } 0 \leq i < \frac{N}{m} + \frac{N}{n} + \frac{2n}{m} + 1.$$

This requires less than

$$\left(1 + \frac{n}{m}\right) N + \frac{2n^2}{m} + 2n \tag{2}$$

multiplications of B times a vector. However, for every y_ν, the νth columns of the sequence matrices $a^{(i)}$, namely $x^{\mathrm{tr}} B^i y_\nu$, can be computed simultaneously, yielding a coarse-grain parallelization. Alternately, one may for each i perform the products $B \cdot B^{i-1} y_\nu$ in parallel, as Coppersmith does, which is finer grain and requires synchronization for each i. Note that the products $x^{\mathrm{tr}} \cdot (B^i y_\nu)$ additionally require for all ν some

$$O(\,(m+n)\,N^2\,)$$

arithmetic operations in $\mathbf{K}$, if done sequentially.

Step C2: Let $D = \lceil N/n \rceil$, $S = n(D+1)$, which is bounded as $N + n \le S < N + 2n$, and let $E = \lceil S/m \rceil$, $R = mE$, which are bounded as $S \le R$ and $E < N/m + 2n/m + 1$. Compute a non-zero solution to the linear homogeneous $R \times S$ linear system (of block Toeplitz structure)

$$\begin{bmatrix} a^{(D)} & | \ \cdots \ | & & | \ a^{(1)} \ | & a^{(0)} \\ \hline a^{(D+1)} & | \ a^{(D)} \ | & & | \ a^{(2)} \ | & a^{(1)} \\ \hline \vdots & | & \ddots & | & \vdots \\ \hline a^{(D+E-1)} & | \ \cdots \ | & & | & a^{(E-1)} \end{bmatrix} \begin{bmatrix} c^{(D)} \\ \hline c^{(D-1)} \\ \hline \vdots \\ \hline c^{(0)} \end{bmatrix} = 0^R, \quad c^{(i)} = \begin{bmatrix} c_1^{(i)} \\ \vdots \\ c_n^{(i)} \end{bmatrix} \in \mathbf{K}^n. \tag{3}$$

Note that

$$D + E < \frac{N}{n} + 1 + \frac{N}{m} + \frac{2n}{m} + 1,$$

which bounds the length of the sequence $a^{(i)}$. Define the generating polynomial with (right-hand-side) vector coefficients

$$f(\lambda) = \lambda^L y c^{(L)} + \lambda^{L+1} y c^{(L+1)} + \cdots + \lambda^D y c^{(D)}, \; c^{(\ell)} = 0^n \text{ for } 0 \le \ell < L, \; c^{(L)} \ne 0^n.$$

Coppersmith in his paper computes such a non-zero vector polynomial by his generalization of the Berlekamp/Massey algorithm to polynomials with matrix coefficients. In any case, we need to have

$$x^{\mathrm{tr}} B^j f(B) = 0^m \quad \text{for all } 0 \le j \le E - 1.$$

As we will argue later, with high probability the projections by x^{tr} do not introduce any additional linear dependence, so that

$$f(B) = B^L y c^{(L)} + B^{L+1} y c^{(L+1)} + \cdots + B^D y c^{(D)} = 0^N. \tag{4}$$

Step C3: Compute

$$\widehat{w} = z c^{(L)} + B z c^{(L+1)} + \cdots + B^{D-L} z c^{(D)}.$$

This requires at most $D - L$ additional multiplications of B times a vector (using a Horner evaluation scheme). One may argue as above that $\widehat{w} \ne 0^N$ with probability at least $1 - 1/\mathrm{card}(\mathbf{K})$ (see also proof of Theorem 1 in §4). Suppose now that $\widehat{w} \ne 0^N$. Finally, determine the first integer i such that $B^i \widehat{w} = 0^N$ and return $w = B^{i-1}\widehat{w}$. By (4), this should happen, with high probability, for an integer $i \le L + 1$. At most $L + 1$ more multiplications of B by a vector are required. Altogether, this step performs

$$D + 1 < \frac{N}{n} + 2 \tag{5}$$

multiplications of B by a vector, and additionally $O(N^2)$ arithmetic operations in $\mathbf{K}$ are required to compute $z c^{(i)}$ for $L \le i \le D$ and add the $D - L + 1$ vectors in the Horner scheme.

Coppersmith's paper raises two distinct problems:

1. The efficient computation of a non-trivial solution to (3). He proposes a clever generalization of the Berlekamp/Massey algorithm to linearly generated sequences of matrices. Although one can define the notion of a minimum generator, a proof that the algorithm produces it has so far eluded us. However, we may proceed directly by computing a non-trivial solution of our system by either a method for Toeplitz-like matrices or by the Wiedemann algorithm itself and by using a fast polynomial (over $\mathbf{K}$) multiplication algorithm (see §5 and §6).

2. The probabilistic analysis, in particular the fact that with high probability the found polynomial $f(\lambda)$ satisfies (4). We will show this to be true at least in the case that the minimum polynomial f^B of the coefficient matrix B has degree $\deg(f^B) = \operatorname{rank}(B) + 1$. Fortunately, by certain randomizations this condition can be enforced for any matrix B (Kaltofen and Saunders 1991; see also the proof of Theorem 2 in §5). Let us consider, e.g., solving a non-singular system $x = A^{-1}b$. We then can randomize $\widetilde{A} = AVG$ where V is random unit triangular Toeplitz matrix and G is random diagonal matrix, and execute the block Wiedemann method on the $(N+1) \times (N+1)$ matrix

$$B = \left[\begin{array}{c|c} \widetilde{A} & b \\ \hline 0^{1\times N} & 0 \end{array}\right].$$

Note that $\widetilde{A}$ has with high probability n distinct eigenvalues.

4 Probabilistic analysis

We now justify Coppersmith's block version of the Wiedemann We will prove the following theorem.

Theorem 1. *Let $\mathbf{K}$ be a finite field, and let $B \in \mathbf{K}^{N\times N}$ be a singular matrix whose minimal polynomial f^B has degree*

$$\deg(f^B) = \operatorname{rank}(B) + 1.$$

Suppose the vector blocks $x^{\mathrm{tr}} \in \mathbf{K}^{m\times N}$ and $z \in \mathbf{K}^{N\times n}$ are chosen at random. Suppose $\widehat{w} \in \mathbf{K}^N$ is computed by steps (C1)–(C3) of §3. Then with probability no less than

$$1 - \frac{2\operatorname{rank}(B)+1}{\operatorname{card}(\mathbf{K})} \geq 1 - \frac{2N-1}{\operatorname{card}(\mathbf{K})}$$

we have

$$\widehat{w} \neq 0^N, \quad B^{L+1}\widehat{w} = 0^N \text{ for some integer } L \leq \lceil N/n \rceil.$$

The key property for the algorithm to succeed is equation (4). We will prove (4) first if the entries in x and z are indeterminates $\xi_{\iota,\mu}$ and $\zeta_{\iota,\nu}$, where $1 \le \mu \le m$, $1 \le \nu \le n$, and $1 \le \iota \le N$. In this case, the algorithm is performed over the multivariate rational function field over K,

$$\mathsf{L} = \mathsf{K}(\xi_{1,1}, \ldots, \xi_{N,m}, \zeta_{1,1}, \ldots, \zeta_{N,n}).$$

In order to distinguish when the algorithm is performed over K and when over L, we will write $\mathcal{X}$ and $\mathcal{Z}$ for the undetermined x and z and

$$\mathcal{Y} = B\mathcal{Z} \in \mathsf{L}^{N \times n}, \quad \mathcal{A}^{(i)} = \mathcal{X}^{\mathrm{tr}} B^i \mathcal{Y} \in \mathsf{L}^{m \times n}.$$

The equation (4) is equivalent to the solution vector c of (3) satisfying the following block Krylov system:

$$[\, B^{D+1}z \mid \ldots \mid B^2 z \mid Bz \,] \begin{bmatrix} c^{(D)} \\ \hline c^{(D-1)} \\ \hline \vdots \\ \hline c^{(0)} \end{bmatrix} = 0. \tag{6}$$

Clearly, any solution of (6) also solves (3). We first state that generically, i.e., over L, no other solutions to (3) exists. We will prove this fact later using Proposition 1 stated below.

Proposition 2. *Suppose that the minimum polynomial f^B of B has the degree*

$$\deg(f^B) = \min\{N, \mathrm{rank}(B) + 1\}.$$

Then for $D = \lceil N/n \rceil$ and $E = \lceil n(D+1)/m \rceil$ we have the rank equalities

$$\mathrm{rank}\left(\begin{bmatrix} \mathcal{A}^{(D)} & \ldots & & \mathcal{A}^{(1)} & \mathcal{A}^{(0)} \\ \hline \mathcal{A}^{(D+1)} & \mathcal{A}^{(D)} & & \mathcal{A}^{(2)} & \mathcal{A}^{(1)} \\ \hline \vdots & & \ddots & & \vdots \\ \hline \mathcal{A}^{(D+E-1)} & \ldots & & & \mathcal{A}^{(E-1)} \end{bmatrix} \right)$$

$$= \mathrm{rank}([\, B^D \mathcal{Y} \mid B^{D-1}\mathcal{Y} \mid \ldots \mid B\mathcal{Y} \mid \mathcal{Y} \,])$$

$$= \mathrm{rank}(B) = \begin{cases} N & \text{if } B \text{ is non-singular,} \\ \deg(f^B) - 1 & \text{if } B \text{ is singular.} \end{cases}$$

The proof of this proposition is based on its validity for $m = n = 1$, which we shall formulate as our first proposition. We will denote our generic sequence by

$$\alpha^{(i)} = \mathcal{A}^{(i)}_{1,1} = \mathcal{X}_1^{\mathrm{tr}} B^{i+1} \mathcal{Z}_1 \in \mathsf{L}, \quad \text{for } i \ge 0.$$

Proposition 1. *Let $M' \geq M \geq N$. Define*

$$\mathcal{T} = \begin{bmatrix} \alpha^{(M)} & \cdots & \alpha^{(1)} & \alpha^{(0)} \\ \alpha^{(M+1)} & \alpha^{(M)} & \alpha^{(2)} & \alpha^{(1)} \\ \vdots & \ddots & & \vdots \\ \alpha^{(M+M'-1)} & \cdots & & \alpha^{(M'-1)} \end{bmatrix} \in \mathsf{L}^{M' \times (M+1)}$$

and

$$\mathcal{K} = [\, B^M \mathcal{Z}_1 \mid \ldots \mid B^2 \mathcal{Z}_1 \mid B \mathcal{Z}_1 \,] \in \mathsf{L}^{N \times M}.$$

Then

$$\mathrm{rank}(\mathcal{T}) = \mathrm{rank}(\mathcal{K}) = \begin{cases} \deg(f^B) & \text{if } B \text{ is non-singular,} \\ \deg(f^B) - 1 & \text{if } B \text{ is singular.} \end{cases}$$

Proof. The argument can be deduced from the probabilistic analysis of Wiedemann (1986, §V and §VI). Since $f^B(\lambda)$ linearly generates the sequence

$$\{B^i \mathcal{Z}_1\}_{i=0}^{\infty}, \quad \text{where } B^i \mathcal{Z}_1 \in \mathsf{L}^N,$$

we must have the rank inequality

$$\mathrm{rank}([\, B^M \mathcal{Z}_1 \mid \ldots \mid B \mathcal{Z}_1 \mid \mathcal{Z}_1 \,]) \leq \deg(f^B)$$

for any vector $\mathcal{Z}_1$ and any integer $M \geq N$. Moreover, there exists a specialization of $\mathcal{Z}_1$ to a vector $z_1 \in \mathsf{K}^N$ such $f^B(\lambda)$ is the minimum linear generating polynomial of the sequence

$$\{B^i z_1\}_{i=0}^{\infty}, \quad \text{where } B^i z_1 \in \mathsf{K}^N,$$

hence

$$\mathrm{rank}([\, B^N z_1 \mid \ldots \mid B z_1 \mid z_1 \,]) = \deg(f^B),$$

and therefore generically the rank cannot be lower. The existence of such a vector z_1 follows, e.g., by considering the rational canonical form (Frobenius form) of B. Certainly, for B in companion form

$$B = \begin{bmatrix} 0 & 1 & 0 & \cdots & 0 \\ 0 & 0 & 1 & \cdots & 0 \\ \vdots & \vdots & & \ddots & \vdots \\ 0 & 0 & \cdots & & 1 \\ b_{N,1} & b_{N,2} & \cdots & & b_{N,N} \end{bmatrix} \text{ we may choose } z_1 = \begin{bmatrix} 0 \\ 0 \\ \vdots \\ 0 \\ 1 \end{bmatrix},$$

and this argument extends to the block diagonal shape with companion blocks of the rational canonical form. If B is non-singular, the minimum generating polynomial of

$$\{B^i \mathcal{Y}_1\}_{i=0}^{\infty} = \{B^{i+1} \mathcal{Z}_1\}_{i=0}^{\infty}$$

does not change, while for singular B the minimum generating polynomial is $f^B(\lambda)/\lambda$, thus the rank drops by 1. We define this polynomial as

$$f_-^B(\lambda) = \begin{cases} f^B(\lambda) & \text{if } B \text{ is non-singular,} \\ f^B(\lambda)/\lambda & \text{if } B \text{ is singular.} \end{cases}$$

So far, we have shown that $\operatorname{rank}(\mathcal{K}) = \deg(f_-^B)$.

Second, we need to prove that $\operatorname{rank}(\mathcal{T}) = \operatorname{rank}(\mathcal{K})$. The proof of this is very similar. It follows from Wiedemann (1986, §VI) that there exists a vector $x_1 \in \mathsf{K}^N$ such that the sequence

$$\{x_1^{\text{tr}} B^{i+1} z_1\}_{i=0}^{\infty}$$

has the same minimum generator as

$$\{B^{i+1} z_1\}_{i=0}^{\infty},$$

which is $f_-^B(\lambda)$. Furthermore, $f_-^B(\lambda)$ is already a generating polynomial of the generic sequence

$$\{\alpha^{(i)}\}_{i=0}^{\infty}, \tag{7}$$

and must therefore also be minimal for that sequence, because a specialization is.

We finally argue that the rank of $\mathcal{T}$ is the degree of the minimum generator of (7), namely that $\operatorname{rank}(\mathcal{T}) = \deg(f_-^B)$. Consider any non-zero solution in L^{M+1} of

$$\mathcal{T}\gamma = \begin{bmatrix} \alpha^{(M)} & \cdots & \alpha^{(1)} & \alpha^{(0)} \\ \alpha^{(M+1)} & \alpha^{(M)} & \alpha^{(2)} & \alpha^{(1)} \\ \vdots & & \ddots & \vdots \\ \alpha^{(M+M'-1)} & \cdots & & \alpha^{(M'-1)} \end{bmatrix} \begin{bmatrix} \gamma^{(M)} \\ \gamma^{(M-1)} \\ \vdots \\ \gamma^{(0)} \end{bmatrix} = 0^{M'}. \tag{8}$$

Then for all $j = 0, \ldots, M-1 \le M'-1$

$$\alpha^{(M+j)}\gamma^{(M)} + \cdots + \alpha^{(j)}\gamma^{(0)} = 0,$$

hence the polynomial

$$\varphi(\lambda) = \gamma^{(M)}\lambda^M + \cdots + \gamma^{(1)}\lambda + \gamma^{(0)} \in \mathsf{L}[\lambda]$$

generates the entire sequence (7) (Lemma 1 in §2). This implies that f_-^B divides φ, so φ is in the linear span over L of

$$f_-^B(\lambda), \lambda f_-^B(\lambda), \lambda^2 f_-^B(\lambda), \ldots, \lambda^\delta f_-^B(\lambda), \quad \text{where } \delta = M - \deg(f_-^B).$$

Also, any coefficient vector of a polynomial in that linear span generates the sequence, thus solves (8). Therefore the rank of $\mathcal{T}$,

$$M + 1 - \text{the dimension of kernel}(\mathcal{T}),$$

is equal to $\deg(f_-^B)$. ⊠

Proof of Proposition 2. Consider the specialization

$$\mathcal{Z}' = [\,\mathcal{Z}_1 \mid \underbrace{B^{D+1}\mathcal{Z}_1}_{\mathcal{Z}'_2} \mid \underbrace{B^{2(D+1)}\mathcal{Z}_1}_{\mathcal{Z}'_3} \mid \ldots \mid \underbrace{B^{(n-1)(D+1)}\mathcal{Z}_1}_{\mathcal{Z}'_n}\,].$$

Then the set of columns in

$$[\,B^D\mathcal{Z}' \mid B^{D-1}\mathcal{Z}' \mid \ldots \mid \mathcal{Z}'\,]$$

is equal to

$$\{\mathcal{Z}_1, B\mathcal{Z}_1, B^2\mathcal{Z}_1, \ldots, B^{n(D+1)-1}\mathcal{Z}_1\}.$$

Since $n(D+1) > N$, this set has rank equal $\deg(f^B)$, as is argued in the proof of Propostition 1. Therefore the "more generic" matrix

$$[\,B^D\mathcal{Z} \mid B^{D-1}\mathcal{Z} \mid \ldots \mid \mathcal{Z}\,]$$

also has rank greater equal $\deg(f^B)$. Now define

$$\mathcal{K}^{\boxplus} = [\,B^D\mathcal{Y} \mid B^{D-1}\mathcal{Y} \mid \ldots \mid \mathcal{Y}\,] = B \cdot [\,B^D\mathcal{Z} \mid B^{D-1}\mathcal{Z} \mid \ldots \mid \mathcal{Z}\,],$$

which thus satisfies $\operatorname{rank}(\mathcal{K}^{\boxplus}) \leq \operatorname{rank}(B)$. If B is non-singular, the matrix $\mathcal{K}^{\boxplus}$ actually has full rank N, since by assumption $\deg(f^B) = N$. From Proposition 1 we further get for a singular B that

$$\deg(f^B) - 1 = \operatorname{rank}([\,B^{D+1}\mathcal{Z}' \mid B^D\mathcal{Z}' \mid \ldots \mid B\mathcal{Z}'\,]) \leq \operatorname{rank}(\mathcal{K}^{\boxplus}),$$

hence

$$\deg(f^B) - 1 \leq \operatorname{rank}(\mathcal{K}^{\boxplus}) \leq \operatorname{rank}(B),$$

which implies by the assumption of the theorem that $\operatorname{rank}(\mathcal{K}^{\boxplus}) = \operatorname{rank}(B)$. Furthermore, if B is singular and $\deg(f^B) = N$ it follows that $\operatorname{rank}(B) = N-1 = \deg(f^B) - 1$.

We will use a similar specialization for the columns of $\mathcal{X}$ to establish that the rank of

$$\mathcal{T}^{\boxplus} = \left[\begin{array}{c|c|c|c|c} \mathcal{A}^{(D)} & \ldots & & \mathcal{A}^{(1)} & \mathcal{A}^{(0)} \\ \hline \mathcal{A}^{(D+1)} & \mathcal{A}^{(D)} & & \mathcal{A}^{(2)} & \mathcal{A}^{(1)} \\ \hline \vdots & & \ddots & & \vdots \\ \hline \mathcal{A}^{(D+E-1)} & \ldots & & & \mathcal{A}^{(E-1)} \end{array}\right] \tag{9}$$

agrees with the rank of $\mathcal{T} \in \mathsf{L}^{M' \times (M+1)}$ of Proposition 1 with the dimensions

$$M = n(D+1) - 1 = S - 1 \geq N$$

and

$$M' = mE = R > S - 1.$$

Consider the specialization $\mathcal{Z}'$ given above, and

$$\mathcal{X}' = [\mathcal{X}_1 \mid \underbrace{B^E \mathcal{X}_1}_{\mathcal{X}'_2} \mid \underbrace{B^{2E} \mathcal{X}_1}_{\mathcal{X}'_3} \mid \ldots \mid \underbrace{B^{(m-1)E} \mathcal{X}_1}_{\mathcal{X}'_n}].$$

Then with

$$\mathcal{A}'^{(i)} = \mathcal{X}'^{\text{tr}} B^{i+1} \mathcal{Z}'$$

there exist permutation matrices $P \in \{0,1\}^{R \times R}$ and $Q \in \{0,1\}^{S \times S}$ such that

$$PTQ = \left[\begin{array}{c|c|c|c|c}
\mathcal{A}'^{(D)} & \ldots & & \mathcal{A}'^{(1)} & \mathcal{A}'^{(0)} \\ \hline
\mathcal{A}'^{(D+1)} & \mathcal{A}'^{(D)} & & \mathcal{A}'^{(2)} & \mathcal{A}'^{(1)} \\ \hline
\vdots & & \ddots & & \vdots \\ \hline
\mathcal{A}'^{(D+E-1)} & \ldots & & & \mathcal{A}'^{(E-1)}
\end{array}\right].$$

The row and column permutations move the entry

$$\begin{aligned}
\mathcal{A}'^{(D+I-J)}_{i,j} &= \mathcal{X}'^{\text{tr}}_i B^{D+I-J} B \mathcal{Z}'_j \\
&= \mathcal{X}^{\text{tr}}_1 B^{(i-1)E} B^{D+1+I-J} B^{(j-1)(D+1)} \mathcal{Z}_1
\end{aligned}$$

in the right hand side block Toeplitz matrix, which is in row $mI + i$, where $0 \le I < E$ and $1 \le i \le m$, and column $nJ + j$, where $0 \le J < D + 1$ and $1 \le j \le n$, to row $E(i-1) + I + 1$ and column $(D+1)(n-j) + J + 1$ in $\mathcal{T}$, namely

$$\begin{aligned}
\mathcal{T}_{E(i-1)+I+1,(D+1)(n-j)+J+1} &= \alpha^{(\,M \;+\; E(i-1)+I+1 \;-\; ((D+1)(n-j)+J+1)\,)} \\
&= \mathcal{X}^{\text{tr}}_1 B^{n(D+1)+E(i-1)+I+(D+1)(j-n)-J} \mathcal{Z}_1.
\end{aligned}$$

Therefore, the rank of $\mathcal{T}^{\boxplus}$ is no less than the rank of $\mathcal{T}$ with the given dimensions, which by Proposition 1 and the assumptions is equal to $\deg(f^B) = N$ for non-singular B, and $\deg(f^B) - 1$ for singular B. Since the kernel of $\mathcal{K}^{\boxplus}$ is contained in the kernel of $\mathcal{T}^{\boxplus}$, the rank cannot be more. ⊠

Proof of Theorem 1. Let

$$\Delta(\xi_{1,1}, \ldots, \xi_{N,m}, \zeta_{1,1}, \ldots, \zeta_{N,n})$$

be a non-zero maximal minor of $\mathcal{T}^{\boxplus}$ in (9). Then for all matrices x and z with

$$\Delta(x_{1,1}, \ldots, x_{N,m}, z_{1,1}, \ldots, z_{N,n}) \neq 0$$

any solution to (3) must also solve (6), because the ranks of both coefficient matrices will be $\deg(f^B) - 1$. Hence

$$Bzc^{(0)} + B^2zc^{(1)} + \cdots + B^{D+1}zc^{(D)} = B^{L+1}\widehat{w} = 0^N$$

for $0 \leq L \leq D$ such that $c^{(L)} \neq 0$ and $c^{(\ell)} = 0$ for $0 \leq \ell < L$. By a lemma of Zippel (1979)/Schwartz (1980) the probability of hitting a zero of Δ is no more than $\deg(\Delta)/\mathrm{card}(\mathsf{K}) \leq 2\,\mathrm{rank}(B)/\mathrm{card}(\mathsf{K})$.

It remains to estimate the probability that $\widehat{w} \neq 0$. The argument, by Coppersmith, is as that for step W3. For a matrix $y = Bz \in \mathsf{K}^{N\times n}$ consider the equivalence class

$$\{z' \in \mathsf{K}^{N\times n} \mid y = Bz' = Bz\} \tag{10}$$

of $\mathsf{K}^{N\times n}$. Then for each member in that class

$$\begin{aligned}\widehat{w}' &= z'c^{(L)} + Bz'c^{(L+1)} + \cdots + B^{(D-L)}z'c^{(D)} \\ &= \underbrace{zc^{(L)} + Bzc^{(L+1)} + \cdots + B^{(D-L)}zc^{(D)}}_{\widehat{w}} + (z'-z)c^{(L)},\end{aligned}$$

where

$$z' - z = [w_1 \mid w_2 \mid \ldots \mid w_n] \quad \text{with } w_\nu \in \mathrm{kernel}(B) \text{ for all } 1 \leq \nu \leq n.$$

Since, given $c^{(L)} \in \mathsf{K}^n \setminus \{0^n\}$, the linear span

$$c_1^{(L)}w_1 + \cdots + c_n^{(L)}w_n$$

uniformly samples $\mathrm{kernel}(B)$ for randomly chosen $w_\nu \in \mathrm{kernel}(B)$, at most a fraction of $1/\mathrm{card}(\mathsf{K})$ matrices in the set (10) can give $-\widehat{w}$ as that linear combination and thus lead to $\widehat{w}' = 0$. Therefore, the probability that $\widehat{w} = 0$ is no more than $1/\mathrm{card}(\mathsf{K})$.

Summing both estimates bounds the probability of failure. ⊠

5 Algorithms and their running times

The block Wiedemann method of §3 is used to solve both non-singular and singular sparse linear systems, i.e., linear systems with an efficient way to multiply the coefficient matrix by any vector. The method is randomized and can be executed sequentially or in parallel. Especially in the latter form the method becomes very efficient. We now present several algorithms that are based on the block Wiedemann algorithm of §3. One main point is that we are able to give both explicit expected running times and estimates on the success probability of the randomizations. We have the following theorem, which focuses on the sequential performance of the blocking. A corollary considering the parallel costs is given below.

Theorem 2. *Let $B \in \mathbf{K}^{N \times N}$ be a singular matrix and $1 \leq m, n \leq N$. Then one can compute a solution vector $w \in \mathbf{K}^N \setminus \{0^N\}$ with $Bw = 0^N$ in no more than*

$$\left\lfloor \left(1 + \frac{n}{m} + \frac{1}{n}\right) N + \frac{2n^2}{m} + 2n + 2 \right\rfloor$$

multiplications of B times a vector in $\mathbf{K}^N$, and an additional

$$O((m+n)N^2 \log N \operatorname{loglog} N)$$

arithmetic operations in $\mathbf{K}$. The algorithm selects no more than $(m+n+5)N$ random elements in $\mathbf{K}$ and succeeds to produce a solution with probability no less than

$$1 - \frac{3/2\,(N^2+N)}{\operatorname{card}(\mathbf{K})}.$$

The algorithm requires an additional $O((m+n)N)$ amount of storage for field elements in $\mathbf{K}$.

Proof. Consider the perturbed matrix

$$\widetilde{B} = UBVG$$

where $U \in \mathbf{K}^{N \times N}$ is a random unit upper triangular Toeplitz matrix, $V \in \mathbf{K}^{N \times N}$ is a random unit lower triangular Toeplitz matrix, and $G \in \mathbf{K}^{N \times N}$ is a random non-singular diagonal matrix. Then with probability of at least

$$1 - \frac{3(N-1)N/2}{\operatorname{card}(\mathbf{K})}$$

for the minimum polynomial $f^{\widetilde{B}}$ of $\widetilde{B}$ we have

$$\deg(f^{\widetilde{B}}) = \operatorname{rank}(\widetilde{B}) + 1$$

(Kaltofen and Saunders 1991, Theorem 2 and Lemma 2). Also, for a vector $b \in \mathbf{K}^N$ the product $\widetilde{B}b$ can be computed by one multiplication of B by a vector, and an additional $O(N \log N \operatorname{loglog} N)$ arithmetic operations in $\mathbf{K}$. We remark that the use of Beneš networks (Wiedemann 1986, §V) can reduce the latter complexity by the $\operatorname{loglog} N$ factor at the cost of requiring $O(N \log N)$ random field elements.

Now, the matrix $\widetilde{B}$ satisfies the assumptions of Theorem 1, and we can find a non-zero solution $\widetilde{w} \in \mathbf{K}^N \setminus \{0^N\}$ to $\widetilde{B}\,\widetilde{w} = 0^N$. Thus $w = VG\widetilde{w} \neq 0^N$ solves $Bw = 0^N$. By (2) and (5) the method multiplies $\widetilde{B}$ by a vector no more than

$$\left(1 + \frac{n}{m} + \frac{1}{n}\right) N + \frac{2n^2}{m} + 2n + 2$$

many times. The extra work in terms of arithmetic operations in $\mathbf{K}$ is $O((m+n)N^2)$ plus the work it takes to solve (3). Suppose then that we compute a non-zero solution to (3) by the standard Wiedemann method, e.g., by the algorithm described in §3.

The coefficient matrix

$$A = \left[\begin{array}{c|c|c|c|c} a^{(D)} & \ldots & & a^{(1)} & a^{(0)} \\ \hline a^{(D+1)} & a^{(D)} & & a^{(2)} & a^{(1)} \\ \hline \vdots & & \ddots & & \vdots \\ \hline a^{(D+E-1)} & \ldots & & & a^{(E-1)} \end{array}\right]$$

of (3) is not square, which requires some modification. One could, e.g., pre-multiply A by a random unit upper triangular $S \times R$ Toeplitz matrix, which is a rank preserving operation with high probability. However, by inspection of the proof of Propostion 2 we see that we may drop the rows in position R, $R-m, \ldots, R-(R-S-1)m$ without affecting the probabilistic rank estimates. The Wiedemann algorithm requires $\leq 3R$ multiplications of A times a vector and $O(R^2)$ arithmetic operations. We are left with the problem of efficiently performing the matrix times vector multiplication

$$A \cdot \left[\begin{array}{c} b^{(0)} \\ \hline \vdots \\ \hline b^{(D)} \end{array}\right] = \left[\begin{array}{c} b'^{(0)} \\ \hline \vdots \\ \hline b'^{(D)} \end{array}\right], \quad \text{where } b^{(i)}, b'^{(i)} \in \mathbf{K}^n. \tag{11}$$

Consider the polynomial multiplication

$$\begin{aligned}(a^{(0)} + a^{(1)}\lambda + \cdots + a^{(D+E-1)}\lambda^{D+E-1}) \cdot (b^{(0)} + b^{(1)}\lambda + \cdots + b^{(D)}\lambda^{D}) \\ = \cdots + b'^{(0)}\lambda^{D} + b'^{(1)}\lambda^{D+1} + \cdots + b'^{(D)}\lambda^{D+E-1} + \cdots\end{aligned}$$

with non-commuting coefficients in the algebras $\mathbf{K}^{m\times n}$ and $\mathbf{K}^n$. By the results in (Kaltofen and Cantor 1991) the product can be found in

$$O(D' \log D' \operatorname{loglog} D'), \quad \text{where } D' = D + E,$$

algebra operations, i.e., additions and subtractions in $\mathbf{K}^{m\times n}$ and $\mathbf{K}^n$, and $O(D' \times \log D')$ multiplications of $m \times n$ matrices by vectors in $\mathbf{K}^n$. These are

$$O((m+n)N \log N \operatorname{loglog} N)$$

arithmetic operations in $\mathbf{K}$ for computing Ab. Alternately, we could have re-arranged the rows and columns of A to obtain an $m \times n$ block matrix with $(D+1) \times E$ Toeplitz blocks. ⊠

Theorem 2 can be employed to solve non-singular systems as outlined in the last paragraph of §3. We shall formulate the result not in terms of the block sizes m and n, but in terms of the quantity

$$\epsilon = \frac{n}{m} + \frac{1}{n}.$$

For suitable constant block sizes ϵ can be made arbitrarily close to 0. Thus we have the following sequential complexity result.

Corollary 1. *Let $B \in \mathbf{K}^{N \times N}$ be a non-singular matrix and let $\epsilon > 0$ be fixed. Then one can compute the solution vector $w = B^{-1}b$ with $b \in \mathbf{K}^N$ in no more than*

$$(1+\epsilon)N + O(1)$$

multiplications of B times a vector in $\mathbf{K}^N$, and an additional

$$O(N^2 \log N \operatorname{loglog} N)$$

arithmetic operations in $\mathbf{K}$. The algorithm selects $O(N)$ random elements in $\mathbf{K}$ and succeeds to produce the solution with probability no less than

$$1 - \frac{N^2 + 4N + 3}{\operatorname{card}(\mathbf{K})}.$$

The algorithm requires an additional $O(N)$ amount of storage for field elements in $\mathbf{K}$. Note that here all big-O estimates depend on ϵ.

Of course, the main application of blocking is to compute the sequence of matrices $a^{(i)}$ in parallel. In order to make the statement of the next corollary simpler, we suppose that $m = n \approx \sqrt{N}$ and that we have n (loosely linked) parallel processors.

Corollary 2. *On $\lceil \sqrt{N} \rceil$ processors one may compute using $O(N\sqrt{N})$ random elements in $\mathbf{K}$ a solution to the linear system $Bw = b$, where $B \in \mathbf{K}^{N \times N}$ and $b \in \mathbf{K}^N$, in $O(\sqrt{N})$ (parallel) multiplications of B times vectors, $O(N^2)$ (parallel) arithmetic operations in $\mathbf{K}$, and an additional $O(N^2\sqrt{N} \log N \operatorname{loglog} N)$ sequential arithmetic operations in $\mathbf{K}$.*

Note that we may also solve the arising block-Toeplitz system (3) in parallel, which improves on the sequential operation count $O(N^2\sqrt{N} \log N \operatorname{loglog} N)$ in the above corollary by a factor of $\sqrt{N}$ as follows. When using the adaption of Wiedemann's method proposed in the proof of Theorem 2, we must first parallelize the computation of (11). For this we employ a parallel version of the algorithm by Cantor and Kaltofen (1991), which with $n = m \approx \sqrt{N}$ processors can compute the matrix times vector product (11) in $O(N \log N \operatorname{loglog} N)$ arithmetic operations in $\mathbf{K}$; this is because the Cantor/Kaltofen algorithm for polynomial multiplication constructs a parallel circuit that is actually of depth $O(\log N)$. Second, we must also parallelize the Berlekamp/Massey step in Wiedemann's algorithm (Step W2 in §3). Again, we can appeal to the parallel implementation of the extended Euclidean algorithm on a systolic array (Brent and Kung 1983), which with $\lceil \sqrt{N} \rceil$ processors finds the needed linear recurrence in $O(N\sqrt{N})$ arithmetic steps (see also Dornstetter 1987). The operations necessary for the evaluation of the generating polynomial at the matrix (see Step W3 in §3) are again parallelizable by using the parallel method for computing products such as (11) discussed before. Altogether, we require with $\lceil \sqrt{N} \rceil$ processors $O(N^2 \log N \operatorname{loglog} N)$ (parallel) arithmetic operations in $\mathbf{K}$ for the solution of (3). However, all these substeps utilize a much more fine grain parallelism than does the parallel computation of the sequence $a^{(i)}$ of Step C1 in §2.

6 Conclusion

Our main contribution in this paper is to give a theoretical basis for the block generalization of the Wiedemann method. We could prove our algorithm for sufficiently large fields and by using certain perturbations of the input matrix. The algorithm may still be valid without the assumptions on the degree of the minimum polynomial. However, Coppersmith also notes that for certain "pathological" cases the straight-forward algorithm might fail to compute a solution. Also, Coppersmith's application to integer factoring has the smallest coefficient field $\mathsf{K} = \mathsf{F}_2$. However, in that situation, Propostion 2 could be relaxed. If the rank of (3) were one or two less than the rank of (6), with probability 1/2 or 1/4 we still would find a solution to (6). For very large finite fields such a rank deficiency would make the problem quite infeasible.

Our algorithms are formulated for finite fields only, but it is not difficult to extend them to fields such as the rational numbers and functions by the use of Chinese remaindering, interpolation, and p-adic lifting (McClellan 1973, Moenck and Carter 1979).

We have used the standard Wiedemann method for analyzing the complexity of computing a solution to (3). There is a slightly faster way of deriving a solution, based on the theory of Toeplitz-like matrices, i.e., matrices with small displacement rank (Kailath et al. 1979). Then it is possible to compute a solution to (3) in $O((m+n)N^2)$ arithmetic operations in K, thus saving the $\log N \operatorname{loglog} N$ factor for Step C2, using the generalized Levinson-Trench algorithm for inverting a matrix of small displacement rank (the condition that all possible leading principal minors are non-zero can be enforced by multiplying with random triangular Toeplitz matrices). Incidentally, Coppersmith's generalized Berlekamp/Massey method has the same asymptotic complexity.

Lobo has implemented several versions of the block Wiedemann algorithm in the programming language C for $\mathsf{K} = \mathsf{F}_p$ and executed it using simultaneously 4 Sun Sparc 2 processors, each rated 28.5MIPS. For $p = 32749$ he can solve a 20K×20K system with 1.32M non-zero entries in about 60 CPU hours. The details of this experiment will be published in a forthcoming paper.

Acknowledgement: Thanks to Austin Lobo for discussions on the theory and implementation of the block Wiedemann method, and to the referee for his comments.

Literature Cited

Brent, R. P. and Kung, H. T., "Systolic VLSI arrays for linear-time GCD computation," *Proc. VLSI '83*, pp. 145–154 (1983).

Cantor, D. G. and Kaltofen, E., "On fast multiplication of polynomials over arbitrary algebras," *Acta Inform.* **28**/7, pp. 693–701 (1991).

Coppersmith, D., "Solving linear equations over GF(2) via block Wiedemann algorithm," *Math. Comput.*, p. to appear (1992).

Dornstetter, J. L., "On the equivalence between Berlekamp's and Euclid's algorithms," *IEEE Trans. Inf. Theory* IT-**33**/3, pp. 428–431 (1987).

Kailath, T., Kung, S.-Y., and Morf, M., "Displacement ranks of matrices and linear equations," *J. Math. Analysis Applications* **68**, pp. 395–407 (1979).

Kaltofen, E. and Saunders, B. D., "On Wiedemann's method of solving sparse linear systems," in *Proc. AAECC-9*, Springer Lect. Notes Comput. Sci. **539**; pp. 29–38, 1991.

Lenstra, A. K., Lenstra, H. W., Manasse, M. S., and Pollard, J. M., "The number field sieve," *Proc. 22nd Annual ACM Symp. Theory Comp.*, pp. 564–572 (1990).

Massey, J. L., "Shift-register synthesis and BCH decoding," *IEEE Trans. Inf. Theory* **IT-15**, pp. 122–127 (1969).

McClellan, M. T., "The exact solution of systems of linear equations with polynomial coefficients," *J. ACM* **20**, pp. 563–588 (1973).

Moenck, R. T. and Carter, J. H., "Approximate algorithms to derive exact solutions to systems of linear equations," *Proc. EUROSAM '79, Springer Lec. Notes Comp. Sci.* **72**, pp. 65–73 (1979).

Schwartz, J. T., "Fast probabilistic algorithms for verification of polynomial identities," *J. ACM* **27**, pp. 701–717 (1980).

Wiedemann, D., "Solving sparse linear equations over finite fields," *IEEE Trans. Inf. Theory* **IT-32**, pp. 54–62 (1986).

Zippel, R., "Probabilistic algorithms for sparse polynomials," *Proc. EUROSAM '79, Springer Lec. Notes Comp. Sci.* **72**, pp. 216–226 (1979).

Relations among Lie Formal Series and Construction of Symplectic Integrators

P.-V. Koseleff

Aleph et Géode
Centre de Mathématiques, École Polytechnique
91128 Palaiseau
e-mail : koseleff@polytechnique.fr

Abstract. Symplectic integrators are numerical integration schemes for hamiltonian systems. The integration step is an explicit symplectic map. We find symplectic integrators using universal exponential identities or relations among formal Lie series. We give here general methods to compute such identities in a free Lie algebra. We recover by these methods all the previously known symplectic integrators and some new ones. We list all possible solutions for integrators of low order.

1 INTRODUCTION

Lie series and Lie transformations have found many applications, particularly in celestial mechanics (see [3]) or in hamiltonian perturbation theory (see for example [2, 4, 6]). These techniques have the advantage of providing explicit approximating systems that are also hamiltonian.

In hamiltonian mechanics, it is often important to know the time evolution mapping, that is to say the position of the solution after a certain given time. In celestial mechanics, long integrations have mostly used high order multistep integration methods. A disadvantage of such methods is that the error in position grows quadratically in time or linearly with symmetric integrators.

For very long time integration, there has been recently a development of numerical methods preserving the symplectic structure (see for example [7, 14, 15, 16]), which seem to be more efficient with respect to the computational cost.

Symplectic integrators may be seen as the time evolution mapping of a slightly perturbed Hamiltonian, that is to say as a Lie transformation that can be represented either by an exponential, a product of increasing order single exponentials or a proper Lie transformation. Constructing explicit high order symplectic integrators requires the manipulation of formal identities like exponential identities.

In section **2.**, we give some general methods to manipulate formal Lie series and Lie algebra automorphisms. We recall some theorems related to exponential identities and give explicit methods to compute them. They make use the Lyndon basis, which is particularly adapted to this problem.

In section **3.**, we recall first some definitions of the Hamilton formalism. Then we show how the algorithms described in section **2.** provide symplectic integrators. The idea of such constructions originates in Forest & Ruth ([7]) or more recently Yoshida ([16]). Our approach in this paper is to combine the use of proper Lie transforms and exponentials. This avoids many unnecessary direct calculations of exponential indentities. At the end we propose some improvement in the case when the Hamiltonian is seperated into kinetic and potential energies.

All the algorithms described in the present paper have been implemented using Axiom (NAG) running on IBM-RS/6000-550.

2 LIE ALGEBRAIC FORMALISM

In hamiltonian mechanics, the use of Lie methods or Lie transformations is efficient when it becomes easy to manipulate Lie polynomials and to express exponential identities like the Baker-Campbell-Hausdorff formula. Our aim in this section is to give general methods for the computation of such identities.

These identities are universal Lie algebraic identities, that is to say they do not depend on the Lie algebra we work in or the Lie bracket we use. We work in free Lie algebras and with formal Lie series, neglecting all the convergence problems that can appear with analytical functions for example.

We will use the Lyndon basis for the formal computations but all the identities can be later evaluated in any Lie algebra.

2.1 Definitions

In this paper X will denote an alphabet, that is to say an ordered set (possibly endless).

R is a ring which contains the rational numbers $\mathbb{Q}$.

X^* is the free monoid generated by X. X^* is totally ordered with the lexicographic order.

$M(X)$ is the free magma generated by X. Having defined $M_1(X)$ as X, we define $M_n(X)$ by induction on n:

$$M_n(X) = \bigcup_{p+q=n} M_p \times M_q \quad \text{and} \quad M(X) = \bigcup_{n\geq 1} M_n(X). \tag{1}$$

$\mathcal{A}_R(X)$ is the associative algebra, that is to say the R-algebra of X^*.

A Lie algebra is an algebra in which the multiplication law $[,]$ is bilinear, alternate and satisfies the Jacobi identity:

$$[a,[b,c]] + [b,[c,a]] + [c,[a,b]] = 0. \tag{2}$$

$L_R(X)$ or $L(X)$ is the free Lie algebra on X. It is defined as the quotient of the R-algebra of $M(X)$ by the ideal generated by the elements (u,u) and $(u,(v,w)) + (v,(w,u)) + (w,(u,v))$.

An element of $M(X)$ considered as element of $L(X)$ will be called a Lie monomial. $L_n(X)$ is the free module generated by those of length n. Thus $L(X)$

is graded by the length denoted by $|x|$ for $x \in M(X)$. If $|X| = q < \infty$ we have Witt's formula (see [1, 9, 10]):

$$\sum_{d|n} d \dim L_d(X) = q^n. \tag{3}$$

2.2 Formal Lie series

Given a weighted alphabet X in which each letter a has an integer weight $|a|_*$, we take as graduation for $L(X)$ the weight $||_*$ which is defined as the unique extension of the weight in X. We call $\tilde{L}_n(X)$ (resp. $\tilde{\mathcal{A}}_n(X)$) the submodule of $L(X)$ (resp. $\mathcal{A}(X)$) generated by the elements of weight n. We define the formal Lie series $\tilde{L}(X)$ and $\tilde{\mathcal{A}}(X)$ as

$$\tilde{L}(X) = \prod_{n\geq 0} \tilde{L}_n(X) \quad \text{and} \quad \tilde{\mathcal{A}}(X) = \prod_{n\geq 0} \tilde{\mathcal{A}}_n(X). \tag{4}$$

We will write $x \in \tilde{L}(X)$ as a series $\sum_{n\geq 0} x_n$. $\tilde{L}(X)$ is a complete Lie algebra with the Lie bracket

$$[x, y] = \sum_{n\geq 0} \sum_{p+q=n} [x_p, y_q]. \tag{5}$$

Denoting by $\tilde{L}(X)^+$ (resp. $\tilde{\mathcal{A}}(X)^+$) the ideal of $\tilde{L}(X)$ (resp. $\tilde{\mathcal{A}}(X)$) generated by the elements of positive weight, we can define the exponential and the logarithm as

$$\begin{array}{ll} \exp : \tilde{\mathcal{A}}(X)^+ \to 1 + \tilde{\mathcal{A}}(X)^+ & \log : 1 + \tilde{\mathcal{A}}(X)^+ \to \tilde{\mathcal{A}}(X)^+ \\ \quad x \mapsto \sum_{n\geq 0} \dfrac{x^n}{n!} & \quad x \mapsto -\sum_{n\geq 1} \dfrac{(1-x)^n}{n!}. \end{array} \tag{6}$$

They are mutually reciprocal functions and we have (see [1, Ch. II, §5]) the

Theorem 1 (Campbell-Hausdorff). *If $x, y \in \tilde{L}(X)^+$ then*

$$\log[\exp(x)\exp(y)] \in \tilde{L}(X)^+. \tag{7}$$

More precisely, we have the following

Lemma 2. *Given $x, y \in \tilde{L}(X)^+$, we have*

- $\exp(x)\exp(y) = \exp(z)$ *where* $z \in \tilde{L}(X)^+$,
- $z = z_p + z_q + \sum_{n\geq p+q} z_n$ *and* $z_p + z_q = x_p + y_q$,
- $z_m \in \tilde{L}_m(x_p, \ldots, x_m, y_q, \ldots, y_m)$ *for* $m \geq p+q$.

Using the preceding lemmas we deduce (see [12]) the

Proposition 3 (Factored product expansion). *Given $k \in \tilde{L}(X)^+$, there is a unique series $g \in \tilde{L}(X)^+$ such that*

$$\exp(\textstyle\sum_{n\geq 1} k_n) = \cdots \exp(g_n) \cdots \exp(g_1). \tag{8}$$

The above proposition is proved by induction, constructing $g \in \tilde{L}(X)$ and $k^{(p)} \in \prod_{n>p} \tilde{L}_n(X)$ such that, for each $p \geq 1$,

$$\exp(k) = \exp(k^{(p)}) \exp(g_p) \cdots \exp(g_1). \tag{9}$$

2.3 Lie series automorphisms

We denote for x in $\tilde{L}(X)$ by $L(x)$ or L_x, the Lie operator $L_x y = [x, y]$. From the Jacobi identity (2) we have $[L_x, L_y] = L_{[x,y]}$, in which $[,]$ denotes the commutator. The set of L_x is a Lie algebra that we call the adjoint Lie algebra. For any Lie series automorphisms T, we have by definition $[Tf, Tg] = T[f, g]$. The Lie series automorphisms act on the adjoint Lie algebra by

$$TL_fT^{-1} = L_{Tf}. \tag{10}$$

Let us give now some example of Lie transformations that play an important role in hamiltonian mechanics.

The exponential. Given $x \in \tilde{L}(X)^+$, we consider $\exp(L_x)$ defined as

$$\exp(L_x)y = \sum_{i\geq 0} \frac{L_x^i}{i!}y. \tag{11}$$

From the Jacobi identity (2), we have by induction on $k \geq 0$, for any $f, g, h \in \tilde{L}(X)^+$

$$L_f^k[g,h] = \sum_{i=0}^{k} \binom{k}{i} \left[L_f^i g, L_f^{k-i}h\right].$$

We therefore deduce that

$$\begin{aligned} \exp(L_f)[g,h] &= \sum_{n\geq 0} \frac{1}{n!} \sum_{p=0}^{n} \binom{n}{p} \left[L_f^p g, L_f^{n-p}h\right] \\ &= \sum_{p+q\geq 0} \tfrac{1}{p+q!}\tfrac{p+q!}{p!q!} \left[L_f^p g, L_f^q h\right] = [\exp(L_f)g, \exp(L_f)h]. \end{aligned} \tag{12}$$

Thus, for each $x \in \tilde{L}(X)$, $\exp(L_x)$ is a Lie algebra automorphism. From the Campbell-Hausdorff theorem (1), the set of all $\exp(L_x)$ is a group $\mathbf{G}$ that we will call the Lie transformations group.

The Lie transform. Using (10) and $\dfrac{d}{dt}\exp(tL_x) = L_x \exp(tL_x)$, we obtain the following identity:

$$\frac{d}{dt}\exp(tL_x)\exp(tL_y) = L\left[x + \exp(tL_x)y\right]\exp(tL_x)\exp(tL_y). \tag{13}$$

The map $S(t) = \exp(tL_x)\exp(tL_y) \in \mathbf{G}$ is the solution of (13). For a given z, the solution of $\frac{d}{dt}S = L_z S$ is not necessarily $\exp(tL_z)$ as z may depend on t, but is known as the inverse Lie transformation associated to $\int_t z dt$. For $w = \sum_{n\geq 1} t^n w_n$, T_w and T_w^{-1} are the solution ([3]) of

$$\frac{d}{dt}T_w = -T_w L_{\frac{dw}{dt}} \quad \text{and} \quad \frac{d}{dt}T^{-1}w = L_{\frac{dw}{dt}} T_t^{-1}. \tag{14}$$

For $g \in \tilde{L}(X)$, we have (see [2, 3]) $G = \sum_{n\geq 0} G_n = T_w^{-1} g$ where

$$G_0 = g_0, \quad G_{0,n} = g_n, \quad G_{p,q} = \sum_{k=1}^{p} \tfrac{k}{p}[w_k, G_{p-k,q}], \quad G_n = \sum_{p=0}^{n} G_{p,n-p}. \quad (15)$$

The Dragt-Finn transform. There is another transformation that plays an important role in the Lie transformations theory that is called the Dragt-Finn transform and is an infinite product of exponential maps (see [4]).

Given $g = \sum_{n\geq 1} g_n$, we define M_g and M_g^{-1} as

$$M_g = \exp(-L_{g_1}) \cdots \exp(-L_{g_n}) \cdots \quad \text{and} \quad M_g^{-1} = \cdots \exp(L_{g_n}) \cdots \exp(L_{g_1}) \quad (16)$$

Using proposition 3, we express the exponential of a Lie operator as a Dragt-Finn transform.

2.4 Relations between the exponential and the Lie transform.

The three above transformations are totally defined by generating series which satisfy the following :

Proposition 4. *Given $w, k, g \in \tilde{L}(X)^+$, there exist*

— $k' \in \tilde{L}(X)^+$ *with* $k'_n - w_n \in L_{\mathbb{Q}}(w_1, \ldots w_{n-1})$ *such that* $\exp(L_{k'}) = T_w^{-1}$,
— $g' \in \tilde{L}(X)^+$ *with* $g'_n - k_n \in L_{\mathbb{Q}}(k_1, \ldots k_{n-1})$ *such that* $M_g^{-1} = \exp(L_k)$,
— $w' \in \tilde{L}(X)^+$ *with* $w'_n - g_n \in L_{\mathbb{Q}}(g_1, \ldots g_{n-1})$ *such that* $T_w^{-1} = M_g^{-1}$.

We first prove the third part of the above proposition.
Given $M_g^{-1}(t) = \cdots \exp(t^n L_{g_n}) \cdots \exp(tL_{g_1})$, we have using (10)

$$\begin{aligned}
\frac{d}{dt} M_g &= \sum_{n\geq 1} \left[e^{-tL_{g_3}} \cdots e^{-t^{n-1}L_{g_{n-1}}}\right] \left[\frac{d}{dt}\left[e^{-t^n L_{g_n}}\right]\right] \left[e^{-t^{n+1}L_{g_{n+1}}} \cdots\right] \\
&= \sum_{n\geq 1} \left[e^{-tL_{g_3}} \cdots e^{-t^{n-1}L_{g_{n-1}}}\right] \left[-nt^{n-1} L_{g_n} e^{-t^n L_{g_n}}\right] \left[e^{-t^{n+1}L_{g_{n+1}}} \cdots\right] \\
&= M_g \sum_{n\geq 1} M_g^{-1} \left[e^{-tL_{g_3}} \cdots e^{-t^{n-1}L_{g_{n-1}}}\right] \left[-nt^{n-1} L_{g_n}\right] \left[e^{-t^n L_{g_n}} \cdots\right] \\
&= M_g \sum_{n\geq 1} \left[\cdots e^{t^n L_{g_n}}\right] \left[-nt^{n-1} L_{g_n}\right] \left[e^{-t^n L_{g_n}} \cdots\right] \\
&= M_g L \left[\sum_{n\geq 1} -nt^{n-1} \left[\cdots e^{t^n L_{g_n}}\right] g_n\right]. \qquad (17)
\end{aligned}$$

On the other hand, from (14), we have $\frac{d}{dt}T_w = -T_w L_{\frac{dw}{dt}}$. With the initial conditions $T_w(0) = M_g(0) = Id$, we deduce that $T_w = M_g$ iff

$$\frac{dw}{dt} = \sum_{n\geq 1} t^{n-1}\left[\sum_{k=1}^{n} k \sum_{\substack{(k+1)m_{k+1}+\cdots \\ +(n-k)m_{n-k}=n-k}} \frac{L_{g_{n-k}}^{m_{n-k}}\cdots L_{g_{k+1}}^{m_{k+1}}}{m_{k+1}!\cdots m_{n-k}!} g_k\right], \tag{18}$$

or equivalently

$$w_n = \sum_{k=1}^{n} \frac{k}{n} \sum_{\substack{(k+1)m_{k+1}+\cdots \\ +(n-k)m_{n-k}=n-k}} \frac{L_{g_{n-k}}^{m_{n-k}}\cdots L_{g_{k+1}}^{m_{k+1}}}{m_{k+1}!\cdots m_{n-k}!} g_k = g_n + G_n \tag{19}$$

in which $G_n \in L(g_1,\ldots,g_{n-1})$.

Using the proposition 3, one proves the existence of $g = \sum_{n\geq 1} g_n$ such that

$$\exp(\textstyle\sum_{n\geq 1} L_{k_n}) = \cdots \exp(L_{g_n})\cdots \exp(L_{g_1}), \tag{20}$$

in which $g_n = k_n + K_n$ and $K_n \in L_{\mathbb{Q}}(k_1,\ldots,k_{n-1})$. Combining (19) and (20) we deduce (4).

⁂

We deduce in passing that any Lie transformation $T \in \mathbf{G}$ may be expressed as an exponential of a Lie operator or as an infinite product of single exponentials or as a proper Lie transform. The use of a representation depends deeply on the result we look for. For example, if we have to compose transformations, it is much easier to consider Lie transforms because their product is a Lie transform whose generating function appears easily from (13).

2.5 Computing the relations.

In order to compute the relations between the generating series we try to solve at each order

$$\exp(L_k)a = T_w^{-1}a. \tag{21}$$

We first have the following lemma resulting from the expansion of the exponential or the Lie transform (15)

Lemma 5. *Let $T = \exp(L_k)$ (resp. $T = T_w^{-1}$) and $A = \sum_n A_n = Ta$, then for each $n \geq 0$, $A_n = -[a,k_n] + K_n$ (resp. $A_n = -[a,w_n] + W_n$) in which $K_n \in L(a,k_1,\ldots k_{n-1})$ (resp. $W_n \in L(a,w_1,\ldots w_{n-1})$).*

In order to compute the relations, we need a basis of the free Lie algebra $L(X)$. We propose here the use of the Lyndon basis that has many useful properties.

Basis of the free Lie Algebra. Given $w = uv$ a word of X^*, we will say that u and v are left factor and right factor respectively. For any $u, v \in X^*$, uv and vu are said conjugate and therefore X^* is divided in conjugate classes. The minimal element of each conjugate class is called a Lyndon word. The set $\mathbf{L}$ of Lyndon words satisfies many properties like (see [10])

Lemma 6. *$w \in \mathbf{L}$ iff there exist $u < v \in \mathbf{L}$ such that $w = uv$.*

Unfortunately a Lyndon word may be decomposed in many ways. For example $aabb = a.abb = aab.b$. For $l < m$ and $lm \in \mathbf{L}$, we call $\sigma(lm) = (l, m)$ the standard factorization of lm when $|m|$ is maximal. Therefore, we define a one-to-one correspondence between $\mathbf{L}$ and the Lyndon brackets $\mathcal{L} \subset M(X)$. We define

- $\Lambda : \mathbf{L} \to \mathcal{L} \subset M(X)$ by: $\Lambda(a) = a$ if $a \in X$ otherwise $\Lambda(lm) = (\Lambda(l), \Lambda(m))$ if $\sigma(lm) = (l, m)$.
- $\delta : M(X) \to X^*$ as the canonical application of unparenthesing.

Denoting $\mathcal{L} \cap M_n(X)$ by $\mathcal{L}_n(X)$, we have (see [10]) the following

Theorem 7. *$\mathcal{L}_n(X)$ is a basis of $L_n(X)$.*

Construction of the basis. One of the main properties is that (see [10])

Lemma 8. *if $u < v \in \mathcal{L}$ then $[u, v] \in \mathcal{L}$ iff $|u| = 1$ or $u = [u_1, u_2] < v \leq u_2$.*

One can now built the Lyndon basis as follows:

$$\mathcal{L}_1(X) = X,$$

$$\mathcal{L}_2(X) = \{[x, y]; x < y \in \mathcal{L}_1\},$$

$$\mathcal{L}_n(X) = \bigcup_{x \in \mathcal{L}_1(X)} \{[x, y]; y \in \mathcal{L}_{n-1}(X), x < y\} \tag{22}$$

$$\bigcup \bigcup_{p=2}^{n} \bigcup_{x \in \mathcal{L}_p(X)} \{[x, y]; y \in \mathcal{L}_{n-p}(X), [x_1, x_2] = x < y \leq x_2\}. \tag{23}$$

If $\mathcal{L}_1$ is well sorted then the $\bigcup$ in (23) are disjoint and the results are sorted. The knowledge of the basis is actually not very useful to work in the free Lie algebra except in some specific cases that will occur in the last section.

Decomposition onto the basis. We give here an algorithm for writing a Lie polynomial in terms of Lyndon brackets. This algorithm (see [10]) is also a proof that the Lyndon brackets generate the free Lie algebra.

Given two polynomials that are linear combinations of Lyndon brackets $p = \sum_i \alpha_i p_i$ and $q = \sum_i \beta_i q_i$, we have $p * q = \sum_{i,j} \alpha_i \beta_j \mathtt{mult}(p_i, q_j)$ in which $\mathtt{mult}(a, b)$ denotes the decomposition of $[a, b]$ onto the Lyndon basis. The point is to know how to multiply two Lyndon brackets. We will use the algorithm proposed by Perrin [10]. Implementations have been realized by Petitot [11] and myself more recently.

if $u = v$ **then** $\texttt{mult}(u, v) := 0$
else if $v < u$ **then** $\texttt{mult}(u, v) := -\texttt{mult}(v, u)$
if $|u| = 1$ **then** $\texttt{mult}(u, v) := 1.[u, v]$
else $u = [u_1, u_2]$ **if** $v \le u_2$ **then** $\texttt{mult}(u, v) := 1.[u, v]$
else $\texttt{mult}(u, v) := 1.u_1 * \texttt{mult}(u_2, v) + \texttt{mult}(u_1, v) * 1.u_2$.

2.6 Solving triangular systems

We deduce from the lemma 8 the following useful

Lemma 9. *Let* $X = \{a\} \bigcup X_1$ *where* $a < b$ *for each* $b \in X_1$. *Then for each* $x = \sum_i \alpha_i x_i \in L(X_1)$, *the decomposition of* $[a, x]$ *onto the Lyndon basis of* $\mathcal{L}(X)$ *is* $[a, x] = \sum_i \alpha_i [a, x_i]$.

This lemma shows that the injective mapping

$$\mathcal{L}(X_1) \rightarrow \mathcal{L}(\{a\} \bigcup X_1) \tag{24}$$

$$x \mapsto (a, x) \tag{25}$$

may be extended to $L_a : L(X_1) \rightarrow L(\{a\} \bigcup X_1)$. This property is specific to the Lyndon basis.

⋆⋆⋆

With this lemma, (21) becomes a triangular system in form

$$\begin{cases} [a, k_1] = [a, w_1] \\ [a, k_2] + K_2 = [a, w_2] + W_2 \\ \vdots \quad \vdots \quad \vdots \\ [a, k_n] + K_n = [a, w_n] + W_n \end{cases}, \tag{26}$$

in which $K_i - W_i \in L_{\mathbb{Q}}(a, k_1, \cdots, k_{i-1}, w_1, \cdots, w_{i-1})$. This system can be solved by successive evaluations of the k_i's (resp. w_i's) using the the factorization lemma 9. The relation (19) cannot be generated using equation (21) and the Hall basis.

⋆⋆⋆

This method may be used in order to express any $T \in \mathbf{G}$ as an exponential or as the inverse of a Lie transform. Let $A = Ta = \sum_{n \ge 0} A_n$. We formally compute for example $E = \exp(L_k)a = \sum_{n \ge 0} E_n \in \tilde{L}(a, k_1, \ldots, k_n, \ldots)$. We therefore obtain the following triangular system

$$\begin{cases} A_0 = E_0 = a \\ A_1 = E_1 = -[a, k_1] \\ A_2 = E_2 = -[a, k_2] + R_2 \\ \vdots \quad \vdots \quad \vdots \\ A_n = E_n = -[a, k_1] + R_n \end{cases} \tag{27}$$

that may be solved by the factorization lemma (9) as the system is consistent.

This method can also be applied for the computing of any relation between generating series in **G** like the Baker-Campbell-Hausdorff formula. For example we found at the order 6

$$\begin{aligned}\log[\exp x \exp y] = {} & x + y - \tfrac{1}{2}[x, y] + \tfrac{1}{12}[x, [x, y]] + \tfrac{1}{12}[[x, y], y] \\ & - \tfrac{1}{24}[x, [[x, y], y]] - \tfrac{1}{720}[x, [x, [x, [x, y]]]] + \tfrac{1}{180}[x, [x, [[x, y], y]]] \\ & + \tfrac{1}{180}[x, [[[x, y], y], y]] + \tfrac{1}{120}[[x, y], [[x, y], y]] + \tfrac{1}{360}[[x, [x, y]], [x, y]] \\ & - \tfrac{1}{720}[[[[x, y], y], y], y] + \tfrac{1}{1440}[x, [x, [x, [[x, y], y]]]] \\ & - \tfrac{1}{360}[x, [x, [[[x, y], y], y]]] - \tfrac{1}{240}[x, [[x, y], [[x, y], y]]] \\ & - \tfrac{1}{720}[x, [[x, [x, y]], [x, y]]] + \tfrac{1}{1440}[x, [[[[x, y], y], y], y]]\end{aligned}$$

Given $w = \sum_{n\geq 1} w_n$, we have at the order 6 $\exp(L_k) = T_w^{-1}$ in which

$$\begin{aligned}k = {} & w_1 + w_2 + w_3 - \tfrac{1}{6}[w_1, w_2] + w_4 - \tfrac{1}{4}[w_1, w_3] + w_5 - \tfrac{3}{10}[w_1, w_4] - \tfrac{1}{10}[w_2, w_3] \\ & + \tfrac{1}{120}[w_1, [w_1, w_3]] + \tfrac{1}{60}[[w_1, w_2], w_2] + \tfrac{1}{360}[w_1, [w_1, [w_1, w_2]]] + w_6 \\ & - \tfrac{1}{3}[w_1, w_5] - \tfrac{1}{6}[w_2, w_4] + \tfrac{1}{60}[w_1, [w_1, w_4]] + \tfrac{1}{30}[w_1, [w_2, w_3]] \\ & + \tfrac{1}{24}[[w_1, w_3], w_2] + \tfrac{1}{240}[w_1, [w_1, [w_1, w_3]]] - \tfrac{1}{180}[w_1, [[w_1, w_2], w_2]]\end{aligned}$$

3 HAMILTONIAN FORMALISM

An hamiltonian system is the given of a phase space E which can be identified to $\mathbb{R}^{2n}$, a set of variables

$$(q, p) = (q_1, \ldots, q_n, p_1, \ldots, p_n) = (z_1, \ldots, z_{2n}), \tag{28}$$

and an Hamiltonian $h = h(p, q, t)$. We consider the system of differential equations

$$\dot{p}_i = -\frac{\partial h}{\partial q_i}, \dot{q}_i = \frac{\partial h}{\partial p_i}, \quad 1 \leq i \leq n, \tag{29}$$

where $\dot{z} = \frac{dz}{dt}$ denotes the total time derivative. Introducing the Poisson bracket

$$\{f, g\} = \sum_{i=1}^{n} \frac{\partial f}{\partial p_i}\frac{\partial g}{\partial q_i} - \frac{\partial g}{\partial p_i}\frac{\partial f}{\partial q_i} \tag{30}$$

that turns the set of smooth functions on E onto a Lie algebra, (29) becomes

$$\dot{z}_i = \{z_i, h\} = -L_h z_i, \quad 1 \leq i \leq 2n, \tag{31}$$

and for any function f on the phase space we get

$$\dot{f} = \tfrac{\partial f}{\partial t} + \{f, h\} = -L_h f + \tfrac{\partial f}{\partial t} \tag{32}$$

along the trajectories. In particular, if h is not time-dependent, it is a first integral of the system.

From the Jacobi identity (2), we deduce that $[L_f, L_g] = L_{\{f,g\}}$, where $[,]$ denotes the commutator.

A transformation on the phase space E is said canonical if it preserves the Poisson brackets. Such transformations are also called symplectic as their Jacobians belong to the symplectic group. One extends the canonical transformations on the functions on the phase space by $Tf(z) = f(T(z))$. Canonical transformations act on the Lie algebra of the Lie operators by $TL_fT^{-1} = L_{Tf}$.

3.1 Time-evolution mapping

A canonical transformation appearing in hamiltonian mechanics is the time-evolution mapping $S_h(t) : z \mapsto z(t)$. From (31), $S_h(t)$ is the solution of the differential equation

$$\frac{d}{dt}S_h(t) = -S_h(t)L_h, S_h(0) = Id. \tag{33}$$

If h is not time-dependent we have $S_h(t) = e^{-tL_h}$ and a formal solution of (31) is given by its Taylor series, called Lie series

$$z(t) = \sum_{n\geq 0} t^n \frac{L_h^n}{n!} z. \tag{34}$$

For example, if $h = \frac{\omega}{2}(p^2 + q^2)$, then

$$S_h(t) = \begin{pmatrix} cos(\omega t) & -\sin(\omega t) \\ sin(\omega t) & \cos(\omega t) \end{pmatrix}.$$

If h is time-dependent, say for example $h = \sum_{n\geq 0} t^n h_n$, then $S_h(t)z$ may be written as a Lie series and from proposition (4), there exists $k = \sum_{n\geq 0} t^n k_n$ such that $S_h(t) = e^{-tL_k}$. Furthermore, we have an algorithm to compute k. k is an invariant function of the system but is not the Hamiltonian governing the system.

3.2 Discrete integration

h is not necessarly as simple as above and it may be quit hard to calculate $S_h(t)$ so we try to calculate truncated Lie series or approximating solutions by discrete integration.

The Euler scheme to integrate (29) on the path τ is the mapping

$$\begin{pmatrix} q \\ p \end{pmatrix} \to \begin{pmatrix} q_\tau \\ p_\tau \end{pmatrix} = \begin{pmatrix} q \\ p \end{pmatrix} + \tau \begin{pmatrix} \frac{\partial h}{\partial p} \\ -\frac{\partial h}{\partial q} \end{pmatrix} \tag{35}$$

which is generally not symplectic. When $h = \frac{\omega}{2}(p^2 + q^2)$ we obtain the mapping

$$\begin{pmatrix} p_\tau \\ q_\tau \end{pmatrix} = \begin{pmatrix} 1 & -\omega\tau \\ \omega\tau & 1 \end{pmatrix} \begin{pmatrix} p \\ q \end{pmatrix} \tag{36}$$

which is not symplectic since its determinant equals $1+\omega^2\tau^2$. Furthermore at each step τ, the energy grows by a factor $1+\omega^2\tau^2$. As the total energy is a first integral for the hamiltonian system, it is obvious that the difference between the exact solution and the discrete solution (36) grows secularly. Moreover the product $S(t)$ of the two symplectic transformations

$$\begin{cases} p \longrightarrow p - \tau\dfrac{\partial H}{\partial q} \\ q \longrightarrow q \end{cases} \quad \text{and} \quad \begin{cases} p \longrightarrow p \\ p \longrightarrow p + \tau\dfrac{\partial H}{\partial p} \end{cases} . \tag{37}$$

is symplectic. With $h = \frac{\omega}{2}\left(p^2+q^2\right)$, the previous operator becomes

$$S(\tau) = \begin{pmatrix} 1-\omega^2\tau^2 & \omega\tau \\ -\omega\tau & 1 \end{pmatrix}. \tag{38}$$

3.3 Symplectic schemes

The case when $h = T(p) + V(q)$ is a sum of a kinetic energy and a potential plays an important role. Let us denote by $S_T(t) = e^{-tL_T}$ and $S_V(t) = e^{-tL_V}$ the time-evolution mappings associated to T and V respectively. As $\{V,p\}$ depends on q and $\{T,q\}$ depends on p, we have

$$S_T(t)p = p, S_T(t)q = q + t\frac{\partial T}{\partial p}, S_V(t)p = p - t\frac{\partial V}{\partial q}, S_V(t)q = q. \tag{39}$$

On the other hand, we have through the Baker-Campbell-Hausdorff formula

$$S(t) = S_T(t)S_V(t) = e^{-t(L_T+L_V)+\frac{t^2}{2}L_{\{T,V\}}+o(t^2)} = S_h(t) + o(t) \tag{40}$$

that is to say $S(t)$ is a linear symplectic map close to the time evolution mapping.

The main idea is to construct for given n and k, $S^{(n)}(t) = S_h(t) + o(t^k)$ as

$$S^{(n)}(t) = S_T(c_1t)S_V(d_1t)\cdots S_T(c_nt)S_V(d_nt). \tag{41}$$

For a given integration step, we then obtain a series of transformations

$$q^0 = q, p^0 = p, p^{i+1} = p^i - d_i\frac{\partial V}{\partial q}(q^i), q^{i+1} = q^i + c_i\frac{\partial T}{\partial p}(p^{i+1}) \tag{42}$$

and $p_\tau = p^n, q_\tau = q^n$ are computed after n evaluations of $\dfrac{\partial T}{\partial p}$ and $\dfrac{\partial V}{\partial q}$.

3.4 Symplectic Integrators

Let us consider an Hamiltonian $H = A + B$, the two maps $S_A(t) = e^{-tL_A}$ and $S_B(t) = e^{-tL_B}$, and a given integer k, one seeks a minimal set of coefficients $c_1, \ldots, c_n, d_1, \ldots d_n$, such that

$$S_A(c_1 t) S_B(d_1 t) \cdots S_A(c_n t) S_B(d_n t) = e^{-tL_H} + o(t^k). \tag{43}$$

Direct method: This can be reformulated, looking for $c_1, \ldots, c_n, d_1, \ldots d_n$, such that

$$S_A(c_1 t) S_B(d_1 t) \cdots S_A(c_n t) S_B(d_n t) z = e^{-tL_H} z + o(t^k). \tag{44}$$

At each order p, let us consider $\tilde{L}_p(z, A, B)$, in which $|z|_* = 0$, $|A|_* = |B|_* = 1$. The part of order t^p of the left hand side of (43) belongs to the subspace of $\tilde{L}_{p+1}(z, A, B)$ generated by the Lie monomials in which z appears once. The dimension of this subspace is 2^p.

Invariant function: The problem (43) is equivalent to the finding of an invariant function $K(t) = H + o(t^{k-1})$ such that

$$S_A(c_1 t) S_B(d_1 t) \cdots S_A(c_n t) S_B(d_n t) = e^{-tK}. \tag{45}$$

Using the Baker-Campbell-Hausdorff formula, one express $K = \sum_{n \geq 0} t^n K_{n+1}$ onto $\tilde{L}(A, B)$, in which $|A|_* = |B|_* = 1$. K_p belongs to $\tilde{L}_p(A, B)$ which dimension is given by Witt's formula.

Perturbed Hamiltonian: The problem (43) is also equivalent to the finding of an Hamiltonian $W(t) = H + o(t^{k-1})$ such that

$$S_A(c_1 t) S_B(d_1 t) \cdots S_A(c_n t) S_B(d_n t) = S_W. \tag{46}$$

Using (13) and (14), one can express $W(t) = \sum_{n \geq 0} t^n W_{n+1}$ in $\tilde{L}(A, B)$. With this method, there is no need to calculate any exponential identity. We have to bear in mind that $S_W(t) = T_{\tilde{W}}$ in which $\frac{d\tilde{W}}{dt} = W$.

⁂

Using the third method, one gets the four first symplectic integrators.

- For $k = 1$ one has two linear equations and a solution is $c_1 = d_1 = 1$
- For $k = 2$ the solution is reached for $n = 2$ and $c_1 = c_2 = \frac{1}{2}, d_1 = 1$. We thus obtain the second-order symplectic integrator $S_2 = S_A(\frac{t}{2}) S_B(t) S_A(\frac{t}{2})$.
- For $k = 3$, we get after reduction

$$d_2 + 2c_1 = 1, d_1 - 2c_1 = 0, c_3 + c_1 = \tfrac{1}{2}, c_2 = \tfrac{1}{2}, c_1^2 - \tfrac{1}{2}c_1 + \tfrac{1}{12} = 0. \tag{47}$$

We thus have 2 complex solutions involving 5 factors. A real solution with 6 factors is for example (see also [7])

$$c_1 = -c_2 = \tfrac{2}{3}, d_3 = \tfrac{1}{24}, d_2 = -\tfrac{3}{4}, d_1 = -\tfrac{7}{24}, c_3 = 1. \tag{48}$$

- For $k=4$, we get after reduction by some Gröbner package implemented in Axiom

$$\left\{\begin{array}{r} d_3+288\,c_1^4-312\,c_1^3+96\,c_1^2-14\,c_1+\frac{1}{2}=0\\ d_2-288\,c_1^4+312\,c_1^3-96\,c_1^2+16\,c_1-\frac{3}{2}=0\\ d_1-2\,c_1=0\\ c_4+144\,c_1^4-156\,c_1^3+48\,c_1^2-7\,c_1+\frac{1}{4}=0\\ c_3+c_1-\frac{1}{2}=0\\ c_2-144\,c_1^4+156\,c_1^3-48\,c_1^2+7\,c_1-\frac{3}{4}=0\\ c_1^5-\frac{5}{4}\,c_1^4+\frac{13}{24}\,c_1^3-\frac{1}{8}\,c_1^2+\frac{1}{64}\,c_1-\frac{1}{1152}=0 \end{array}\right. . \qquad (49)$$

The last polynomial is factorized over $\mathbb{Q}$ in $(c_1^2-\frac{1}{4}\,c_1+\frac{1}{24})(c_1^3-c_1^2+\frac{1}{4}\,c_1-\frac{1}{48})$. Taking c_1 as one of the complex roots of $c_1^2-\frac{1}{4}\,c_1+\frac{1}{24}=0$, one finds

$$d_3+2\,c_1=\tfrac{1}{2}, d_2=\tfrac{1}{2}, d_1=2\,c_1, c_4+c_1=\tfrac{1}{4}, c_3+c_1=\tfrac{1}{2}, c_2-c_1=\tfrac{1}{4} \qquad (50)$$

or equivalently

$$c_4=\bar{c_1}, c_3=\bar{c_2}=\tfrac{1}{2}-c_1, d_3=\bar{d_1}=2\bar{c_1}, d_2=\tfrac{1}{2}. \qquad (51)$$

Taking c_1 as root of $c_1^3-c_1^2+\frac{1}{4}\,c_1-\frac{1}{48}=0$, for example $c_1=\frac{1}{2(\sqrt[3]{2}-1)}$, one gets

$$d_3=d_1=2c_1, d_2=1-4c_1, c_4=c_1, c_2=c_3=\tfrac{1}{2}-c_1. \qquad (52)$$

We found all the solutions for this integrator. The 3 above were already known.
- This method cannot be applied for $k=6$ as the set of equations is too big. There is no integrator with less than 13 factors, as tested with the Macaulay package.

The real valued integrators for $k=2$ or 4 are reversible, that means $S(-t)=S^{-1}(t)$.

3.5 Reversible Integrators

Representing a reversible integrator $S(t)$ by an exponential $\exp(-tL_K)$, we deduce that $K(t)=K(-t)$. Looking for reversible integrators, we can deduce from the Campbell-Hausdorff formula the

Lemma 10 ([13]). *If $S_{2k}(t)$ is a reversible symplectic integrator of order $2k$, then*

$$S(t)=S_{2k}\left(\frac{1}{2-\sqrt[2k+1]{2}}t\right)S_{2k}\left(-\frac{\sqrt[2k+1]{2}}{2-\sqrt[2k+1]{2}}t\right)S_{2k}\left(\frac{1}{2-\sqrt[2k+1]{2}}t\right)$$

is a reversible symplectic integrator of order $2k+2$.

This lemma allows us to built reversible symplectic integrators of order $2k$ as products of $2.3^{k-1}+1$ single operators S_A or S_B. With this method we should find a sixth-order integrator as a product of 19 operators.

⁂

One can try to find directly reversible integrators looking for

$$S_R^{(n)}(t) = S_A(c_n t)S_B(d_n t)\cdots S_A(c_1 t)S_B(d_n t)S_A(c_0 t)S_B(d_1 t)S_A(c_1 t)\cdots$$
$$S_B(d_n t)S_A(c_n t)$$

that we can express as an exponential or a Lie transform.

Representing $S_R^{(n)}$ as an exponential e^{-tL_K} has the advantage that $K(t) = K(-t)$. Moreover we have the following lemma resulting from (4)

Lemma 11. *If* $S_R^{(n)} = e^{-tL_K} = T_{tW}$ *with* $h = K + o(t^{2k-2}) = W + o(t^{2k-2})$, *then* $W = h + o(t^{2k-1})$.

As $S_R^{(n)}$ is reversible, $K(t) = A + B + \sum_{n\geq k} t^{2n} K_{2n+1}$. From the proposition (4), we have $W(t) = A + B + \sum_{n\geq 2k-1} t^n W_{n+1}$ in which $2kW_{2k} = K_{2k} + R_{2k}$ and $R_{2k} \in \tilde{L}_{\mathbb{Q}[c_0,\ldots,c_n,d_1,\ldots,d_n]}\{K_1,\ldots,K_{2k-1}\}$. As $K_2 = K_3 = \cdots = K_{2k} = 0$, we deduce that $W_{2k} = 0$.

This lemma proves that there is no need to consider odd terms of the Hamiltonian obtained with reversible integrators.

- For $k = 4$, one finds 3 reversible integrators obtained with the direct method.
- For $k = 6$, one proves that there is no solutions for $n < 8$. For $n = 8$, one sees, using the Hilbert function implemented in Macaulay, that the variety of solutions in $\mathbb{Z}/p\mathbb{Z}$ ($p = 31991$) is constituted of 39 points. There is at most 39 algebraic solutions over $\mathbb{Q}$.

⁂

Another solution has been proposed by Yoshida [16] consisting in the finding of reversible integrators as reversible product of second-order integrators S_2. We look for

$$S^{(n)}(t) = S_2(c_n t)\cdots S_2(c_1 t)S_2(c_0 t)S_2(c_1 t)\cdots S_2(c_n t) = e^{-tL_{K^{(n)}}}. \tag{53}$$

Since S_2 is reversible we have $S_2(t) = \exp(-tL_k)$ where

$$K = K_1 + t^2 K_3 + t^4 K_5 + t^6 K_7 + t^8 K_9 + o(t^{10}). \tag{54}$$

We work now on the free Lie algebra on the alphabet $X = \{K_1, K_3, K_5, K_7, K_9\}$ in which $|K_i|_* = i$. We therefore have

$$K^{(0)} = c_0 K_1 + t^2 c_0^3 K_3 + t^4 c_0^5 K_5 + t^6 c_0^7 K_7 + t^8 c_0^9 K_9 + o(t^{10}). \tag{55}$$

Bearing in mind that $K^{(n)}$ has only odd terms and belongs to $\tilde{L}(X)$, we will compute:

— the Lyndon basis of $L_{2n+1}(X)$ denoted by $K_{1,1}, K_{3,1}, \ldots, K_{5,1}, \ldots, K_{7,1}, \ldots, K_{9,1}, \ldots$

— $\log\left[S_2(xt)e^{-tL_{\tilde{H}}}S_2(xt)\right]$ at a given order k in which $\tilde{H} = \sum_{i,j} a_{i,j}K_{i,j}$ is generic.

- For $k = 4$, we find the real valued reversible integrator previously found by the direct method or using the lemma (10).
- For $k = 6$, we have four equations with four unknowns $c_0, \ldots, c_3$. The solution is obtained after eliminations with

$$\begin{aligned}P_0(c_0) = {} & c_0^{39} + 4\, c_0^{38} - 18\, c_0^{37} - \tfrac{232}{3}\, c_0^{36} + \tfrac{6469}{45}\, c_0^{35} + \tfrac{8108}{15}\, c_0^{34} - \tfrac{82144}{135}\, c_0^{33} - \\ & \tfrac{239008}{135}\, c_0^{32} + \tfrac{870652}{675}\, c_0^{31} + \tfrac{5898416}{2025}\, c_0^{30} - \tfrac{618824}{675}\, c_0^{29} - \tfrac{5158016}{2025}\, c_0^{28} + \\ & \tfrac{2525372}{30375}\, c_0^{27} + \tfrac{32135888}{30375}\, c_0^{26} - \tfrac{1377776}{10125}\, c_0^{25} - \tfrac{33361568}{91125}\, c_0^{24} + \tfrac{536566}{10125}\, c_0^{23} + \\ & \tfrac{35651416}{455625}\, c_0^{22} - \tfrac{19660868}{1366875}\, c_0^{21} - \tfrac{8051504}{455625}\, c_0^{20} + \tfrac{5636474}{1366875}\, c_0^{19} + \tfrac{11313208}{4100625}\, c_0^{18} - \\ & \tfrac{17674448}{20503125}\, c_0^{17} - \tfrac{8733536}{20503125}\, c_0^{16} + \tfrac{1302268}{6834375}\, c_0^{15} + \tfrac{87632}{2460375}\, c_0^{14} - \tfrac{624184}{20503125}\, c_0^{13} + \\ & \tfrac{288448}{922640625}\, c_0^{12} + \tfrac{3333844}{922640625}\, c_0^{11} - \tfrac{716752}{922640625}\, c_0^{10} - \tfrac{127664}{553584375}\, c_0^{9} + \tfrac{143264}{922640625}\, c_0^{8} - \\ & \tfrac{136499}{4613203125}\, c_0^{7} - \tfrac{19996}{8303765625}\, c_0^{6} + \tfrac{117142}{41518828125}\, c_0^{5} - \tfrac{33848}{41518828125}\, c_0^{4} + \\ & \tfrac{17431}{124556484375}\, c_0^{3} - \tfrac{9668}{622782421875}\, c_0^{2} + \tfrac{656}{622782421875}\, c_0 - \tfrac{64}{1868347265625} = 0\end{aligned}$$

  and $c_1 = P_1(c_0), c_2 = P_2(c_0), c_3 = P_3(c_0)$ where P_1, P_2, P_3 are polynomials of degree 38. P_0 is irreducible over $\mathbb{Q}$ and has only three real roots. Thus, we find 36 complex integrators and 3 real valued. All the solutions are reached with this method as there is at most 39 solutions.
- For $k = 8$, Yoshida ([16]) has found 5 real valued integrators using numerical methods. These integrators involve 31 single integrators S_A or S_B. We proved, using standard basis computed with Macaulay, that these integrators are not products of 5 fourth-order symplectic integrators.

3.6 Special cases

Most of the times, when $h = T(p) + V(q)$, the kinetic energy is just a quadratic form in p. That means that $\{T, V\}$ is of degree one in p, $\{\{T, V\}, V\}$ depends only on q and $\{\{\{T, V\}, V\}, V\} = 0$. We can therefore hope to find symplectic integrators of order 4 or 6 with less terms.

Unfortunately, there is no integrator of order 4 using less than 7 terms.

As $\{\{T, V\}, V\}$ depends only on q, $V_1 = \alpha V + t^2\beta\{\{T, V\}, V\}$ depends only on q and t for any α, β and we have like above

$$e^{-tL_{V_1}}p = p - t\frac{\partial V_1}{\partial q} \quad \text{and} \quad e^{-tL_{V_1}}q = q. \tag{56}$$

Denoting $e^{-t(\alpha L_V + \beta t^2 L_{\{\{T,V\},V\}})}$ by $S_{\alpha,\beta}(t)$ we look now for integrators $S^{(n)}$ as product of

$$S_{c_n,z_n}(t)S_T(d_nt)\cdots S_{c_1,z_1}(t)S_T(d_0t)S_{c_1,z_1}(t)\cdots S_T(d_nt)S_{c_n,z_n}(t) \tag{57}$$

or

$$S_T(d_n t) S_{c_n, z_n}(t) \cdots S_T(d_1 t) S_{c_0, z_0}(t) S_T(d_1 t) \cdots S_{c_n, z_n}(t) S_T(d_n t) \qquad (58)$$

With this method we found an integrator of order 4 as a product of 5 factors and an integrator of order 6 as product of 9 factors.

- For the fourth-order integrator, we find as general solution

$$z_1 + \tfrac{1}{2}\, z_0 + \tfrac{1}{48} = 0, d_1 - \tfrac{1}{2} = 0, c_0 - \tfrac{2}{3} = 0, c_1 - \tfrac{1}{6} = 0 \qquad (59)$$

  and we can take $z_1 = 0$, $z_0 = -\frac{1}{24}$, $d_1 = \frac{1}{2}, c_0 = \frac{2}{3}, c_1 = \frac{1}{6}$.
- For the sixth-order integrator, we find

$$\begin{gathered} z_2 = 0, z_1 - \tfrac{15}{16}\, c_2^2 + \tfrac{1}{4}\, c_2 - \tfrac{1}{96} = 0, z_0 - \tfrac{3}{8}\, c_2^2 - \tfrac{1}{4}\, c_2 + \tfrac{1}{48} = 0, \\ d_2 + 3\, c_2 - \tfrac{1}{2} = 0, d_1 - 3\, c_2 = 0, c_0 - 30\, c_2^2 + 12\, c_2 - 1 = 0, \\ c_1 + 15\, c_2^2 - 5\, c_2 = 0, c_2^3 - \tfrac{1}{2}\, c_2^2 + \tfrac{1}{18}\, c_2 - \tfrac{1}{540} = 0 \end{gathered}$$

Moreover the evaluation of $\frac{\partial V_1}{\partial q}$ requires 2 evaluations so the above integrators require in fact 6 evaluations and 12 evaluations at each step respectively, which is an improvement with respect to the integrators found in the general case.

***

In table 1., we give for each order and each method, the number of polynomial equations to solve, the number of operators involved in the integrators and the number of solutions. For the first column, D means *using the direct method (44)*, L means *using the Lie transformations or the exponential (46) or (45)*, RL means *looking for reversible integrators using Lie transformations (lemma (11))*, S_2 means *looking for reversible products of operators* S_2. In the column "solutions", when two numbers appear, the first one is the number of solutions and the second one the number of real solutions.

For example, in the row corresponding to the order 4, the set of algebraic equalities contains 30 polynomials (sum in the column) with the method D, 8 polynomials with the method L, 6 polynomials with the method RL and 2 polynomials with the method S_2. In the same row, one finds solutions that give integrators of order 4 as product of 7 operators. One finds 5 solutions with the method L (1 real solution), and 3 with the other methods. In the row corresponding to the order 3, there are 3 integrators as product of 5 operators but none is real valued and there is an infinity of real valued integrators as product of 6 single operators.

4 Conclusion

We showed in this paper that there are exactly 5 fourth-order symplectic integrators involving 7 operators. Three of them are known (see [7, 16]). We showed, that there are exactly 39 reversible sixth-order symplectic integrators involving

Order	Polynomials				Operators			Solutions		
			Reversible			Reversible			Reversible	
	D	L	RL	S_2	L	RL	S_2	L	RL	S_2
1	2	2	2		1			1		
2	4	1		1	3	3	3	1		
3	8	2	4		5			3		
					6			∞		
4	16	3		1	7	7	7	5-1	3-1	3-1
5	32	6	6		11?			46?		
6	64	9		2	≥ 13	15	15		39	39-3
7	128	18	18							
8	256	30		4			31			?-5
9	512	56	56							
10	1024	99		8						

Table 1. — Symplectic Integrators —

15 operators. All of them are reversible products of second-order integrators. Three of them were known ([16]). In ([16]), Yoshida has found 5 eight-order integrators involving 31 operators: here we proved that there are not reversible products of 5 fourth-order integrators.

We have shown why the Lyndon basis is particularly adapted for computing the relations between Lie transforms and exponentials. We also found integrators in the case when $h = T(p) + V(q)$ and $T(p)$ is a quadratic form in p.

Bibliography

[1] Bourbaki, N.: Groupes et algèbres de Lie, Éléments de Mathématiques, Hermann, Paris, 1972.

[2] Cary, J.R.: Lie Transform Perturbation Theory for Hamiltonian Systems, in Physics Reports, North-Holland Publishing Company **79-2** (1981), 129–159.

[3] Deprit, A.: Canonical transformations depending on a small parameter, Cel. Mech. **1** (1969), 12–30.

[4] Dragt, A. J., Finn, J. M.: Lie Series and invariant functions for analytic symplectic maps, J. Math. Physic **17** (1976), 2215–2227.

[5] Dragt, A. J., Healy, L. M.: Concatenation of Lie Algebraic Maps, in Lie Methods in Optics II, Lec. Notes in Physics **352** (1988).

[6] Finn, J. M.: Lie Series: a Perspective, Local and Global Methods of nonlinear Dynamics, Lec Notes in Physics **252** (1984), 63–86.

[7] Forest, E., Ruth, D.: Fourth-Order Symplectic Integration, Physica D **43** (1990), 105–117.

[8] Koseleff, P.-V., Thèse de troisième cycle, École Polytechnique, (to appear).

[9] Michel, J.: Bases des Algèbres de Lie Libres, Étude des coefficients de la formule de Campbell-Hausdorff, Thèse, Orsay, 1974.

[10] Perrin, D.: Factorization of free monoids, in Lothaire M., Combinatorics On Words, Chap. 5, Addison-Wesley (1983).

[11] Petitot, M.: Algèbre non commutative en Scratchpad : application au problème de la réalisation minimale analytique, Thèse, Université de Lille I, 1991.

[12] Steinberg, S.: Lie Series, Lie Transformations, and their Applications, in Lie Methods in Optics, Lec. Notes in Physics **250** (1985).

[13] Suzuki M.: Fractal Decomposition of Exponential Operators with Applications to Many-Body Theories and Monte Carlo Simulations, Ph. Letters A **146** (1990), 319–323.

[14] Wisdom J., Holman, M.: Symplectic Maps for the N-Body Problem, The Astr. J. **102(4)** (1991), 1528–1538.

[15] Yoshida, H.: Conserved Quantities of Symplectic Integrators for Hamiltonian Systems, Physica D (1990).

[16] Yoshida, H.: Construction Of Higher Order Symplectic Integrators, Ph. Letters A **150**, (1990), 262–268.

Exponential sums as discrete Fourier transform with invariant phase functions

Gilles Lachaud
Laboratoire de Mathématiques Discrètes du C.N.R.S.
Luminy Case 930, 13288 Marseille cedex 9, France

Abstract

We give estimates for exponential sums over finite fields in several variables. We study the case where the phase is either quadratic or more generally invariant under the action of a finite group. The bounds obtained are better than the general ones; they imply some estimates for certain sums in one variable, and for the number of solutions of the trace equation $Tr(x^d + vx) = 0$. In an appendix we discuss the link between exponential sums and bent functions.

1 Introduction

Exponential sums play in finite mathematics the role of the oscillating integrals in applied mathematics. They are naturally linked with the theory of equations over finite fields, as it has been noticed since the early nineteenth century, through the works of Lagrange and Gauss; and as such, they are used nowadays as a technical tool as soon as one studies linear recurring sequences (feedback shift registers) or difference sets and bent functions (cf. section 9). Exponential sums also have applications to the computations of the weights of linear codes : we refer for instance to [12], [13], [21] for these applications. In particular the inequalities given in section 8 are applied by Wolfmann (cf. [22]) in order to derive some bounds for the weights of certain families of binary linear codes.

Let q be a power of a prime p, and let $\mathbf{F}_q$ be the field with q elements. If $x \in \mathbf{F}_q$ let

$$\psi(x) = exp(\frac{2i\pi}{p} Tr_{\mathbf{F}_q/\mathbf{F}_p} x),$$

where $Tr_{\mathbf{F}_q/\mathbf{F}_p} x$ is the trace of x in the extension $\mathbf{F}_q/\mathbf{F}_p$. We write $V = \mathbf{F}_q^n$ and we note $x.y$ the usual scalar product of two elements x and y in V. If F is a function defined on V, the *Fourier transform* of F is

$$\hat{F}(v) = \sum_{x \in V} F(x)\overline{\psi}(v.x).$$

We shall confine here to the case where F comes from a phase function, that is

$$F(x) = \psi(f(x))$$

where *the phase function* $f = f(X_1, \ldots, X_n) \in \mathbf{F}_q[X_1, \ldots, X_n]$ is a polynomial $\neq 0$ with n variables; hence it is better to introduce the *exponential sum* associated to f and $v \in V$:

$$S(f, v = \sum_{x \in V} \psi(f(x) - v.x).$$

This is the discrete analog of the *oscillating integral* in the theory of the Fourier transform in the usual euclidean space :

$$I(f, v) = \int_{x \in V} a(x)\ e^{2i\pi(f(x) - v.x)},$$

where a is any smooth amplitude function with compact support. This topic is classical:for instance we recall in section 2 the results on quadratic phases, and in section 9 we connect exponential sums to the theory of bent functions. There is nothing essentially new in these sections, except perhaps in the formulation. On the other hand, apart from those attributed to somebody else, the theorems and corollaries of sections 3 to 8 are new.

2 Quadratic phases

Our purpose is to give bounds for exponential sums if we assume that the phase function satisfies certain invariance properties. The most classical instances of invariant functions are quadratic forms : they are invariant under their orthogonal group. But in that case, it is possible to compute explicitely the exponential sum defined by a quadratic phase function. In order to take care of characteristic two, we define a quadratic form as a function $f : V \to \mathbf{F}_q$ satisfying

$$f(\lambda x) = \lambda^2 f(x) \; if \lambda \in \mathbf{F}_q \; and \; x \in V,$$

$$f(x+y) = f(x) + f(y) + B(x,y) \; if \; x \; and \; y \in V,$$

where B is a symmetric bilinear form on V. Let

$$V^0 = \{v \in V \mid B(v,x) = 0, x \in V\}$$

the subspace which is the orthogonal of V, and let $c(f) = corank(B) = dim\ V^0$. The form f is nondegenerate if and only if $V^0 = \{0\}$. If the characteristic p is odd, then the restriction of f to V^0 is identically zero. If the characteristic p is even and if f is nondegenerate, the dimension of V is even.

Theorem. *Let f be a quadratic form on V.*
a) *if $v \in V$, then*

$$\mid S(f,v) \mid^2 = q^n \sum_{x \in V^0} \psi(f(x) - v.x);$$

in particular if the restriction of f to V^0 is identically zero, then

$$\mid S(f,v) \mid \leq q^{\frac{n+c(f)}{2}};$$

b) If f is identically zero on V^0, then

$$\mid S(f,0) \mid^2 = q^{n+c(f)};$$

c) Assume $S(f,0) \neq 0$. Let $r : V \to V$ be the linear operator with kernel V^0 such that $B(x,y) = r(x).y$; if there is $u \in V$ such that $r(u) = v$, then

$$S(f,v) = \psi(f(u) - v.u) S(f,0);$$

if there is no such u, then $S(f,v) = 0$.

This is an analog of the *stationary phase method.* We recover here in the case of finite fields a general formula of Weil [20], théorème 2; this meets also some results of Carlitz [2]. There are analogous results for hermitian forms.

3 Invariant phases

We now give bounds on exponential sums when f is invariant under the action of a finite group : such phase functions have already been considered by Mordell [15] and Ono [16]. Let there be given a linear representation ρ of a finite group G in V, and assume that the function f is nonzero and invariant under G, that is satisfies

$$f(\rho(g)x) = f(x)$$

for every $x \in V$ and every $g \in G$.We set

$$G(v) = \{g \in G \mid \rho(g)v = v\}.$$

Theorem. *If $v \neq 0$ and if f is invariant under G, then*

(1) $$| S(f,v) | \leq \left(\frac{\#G(v)}{\#G}\right)^{\frac{1}{2}} q^n.$$

If f is homogeneous of degree d, we take

$$G = \{t \in \mathbf{F}_q^\times \mid t^d = 1\},$$

then $G(v) = \{1\}$ and $\#G = (d, q-1)$,whence :

Corollary. *If f is homogeneous of degree $d \geq 2$ and if $v \neq 0$, then*

(2) $$| S(f,v) | \leq \frac{q^n}{\delta^{\frac{1}{2}}},$$

where $\delta = (d, q-1)$.

4 Application : sums in one variable I

If we apply the preceding results to the binary form

$$f(x_1, x_2) = x_1^d - x_1^d,$$

where $d \geq 2$, we get estimates on the sum

$$S(ax^d, v) = \sum_{x \in \mathbf{F}_q} \psi(ax^d + vx).$$

For that sum, the Weil inequality holds :

(3) $$| S(ax^d, v) | \leq (d-1) q^{\frac{1}{2}}.$$

If $v = 0$, the inequality due to Hardy and Littlewood [6] if $q = p$ and to Hua and Vandiver [7] in any case is the following :

$$(4) \qquad | S(ax^d, 0) | \leq (\delta - 1)q^{\frac{1}{2}},$$

where $\delta = (d, q-1)$. In fact, the mean value of these sums is much smaller, since

$$\sum_{a \in \mathbf{F}_q} | S(ax^d, v) |^2 = q(q - d + 1).$$

Theorem. *If $a \in \mathbf{F}_q^\times$ and if $v \in V$, $v \neq 0$, then*

$$(5) \qquad | S(ax^d, v) | \leq \frac{q}{\delta^{\frac{1}{2}}}.$$

This inequality generalizes that of Akulinicev [1].

Another example can be given for exponential sums where the phase is a *norm form*. Let there be given a tower of fields :

$$\mathbf{F}_p \subset \mathbf{F}_q \subset \mathbf{F}_Q,$$

and consider the norm form of the extension $\mathbf{F}_Q/\mathbf{F}_q$:

$$N(x) = N_{\mathbf{F}_Q/\mathbf{F}_q}(x) = x_\nu, \nu = \frac{Q-1}{q-1}, x \in \mathbf{F}_Q.$$

Take $V = \mathbf{F}_Q$, and $N = N_{\mathbf{F}_Q/\mathbf{F}_q}$ as phase function, in such a way that

$$S(N, v) = \sum_{x \in \mathbf{F}_Q} \psi(x^\nu + vx);$$

then $S(N, 0) = 1 - \nu$.

Corollary. *If $v \neq 0$, then*

$$(6) \qquad | \sum_{x \in \mathbf{F}_Q} \psi(x^\nu + vx) | \leq \left(\frac{q-1}{Q-1}\right)^{\frac{1}{2}} Q.$$

As a third example, we use the *determinant* as phase function. Let $V = M(m, \mathbf{F}_q)$ be the vector space of square matrices of size m. Take

$$f(x) = det(x), x \in M(m, \mathbf{F}_q).$$

Let

$$\phi(q, m) = \prod_{i=1}^{m} \left(1 - \frac{1}{q^i}\right).$$

Corollary. *If $v \in M(m, \mathbf{F}_q)$ is inversible, then*

(7) $$\left| \sum_{x \in M(m,\mathbf{F}_q)} \psi(det(x) + v.x) \right| \leq \left(\frac{q-1}{\phi(q,m)} \right)^{\frac{1}{2}} q^{\frac{m^2}{2}}.$$

If $q = 2$ then $\phi(2, m) \geq 0.288\ldots$, hence

$$\left| \sum_{x \in M(m,\mathbf{F}_2)} \psi(det(x) + v.x) \right| \leq c\, 2^{\frac{m^2}{2}},$$

with $c = 1.86084\ldots$

5 Semi-invariant phases

As long as one studies general exponential sums in several variables, the bounds are not sharp. The general inequality of [5], which is a generalization of the Carlitz-Uchiyama inequality, is the following :

Theorem (Deligne). *If $f \in \mathbf{F}_q[X_1, \ldots, X_n]$ is a polynomial function such that $f \neq g^p - g + c^{st}$ for any polynomial $g \in \mathbf{F}_q[X_1, \ldots, X_n]$ then*

(8) $$| S(f, v) | \leq (d-1) q^{n-\frac{1}{2}}.$$

We have also (cf.[10]) :

Theorem (Katz). *If $f \in \mathbf{F}_q[X_1, \ldots, X_n]$ is a polynomial function such that the hypersurface $f = c^{st}$ is generically irreducible, then*

$$| S(f, v) | \leq A\, q^{n-1}.$$

The defect of this inequality is that the constant A is not explicit.

One can improve these results when one consider a slightly more general situation than in section 3. Let there be given this time a linear representation $\rho : G \rightarrow GL(V)$ of a finite group G in V, and a character $\omega : G \rightarrow \mathbf{F}_q^{\times}$. We assume now that the function f is *semi-invariant* under G of weight ω, i.e. satisfies

$$f(\rho(g)x) = \omega(g) f(x)$$

for every $x \in V$ and every $g \in G$. For $v \in V$ we set

$$G(1, v) = \{g \in G \mid \omega(g) = 1 \text{ and } \rho(g)v = v\}.$$

Let W be a stable subspace of G : $\rho(G)W \subset W$. The behaviour of $S(f, v)$ depends for $v \in W$ on the size of the algebraic set $X(W)$ in $V \times V$ defined by the system

$$f(x) = f(y),$$

$$w.x = w.y \; for \; w \in W.$$

Theorem. *If $v \in W$, then*

$$(9) \qquad | S(f,v) |^2 \leq \left(\frac{\#G(1,v)}{\#G}\right) \left(q^{\dim W+1} \#X(W) - q^{2n}\right)$$

The simplest instance of semi-invariant functions are still the homogeneous functions of degree d : in this case we choose a non-zero $v \in V$, take $G = \mathbf{F}_q^{\times}$, $W = \mathbf{F}_q v$, and

$$\rho(t)w = tw, \omega(t) = t^d$$

for $t \in G$ and $w \in W$. Here the algebraic set $X(W)$ is the hypersurface

$$X(v) : f(x) = f(y), v.x = v.y.$$

We obtain a bound on the size of $X(v)$ with the help of the following result (cf. [19]) which is an answer to a question of Tsfasman :

Theorem (Serre). *Let F be a non-zero homogeneous form with $N \geq 2$ variables of degree $d \leq q$; if*

$$S = \{x \in \mathbf{F}_q^N \mid F(x) = 0\},$$

then

$$\#S - q^{N-1} \leq (d-1)(q-1)q^{N-2}.$$

From the preceding we deduce :

Theorem. *Assume $n \geq 2$ and $2 \leq d \leq q-1$; if f is homogeneous of degree d and if $X(v)$ is of codimension 2, then*

$$(10) \qquad | S(f,v) | \leq (d-1)^{\frac{1}{2}} q^{n-\frac{1}{2}}$$

$$(11) \qquad | S(f,0) | \leq \delta^{-\frac{1}{2}} (d-1)^{\frac{1}{2}} q^{n-\frac{1}{2}}.$$

Under the assumptions which have been made, the inequality (10) is better than (8) in any case, and (11) is better than (8) if $\delta > 1$.

The following inequality is derived from the preceding ones, but the right hand side is independent of d.

Corollary. *Assume $d \geq 2$ and $n \geq 2$. If $d \mid q-1$, and if codim $X(v) = 2$, then*

$$(12) \qquad | S(f,v) | \leq q^{n-\frac{1}{4}}.$$

6 Application : sums in one variable II

We now apply like in section 4 the results of section 5 to sums in one variable.

Theorem. *If* $2 \le d \le q-1$, *if* d *is not a power of* p, *if* $a \in \mathbf{F}_q^\times$, *and if* $v \in \mathbf{F}_q^\times$, *then*

(13) $$| S(ax^d, v) | \le (d-1)^{\frac{1}{4}} q^{\frac{3}{4}},$$

and

(14) $$| S(ax^d, 0) | \le \delta^{\frac{1}{4}} (d-1)^{\frac{1}{4}} q^{\frac{3}{4}}.$$

This theorem is an explicit version of the rough estimate of Davenport and Heilbronn [3] used in the *circle method* :

$$| S(ax^d, v) | = O(q^{\frac{3}{4}})$$

if $q \to \infty$ and d is fixed. If $v \ne 0$, the inequality (13) generalizes an inequality of Karacuba [9]; it is not trivial if $2 \le d \le q-1$. If $d > q^{\frac{1}{3}}$, this inequality is better than Weil inequality (3) in this case. If $v = 0$, the inequality (14) is better than the one of Hardy-Littlewood (4) if $d > 1 + \sqrt{q}$.

A combination of (5) and (13) leads to the following :

Corollary. *if* $d \ge 2$ *and if* $d \mid q-1$, *if* $a \in \mathbf{F}_q^\times$ *and if* $v \in \mathbf{F}_q^\times$, *then*

(15) $$| S(ax^d, v) | \le q^{\frac{5}{6}}.$$

7 Nondegenerate forms

A polynomial $f = f(X_1, \ldots, X_n) \in \mathbf{F}_q[X_1, \ldots, X_n]$ is *nondegenerate* if its discriminant is $\ne 0$. This is so if and only if

$$\textit{grad } f(x) = 0 \textit{ and } f(x) = 0 \Rightarrow x = 0;$$

this means that the hypersurface defined by f in the projective space $\mathbf{P}^n(\mathbf{F}_q)$ is smooth.

Let

$$b_n(d) = (d-1)\frac{(d-1)^n - (-1)^n}{d};$$

this is the (corrected) *Betti number* of smooth hypersurfaces of degree d in $\mathbf{P}^n$, which is also the degree of the numerator of the zeta functions of these hypersurfaces.

For nondegenerate phases, the major inequality (cf. [4], [18]) is as follows.

Theorem (Deligne). *If $f \in \mathbf{F}_q[X_1, \ldots, X_n]$ is nondegenerate and homogeneous of degree d, and if $(d,p) = 1$, then*

$$(16) \qquad | S(f,v) | \leq (d-1)^n \, q^{\frac{n}{2}}.$$

From Deligne theorem on Weil conjectures ([4]) and 9, we get :

Theorem. *Assume $n \geq 2$, $2 \leq d \leq q-1$, and $(d,p) = 1$. Assume moreover that $f \in \mathbf{F}_q[X_1, \ldots, X_n]$ is nondegenerate and homogeneous of degree d; then*

$$(17) \qquad | S(f,0) | \leq \delta^{\frac{1}{2}} \, b_{2n-1}(d)^{\frac{1}{2}} \, q^{\frac{n}{2}}.$$

This theorem implies

$$(18) \qquad | S(f,0) | \leq \delta^{\frac{1}{2}} \, (d-1)^{n-\frac{1}{2}} \, q^{\frac{n}{2}};$$

This is slightly better than (16) if $\delta < d$.

Theorem. *Assume f nondegenerate and homogeneous of degree d; if the varieties $f = 0$ and $v.x = 0$ are mutually transverse in V, then*

$$(19) \qquad | S(f,v) | \leq b_{2n-2}(d)^{\frac{1}{2}} \, q^{\frac{n}{2}+\frac{1}{4}}.$$

The preceding result is generic in some sense, since the varieties $f = 0$ and $v.x = 0$ are mutually transverse in V if and only if v does not belong to a certain variety, thanks to the elimination theorem. The inequality (19) implies

$$(20) \qquad | S(f,v) | \leq (d-1)^{n-1} \, q^{\frac{n}{2}+\frac{1}{4}},$$

which is better than (16) if and only if $d \geq 1 + q^{\frac{1}{4}}$.

Corollary. *Assume $d \geq 2$ and $v_1^d - v_2^d \neq 0$; then*

$$(21) \qquad | S(ax^d, v_1)\bar{S}(ax^d, v_2) | \leq (d-1) \, q^{\frac{5}{4}}.$$

This has to be compared to the estimates given in section 4 and 6 !

8 The trace equation

Assume $q > 2$, and $2 \leq d \leq q-1$. The preceding results on $S(f,v)$ give bounds on the number of points on the affine *Artin-Schreier variety* :

$$X : y^p - y = f(x),$$

since

$$\#X(\mathbf{F}_q) = q + \sum_{c \in \mathbf{F}_q} S(cf, cv);$$

for instance we get from (13), with the hypotheses made there :

(22) $$| \#X(\mathbf{F}_q) - q | \leq (p-1)(d-1)^{\frac{1}{4}} q^{\frac{3}{4}}.$$

The number $T(ax^d, v)$ of solutions $x \in \mathbf{F}_q$ of the equation

$$Tr_{\mathbf{F}_q/\mathbf{F}_p}(ax^d + vx) = 0$$

satisfies

$$pT(ax^d, v) = \#X(\mathbf{F}_q);$$

we thus get

(23) $$| T(ax^d, v) - \frac{q}{p} | \leq \frac{p-1}{p} (d-1)^{\frac{1}{4}} q^{\frac{1}{4}}$$

and this gives information on the weights of some families of binary linear codes [22].

9 Appendix : bent functions

We show here how exponential sums link to the theory of bent functions. Let $R(V)$ be the space of all functions on $V = \mathbf{F}_q^n$, and for $f \in R(V)$ let

$$\| S(f) \| = Max_{v \in V} | S(f, v) | .$$

Define

$$\mu_q(n) = Min_{f \in R(V)} \| S(f) \| .$$

Definition. *We say that* $f \in \mathbf{F}_q^n$ is a bent function *if* $\| S(f) \| = \mu_q(n)$.

We have

$$\mu_q(n) = \mu_p(nt) \; if \; q = p^t,$$

since every function on $\mathbf{F}_p^{nt}$ can be written as the trace of a function on $\mathbf{F}_q^n$.

Proposition. a) *if q is odd or if q is even and n is even then*

$$\mu_q(n) = q^{\frac{n}{2}};$$

b) if q is even and n is odd then

$$q^{\frac{n}{2}} \leq \mu_q(n) \leq \sqrt{2}\, q^{\frac{n}{2}};$$

c) if $\| S(f) \| = q^{\frac{n}{2}}$, then f is a bent function.

Since $\mu_q(n) = 2^{\frac{n}{2}}$ if $q = 2$ and n is even, our definition of bent functions generalize the one given in that case in [14], Ch. 14. Bent functions in characteristic $\neq 2$ have already been considered in [8] and [11].

The theorem of section 2 shows that quadratic forms of maximal rank are bent if q is odd or if q is even and n is even.

If q is even and n is odd, notice that we have

$$\mu_2(2k+1) = 2^{k+1} \; for \; 0 \leq k \leq 3,$$

that one deduces from the results of Patterson and Wiedemann [17] that

$$\mu_2(15) \leq 216 < 2^8 = 256,$$

and that one may conjecture

$$\mu_2(n) \sim 2^{n/2} \; if \; n \to \infty.$$

Remark. Assume $q = 2$. Recall that the *distance* $d(f,g)$ between two functions f and g defined on V is the number of $x \in V$ for which $f(x) \neq g(x)$. Let A be the subspace of affine functions on V; every $g \in A$ writes $g(x) = v.x$ or $g(x) = v.x + 1$ with $v \in V$; if we define

$$r(f) = Min_{g \in A} d(f,g),$$

it is not hard to see that

$$r(f) = 2^{n-1} - \frac{1}{2} \parallel S(f) \parallel .$$

Now let

$$r_q(n) = Max_{f \in R(V)} r(f),$$

the preceding relation shows that

$$r_q(n) = 2^{n-1} - \frac{1}{2}\mu_q(n),$$

and *f is a bent function if and only if $r(f) = r_q(n)$* : this is why bent functions are named *maximally nonlinear* in russian.

References

[1] Akulinicev, N.M., *Estimates for rational trigonometric sums of a special type*, Dokl. Akad. Nauk SSSR **161 (1965)**, 743-745; = Soviet Math. Dokl. **6** (1965), 480-482.

[2] Carlitz, L., *Evaluation of some exponential sums over a finite field*, Math. Nachr. **96** (1980), 319-339.

[3] Davenport, H., Heilbronn, H., *On an exponential sum*, Proc. London Math. Soc. (2), **41** (1936), 449-453; = Davenport, H., Collected Works, vol. IV, n° 14, 1524-1528.

[4] Deligne, P., *La conjecture de Weil I*, Publ. Math. I.H.E.S. **43** (1974), 273-307.

[5] Deligne, P., *Applications de la formule des traces aux sommes trigonométriques*, p. 168-232 in Séminaire de Géométrie Algébrique du Bois-Marie SGA 4 1/2, Lect. Notes in Math. **569**, Springer, Berlin 1977.

[6] Hardy, G.H., Littlewood, J.E., *Some problems of partitio numerorum IV. The singular series in Warings problem and the value of the number G(k)*, Math. Z. **12** (1922), 161-188; = Hardy, G.H., Coll. Works, vol. 1, 478-505, Oxford University Press, 1966.

[7] Hua, L.K., Vandiver, H.S., *On the existence of solutions of certain equations in a finite field*, Proc. Nat. Acad. Sci. U.S.A. **34** (1948), 258-263.

[8] Ipatov, V.P., Kalametdinov, B. Zh., *Ensembles of p-ary sequences of bent functions*, Izv. Vyssh. Uchebn. Zaved. Mat. (1988), n° 3, 26-32, 86; = Soviet Math. (Iz. VUZ) **32** (1988),n° 3, 34-43.

[9] Karacuba, A.A., *Estimates of complete trigonometric sums*, Mat. Zametki **1** (1967), 199-208; engl. transl., Math. Notes **1** (1967), 133-139.

[10] Katz, N.M, *sommes exponentielles*, Astérisque **79**, S.M.F., Paris, 1980.

[11] Kumar, P. V., Scholtz, R.A., Welch, L.R., *Generalized bent functions and their properties*, J. combinatorial Theory Ser. A **40** (1985), 90-107.

[12] Lachaud, G., *Artin-Schreier curves, exponential sums, and the Carlitz-Uchiyama bound for geometric codes*, Journal of Number Theory **39** (1991), p. 18-40.

[13] Lachaud, G., *Artin-Schreier curves, exponential sums, and coding theory*, Theoretical Computer Science **94** (1992), p.295-310.

[14] MacWilliams, F.J., Sloane, N.J.A., *The Theory of Error-Correcting Codes*,North-Holland, Amsterdam, 1977.

[15] Mordell, L.J., *On a sum analogous to a Gausss sum*, Quart. J. of Math. **3** (1932), 161-167.

[16] Ono, T., *A remark on Gaussian sums and algebraic groups*, J. Math. Kyoto Univ. **13** (1973), 139-142.

[17] Patterson, N.J., Wiedemann, D.H., *The covering radius of the* $(215, 16)$ *Reed-Muller code is at least* 16276, IEEE Trans. Inform. Theory **29** (1983), 354-356.

[18] Serre, J-P., *Majoration de sommes exponentielles*, Journées Arith. Caen, Astérisque **41-42** (1977), 111-126; = Oeuvres, t.III, $n°$ 111, 368-383, Springer, Berlin, 1986.

[19] Serre, J-P., *Lettre à M. Tsfasman*, Journées Arithmétiques de Luminy, Astérisque **198-200** (1992), p.351-353.

[20] Weil, A., *Sur certains groupes d'opérateurs unitaires*, Acta Math. **111** (1964), 143-211; = Coll. papers, III, [1964b].

[21] Wolfmann, J., *New bounds on cyclic codes from algebraic curves*, Lect. Notes in Comp. Sci., vol. 388, Springer, New York, 1989, p. 47-62.

[22] Wolfmann, J., *Polynomial description of binary linear codes and related properties*, AAECC **2** (1991).

Application of Finite Fields to Memory Interleaving

Abraham Lempel[1] and Gadiel Seroussi

Hewlett-Packard Laboratories
1501 Page Mill Road
Palo Alto, CA 94304

[1] Also, Technion - Israel Institute of Technology, Haifa, Israel.

Abstract. A hashing scheme for memory interleaving, based on the properties of finite field exponentiation is presented. The scheme maps sequences of l-bit addresses to sequences of m-bit memory module numbers, with $m<l$. For input sequences that are in arithmetic progression, the output sequence has a provably uniform distribution on the average, and no "pathologies" for a prescribed range of strides in the input sequence. We prove bounds on the lengths of runs in the output sequence, and prove the surprising result that when m/l is bounded away from both 0 and 1, the run length can be bounded by a *constant*. The proposed scheme is highly amenable to fast systolic implementation.

1 Introduction

With the continuous advances in semiconductor technology and architecture design, computer CPU clock rates have increased dramatically in recent years, in a trend that is expected to continue in the near future. Memory chips, on the other hand, have seen significant but less dramatic increases in speed. In order to take full advantage of a powerful CPU, a computer memory system must be able to sustain data transfer rates matched to the CPU processing capabilities. One solution to this problem is offered by *memory interleaving* [1]-[5]. In an interleaved system, a main memory of size 2^l is divided into M *modules*, where M is a positive integer (usually, $M=2^m$ for some integer m such that $0<m<l$, l being the number of bits in a main memory address). Each main memory address is mapped to a module, and to an address within that module. Such a mapping is called a *hashing scheme*. Clearly, the mapping must be one-to-one.

Ideally, memory addresses would be distributed among the modules so that the CPU would access each module exactly once in M consecutive clock cycles. This would allow each module to run at a clock rate M times slower than the CPU rate. In practice, however, the sequence of memory addresses accessed by a computer program is unpredictable at computer design time, and it varies widely among different computer programs. Therefore, for any hashing scheme, there will be "pathological" address sequences that would map, in the worst case, all the accesses to the same memory module. This situation can be alleviated by buffering the accesses to the modules. A hashing scheme is then required to distribute addresses uniformly "on the average". To this end, the hashing scheme must perform well for the most common memory access patterns. An important instance of these are arithmetic progressions of the form $a_0, a_0+s, a_0+2s, \cdots, a_0+ks, \cdots$. Here, a_0 is a *base address*, and s is a fixed *stride*. Arithmetic progressions occur when a program is sequentially accessing the elements of an array of fixed size data objects. The base address is the address of the beginning of the array, and the stride is the size of the data objects.

In addition to a uniform address distribution on the average, it is also desirable to limit the number of times a given memory module can be accessed in consecutive cycles. Hence, we would like to bound the lengths of *runs* in the output sequence.

In this paper, we present a memory hashing scheme that performs provably well for *all* stride values up to a designed upper bound. For values above the upper bound, the set of "pathological" stride values is completely characterized. The proposed hashing scheme, which is based on finite field exponentiation, is described in Section 2. There, we prove the main mathematical properties of the scheme, and bound the lengths of runs in the output sequece. We prove that, when m/l is bounded away from both 0 and 1, the run length can be bounded by a *constant* independent of l and m. The computations involved in the hashing scheme are amenable to pipelining and, thus, to a fast systolic implementation. This is described in section 3.

2 The Hashing Scheme

Let $F = GF(2^n)$, for $n \geq 1$. In the sequel, we shall make free use of well known properties of F as presented, for instance, in [6],[7].

We are now ready to describe our hashing scheme. Let l denote the length of a main memory address, and let $M = 2^m$ denote the number of interleaved memory modules. Let S denote the maximum expected stride value (i.e. the maximum chosen stride value for which the performance of the hashing scheme is guaranteed), and let n be a positive integer such that:

$$m \leq n \leq l,$$

$$S < 2^n - 1,$$

and $p = 2^n - 1$ is a prime number (a *Mersenne* prime).[1] Choose a primitive

1. While the question of existence of an infinite sequence of Mersenne primes is, of course, an open one, for practical values of l, n can be taken from the list {3, 5, 7, 13, 17, 19, 31}. Also, most of the properties of the proposed hashing scheme remain valid even for composite $2^n - 1$ if we define p to be the smallest prime divisor of $2^n - 1$, and we take $S < p$.

element $\alpha \in F$. Let a denote an address in the range $0 \le a \le 2^l - 1$. The address a is mapped to a memory module number B and an internal address C (within module B) whose binary representations are obtained as follows:

1. Compute the n-tuple $A = (A(0)\, A(1) \cdots A(n\text{-}1)) \triangleq \alpha^a$, $A(i) \in \{0,1\}$.

2. Set $B = (A(0)\, A(1) \cdots A(m\text{-}1))$.

3. Set $D = (A(m)\, A(m+1) \cdots A(n\text{-}1))$, and compute the $(l-n+1)$-tuple

$$E = \left\lfloor \frac{a}{p} \right\rfloor .$$

Then, $C = (D \mid E)$ (the binary concatenation of D and E).

In Step 1 above, one needs to know the basis used to represent elements of F as binary n-tuples. For the time being, we assume that an arbitrary basis is used. Later, we will show how some of the properties of the hashing scheme depend on the choice of basis.

Proposition 1. *The mapping $a \to (B,C)$ is one-to-one.*

Proof. Let $a' = a \bmod p$. Then, due to the uniqueness of quotient and remainder in integer division, the mapping $a \to (a',E)$ is one-to-one. Since $\alpha^p = 1$, we have $\alpha^a = \alpha^{a'}$. Therefore, we have $A = \alpha^{a'}$. The latter is a one-to-one mapping between the set of integers $\{0,1, \cdots, p\text{-}1\}$ and the set of nonzero binary n-tuples. Hence, the mapping $a \to (A,E)$ is one-to-one. Since (B,C) is just a re-partitioning of the bits in (A,E), the mapping $a \to (B,C)$ is one-to-one. □

Note that the maximum possible value of E, given by $\bar{E} = (2^l - 1)/(2^n - 1)$, requires $l-n+1$ bits in its binary representation, as it is bounded by $2^{l-n} < \bar{E} < 2^{l+1-n}$. Thus, the transformed address (B,C) requires $l+1$ bits. However, the overhead of this extra bit can be eliminated by either

(i) limiting the input address range to $0 \le a \le 2^l - 2^{l-n} - 1$, and using the highest 2^{l-n} addresses for other purposes (e.g. I/O), or

(ii) using a different mapping for the "high" addresses in the range $2^l - 2^{l-n} - 1 < a \le 2^l - 1$. This can be easily achieved by noting that $\alpha^a \neq 0$ for all a. Therefore, pairs (A,E) with $A = 0$ (and the corresponding pairs (B,C)) never occur as outputs of the address mapping. On the other hand, all "high" addresses are of the form $[1^n \mid X]$, where 1^n denotes the all-ones n-tuple, and X is an arbitrary $(l-n)$-tuple. Thus, we can map $a = [1^n \mid X]$ to $(A,E) = (0^n, X)$, and the corresponding (B,C) pair. In the sequel, we assume that approach (i) is taken.

Consider a sequence of addresses $a_0, a_0+s, a_0+2s, \cdots, a_0+ks, \cdots$, with $1 \le s < p$. Denote $a_k = a_0 + ks$, and let A_k and B_k denote the quantities corresponding to a_k in the mapping described above, i.e. $A_k = \alpha^{a_k}$, and B_k consists of the m least significant bits of A_k. Furthermore, let **A** and **B** denote the sequences $\mathbf{A} = \{A_k\}_{k=0}^{p-1}$, $\mathbf{B} = \{B_k\}_{k=0}^{p-1}$. Clearly, **A** and **B** depend on the stride value s and the base address a_0.

Proposition 2. *For any base address a_0 and stride s, $1 \le s < p$,* **A** *consists of all the nonzero binary n-tuples.*

Proof. Since p is a prime number, and $1 \le s < p$, we have $\gcd(s,p)=1$. Hence, $\beta = \alpha^s$ is a primitive element of F. Now,

$$A_k = \alpha^{a_k} = \alpha^{a_0}\alpha^{ks} = \alpha^{a_0}\beta^k.$$

Let $\gamma = \alpha^{a_0}$. Then, we have

$$A_k = \gamma\,\beta^k, \quad 0 \le k \le p\text{-}1.$$

Since β is primitive and $\gamma \neq 0$, A_k runs through all nonzero elements of F. □

Notice that the requirement $1 \le s < p$ guarantees that $\gcd(s,p)=1$ and, hence, the validity of Proposition 2. For $s \ge p$, Proposition 2 is still valid whenever s is not a multiple of p. If s is a multiple of p, then all n-tuples in **A** are identical, resulting in repeated accesses to the same memory module. The following propositions are true for all stride values s, $1 \le s < p$, and all values of the base address a_0.

Proposition 3. *Let N_B denote the number of occurrences of a given m-tuple B in the sequence* **B**. *Then, we have*

$$N_B = \begin{cases} 2^{n-m} - 1 & B = 0, \\ 2^{n-m} & B \neq 0. \end{cases}$$

Proof. This is a direct consequence of Proposition 2, and of the fact that B_k is the m-bit prefix of A_k. Every nonzero m-tuple can be completed in 2^{n-m} different ways to a nonzero n-tuple. The zero m-tuple can be completed in $2^{n-m} - 1$ different ways to a nonzero n-tuple. □

Proposition 3 leads to the following.

Proposition 4. The average (cyclic) distance between two consecutive occurrences of a given m-tuple B in **B** is given by

$$\delta_B = \begin{cases} 2^m + \dfrac{2^m - 1}{2^{n-m} - 1} & B = 0, \\ \\ 2^m - \dfrac{1}{2^{n-m}} & B \neq 0. \end{cases}$$

Proof. The number of m-tuples in **B** is $p = 2^n - 1$. Hence, the average (cyclic) distance between two consecutive occurrences of a given m-tuple B in **B** is given by $\delta_B = (2^n - 1)/N_B$. The claim of the proposition now follows by using the value of N_B from Proposition 2, after some simple algebraic simplifications. □

Propositions 3 and 4 show that the distribution of addresses to interleaved modules is, on the average, close to uniform, regardless of the stride value. The slight deviation from uniformity due to the lower frequency of occurrence of the all-zero m-tuple is small when n is sufficiently large with respect to m, which is the case in most practical applications. As an example of the foregoing discussion, Table 1 presents numerical values of N_B and δ_B for some practical values of n and m.

Table 1. Numerical examples

		N_B		δ_B	
n	m	$B=0$	$B\neq0$	$B=0$	$B\neq0$
13	6	127	128	64.496	63.992
17	7	1023	1024	128.124	127.999
19	8	2047	2048	256.125	255.999

The previous claims characterize the distribution of m-tuples over one period of the sequence **B**, of length p. In the following claims, we bound the repetition rates of m-tuples over much shorter segments of the sequence.

Proposition 5. *Let* $\mathbf{B}_k$ *denote the* $m \times n$ *matrix whose columns are the* n *consecutive m-tuples* $B_k, B_{k+1}, \cdots, B_{k+n-1}$ *from the sequence* **B**. *Then,* $\mathbf{B}_k$ *has rank m.*

Proof. Consider the matrix $\mathbf{A}_k$ with columns $A_k, A_{k+1}, \cdots, A_{k+n-1}$. These columns represent n consecutive powers of α and, hence, are linearly independent. Therefore, any subset of m rows of $\mathbf{A}_k$ must be linearly independent. In particular, the rows of $\mathbf{B}_k$ are such a subset, and, thus, $\mathbf{B}_k$ has rank m. □

A *run* of length t in **B** is a segment $B_k, B_{k+1}, \cdots, B_{k+t-1}$ with $B_k = B_{k+1} = \cdots = B_{k+t-1}$. The following claims are immediate consequences of Proposition 5:

Proposition 6. *No m-tuple B can occur more than $n-m+1$ times in a segment of length n of the sequence* **B**.

Proposition 7. **B** *does not contain any run of length greater than $n-m+1$.*

The bound in Proposition 7 is tight when a standard basis $\mathbf{b} = (1\ \alpha\ \alpha^2\ \cdots\ \alpha^{n-1})$ is used, as shown in the following.

Proposition 8. If F is represented using the standard basis **b**, then, for $s = 1$, and any base address a_0, there is a run of length $n-m+1$ starting at B_k, where k is such that $\alpha^{a_0+k} = \alpha^m(1+\alpha)^{-1}$.

Proof. For the given values of s and k, and $i \geq 1$, we have

$$A_{k+i} = \alpha^{a_{k+i}} = \alpha^{a_0+k+i} = \frac{\alpha^{m+i}}{1+\alpha} = \frac{\alpha^{m+i-1}}{1+\alpha} + \alpha^{m+i-1}$$

$$= \cdots = \frac{\alpha^m}{1+\alpha} + \alpha^m + \alpha^{m+2} + \cdots + \alpha^{m+i-1}$$

$$= A_k + \alpha^m + \alpha^{m+2} + \cdots + \alpha^{m+i-1}.$$

Hence, for $0 \leq i \leq m-n$, A_{k+i} differs from A_k by a sum of powers of α from the set $\{\alpha^m, \alpha^{m+1}, \cdots, \alpha^{n-1}\}$. In standard basis representation, this implies that A_{k+i} differs from A_k only in the last $n-m$ coordinates. Therefore, we have $B_k = B_{k+1} = \cdots = B_{k+m-n}$. □

Proposition 7 can be significantly improved if we allow for non-standard bases, as shown bellow.

Theorem 1. *Let $r = \lceil (2n-1)/m \rceil$. Then, there exists a basis $\Omega = (\omega_0, \omega_1, \cdots, \omega_{n-1})$ of F over GF(2) such that, when elements of F are represented with respect to Ω, the sequence* **B** *does not contain any run of length greater than r for any stride value s.*

Before proceeding to prove Theorem 1, we consider an example. Let $n = 13$ and $m = 6$. Proposition 7 guarantees that, for any stride s and for any basis of F, **B** does not contain any run of length greater than 8. Theorem 1 guarantees that there is a basis of F such that for any stride s, **B** does not contain any run of length greater than 5. In an exhaustive computer search it was found that, in actuality, there are bases of F that limit the maximum run length to 4. Asymptotically, if the ratio n/m is bounded away from both 0 and 1, Theorem 1 bounds the maximum run length from above by a *constant* approximately equal to $2n/m$, while the maximum run length in Proposition 7 grows linearly with n.

The following definitions and lemma are needed for the proof of Theorem 1.

Let S be a linear subspace of $F=GF(2^n)$. A *geometric progression* of length r in S is a sequence of elements $\gamma, \gamma\beta, \cdots, \gamma\beta^{r-1}$, with $\gamma\beta^i \in S$, $0 \le i \le r-1$, $\gamma, \beta \in F$, $\gamma \ne 0$, and β not in any proper subfield of F. Let $L(S)$ denote the length of the longest geometric progression in S.

Lemma 1. *Let k be an integer such that $1 \le k < n$, and let $R(n,k) = \dfrac{2n-1}{n-k}$. Then, there exists a k-dimensional subspace S of F such that $L(S) < R(n,k)$.*

Proof. Let $r \ge 2$ be an integer, and let $Q(n,k,r)$ denote the number of k-dimensional subspaces of F that contain geometric progressions of length r. Clearly, we have $Q(n,k,r)=0$ for $r>k$, since any $k+1$ consecutive powers of a proper element $\beta \in F$ are linearly independent, and, thus, cannot be contained in a k-dimensional space. Hence, we may assume $r \le k$. Let S be a subspace containing a geometric progression $\gamma, \gamma\beta, \cdots, \gamma\beta^{r-1}$. Since the elements of the geometric progression are linearly independent, they can be completed to form a basis $\Gamma = (\gamma, \gamma\beta, \cdots, \gamma\beta^{r-1}, \delta_r, \delta_{r+1}, \cdots, \delta_{k-1})$ of S. In order to define a geometric progression of length $r \ge 2$, we have 2^n-1 choices for γ and at most 2^n-2 choices for β. However, if the geometric progression of length r defined by γ and β is contained in S, so is the one defined by $\gamma\beta^{r-1}$ and β^{-1}. The number of choices for $\delta_r, \cdots, \delta_{k-1}$ is $(2^n-2^r)(2^n-2^{r+1}) \cdots (2^n-2^{k-1})$. On the other hand, if Γ spans S, so does $\Gamma' = (\gamma, \gamma\beta, \cdots, \gamma\beta^{r-1}, \delta'_r, \delta'_{r+1}, \cdots, \delta'_{k-1})$, which can be written as

$$\Gamma' = \Gamma M,$$

where M is a nonsingular matrix with the first r columns being the first r unit vectors, and the last k-r vectors forming a $k \times (k\text{-}r)$ matrix of rank k-r. The number of such matrices is $(2^k-2^r)(2^k-2^{r+1}) \cdots (2^k-2^{k-1})$. It follows from the above discussion that

$$Q(n,k,r) \le \frac{(2^n-1)(2^n-2)}{2} \; \frac{(2^n-2^r)(2^n-2^{r+1}) \cdots (2^n-2^{k-1})}{(2^k-2^r)(2^k-2^{r+1}) \cdots (2^k-2^{k-1})}. \tag{3}$$

Let $P(n,k)$ denote the number of k-dimensional subspaces of dimension k of F. It is well known [6] that

$$P(n,k) = \frac{(2^n-1)(2^n-2^1) \cdots (2^n-2^{k-1})}{(2^k-1)(2^k-2^1) \cdots (2^k-2^{k-1})}. \tag{4}$$

Clearly, if $P(n,k)/Q(n,k,r) \ge 1$, then there is a k-dimensional space S that does not contain any geometric progression of length r. Now, from (3) and (4) we have

$$\frac{P(n,k)}{Q(n,k,r)} \geq \frac{2}{(2^n-1)(2^n-2)} \prod_{i=0}^{r-1} \frac{(2^n-2^i)}{(2^k-2^i)}$$

$$> \frac{2}{(2^n-1)(2^n-2)} (2^{n-k})^r > 2^{(n-k)r-(2n-1)}. \qquad (5)$$

It follows from (5) that $P(n,k)/Q(n,k,r) > 1$ when $(n-k)r-(2n-1) \geq 0$, or equivalently, when $r \geq R(n,k)$. □

Proof of Theorem 1. Let $\Omega = (\omega_0, \omega_1, \cdots, \omega_{n-1})$ be the basis used to represent elements of F. Assume there is a run of length $r+1$ starting at B_k for some stride value s. Let $\beta = \alpha^s$. Then, for some $\gamma' \in F$, the elements $\gamma', \gamma'\beta, \gamma'\beta^2, \cdots, \gamma'\beta^r$ are identical in their first m coordinates. This implies that, for $\gamma = \gamma'(1+\beta)$, the elements $\gamma, \gamma\beta, \cdots, \gamma\beta^{r-1}$ are zero in their first m coordinates. Hence, the geometric progression $\gamma, \gamma\beta, \cdots, \gamma\beta^{r-1}$, of length r, is contained in the $(n-m)$-dimensional subspace S spanned by $\omega_m, \omega_{m+1}, \cdots, \omega_{n-1}$. By Lemma 1, there is a choice of $\omega_m, \omega_{m+1}, \cdots, \omega_{n-1}$ such that the longest geometric progression in S is of length $r' < R(n, n-m)$. Hence, for this choice of $\omega_m, \omega_{m+1}, \cdots, \omega_{n-1}$, if $r \geq R(n,n-m) = (2n-1)/m$, there are no arithmetic progressions of length r in S, and there are no runs of length greater than r in **B**. Notice that, other than the obvious requirement that they form a basis of F together with $\omega_m, \omega_{m+1}, \cdots, \omega_{n-1}$, there are no constraints on the choice of $\omega_0, \omega_1, \cdots, \omega_{m-1}$. □

3 Systolic computation

The main component of the hashing computation described in Section 2 is the exponentiation operation $A = \alpha^a$. Let $a' = a \bmod p$, and let $(a(0)\ a(1)\ \cdots\ a(n-1))$ denote the binary expansion of a'. Then, we have

$$A = \alpha^a = \alpha^{a'} = \alpha^{\sum_{i=0}^{n-1} 2^i a(i)} = \prod_{i=0}^{n-1} \alpha^{2^i a(i)}.$$

This computation is amenable to a pipeline implementation, as illustrated in Figure 2 for $n=4$. The schematic symbols used in the figures are defined in Figure 1.

The constant multipliers α^{2^i} are linear operators on their n-bit inputs, and, thus, are implemented as XOR arrays, i.e. each bit of the output is a XOR combination of a subset of the n input bits. The computation in Figure 2 proceeds from top to bottom, taking n clock cycles to complete. However, each stage is used in only one clock cycle per exponentiation operation. Thus, the pipeline can process n different exponentiation operations at any given clock cycle. This arrangement is shown in the systolic array of Figure 3, where $a(j)_i$ denotes the j-th bit of the i-th input argument a_i'. The systolic array inputs one address and outputs one exponentiation result per clock cycle, with the clock rate being limited only by the propagation delays of one constant multiplier and one multiplexer.

The systolic array of Figure 3 has a latency delay of n clock cycles. This delay can be reduced to $\lfloor n/k \rfloor$ by processing k bits of the input per pipeline stage. Each stage would then include 2^k constant multipliers, and a 2^k:1 multiplexer. This is shown in Figure 4 for $n=4$ and $k=2$.

The hashing scheme also requires the computation of $a'=a \bmod p$ and $E=\lfloor a/p \rfloor$. Because of the special form of the prime $p=2^n-1$, this computation is much simpler that a full-fledged integer division.

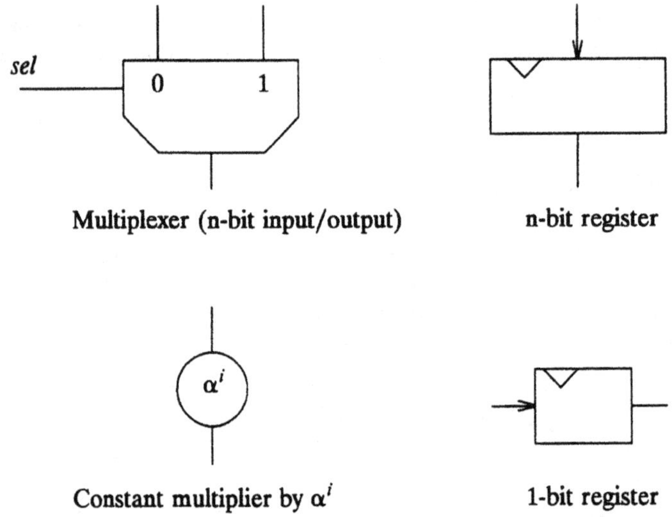

Fig. 1: Schematic symbols

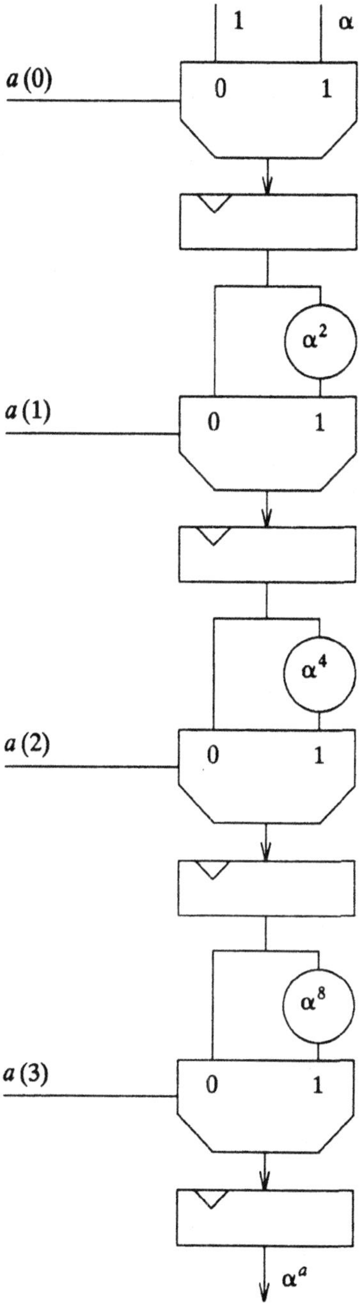

Fig. 2: Basic exponentiation circuit

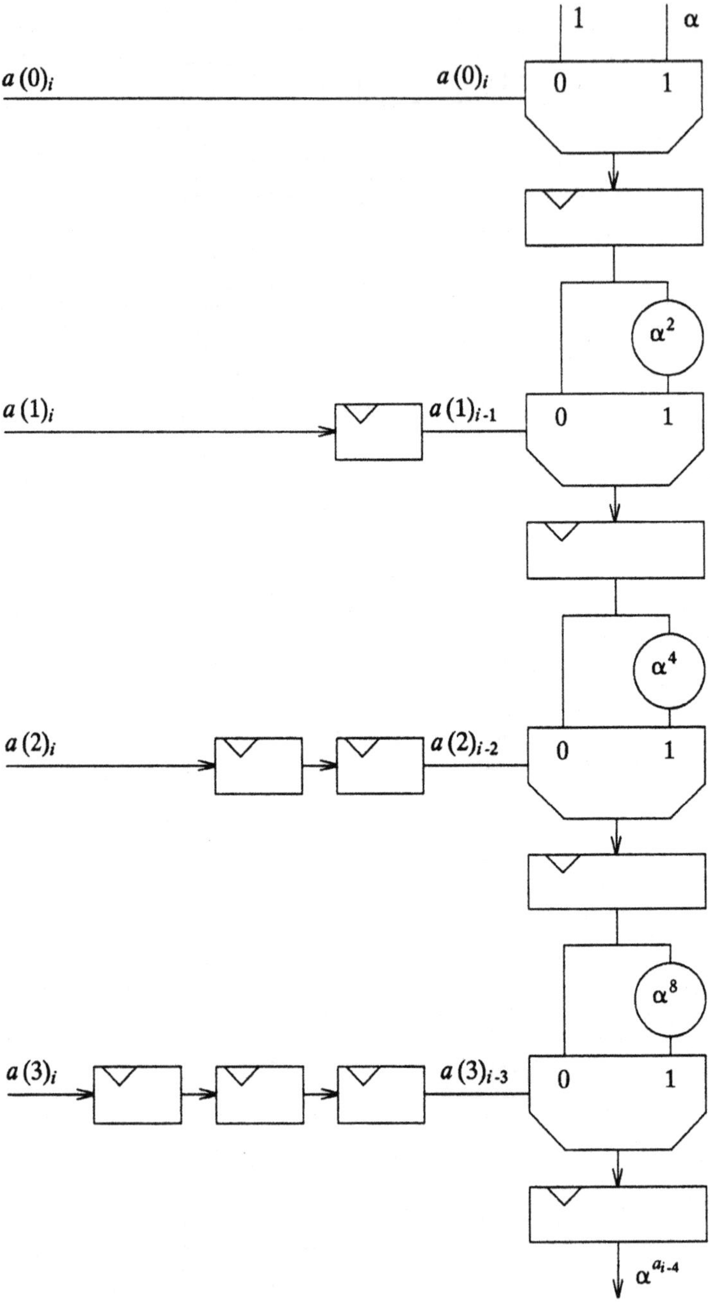

Fig. 3: Systolic array for exponentiation

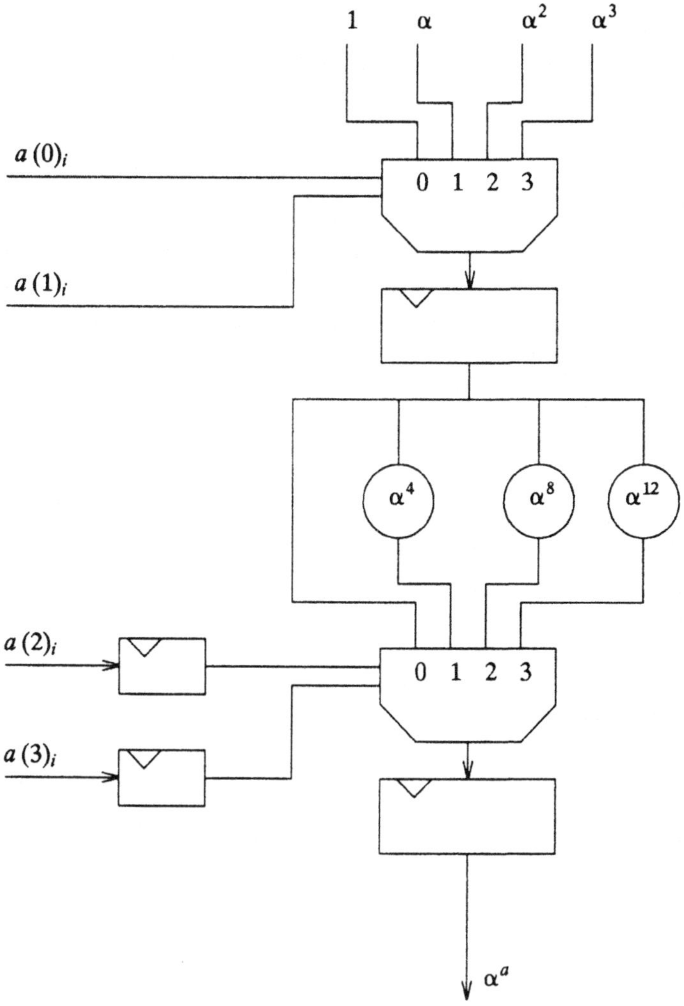

Fig. 4: Exponentiation array with shortened pipeline

Acknowledgment. We are grateful to Bob Rau and Bill Worley of HP Labs for introducing us to the problem, and for stimulating discussions.

References

1. D. H. Lawrie and C. R. Vora: The Prime Memory System for Array Access. IEEE Transactions on Computers, Vol. TC-31, No. 5, pp. 435-442, May 1982.
2. J. M. Frailong, W. Jalby and J. Lenfant: XOR-Schemes: A Flexible Data Organization in Parallel Memories. Proceedings of the 1985 International Conference on Parallel Processing, pp. 276-283, August 1985.
3. A. Norton and E. Melton: A Class of Boolean Linear Transformations for Conflict-Free Power-of-Two Stride Access. Proceedings of the 1987 International Conference on Parallel Processing, pp. 247-254, 1987.
4. D. Lee: Scrambled Storage for Parallel Memory Systems. Proceedings of the 1988 International Symposium on Computer Architecture, pp. 232-239, 1988.
5. B. R. Rau, M. S. Schlansker and D. W. L. Yen: The Cydra 5 Stride-Insensitive Memory System. Proceedings of the 1989 International Conference on Parallel Processing, Vol. 1 pp. 242-246, August 8-12, 1989.
6. F. J. MacWilliams, N. J. A. Sloane: *The Theory of Error-Correcting Codes*, North-Holland, Amsterdam, 1977.
7. R. Lidl and H. Niederreiter: *Finite Fields*, Encyclopedia of Mathematics and its Applications, Gian-Carlo Rota, Editor, Addison-Wesley, Reading, MA, 1983.

An Elementary Proof of a Partial Improvement to the Ax-Katz Theorem

O. Moreno[1]* and C.J. Moreno[2]**

[1] University of Puerto Rico Rio Piedras, PR 00931 USA
Email: o_moreno@uprenet.bitnet,
moreno@sun386-gauss.uprr.pr
[2] Baruch, CUNY Box 545 N. Salem, NY 10560 USA
Email: carlos@kronecker.baruch.cuny.edu

Abstract. Moreno and Moreno have recently given a result that in many cases improves upon the *Ax-Katz* theorem. We will presently give an elementary and self contained proof of this and also give a corresponding improvement upon a result of Adolphson and Sperber.

1 Introduction.

In connection with his work on quasi-algebraically closed fields, Artin was led to make the following conjecture: If $F = F(x_1, \ldots, x_n)$ is a homogeneous polynomial of total degree d over k, a finite field having $q = p^f$ elements, and $n > d$, then F has a nontrivial zero (It seems that Dickson had also made the same conjecture a number of years earlier [2]). C. Chevalley proved this in [4] and even showed the hypothesis of homogeneity could be replaced by the weaker assumption of no constant term. E. Warning in [14], using a lemma of Chevalley, showed that even without this last assumption the characteristic p of k divides $N(F)$, the number of zeros of F (counting the trivial zero if F has no constant term).

By using an idea of B. Dwork [5], J. Ax [3] greatly improved the theorem of Warning. He proved that if b is the least nonnegative integer such that $b \geq (n-d)/d$, then q^b divides $N(F)$. As a corollary, Ax obtained the following consequence for a system of polynomials: Let $F_i(x_1, \ldots, x_n)$ $(i = 1, \ldots, r)$ be polynomials over k of degree d_i, and λ be the least nonnegative integer which is

$$\geq \frac{n - \sum_{i=1}^{r} d_i}{\sum_{i=1}^{r} d_i}.$$

Then q^λ divides $N(F_1 = 0, \ldots, F_r = 0)$. This corollary, however, could be improved. In 1971, N. M. Katz [6] proved:

Theorem 0. *Let $F_1, \ldots, F_r$ be polynomials in n variables with coefficients in k, a finite field with $q = p^f$ elements. Let μ be:*

$$\mu = \left\lceil \frac{n - \sum_{i=1}^{r} \deg(F_i)}{\max_{1 \leq i \leq r}\{\deg(F_i)\}} \right\rceil,$$

* Work partially supported by NSF grants RII-9014056, component IV of the EPSCoR of Puerto Rico Grant and the ARO grant for Cornell MSI.
** And grants DMS-8711566 and DMS-8712742 from the National Science Foundation.

where $\lceil x \rceil$ denotes the least integer which is $\geq x$. Then q^{μ} divides $N(F_1 = 0, \ldots, F_r = 0)$.

Please note that in the Ax-Katz result, the divisibility depends on the degree of the equations involved. In [11] Moreno and Moreno gave an improvement to the Ax-Katz result (theorem 0-1 below) which in many cases gives an improvement over theorem 0 above, and in those cases the divisibility depends in a fundamental way on the maximal p-weight (as defined below) of the degrees of the terms of the equations. Theorem 1, which is the main theorem in [11] invokes theorem 0 above and we present here a self contained proof of said theorem 1. Our method is furthermore quite elementary, and uses among other things only a simple version of Stickelberger's theorem which is readily established.

Definition 1. The p-weight degree of a monomial $x^{\underline{d}} = x_1^{d_1} x_2^{d_2} \cdots x_n^{d_n}$ is $w_p(x^{\underline{d}}) = \sigma_p(d_1) + \sigma_p(d_2) + \cdots + \sigma_p(d_n)$, where $\sigma_p(d)$ denotes the p-weight of the integer d, i.e., if $d = a_0 + a_1 p + \cdots + a_t p^t$, $\quad 0 \leq a_i < p$, then $\sigma_p(d) = a_0 + a_1 + \cdots + a_t$.

Definition 2. The p-weight degree of a polynomial $F(X_1, \ldots, X_n) = \sum_{\underline{d}} a_{\underline{d}} x^{\underline{d}}$ is $w_p(F) = \max_{x^{\underline{d}},\ a_{\underline{d}} \neq 0} w_p(x^{\underline{d}})$, where the maximum is taken over those monomials which effectively appear in F with a non zero coefficient.

In our proof of theorems 1 below, we follow the elementary and elegant proof of Katz, given by Daqing Wan in [13].

Theorem 1. *Let $F_1, \ldots, F_r$ be polynomials in n variables with coefficients in k, a finite field with $q = p^f$ elements. Let $w_p(F_i)$ be the p-weight degree of F_i and let μ be:*

$$\mu = \left\lceil f \left(\frac{n - \sum_{i=1}^{r} w_p(F_i)}{\max_{1 \leq i \leq r} \{w_p(F_i)\}} \right) \right\rceil .$$

Then p^{μ} divides $N(F_1 = 0, \ldots, F_r = 0)$.

Theorem 0-1. *Let $F_1, \ldots, F_r$ be polynomials in n variables with coefficients in k. Let $w_p(F_i)$ be the p-weight degree of F_i and let μ be:*

$$\mu = \max \left(f \left\lceil \frac{n - \sum_{i=1}^{r} \deg(F_i)}{\max_{1 \leq i \leq r} \{\deg(F_i)\}} \right\rceil , \left\lceil f \left(\frac{n - \sum_{i=1}^{r} w_p(F_i)}{\max_{1 \leq i \leq r} \{w_p(F_i)\}} \right) \right\rceil \right) .$$

We then have that p^{μ} divides $N(F_1 = 0, \ldots, F_r = 0)$.

Corollary. *Let $F(x_1, \ldots, x_n)$ be a polynomial over k, and let $l = w_p(F)$, and $S(F)$ be the exponential sum:*

$$S(F) = \sum_{x_1, \ldots, x_n \in k} \eta(F(x_1, \ldots, x_n)),$$

where η is an additive character defined over k. Let μ be the highest exponent of p-divisibility for $S(F)$, then $\mu \geq f(n/l)$.

The above corollary is to our knowledge new and improves in many cases upon results of Adolphson and Sperber in [1].

The results of this paper were motivated by considerations arising from coding theory, and in particular it is worth mentioning the work of R. McEliece (see [10]). Also we would like to mention our co-authors P. V. Kumar and S. Litsyn (see [7], [9]).On the other hand one of the main points of our paper is the fact of providing an elementary proof of Ax-Katz to coding theory people. This was needed since this theorem has proven to be very useful for this applied field.

2. Proof of theorem 1.

We first introduce some notation from [3], [13]. Let $q = p^f$, Q_p be the p-adic completion of the rationals, and K the unique unramified extension of Q_p of degree f. Then the residue class field of K is k. Let T denote the set of Teichmuller representatives of k in K; let $T^* = T - \{0\}$, the $q-1$ roots of unity. Let ξ be a primitive p-th root of unity. If α is an integer of Q_p, ξ^α is then defined to be ξ^a if a ($\in Z_+$) is congruent to α modulo p. Letting S denote the trace of K over Q_p, we define

$$P(U) = \sum_{m=0}^{q-1} c(m)U^m$$

to be the unique polynomial of degree $q-1$ with coefficients in $K(\xi)$ such that $P(t) = \xi^{S(t)}$ for all $t \in T$. $c(j)$ can be easily determined, that is:

$$c(0) = 1, \quad (1)$$
$$(q-1)c(q-1) = -q, \quad (2)$$
$$(q-1)c(j) = g(j), \ for\ 0 < j < q-1. \quad (3)$$

where $g(j)$ is the Gauss sum defined by

$$g(j) = \sum_{t\in T^*} t^{-1}\xi^{S(j)}.$$

If $0 \le j \le q-1$, let j_i for $i = 0, \ldots, f-1$ be such that $0 \le j_i \le p-1$ and

$$j = \sum_{i=0}^{f-1} j_i p^i;$$

we put

$$\sigma(j) = \sum_{i=0}^{f-1} j_i.$$

Put $\pi = \xi - 1$, then the Stickelberger relation and (1) together imply

$$c(j) \equiv 0 \pmod{\pi^{\sigma(j)}} \quad \text{for} \quad 0 \le j \le q-1.$$

Remark. Simple proofs of the above statements for the case of the prime field are given below in lemmas 1 and 2.

The map $\alpha \to \xi^{S(\alpha)}$ is a nontrivial character of the additive group of the integers of K which is trivial on the maximal ideal of the integers of K. Define

$\eta(x) = P(t)$ for $x \in k$ and t the Teichmuller representative of x, then η is a nontrivial character of the additive group of k. Moreover, if $u \in k$, then

$$\sum_{y_0 \in k} \eta(y_0 u) = \begin{cases} q, & \text{if } u = 0 \\ 0, & \text{otherwise.} \end{cases}$$

It follows that

$$q^r N(F_1, \ldots, F_r) = \sum_x \sum_y \prod_{i=1}^r \eta(y_i F_i(x_1, \ldots, x_n)) = S(F_1, \ldots, F_r)$$

where $x = (x_1, \ldots, x_n) \in k^n$, $y = (y_1, \ldots, y_r) \in k^r$.

The theorem of Ax-Katz, generalizing the Chevalley-Warning theorem, can be split into two very natural and elementary steps: the first consists in reducing the exponential sum $S(F_1, \ldots, F_r)$ with $F_i(x_1, \ldots, x_n)$ a polynomial in n variables with coefficients in k, into an exponential sum over k_p, the prime field (with p elements)

$$\sum_x \sum_y \prod_{i=1}^{fr} \eta(y_i F_i'(x_1, \ldots, x_{nf}))$$

where $x = (x_1, \ldots, x_{nf}) \in k_p^{nf}$, $y = (y_1, \ldots, y_{fr}) \in k_p^{fr}$ and where now F_i' can be grouped in a class consisting of f polynomials with coefficients in k_p, each with degree $\leq l_i$.

We want to replace the above sum $S(F_1, \ldots, F_r)$ with one where the variables take their values in the prime field k_p. Pick a basis of k over k_p:

$$\mu_1, \ldots, \mu_f$$

and represent each variable with values in k in the form

$$x = \sum_{i=1}^f z_i \mu_i.$$

Since the exponent $S(y_i F_i(x_1, \ldots, x_n))$ is the sum of the traces of the monomials which make up F_i, we need only to show the reduction for a typical monomial

$$\alpha y_i x_1^{d_1} \cdots x_n^{d_n}.$$

For this monomial we write

$$\alpha y_i x_1^{d_1} \cdots x_n^{d_n} = \alpha \left(\sum_{t=1}^f {}_i y_t \mu_t \right) \left(\sum_{t=1}^f z_t^{(1)} \mu_t \right)^{d_1} \cdots \left(\sum_{t=1}^f z_t^{(n)} \mu_t \right)^{d_n} \tag{4}$$

$$= \alpha \left(\sum_{t=1}^f {}_i y_t \mu_t \right) \left(\sum_{t=1}^f z_t^{(1)} \mu_t \right)^{\alpha_1 p^{j_1}} \left(\sum_{t=1}^f z_t^{(1)} \mu_t \right)^{\alpha_2 p^{j_2}} \cdots \tag{5}$$

$$= \alpha \left(\sum_{t=1}^f {}_i y_t \mu_t \right) \left(\sum_{t=1}^f z_t^{(1)} \mu_t^{p^{j_1}} \right)^{\alpha_1} \left(\sum_{t=1}^f z_t^{(1)} \mu_t^{p^{j_2}} \right)^{\alpha_2} \cdots, \tag{6}$$

where the $z_t^{(j)}$ take their values in k_p and

$$d_1 = \alpha_1 p^{j_1} + \alpha_2 p^{j_2} + \cdots$$

is the expansion of d_1 in the base p, etc. The conclusion is that the original monomial $y_i x_1^{d_1} x_2^{d_2} \cdots x_n^{d_n}$ of degree $d_1 + d_2 + \cdots + d_n$ has now become a sum of monomials in the new variables $z_t^{(j)}$ of degree

$$l_i = \sum_{t=1}^{n} \sigma(d_t).$$

Using the linearity of S we can thus replace the original sum over k by a sum over k_p with the exact form previously shown.

We have transformed the original exponential sum into another of exactly the same form, but where the number of x's is exactly fn and the number of y's is fr, and the F_i' can be gathered in groups with f members each, each with degree $\leq l_i$. Therefore, it is clear that, given the form of the new sum, in order to find the divisibility by p and complete the proof of theorem 1, all we need to do is prove the Ax-Katz theorem over the ground field (i.e., theorem 1*). In other words theorem 1 now follows if we call upon the case of $R = fr$ and $N = fn$ of theorem 1* below. Also, it turns out that in what follows we can give a particularly simple proof of theorem 1*.

Theorem 1*. *Let k be the prime field with p elements. Let $F_i(x_1, \ldots, x_N)$ be R polynomials, of degree l_i, and μ be:*

$$\mu = \left\lceil \frac{N - \sum_{i=1}^{R} l_i}{\max_{1 \leq i \leq R} l_i} \right\rceil .$$

Then p^μ divides the number of solutions of the system $F_i = 0, \ldots, F_R = 0$ over k.

We give below a simple proof of this version of Ax-Katz theorem, which does not involve Frobenius and uses an elementary case of Stickelberger's theorem which can be proved directly.

Let $K = \mathbf{Q}_p$; T denotes the set of Teichmuller representatives in K of k and $T^\times = T - \{0\}$; these are the $p-1$ roots of unity. For ξ a primitive p-th root of unity, and a p-adic integer α in $\mathbf{Q}_p$, we let ξ^α be defined to be ξ^x, where x is any rational integer congruent to α modulo p.

We need several elementary results about Gauss sums which we now state and prove.

Definition. For an integer j, $0 \leq j \leq p-1$, the Gauss sum is defined by

$$g(j) = \sum_{t \in T^\times} t^{-j} \xi^t.$$

Lemma 1. *There is a unique polynomial*

$$P(U) = \sum_{j=0}^{p-1} c(j) U^j \in K(\xi)[U]$$

such that

$$P(t) = \xi^t, t \in T$$

and whose coefficients are given by

$$c(0) = 1, \tag{7}$$

$$(p-1)c(p-1) = -p, \tag{8}$$

$$(p-1)c(j) = g(j), \ \textit{for } 0 < j < p-1. \tag{9}$$

where $g(j)$ is the Gauss sum.

Proof. The existence and uniqueness of $P(U)$ follows readily by considering the equalities

$$\sum_{m=0}^{p-1} c(j)t^m = \xi^t, \quad t \in T,$$

as a system of p equations in the p unknowns $c(j), \quad 0 \leq j \leq p-1$, which can be solved using the van der Monde determinant. To derive the relation with the Gauss sums we note first that

$$\sum_{t \in T^\times} P(t)t^{-j} = \sum_{t \in T^\times} \xi^t t^{-j} = g(j).$$

On the other hand,

$$\sum_{t \in T^\times} \sum_{m=0}^{p-1} c(m)t^{m-j} = \sum_{m=0}^{p-1} c(m) \sum_{t \in T^\times} t^{m-j}$$

and from the equation

$$\sum_{t \in T^\times} t^{m-j} = \begin{cases} 0 & \text{if } m \neq j \\ (p-1)c(j) & \text{otherwise,} \end{cases}$$

we obtain $(p-1)c(j) = g(j), \quad 0 < j < p-1$. The relation $c(0) = 1$ is clear. Since $g(0) = -1$, a similar argument gives $-1 = (p-1)(c(0) + c(p-1))$ and hence $(p-1)c(p-1) = -p$. This proves the lemma.

Lemma 2. *Let* $\pi = \xi - 1$. *The coefficients of the polynomial* $P(U)$ *satisfy*

$$c(j) \equiv 0 \pmod{\pi^j}, \quad 0 \leq j \leq p-1.$$

Proof. We observe that $g(j)$ is equal to

$$\sum_{t \in T^\times} t^{-j}\xi^t = \sum_{t \in T^\times} t^{-j}(1+\pi)^t \tag{10}$$

$$= \sum_{t \in T^\times} t^{-j} \sum_{i=0}^{\infty} \binom{t}{i} \pi^i, \tag{11}$$

where the binomial coefficient, for any $t \in \mathbf{Q}_p$, is formally defined as the polynomial

$$\binom{t}{i} = \frac{t(t-1)\cdots(t-i+1)}{i!}.$$

From the orthogonality relation

$$\sum_{t \in T^\times} t^a = \begin{cases} p-1 & \text{if } a \equiv 0 \pmod{p-1} \\ 0 & \text{otherwise} \end{cases}$$

we see that if $i < j$, then

$$\sum_{t \in T^\times} t^{-j} \binom{t}{i} = 0,$$

and for $i = j$

$$\sum_{t \in T^\times} t^{-j} \binom{t}{i} = p - 1.$$

If $i > j$, then of course, the polynomial $\binom{t}{i}$ will contain terms with t^j and all these will contribute powers of π higher than π^j. Hence we have

$$\sum_{t \in T^\times} t^{-j} \xi^t \equiv (p-1)\frac{\pi^j}{j!} \bmod \pi^{j+1}.$$

This is Stickelberger's theorem in the case of Gauss sums over the prime field.

Corollary. *With notations as above, we have*

$$g(j)^{p-1} \equiv 0 \mod p^j.$$

Proof. This is a simple consequence of the identity

$$0 = \xi^p - 1 = (\pi + 1)^p = \sum_{i=1}^{p} \binom{p}{i} \pi^i,$$

which is also equivalent to

$$\pi^{p-1} + p \cdot u = 0,$$

where u is the p-adic unit in $\mathbf{Z}_p^\times$ given by

$$u = 1 + \sum_{i=2}^{p-1} \frac{1}{p} \binom{p}{i} \pi^{i-1}.$$

The map $\alpha \mapsto \xi^\alpha$ is a non-trivial character of the additive group of integers of K which is trivial on the maximal ideal of the integers of K. For $x \in k_p$, and $t \in T$ its corresponding Teichmuller representative, we define

$$\psi(x) := P(t);$$

ψ is a non-trivial additive character of k and the basic orthogonality relation is, for $u \in k$,

$$\sum_{y_0 \in k} \psi(y_0 u) = \begin{cases} p & \text{if } u = 0 \\ 0 & \text{otherwise.} \end{cases}$$

From the orthogonality relation we obtain readily

$$p^R N(F_1, \ldots, F_R) = \sum_x \sum_y \prod_{i=1}^R \psi(y_i F_i(x_1, \ldots, x_N))$$

where $x = (x_1, \ldots, x_N) \in k^N, \quad y = (y_1, \ldots, y_R) \in k^R$.

Let

$$F_i = \sum_{w(i) \in W(i)} a(w(i)) x^{w(i)},$$

where $W(i)$ stands for the *exponent set* of the polynomial F_i, that is, the set of integer vectors $w(i) = (w_1(i), \ldots, w_N(i)) \in \mathbf{Z}_+^N$ whose *height*$(w(i))$

$$|w(i)| := w_1(i) + \cdots + w_N(i)$$

is bounded by $|w(i)| \leq l_i$. Here we allow some of the coefficients $a(w(i))$ to be zero.

We have

$$p^R N(F_1, \ldots, F_R) = \sum_x \sum_y \prod_{i=1}^R \prod_{W(i) \in W(i)} \psi(a(w(i)) y_i x^{w(i)}) \tag{12}$$

$$= \sum_{t \in T^N} \sum_{t' \in T^R} \prod_{i=1}^R \prod_{w(i) \in W(i)} P(A(w(i)) t_i' t^{w(i)}) \tag{13}$$

where $A(w(i))$ is the Teichmuller representative of $a(w(i))$, and $t = (t_1, \ldots, t_N) \in T^N$, $t' = (t_1', \ldots, t_R') \in T^R$.

The substitution of the polynomial $P(U)$ into the last equality yields

$$p^R N = \sum_{t \in T^N} \sum_{t' \in T^R} \prod_{i=1}^R \prod_{w(i) \in W(i)} \sum_{m_i(w(i))=0}^{p-1} c(m_i(w(i))) A(w(i))^{m_i(w(i))} \tag{14}$$

$$\cdot\ (t')^{m_i(w(i))} t^{m_i(w(i)) w(i)} \tag{15}$$

$$= \sum_{m \in M} \left\{ \prod_{i=1}^R \prod_{w(i) \in W(i)} A(w(i))^{m_i(w(i))} \right\} \cdot \left\{ \prod_{i=1}^R \prod_{w(i) \in W(i)} c(m_i(w(i))) \right\} \tag{16}$$

$$\cdot \left\{ \sum_{t \in T^N} t^{e(m)} \sum_{t' \in T^R} (t')^{e'(m)} \right\}, \tag{17}$$

where

$$e(m) \in \mathbf{Z}_+^N, \quad e(m) = \sum_{i=1}^{R} \sum_{w(i) \in W(i)} m_i(w(i))w(i)$$

$$e'(m) = (e'_1, \ldots, e'_R) \in \mathbf{Z}_+^R, \quad e'_i = \sum_{w(i) \in W(i)} m_i(w(i)),$$

and M is the set of functions on the R-fold cartesian product of the exponent set

$$m: \quad W(1) \times W(2) \times \cdots \times W(R) \to S \times \cdots \times S$$

with values in the R-fold product of the set $S = \{0, 1, 2, \ldots, p-1\}$:

$$m = (m_1(w(R)), \ldots, m_R(w(R))), \quad 0 \le m_i(w(i)) \le p-1, \ (i = 1, \ldots, R).$$

Since the elements of T are $(p-1)$-st roots of unity, we see easily that

$$\sum_{t \in T^N} t^{e(m)} \sum_{t' \in T^R} (t')^{e'(m)} = 0 \quad \text{if} \quad (p-1)(e(m), e'(m)).$$

Therefore the only terms in the sum $\sum_m$ in (2.) which are not zero, are those whose corresponding vectors $e(m)$ and $e'(m)$ have all their components divisible by $(p-1)$; from now on we assume that this is the case.

For a fixed element $m \in M$, with $(p-1) \mid e'(m)$ and $(p-1) \mid e(m)$, we want to measure the degree of divisibility by q of the corresponding term in (2.). Associated with such an element m we define two numbers:

$$s := \text{number of non-zero entries in } e(m), \tag{18}$$

$$s' := \text{number of non-zero entries in } e'(m), \tag{19}$$

which can be thought of as the Hamming weight of the vectors $e(m)$ and $e'(m)$. It is clear from the definition of s and s' that

$$\sum_{t \in T^N} t^{e(m)} \sum_{t' \in T^R} (t')^{e'(m)} = (p-1)^{s+s'} p^{N+R-s-s'}.$$

This is easily seen by noting that a zero component in the vectors $e(m)$ or $e'(m)$ contributes a p to the sum, and that a non-zero component contributes $p-1$.

From the definition of the integer vector $e(m)$, and the additivity of the *height*, we see that the total degree of t in (2.) satisfies the inequality $(p-1)s \le$ *height*$(e(m))$ and hence

$$s(p-1) \le \sum_{i=1}^{R} \sum_{w(i) \in W(i)} m_i(w(i))|w(i)| \le \sum_{i=1}^{R} \sum_{w(i) \in W(i)} m_i(w(i)) l_i$$

$$= \sum_{i=1}^{R} l_i e'_i.$$

We may suppose that the first s' entries in $e'(m)$ are non-zero:

$$e'_i \neq 0, \quad 1 \leq i \leq s', \quad e'_{s'+1} = \cdots = e'_R = 0.$$

If we let $d = \max\{d_1, \ldots, d_{s'}\}$, then

$$s(p-1) \leq \sum_{i=1}^{s'} l_i e'_i \leq d\left(\sum_{i=1}^{s'} e'_i\right) - \sum_{i=1}^{s'}(d-l_i)e'_i \tag{20}$$

$$\leq d\sum_{i=1}^{s'} e'_i - (p-1)\sum_{i=1}^{s'}(d-l_i), \tag{21}$$

where in the last inequality we used the fact that since each entry in e' is a multiple of $(p-1)$, those which are not zero satisfy $e'_i \geq (p-1), (1 \leq i \leq s'$. Noting that $(p-1) \mid e'_i$, we conclude that

$$\left\lceil \frac{s + s'd - \sum_{i=1}^{s'} l_i}{d} \right\rceil \cdot (p-1) \leq \sum_{i=1}^{s'} e'_i = \sum_{i=1} \sum_{w(i)\in W(i)} m_i(w(i)).$$

From lemma 2. and the fact that p divides π^{p-1}, we obtain that the term in the sum (2.) corresponding to $m \in M$ has

$$\mathrm{ord}_p\left(\prod_{i=0}^{R} \prod_{w(i)\in W(i)} c(m_i(w(i)))\right) = \sum_{i=1}^{R} \sum_{w(i)\in W(i)} \sigma(m_i(w(i)))$$

$$\geq \left\lceil \frac{s - \sum_{i=1}^{s'} l_i}{d} \right\rceil + s'.$$

As we have shown earlier,

$$\mathrm{ord}_p\left(\sum_{t\in T^N} t^{e(m)} \sum_{t'\in T^R} (t')^{e'(m)}\right) = N + R - s - s'.$$

Combining these two estimates and taking the minimum possible lower bound over all terms in the sum (2.), we obtain

$$p^B \mid q^R N(F_1, \ldots, F_R),$$

where

$$B = \inf_{m\in M}\left\{\left\lceil \frac{s - \sum_{i\in L} l_i}{\max_{i\in L} l_i} \right\rceil + N + R - s\right\},$$

where $s = s(m)$ is the Hamming weight of $e(m)$ and $L \subseteq \{1, 2, \ldots, R\}$ is the subset which indexes the non-zero coordinates of $e'(m)$.

To calculate B, we observe that for a fixed subset $L \subseteq \{1, 2, \ldots, R\}$, the inf will occur over those elements $m \in M$ for which $e(m)$ has maximum Hamming weight, i.e. $s = N$. We thus have

$$B = \inf_L \left\{ \left\lceil \frac{N - \sum_{i \in L} l_i}{\max_{i \in L} l_i} \right\rceil + R \right\}.$$

From the assumption

$$N \geq \sum_{i=0}^{R} l_i$$

we have trivially for two subsets $L, L' \subseteq \{1, 2, \ldots, R\}$ with $L' = L \cup \{i'\}$, a disjoint union,

$$\frac{N - \sum_{i \in L} l_i}{\max_{i \in L} l_i} \geq \frac{N - \sum_{i \in L'} l_i}{\max_{i \in L'} l_i}.$$

Therefore we have

$$B = \left(\frac{N - \sum_{i=1}^{R} l_i}{\max_i l_i} \right) + R.$$

This completes the proof of the theorem.

The exponential sum $S(F)$ defined below is an element of Q_p, the p-adic numbers, and the divisibility properties of the exponential sum $(S(F))$ in the corollary below should be considered in the sense of the the p-adic valuation, ord_p, normalized such that $\text{ord}_p\, p = 1$.

Corollary. *Let $F(x_1, \ldots, x_n)$ be a polynomial over k, and let $l = w_p(F)$, and $S(F)$ be the exponential sum:*

$$S(F) = \sum_{x_1, \ldots, x_n \in k} \eta(F(x_1, \ldots, x_n)),$$

where η is an additive character of k. Let μ be the highest exponent ofp-divisibility for $S(F)$, then $\mu \geq f(n/l)$.

The proof of the above corollary follows exactly the same method as that of the above theorem 1.

References

[1] A. Adolphson and S. Sperber, p-adic estimates for exponential sums and the theorem of Chevalley-Warning, *Ann. Scient. E. N. Superior* 4th. series **20** (1987), 545–556.

[2] E. Artin, *Collected Papers*, Springer-Verlag, New York, 1965.

[3] J. Ax, Zeros of polynomials over finite fields, *Amer. J. Math.* **86** (1964), 255–261.

[4] C. Chevalley, Demonstration d'une hypothese de M. Artin, *Abhandlungen aus dem Mathematischen Seminar der Universität Hamburg* **11** (1936), 73–75.

[5] B. Dwork, On the rationality of the zeta function of an algebraic variety, *Amer. J. Math.* **82** (1960), 631–648.

[6] N. M. Katz, On a theorem of Ax, *Amer. J. Math.* **93** (1971), 485–499.

[7] P. V. Kumar and O. Moreno, Prime-Phase Sequences with Periodic Correlation Properties Better Than Binary Sequences, *IEEE Trans. Inform. Theory* Vol. 37 No. 3 (1991), 603–616.

[8] R. Lidl and H. Niederreiter, *Finite Fields, Encyclopedia of Math and its Appl.* **20** Addison-Wesley Publishing Company, 1983.

[9] S. Litsyn, C. J. Moreno and O. Moreno, Divisibility properties and new bounds for exponential sums in one and several variables, submitted to the *AAECC Journal*.

[10] F. J. MacWilliams and N. J. A. Sloane, *The theory of error correcting codes*, North-Holland, Amsterdam, 1977.

[11] Oscar Moreno and Carlos Moreno,*Improvement of the Chevalley Warning and the Ax-Katz Theorems,* accepted in the *American Journal of Math.*

[12] O. Moreno and C. J. Moreno, The MacWilliams-Sloane Conjecture on the tightness of the Carlitz-Uchiyama bound and the weights of duals of BCH codes, submitted to the *IEEE IT Trans.*

[13] Daqing Wan, An elementary proof of a theorem of Katz, *American Journal of Mathematics* **111** (1989), 1–8.

[14] E. Warning, Bermerkung zur vorstehenden Arbeit von Herrn Chevalley, *Abhandlungen aus dem Mathematischen Seminar der Universität Hamburg* **11** (1936), 76–83.

Energy Functions Associated with Error-Correcting Codes

C. Rentería[1] and H. Tapia-Recillas[2]

[1] Escuela Superior de Física y Matemáticas, I.P.N., MEXICO.
[2] Dpto. de Matemáticas, U.A.Metropolitana-I
Apartado Postal 55-534, México 09340, D.F., MEXICO.

Abstract. A function associated with a linear error-correcting code defined over a finite field with q elements, where q is a power of a prime p, is introduced. This function is a generalization of the energy function associated with a linear block code over the field $\mathbb{Z}_p$ as described by Bruck and Blaum in [1]. It is proven to have similar properties. In particular, the Maximum Likelihood Decoding (MLD) problem is shown to be characterized by the maximization of this function.

1 Introduction

The concept of neural network has been attracting the interest of many people because it has properties that are similar to those found in biological and physical systems. The main idea in the proof of the convergence to a stable state of a neural network is by introducing the so-called *energy function* of the neural network and to show that this function is non-decreasing when the state of the network varies as a result of computation. In [1] the concept of error-correcting binary code is related to the concept of neural network, in particular an energy function associated with a binary code is introduced and it is shown that the Maximum Likelihood Decoding (MLD) problem of the code is equivalent to find a global maximum of this energy function. In appendix I of [1] a generalization of these ideas for codes over the finite field $\mathbb{Z}_p$, where p is a prime, is proposed. The main purpose of the present work is to introduce an *energy function* for codes defined over any finite field $GF(q)$, where q is a power of a prime p, and to show that this function has similar properties to the one introduced in [1]. In particular it is shown that the Maximum Likelihood Decoding (MLD) problem of the code is equivalent to finding a maximum of this energy function. Since the codes under consideration are defined over any finite field $GF(q)$, the geometric Goppa codes associated with algebraic projective non-singular curves can be considered. Other results are obtained along the line and several examples of geometric Goppa codes are given to illustrate the ideas presented here.

The paper is organized as follows: in section 2 some results from [1] concerning the energy function associated with linear codes defined over the field $\mathbb{Z}_p$, p a prime, are recalled. In section 3 the *energy function* associated with a linear error-correcting code defined over any finite field $GF(q)$, is introduced. This is done by means of the characters associated with the field $GF(q)$ and the generating matrix of the code. A characterization of the (MLD) problem of the code in terms of the maximization of this energy function is given. In section 4 the energy function associated with the parity check matrix of the code is studied, showing that each local maximum of this energy function corresponds to a codeword and conversely.

2 The Energy Function for Codes over $\mathbb{Z}_p$

In this section the concept of *energy function* associated with a linear code C defined over the field $K = \mathbb{Z}_p$ introduced in [1], as well as a main result concerning this function are recalled.

Let μ_p be the (multiplicative) group of complex p-th roots of unity and let $\xi = exp(2\pi i/p)$. The natural isomorphism of the (additive) group K with the (multiplicative) group μ_p, $b \longrightarrow \xi^b$, can be extended to $K^r \longrightarrow \mu^r$ as $\underline{b} = (b_1, ..., b_r) \longrightarrow \underline{x} = (x_1, ..., x_r)$ where $x_i = \xi^{b_i}$.

Definition 1. Let $A = (a_{ij})$ be a kxn matrix over the field K and let $\underline{\omega} = (\omega_1, ..., \omega_n) \in \mu_p^n$. If $p = 2$ the *polynomial representation* of the matrix A with respect to $\underline{\omega}$ is:

$$E_{\underline{\omega}}^2(\underline{y}) = \sum_{j=1}^{n} \omega_j y_j$$

If $p > 2$, the *polynomial representation* of the matrix A with respect to $\underline{\omega}$ is the following:

$$E_{\underline{\omega}}^p(\underline{y}) = \sum_{j=1}^{n} \sum_{r=1}^{p-1} (\bar{\omega}_j y_j)^r$$

where in both cases $y_j = \prod_{r=1}^{k} x_r^{a_{ij}}$, and the upper bar denotes the complex conjugate.

Now let C be a $[n, k]$-linear code defined over the field $\mathbb{Z}_p$ with generating matrix $G = (g_{ij})$, $i = 1, 2, ..., k$; $j = 1, 2, ..., n$. An information vector $\underline{b} = (b_1, ..., b_k)$ is encoded as the codeword $\underline{c} = (c_1, ..., c_n) = \underline{b} \cdot G$, i.e., $c_j = \sum_{r=1}^{k} b_r g_{rj}$.

Definition 2. The *μ_p-encoding procedure* of the linear code C is $\underline{x} \longrightarrow \underline{y}$ where $\underline{x} = (x_1, ..., x_k)$, $x_i = \xi^{b_i}$ and $\underline{y} = (y_1, ..., y_n)$, $y_j = \xi^{c_j} = \xi^{\sum_{r=1}^{k} b_r g_{rj}}$ $= \prod_{r=1}^{k} \xi^{b_r g_{rj}} = \prod_{r=1}^{k} x^{g_{rj}}$. The encoding procedure is called *systematic* if the generating matrix G is in standard form, i.e., $G = (I_k, A)$.

Definition 3. Let C be a $[n, k]$-linear code over the field $\mathbb{Z}_p$ with generating matrix G. The *energy function* of C with respect to an element $\underline{\omega} \in \mu_p^n$ and the encoding procedure $\underline{x} \longrightarrow \underline{y}$ is the polynomial representation $E_{\underline{\omega}}^p(\underline{y})$ of the generating matrix G with respect to $\underline{\omega}$ as defined above.

Observation 1. Note that $E^2_{\underline{\omega}}(\underline{y}) = \underline{\omega} \cdot \underline{y}$, and $E^p_{\underline{\omega}}(\underline{y}) = \sum_{r=0}^{p-1} \underline{\omega}^{(r)} \cdot \underline{y}^{(r)}$, where $z^{(r)} = (z_1^r, ..., z_n^r)$, (here the centered dot denotes the inner product).

One of the main results concerning the energy function and the Maximum Likelihood Decoding problem of a linear code stated in [1], Theorem 2 and Theorem 4, is the following:

Theorem 4. *Let C be a $[n,k]$-linear code over the field $\mathbb{Z}_p$ with generating matrix G. Let $\underline{x} \longrightarrow \underline{y}$ be the encoding procedure and let $E^p_{\underline{\omega}}$ be the energy function of C with respect to $\underline{\omega} \in \mu_p^n$, as described above. Then the Maximum Likelihood Decoding problem (for the Hamming distance) of $\underline{\omega}$ is equivalent to finding the maximum of $E^p_{\underline{\omega}}$.*

3 The Energy Function for Linear Codes over $GF(q)$

In this section the concept of *energy function* for codes defined over a finite field $GF(q)$, where q is a power of a prime p, is introduced. A similar result relating the MLD problem with the maximization of this function, as theorem 4 above, is proved. In order to illustrate these ideas, some examples are provided with codes arising from non-singular projective algebraic curves, i.e., Goppa codes.

In order to define the energy function we first recall some basic facts about characters of a finite field (see [8], Chap.5 §6 and [14]).

Let $K = GF(q)$ be the finite field with $q = p^m$ elements ($m \geq 1$ and p a prime), and let α be a primitive element of K. Since K is an extension of degree m of the field $GF(p) = \mathbb{Z}_p$, any element $\beta \in K$ can be written as:

$$\beta = \beta_0 + \beta_1\alpha + \cdots + \beta_{m-1}\alpha^{m-1}, \; \beta_i \in GF(p)$$

Equivalently, the element $\beta \in K$ can be identified with the m-tuple $(\beta_0, \beta_1, ..., \beta_{m-1}) \in \mathbb{Z}_p^m$. Let $\xi = exp(2\pi i/p)$ be a primitive p-th root of unity, i.e., a generator of the (multiplicative) group μ_p of p-th roots of unity. For each $\beta = (\beta_0, \beta_1, ..., \beta_{m-1})$ in K, let $\chi : K \longrightarrow \mu_p$ be defined as $\chi_\beta(\gamma) = \xi^{\beta_0\gamma_0 + \cdots + \beta_{m-1}\gamma_{m-1}}$ for $\gamma = (\gamma_0, \gamma_1, ..., \gamma_{m-1}) \in K$. The function χ_β is called a *character* of K. It is readily seen that the character χ_β is a homomorphism from the group $(K, +)$ to the group $(\mu_p, *)$. The set of characters $\Lambda = \{\chi_\beta, \quad \beta \in K\}$ with the product of characters, form a group which is isomorphic to $(K, +)$, via $\beta \longrightarrow \chi_\beta$. An easy but important result about characters that will be useful later is the following:

Lemma 5. *For any non-zero element $\beta \in K$,* $\quad \sum_{\gamma \in K} \chi_\beta(\gamma) = 0$.

Proof. The proof of this result is based on the fact that $\frac{1-\xi^p}{1-\xi} = 0$, (see [8] pag.143). See also [14] pag.14 for an elegant proof.

Now we describe an encoding procedure. If $\underline{b} = (b_1, b_2, ..., b_k) \in K^k$ let $\chi_{\underline{b}} = (\chi_{b_1}, ..., \chi_{b_k}) \in \Lambda^k$ where χ_{b_i} is the character of K determined by the i-th coordinate b_i of the vector $\underline{b}$. Let $A = (a_{ij})$ be a kxn matrix over K, let $\underline{c} = (c_1, ..., c_n) = \underline{b} \cdot A \in K^n$, with $c_j = \sum_{r=1}^{k} b_r a_{rj}$ and let $\psi = (\chi_{c_1}, ..., \chi_{c_n}) \in \Lambda^n$ be the corresponding n-th tuple character determined by the n-tuple vector $\underline{c}$.

Definition 6. Let C be a $[n,k]$-linear code defined over the finite field $K = GF(q)$ with generating matrix $G = (g_{ij})$. The *character encoding procedure* of C is given by the following commutative diagram:

$$\begin{array}{ccc} GF(q)^k & \longrightarrow & \Lambda^k \\ \downarrow & & \downarrow \\ GF(q)^n & \longrightarrow & \Lambda^n \end{array}$$

i.e.,

$$\begin{array}{ccc} \underline{b} = (b_1, ..., b_k) & \longrightarrow & \chi_{\underline{b}} = (\chi_{b_1}, ..., \chi_{b_k}) \\ \downarrow & & \downarrow \\ \underline{c} = (c_1, ..., c_n) & \longrightarrow & \psi_{\underline{c}} = (\chi_{c_1}, ..., \chi_{c_n}) \end{array}$$

where $c_j = \sum_{r=1}^{k} b_k g_{rj}, j = 1, ..., n$

Observe that since χ_β, $\beta \in K$ is a homomorphism, then $\chi_{c_j} = \prod_{r=1}^{k} \chi_{b_r g_{rj}}$. This encoding procedure will be denoted by $\chi \longrightarrow \psi$. Also, it is called *systematic* if the generating matrix G is in standard form (I_k, A).

We now give some examples to illustrate the above encoding procedure. Since most of the linear codes presented here are algebraic-geometric (Goppa) codes, the definition and basic properties of these codes are recalled in the appendix. For more details see for example [5], [6], [7], [9], [13], [15].

1.- If $q = p$ the above character encoding procedure is precisely the encoding procedure $\underline{x} \longrightarrow \underline{y}$ given in [1], and recalled in section 2 above.

2.-Let $K = GF(4) = \{0, 1, \alpha, \alpha^2 = \alpha + 1\}$ be the finite field with 4 elements, and let X be the projective algebraic plane curve over K given by the relation $x^3 + y^3 + z^3 = 0$. It is easy to see that X is a non- singular curve of genus 1 having 9 rational points over K (see [9], [15]): $P_1 = (1, 0, \alpha^2)$, $P_2 = (1, 0, \alpha)$, $P_3 = (1, 0, 1)$, $P_4 = (1, \alpha^2, 0)$, $P_5 = (1, \alpha, 0)$, $P_6 = (1, 1, 0)$, $P_7 = (0, \alpha^2, 1)$, $P_8 = (0, \alpha, 1)$, $Q = (0, 1, 1)$.
Consider the K-rational divisors $D = \sum_{i=1}^{8} P_i$ and $G = 2Q$. From the Riemann-Roch theorem it follows that the K-linear space $L(G) = \{f \in K(X) : \text{div}(f) + G \geq 0\}$ has dimension 2. A basis for this space is given by the rational functions $\{1, x/(y+z)\}$. Thus the geometric Goppa code $C = C(D, G)$ is a [8,2]-linear code with minimal (Hamming) distance 6 and generating matrix:

$$\begin{pmatrix} 1 & 1 & 1 & 1 & 1 & 1 & 1 & 1 \\ \alpha & \alpha^2 & 1 & \alpha & \alpha^2 & 1 & 0 & 0 \end{pmatrix}$$

If $\underline{b} = (b_1, b_2) \in K^2$, then $\underline{c} = (c_1, ..., c_8) = (b_1+\alpha b_2, b_1+\alpha^2 b_2, b_1+b_2, b_1+\alpha b_2, b_1+\alpha^2 b_2, b_1+b_2, b_1, b_2)$, and the character encoding procedure is $\psi = (\chi_{c_1}, ..., \chi_{c_8})$. For instance $\chi_{c_1} = \chi_{b_1+\alpha b_2} = \chi_{b_1}\chi_{\alpha b_2}$. More especifically, if $\underline{b} = (\alpha, \alpha^2)$, then $c_1 = \alpha^2$ and $\chi_{c_1} = \chi_{\alpha^2}$. Since $0 = (0,0), 1 = (1,0), \alpha = (0,1), \alpha^2 = (1,1)$, then $\chi_{\alpha^2} : GF(4) \longrightarrow \mu_2$ is such that $\chi_{\alpha^2}(0) = \chi_{\alpha^2}(\alpha^2) = 1$, $\chi_{\alpha^2}(1) = \chi_{\alpha^2}(\alpha) = -1$

3.- Let $K = GF(4)$ and let X be the projective algebraic plane curve over K given by the relation $x^2y+\alpha y^2z+\alpha^2z^2x = 0$. This curve is non-singular of genus 1 with 9 rational points over K: $P_1 = (1,0,0)$, $P_2 = (0,1,0)$, $P_3 = (0,0,1)$, $P_4 = (1,\alpha,\alpha^2)$, $P_5 = (1,\alpha^2,\alpha)$, $P_6 = (1,1,1)$, $Q_1 = (\alpha,1,1)$, $Q_2 = (1,\alpha,1)$, $Q_3 = (1,1,\alpha)$. Take the divisors $D = \sum_{i=1}^{6} P_i$ and $G = 2Q_1+Q_2$. Then $C(D,G)$ is a [6,3]-linear code over $GF(4)$ (with minimal distance 4). A basis for the K-linear space $L(G)$ is $\{x/(x+y+\alpha z), y/(x+y+\alpha^2 z), \alpha^2 z/(x+y+\alpha^2 z)\}$, and a generating matrix for the code is (see [15]):

$$\begin{pmatrix} 1 & 0 & 0 & 1 & \alpha & \alpha \\ 0 & 1 & 0 & \alpha & 1 & \alpha \\ 0 & 0 & 1 & \alpha & \alpha & 1 \end{pmatrix}$$

If $\underline{b} = (b_1, b_2, b_3) \in K^3$, then $\underline{c} = (c_1, ..., c_6) = \underline{b} \cdot G = (b_1, b_2, b_3, b_1 + \alpha b_2 + \alpha b_3, \alpha b_1 + b_2 + \alpha b_3, \alpha b_1 + \alpha b_2 + b_3)$, and again the encoding procedure is: $\psi = (\chi_{c_1}, ..., \chi_{c_6})$. For instance if $\underline{b} = (1, \alpha, \alpha^2)$ then $\underline{c} = (1, \alpha, \alpha^2, \alpha^2, 1, \alpha)$ and $\psi = (\chi_1, \chi_\alpha, \chi_{\alpha^2}, \chi_{\alpha^2}, \chi_1, \chi_\alpha)$.

Observe that in example 3 the encoding procedure is systematic but not in example 2.

We now define the *energy function* associated with a linear code over any finite field.

Let C be a $[n,k,d]$-linear code defined over the finite field $K = GF(q)$, with q a power of a prime p. Let G be a generating matrix of the code and let $\chi \longrightarrow \psi$ be the character encoding procedure as described above. For an element $\underline{\omega} = (\omega_1, ..., \omega_n) \in \Lambda^n$, let $\underline{w} = (w_1, ..., w_n) \in K^n$ be the corresponding element.

Definition 7. The *energy function* associated with the $[n,k]$-linear code C defined over the finite field $GF(q)$, having generating matrix G, with respect to the element $\underline{w} = (w_1, ..., w_n) \in K^n$ and the above character encoding procedure is:

$$E^q_{\underline{w}}(\psi) = \sum_{j=1}^{n} \sum_{\beta \in K} \overline{\chi_1(\beta w_j)} \chi_1(\beta c_j)$$

where $\underline{c} = (c_1, ..., c_n) = \underline{b} \cdot G$ and $\psi = (\chi_{c_1}, ..., \chi_{c_n})$, for $\underline{b} \in K^k$. Here χ_1 is the character associated with the unit element 1 of the field K, and the upper bar is the complex conjugate.

We now give some examples.

1.- If $q = p$ and $\underline{w} = (w_1, ..., w_n) \in K^n$, for $m \in K = \mathbb{Z}_p$, then $\overline{\chi_1(mw_j)}\chi_1(mc_j) = \overline{w_j} y_j^m$. If $p \neq 2$ the energy function of definition 7 above is precisely the energy function associated with the code C introduced in [1] and recalled in section 2 above. If $p=2$, the energy function of definition 7 and the energy function $E_{\underline{w}}(\underline{y})$ of [1] (see section 2 above) are related as follows: $E^2_{\underline{w}}(\underline{y}) = n + E_{\underline{w}}(\underline{y})$.

2.- Let C be the geometric code in example 2 above, let $\underline{b} = (b_1, b_2) \in K^2$ and let $\underline{w} = (w_1, ..., w_8) \in K^8$. Then $E^4_{\underline{w}}(\psi) = E^4_{\underline{w}}(\chi_{c_1}, ..., \chi_{c_8}) = \sum_{j=1}^{8} \sum_{\beta \in K} \overline{\chi_1(\beta w_j)} \chi_1(\beta c_j)$, where $\underline{c} = (c_1, ..., c_8) = (b_1 + \alpha b_2, b_1 + \alpha^2 b_2, b_1 + b_2, b_1 + \alpha b_2, b_1 + \alpha^2 b_2, b_1 + b_2, b_1, b_2)$. For instance if $\underline{b} = (\alpha, \alpha^2)$ and $\underline{w} = (1, \alpha, \alpha^2, 1, \alpha, \alpha, 0, 0)$, then $E^4_{\underline{w}}(\psi) = 2\sum_{\beta \in K} \overline{\chi_1(\beta)} \chi_1(\beta\alpha^2) + 2\sum_{\beta \in K} \overline{\chi_1(\beta\alpha)} \chi_1(0) + \sum_{\beta \in K} \overline{\chi_1(\beta\alpha^2)} \chi_1(\beta) + \sum_{\beta \in K} \overline{\chi_1(\beta\alpha)} \chi_1(\beta) + \sum_{\beta \in K} \overline{\chi_1(0)} \chi_1(\beta\alpha) + \sum_{\beta \in K} \overline{\chi_1(0)} \chi_1(\beta\alpha^2)$.

3.- Let C be the [6,3,4]-linear code in example 3 above. If $\underline{w} = (w_1, ..., w_6) \in K^6$ and $\underline{b} = (b_1, b_2, b_3) \in K^3$, the corresponding energy function is: $E^4_{\underline{w}}(\psi) = E^4_{\underline{w}}(\chi_{c_1}, ..., \chi_{c_6}) = \sum_{j=1}^{6} \sum_{\beta \in K} \overline{\chi_1(\beta w_j)} \chi_1(\beta c_j)$, where $\underline{c} = (c_1, ..., c_6) = (b_1, b_2, b_3, b_1 + \alpha b_2 + \alpha b_3, \alpha b_1 + b_2 + \alpha b_3, \alpha b_1 + \alpha b_2 + b_3)$. If for instance $\underline{b} = (1, \alpha, \alpha^2)$ then $\underline{c} = (1, \alpha, \alpha^2, \alpha^2, 1, \alpha)$, $\psi = (\chi_1, \chi_\alpha, \chi_{\alpha^2}, \chi_{\alpha^2}, \chi_1, \chi_\alpha)$, and the corresponding energy function is:
$E^4_{\underline{w}}(\psi) = \sum_{\beta \in K} \overline{\chi_1(\beta w_1)} \chi_1(\beta) + \sum_{\beta \in K} \overline{\chi_1(\beta w_2)} \chi_1(\beta\alpha) + \sum_{\beta \in K} \overline{\chi_1(\beta w_3)} \chi_1(\beta\alpha^2) + \sum_{\beta \in K} \overline{\chi_1(\beta w_4)} \chi_1(\beta\alpha^2) + \sum_{\beta \in K} \overline{\chi_1(\beta w_5)} \chi_1(\beta) + \sum_{\beta \in K} \overline{\chi_1(\beta w_6)} \chi_1(\beta\alpha)$

The following result relates the Maximum Likelihood Decoding (MLD) problem, with respect to the Hamming distance, of a code C, with the maximum of the energy function defined above.

Theorem 8. *Let C be a $[n,k]$-linear code over the field $K = GF(q)$. Let $\chi \longrightarrow \psi$ be the character encoding procedure described above and for $\underline{w} = (w_1, ..., w_n) \in K^n$ let $E^q_{\underline{w}}(\psi)$ be the corresponding energy function. Then the (MLD) problem of $\underline{w}$ is equivalent to finding the maximum of $E^q_{\underline{w}}(\psi)$.*

Proof. Since $\chi_1(\gamma)^{-1} = \chi(-\gamma) = \overline{\chi_1(\gamma)}$ for any $\gamma \in K$, $E^q_{\underline{w}}(\psi) = \sum_{j=1}^{n} \sum_{\beta \in K} \chi_1(\beta(c_j - w_j))$. Then $\sum_{\beta \in K} \chi_1(\beta(c_j - w_j)) = q$ if $c_j = w_j$ and equal 0 if $c_j \neq w_j$ since $\sum_{\gamma \in K} \chi_1(\gamma) = 0$ (see lemma 5). From this it follows that $E^q_{\underline{w}}(\psi) = q(\#\{j : c_j = w_j\}) = q(n - d_H(\underline{w}, \underline{c}))$, where d_H denotes the Hamming distance, and the result follows.

Observation 2. The energy function given above can be described in terms of a special case of a Hadamard transform. Recall that if H is a multiplicative group and $f : K \longrightarrow H$ is any function, the Hadamard transform $\hat{f}$ of f is given by $\hat{f}(u) = \sum_{\beta \in K} \chi_u(\beta) f(u)$ (see [8], pag.144)

In our situation if we take the function $f : K \longrightarrow H$ as $f(\gamma) = 1$ for all $\gamma \in K$, then $\hat{f}(c_j - w_j) = \sum_{\beta \in K} \chi_1(\beta(c_j - w_j)) = \sum_{\beta \in K} \overline{\chi_1(\beta w_j)} \chi_1(\beta c_j)$, and $E^q_{\underline{w}}(\psi) = \sum_{j=1}^{n} \hat{f}(c_j - w_j)$.

We now illustrate the above result with some examples.

1.- If $q = p$, theorem 2 and theorem 4 of [1] (recalled above as theorem 4) are obtained from theorem 8.

2.- Let C be the [8,2]-linear geometric code over $GF(4)$ of example 2 above, and let $\underline{w} = (\alpha^2, 1, 1, \alpha, 0, 0, \alpha, \alpha) \in GF(4)^8$. If for instance $\underline{c} = (\alpha^2, 0, 1, \alpha^2, 0, 1, \alpha, \alpha)$, a straightforward calculation shows that $E^4_{\underline{w}}(\psi) = 20$. Note that $d_H(\underline{c}, \underline{w}) = 3$ and $E^4_{\underline{w}}(\psi) = 4(8-3) = 20$. If for instance $\underline{w} = (1, \alpha, \alpha^2, 1, \alpha, \alpha, 0, 0)$ and $\underline{c} = (\alpha^2, 0, 1, \alpha^2, 0, 1, \alpha, \alpha^2)$, a direct calculation shows that $E^4_{\underline{w}}(\psi) = 0$. Note that in this case $d_H(\underline{c}, \underline{w}) = 8$, so that $E^4_{\underline{w}}(\psi) = 4(8-8) = 0$.

3.- Let $K = GF(16)$ and let α be a primitive element (which satisfies $\alpha^4+\alpha^3+1 = 0$). Let X be the projective algebraic curve over K determined by the relation $y^2 + y = x^3 + x + 1$. This curve is non-singular and has 25 rational points over K, $Q = (0, 1, 0)$ being one of them. Consider the K-rational divisors $G = 2Q$ and $D = \sum_{i=1}^{24} P_i$, where the P_i's are the other 24 K-rational points of the curve X. The linear space $L(2Q)$ has dimension 2 and a basis is $\{1, x\}$. The geometric linear code C associated with the triple (X, D, G) has parameters [24,2]. The first row of the 2x24 generating matrix of the code is the vector whose entries are all equal to 1, i.e.,(1, ..., 1), and its second row is $(0, 0, 1, 1, \alpha, \alpha, \alpha^2, \alpha^2, \alpha^3, \alpha^3, \alpha^{12}, \alpha^{12}, \alpha^9, \alpha^9, \alpha^4, \alpha^4, \alpha^{10}, \alpha^{10}, \alpha^5, \alpha^5, \alpha^8, \alpha^8, \alpha^6, \alpha^6)$. Let $\underline{c} = (0, 0, \alpha, \alpha, \alpha^2, \alpha^2, \alpha^3, \alpha^3, \alpha^4, \alpha^4, \alpha^{13}, \alpha^{13}, \alpha^{10}, \alpha^{10}, \alpha^5, \alpha^5, \alpha^{11}, \alpha^{11}, \alpha^6, \alpha^6, \alpha^9, \alpha^9, \alpha^7, \alpha^7)$, which is an element of the code C, and let $\underline{w}$ be the element of K^{24} whose first two coordinates are 1,1 and the other 22 are the same as those of $\underline{c}$, (except the first two, i.e., 0,0). Using a computer program it was shown that the value of the energy function of C with respect to $\underline{w}$, evaluated at $\underline{c}$, is 352. Observe that $d_H(\underline{c}, \underline{w}) = 2$, and hence $16(24-2) = 352$.

The above examples also suggest the following

Corollary 9. *With the same notation as above let* $\underline{0} = (0, ..., 0)$ *and let* $M = \max_{\underline{c} \neq 0} E^q_{\underline{0}}(\psi)$. *Then the minimal (Hamming) distance of the code* C *is* $d = n - \frac{M}{q}$.

Proof. Since $E^q_{\underline{0}}(\psi) = q(n - d_H(\underline{0}, \underline{c})$ and since the $\min_{\underline{c} \neq 0} d_H(\underline{0}, \underline{c})$ occurs at M, the result follows.

Remark. If we take the corresponding element to $\underline{0}$ in the group μ_p^n, i.e.,the element $\underline{1} = (1, ..., 1)$, (all the coordinates equal to 1), then corollary 9 gives the result for the minimal distance of the code, stated in [1], pag.981.

The above corollary may be illustrated with some examples.

1.- Let C be the [8,4]-linear geometric code of example 2 above. A direct calculation shows that $M = \max E^4_{\underline{0}}(\underline{c}) = 8$, where the maximum is taken over all non-zero elements of the code C. Hence $d = 8 - \frac{8}{4} = 6$.

2.- Let C be the [6,3]-linear Goppa code over $GF(4)$ described in example 3 above. A straighforward calculation shows that $M = \max E^4_{\underline{0}}(\underline{c}) = 8$, taken over all non-zero elements of the code C. Then $d = 6 - \frac{8}{4} = 4$. This is in fact a MDS code (see [15]).

4 The Energy Function of the Parity Check Matrix

In [1], theorem 3 it is shown that if $\tilde{E}$ is the energy function associated with the parity check matrix of a binary code, then the elements of the linear code are characterized by the local maximum of the function $\tilde{E}$. In this section using the same techniques as in [1], similar results are obtained for linear codes defined over any finite field.

Let C be a $[n,k]$-linear code over the field $K = GF(q)$ with generating matrix G and parity check matrix H. Without loss of generality we may assume that G is in standard form (I_k, A), where A is a kx$(n-k)$ matrix and I_k is the kxk identity matrix. Hence $H^t = (h_{ij}) = (-A, I_{n-k})^t$. Let $\tilde{E}^q_{\underline{0}}(\psi)$ be the energy function of C associated with the parity check matrix with respect to the zero-vector $\underline{0} = (0,...,0)$.

Lemma 10. *Let $\tilde{E}^q_{\underline{0}}(\psi)$ be as described above. Then $\underline{x} \in C$ if and only if $\tilde{E}^q_{\underline{0}}(\underline{x}) = q(n-k)$.*

Proof. Since $\tilde{E}^q_{\underline{0}}(\psi)$ has $n-k$ terms and each term $\sum_{\beta \in K} \chi_1(\beta v_j) = q$ if and only if $\chi_1(\beta v_j) = 1$ for all $\beta \in K$, the result follows at once (since $\underline{v} \cdot H^t = 0$).

Now a characterization of the elements of the code can be given in terms of the energy function of the parity check matrix.

Theorem 11. *Let C be a $[n,k]$-linear code over $K = GF(q)$ with G, H, and $\tilde{E}^q_{\underline{0}}(\psi)$ as described above. Then $\tilde{E}^q_{\underline{0}}$ has a local maximum at $\underline{x}$ if and only if $\underline{x} \in C$.*

Proof. One direction is given by the above lemma 10. For the converse note that since H^t is in standard form, each term $\chi_1(\beta v_j)$ in $\tilde{E}^q_{\underline{0}}$ has the form $\chi_1(\beta v_j) = [\chi_1(\beta b_1 h_{1j}) \cdots \chi_1(\beta b_k h_{kj})]\chi_1(\beta b_{k+j})$, for $j = 1, 2, ..., n-k$, i.e., in each term $\chi_1(\beta v_j)$, the factor $\chi_1(\beta b_{k+j})$ appears just once, i.e., $\chi_1(\beta b_{k+1})$ appears just in $\chi_1(\beta v_1)$, $\chi_1(\beta b_{k+2})$ appears just in $\chi_1(\beta v_2)$, and so on. Now assume that there is an element $\underline{u}$ that corresponds to a local non-global maximum of the function $\tilde{E}^q_{\underline{0}}$, i.e., $\tilde{E}^q_{\underline{0}}(\underline{u}) = L < M = q(n-k)$. Then since $\tilde{E}^q_{\underline{0}}$ has $n-k$ terms, there is at least one $\sum_{\beta \in K} \chi(\beta u_j)$, not equal to q, but then one summand $\chi(\beta u_j) \neq 1$ for at least one $\beta \in K$. From the above observation on the particular form of this summand, it follows that this term can be made equal to 1 by changing the value of $\underline{u}$, without affecting the value of the other terms. Thus by modifying the original vector $\underline{u}$, each term in $\tilde{E}^q_{\underline{0}}$ can be made to have a value equal to q, so that for this new vector, $\tilde{E}^q_{\underline{0}}$ has value $q(n-k)$, contradicting the fact that $\underline{u}$ is a local maximum.

We give an example. Let C be the geometric code of example 3 above defined over the field $K = GF(4)$. Since the generating matrix is in standard form, the parity check matrix H^t is:

$$\begin{pmatrix} -1 & -\alpha & -\alpha \\ -\alpha & -1 & -\alpha \\ -\alpha & -\alpha & -1 \\ 1 & 0 & 0 \\ 0 & 1 & 0 \\ 0 & 0 & 1 \end{pmatrix}$$

If $\underline{u} = (u_1, u_2, u_3) = (-b_1 - \alpha b_2 - \alpha b_3 + b_4, -\alpha b_1 - b_2 - \alpha b_3 + b_5, -\alpha b_1 - \alpha b_2 - b_3 + b_6)$, the energy function of C with respect to this matrix H^t is: $\tilde{E}^4_{\underline{0}}(\underline{u}) = \sum_{\beta \in K} \chi_1(-\beta b_1)\chi_1(-\beta\alpha b_2)\chi_1(-\beta\alpha b_3)\chi_1(\beta b_4) + \sum_{\beta \in K} \chi_1(-\beta\alpha b_1)\chi_1(-\beta b_2)$ $\chi_1(-\beta\alpha b_3)\chi_1(\beta b_5) + \sum_{\beta \in K} \chi_1(-\beta\alpha b_1)\chi_1(-\beta\alpha b_2)\chi_1(-\beta b_3)\chi_1(\beta b_6)$. Notice that the terms $\chi_1(\beta b_4), \chi_1(\beta b_5), \chi_1(\beta b_6)$ appear just once, respectively, in the first, second and third terms of the energy function. It can be seen that the set of local maxima of the function $\tilde{E}^4_{\underline{0}}(\underline{u})$ is precisely the [6,3,4]-linear Goppa code C.

5 Appendix

In this appendix we recall how Goppa constructed linear codes over finite fields from non-singular algebraic curves (for more details see [5], [6],[7],[9],[13],[15]).

For a non-singular irreducible projective algebraic curve X of genus g over a finite field k, let $k(X)$ be its function field and let $\Omega(X)$ be the linear space of rational differentials of the curve X. For a divisor E on the curve, let $L(E) = \{f \in k(X) : \text{div}(f) + E \geq 0\}$ and let $\Omega(E) = \{\eta \in \Omega(X) : \text{div}(\eta) \geq E\}$. These are finite dimensional k-linear spaces (by the Riemann-Roch theorem), and their dimensions are denoted by $l(E)$ and $i(E)$ respectively. Let $D = P_1 + ... + P_n$ be a divisor on X such that the points P_i's are rationals over k and let G be another rational divisor on X with disjoint support from the support of D. Here we assume that $2g - 2 < \deg(G) < n$.

Let $\alpha : L(G) \longrightarrow k^n$ be the k-linear function defined by $\alpha(f) = (f(P_1), ..., f(P_n))$, and let $\alpha^* : \Omega(G - D) \longrightarrow k^n$ be the k-linear function given by $\alpha^*(\eta) = (\text{res}(\eta, P_1), ..., \text{res}(\eta, P_n))$ (the assumption on the degree of G ensures that these linear functions are injective). The geometric-algebraic (Goppa) codes associated with the triple (X, G, D) are defined as $C(G, D) = \alpha(L(G))$ and $C^*(G, D) = \alpha^*(\Omega(G - D))$. These codes have parameters $[n, l(G), d]$ and $[n, i(G - D), d^*]$, respectively (by the Riemann-Roch theorem, $l(G) = deg(G) - g + 1$ and $i(G - D) = l(K + D - G)$ where K is a canonical divisor on the curve). The minimal distance satisfies $d \geq n - \deg(G)$, $\deg(G) + 2 - 2g \leq d^* \leq \deg(G) + 2 - g$. These codes are dual to each other, i.e. $C(G, D)^{\perp} = C^*(G, D)$. Furthermore, $C(G, D)^{\perp} = C(D - G + (\eta), D)$, where η is a differential on the curve with simple poles and residues 1 at the points P_i's of the support of the divisor D.

Acknowledgment: The authors would like to thank the referees for their helpful suggestions. Author (1) was partially supported by COFAA, IPN and SNI, No.12262, and author (2) was partially supported by SNI, No.4331, México.

References

1. Bruck, J., Blaum, M.: Neural Networks, Error-Correcting Codes, and Polynomials over the Binary n-Cube. IIIE Trans. on Inf. Theory, vol.**35**, No.5 (1989) 976–987
2. Bruck, J., Sanz, J.: A study of neural networks. Int. J. Intelligent Systems, vol.**3** (1988) 59–75
3. Bruck, J., Goodman, J.W.: A generalized convergence theorem for neural networks. IIIE Trans. Inf. Theory, vol.**IT-34** (1988) 1089–1092
4. Hopfield, J.J.: Neural networks and physical systems with emergent collective computational abilities. Proc. Nat. Acad. Sci. (USA) vol.**79** (1986) 269–281
5. Goppa, V.D.: Algebraico-Geometric codes. Math. URSS-Izv **21(1)** (1983) 75–91
6. Goppa, V.D.: *Geometry and Codes.* Mathematics and Its Applications. Kluwer Academic Publishers, (1988)
7. Manin, Y.I., Vlădut, S.G.: Linear codes and modular curves. J.Sov. Math. **30** (6) (1985) 2611–2643
8. MacWilliams, F.J., Sloane, N.J.A.: *The Theory of Error-Correcting Codes.* North-Holland, New York, (1977)
9. Moreno, C. *Algebraic curves over finite fields.* Cambridge U. Press, (1991)
10. Peterson, W.W. and Weldon, W.J. *Error-Correcting Codes.* Cambridge, MA.,MIT Press, (1971)
11. Rentería, C., Tapia-Recillas, H.: A lifting of BCH codes over a hyperelliptic curve. To appear in Marshall Hall Memorial Conference. (J. Wiley & Sons Co.)
12. Rentería, C., Tapia-Recillas, H.: A geometric code on an Artin-Schreier curve. C. Numerantium, vol.**84**, Dec.1991, 3–7, Winnipeg, Canada
13. Stichtenoth, H.: Self-dual Goppa codes. J. of Pure and Applied Algebra **55** (1988) 199–211
14. van Lint, J.H. *Introduction to Coding Theory.* GTM **86**, Springer-Verlag, New York, (1982)
15. van Lint, J.H., van der Geer.: *Introduction to Coding Theory and Algebraic Geometry* Birkhäuser Verlag, (1988)

On Determining All Codes in Semi-Simple Group Rings

Roberta Evans Sabin
Department of Computer Science
Loyola College
Baltimore, Maryland USA
E-Mail: RES@Loyola.edu

Abstract

A group algebra code is an ideal in a group ring, FG. If char(F) $\not|$ order(G), then FG is semi-simple and every such code has an idempotent generator. The group ring and every group algebra code in it are direct sums of disjoint minimal ideals. Thus a list of all possible codes in FG may be produced by first determining idempotent generators of minimal codes which are direct summands of FG. When G is abelian, the task is straightforward and is facilitated by use of the character table for G. When G is non-abelian, however, two-sided ideals, which correspond to the group characters of G, in FG may decompose in multiple ways to one-sided ideals. The decomposition may be varied by altering the matrix representations afforded by the group representations. The number of ways the representation may be altered is limited by the structure of the representation space. With sufficient information about this space, all possible decompositions can be determined. A case study is presented in which a (125,20) two-sided ideal is found to contain 162 disjoint minimal left ideals, each with a distinct weight distribution.

1 Introduction

A group algebra code is an ideal in group ring FG where F is a finite field and G is a finite group. If char(F) $\not|$ $|G| =$ order(G), FG is semi-simple and a principal ideal domain [4]. All codes in FG are direct sums of minimal codes, i.e., ideals which contain no non-trivial subcodes. To list all codes in FG, we may list the primitive idempotent generators of minimal codes and take their sums as code generators. In the abelian case, i.e., when G is abelian, this is a straightforward task. When G is non-abelian, however, the decomposition of FG is not unique and, consequently, the task is much more interesting.

Although few "best" codes have been found among abelian group codes, some recent work has contributed to interest in non-abelian group codes [2]. It is in this spirit that this paper is written.

2 Abelian Codes

If G is abelian and char(F) $\nmid$ |G|, all codes in FG are two-sided ideals and FG decomposes to minimal codes in exactly one way. In the simplest case, $G = \mathbb{Z}_m = \langle x : x^m = 1 \rangle$ and ideals are the familiar cyclic codes. Minimal cyclic codes can be determined by using cyclotomic cosets of the integers mod m to identify the check polynomial, h(x), of the code, and $g(x) = (x^m - 1)/h(x)$ is a generator [6].

Alternatively, the absolutely irreducible group characters of $G = \mathbb{Z}_m$ may be used. The character table for $\mathbb{Z}_m$ provides the m idempotent generators for minimal one-dimensional codes in LG where L is a splitting field for G, in this way: let character $\chi_i : G \rightarrow L$ where $\chi_i(x) = \delta^i$, δ is a primitive m-th root of one. Then

$$\left\{ e_i = |G|^{-1} \sum_{j=0}^{m-1} (\delta^i)^{-j} x^j \ : \ 0 \leq i < m \right\}$$

is the set of mutually orthogonal idempotent generators of minimal codes in LG and $\sum_i e_i = 1$ in LG. (When char(L) = 2, $|G|^{-1} = 1$.)

If F = GF(q), cyclotomic cosets, $C_s^q = \{ s, sq, sq^2, \ldots \}$, partition the integers mod m and indicate which idempotents of minimal codes in LG need be added to yield primitive idempotents in FG [6].

When G is abelian, $G = \mathbb{Z}_{m_1} \times \mathbb{Z}_{m_2} \times \ldots \times \mathbb{Z}_{m_k}$ and FG may likewise be uniquely decomposed. Because of the complexity of operating with polynomials in k indeterminates, it is easiest to arrive at the decomposition by the examination of the character table of G. As with cyclic codes, the table provides the idempotent generators of the $m_1 \cdot m_2 \cdot \ldots \cdot m_k$ minimal one-dimensional codes in LG, where L is a splitting field for G. Then sets of k-tuples, generated in a manner analogous to cyclotomic coset generation, are used to indicate which of these idempotents should be summed to yield idempotents of codes which are minimal in FG [1].

For example, let $\mathbb{Z}_m = \langle x \rangle$ and $\mathbb{Z}_n = \langle y \rangle$, $G = \mathbb{Z}_m \times \mathbb{Z}_n$ and $F = \mathbb{F}_2$ where $2 \nmid mn$. Group characters for G are of the form, $\chi_{ij} : G \rightarrow L$ where $\chi_{ij}(xy) = \delta^i \omega^j$, δ and ω are primitive m-th and n-th roots of one respectively, and $L = GF(2^q)$ is a splitting field for G. Then for every k, l, $0 \leq k < m$, $0 \leq l < n$,

$$e_{kl} = \sum_{j=0}^{n-1} \sum_{i=0}^{m-1} \delta^{-ki}\omega^{-lj}\, x^i y^j$$

generates a minimal, one-dimensional code in LG. Let $\mathcal{N}_{ab} = \{(2^i a \bmod m,\ 2^i b \bmod n) : i \geq 0\}$. For any such ab pair, $\sum_{(k,l) \in \mathcal{N}_{ab}} e_{kl}$ generates a minimal code in FG.

3 Non-Abelian Codes

3.1 Minimal Codes in LG when L is a Splitting Field

It is not possible to generalize the structure of a non-abelian group. Therefore, it is not possible to devise a generalized "polynomial" structure for group ring FG. We will depend on the group characters or, more precisely, the group representations of G for the decomposition of FG.

We begin by generalizing the definition of a group character. Let group G have splitting field L and let $\mathfrak{T}$ be a representation of G over L. Write $\mathfrak{T} : G \to GL_n(L)$, $g \mapsto \mathfrak{T}(g)$. The integer n is the degree of $\mathfrak{T}$; the character χ afforded by $\mathfrak{T}$ is defined

$$\chi : G \to L,\ \chi(g) = \text{trace}\ (\mathfrak{T}(g)).$$

Remark: More properly, $\mathfrak{T}$ is a matrix representation afforded by a representation. For a complete discussion of group representations, we refer the reader to [3], [5], or [7].

Irreducible representations of G over an algebraically closed field are referred to as absolutely irreducible representations. Listing a set of non-equivalent, absolutely irreducible representations of G (over L) provides a character table for G. Only one such table results since equivalent representations afford the same character. When G is abelian, all such absolutely irreducible representations of G over L are of degree 1 [7] and are therefore unique. Thus, we need only speak of group characters for abelian groups. In general, the number of distinct characters for G equals the number of conjugacy classes of G [7, p. 180]. Non-abelian groups have, therefore, at least 1 irreducible representation over L of degree 2 or more.

Representations of G are easily extended to LG. Let $\{\mathfrak{T}_i : 0 \leq i < k\}$ consist of one representative of each set of equivalent, irreducible representations of G over L. Let $\mathfrak{T}_i$ have degree n_i and character χ_i. $\text{Dim(LG)} = \sum_{i=0}^{k-1} n_i^2$. Each $\mathfrak{T}_i$ locates a minimal two-sided ideal in LG with dimension n_i^2 and unique idempotent generator e_i (in the center of G) where

$$e_i = |G|^{-1} \sum_{g \in G} \chi_i(g^{-1})\, g \qquad [5, p. 134].$$

Thus, group LG decomposes uniquely to k two-sided ideals where k is the number of absolutely irreducible representations of G. Minimal two-sided code V_i with idempotent generator e_i (as above) is non-zero only on $\mathcal{T}_i$, i.e., for all $c \in V_i$, $\mathcal{T}_i(c) \neq 0$ and $\mathcal{T}_j(c) = 0$ for all $j \neq i$. We say that V_i *corresponds* to absolutely irreducible representation $\mathcal{T}_i$ (with character χ_i).

If V_i is a minimal two-sided ideal of LG with dimension n_i^2 greater than 1, V_i decomposes to n_i minimal one-sided ideals (each of dimension n_i). Theorem 1, a variation of a scheme defined by Serre [10, pp. 21-24], defines the decomposition. Note that we choose to deal with left ideals, but that Theorem 1 and all statements relating to left ideals could be restated in terms of right ideals.

Theorem 1: Let V be a minimal two-sided ideal in LG, where L is a splitting field for G. If V corresponds to absolutely irreducible representation $\mathcal{T}$ of degree n, then $V = \bigoplus_{i=0}^{n-1} W_i$ where $\forall i$, W_i is a minimal left ideal in LG and W_i is generated by idempotent

$$e_i = \frac{n}{|G|} \sum_{g \in G} \tau_{ii}(g^{-1})\, g$$

where $\tau_{ii}(g^{-1})$ is the (i,i) entry in $\mathcal{T}(g^{-1})$.

Proof: Define a set of projection operators on elements of V:

$$p_i = \frac{n}{|G|} \sum_{g \in G} \tau_{ii}(g^{-1}) \cdot \mathcal{R}(g)$$

where $\mathcal{R}$ is the right regular representation of G, i.e., $\mathcal{R} : G \rightarrow GL(LG)$ and $\mathcal{R}(g)$ is the right-multiplication by g [5]. By Serre [10], $p_i : V \rightarrow W_i$, where W_i is a subspace of V and $V = \bigoplus_{i=0}^{n-1} W_i$. But p_i^{-1} is the transpose of the (unreduced) generating matrix for an ideal in LG with generator $e_i = \frac{n}{|G|} \sum_{g \in G} \tau_{ii}(g^{-1})\, g$. Therefore, this code (the column space of p_i) is W_i. For every $c \in V$, $p_i(c) = c \cdot e_i$. By [10], $p_i : W_i \rightarrow W_i$ is the identity map. For $c \in W_i$, $p_i(c) = c = c \cdot e_i$. Therefore e_i is a right identity in W_i and is idempotent. □

We arrive then at a decomposition of V_i by generating codes with idempotents derived from the diagonal elements of $\mathcal{T}_i(g)$ for all g in G . In fact we could use other projection operators and decompose V by using any subdiagonal of the absolutely irreducible representations. Of greater interest, however, is the fact that the decomposition depends upon the choice of $\mathcal{T}_i$. The ideal V_i corresponds to a set of equivalent absolutely irreducible representations. If two absolutely irreducible representations are equivalent to $\mathcal{T}_i$, the two resulting decompositions of V_i defined by Theorem 1 may not be identical. A particular matrix representation $\mathcal{T}_i$

determines the decomposition of V_i. The search for minimal left ideals (codes) then involves altering the choice of the irreducible representations of G with degree greater than one.

Two representations $\mathfrak{T}_i$ and $\mathfrak{T}'_i$ are equivalent if and only if there exists an invertible matrix P in GL(L) such

$$\mathfrak{T}'_i(g) = P^{-1}\,\mathfrak{T}_i(g)\,P \quad \text{for all } g \in G \qquad [5].$$

What choices of P yield distinct decompositions of V_i? Absolutely irreducible representation $\mathfrak{T}_i$ can be identified with a representation space of dimension n (that is isomorphic to minimal left ideal W_i); $\mathfrak{T}_i$ results from using a particular basis $\{b_1, \ldots, b_n\}$ for the representation space of $\mathfrak{T}_i$. Equivalent representation $\mathfrak{T}'_i$ results from choosing another basis $\{b'_1, \ldots, b'_n\}$ for the representation space. The columns of matrix P are the coefficients of the linear combinations of the original basis vectors needed to represent the new basis vectors, i.e., $P = [\ f_{ij}\ ]$, $1 \leq i,j \leq n$, where for every j, $\sum_{j=1}^{n} f_{ij}\, b_i = b'_j$. If $\mathfrak{T}_i$ is induced from the representation of a subgroup of G, the representation space for $\mathfrak{T}_i$ may be readily identified and alternate bases easily determined.

We illustrate Theorem 1 with the decomposition of a metacyclic group ring. We choose to use metacyclic group rings examples throughout this paper. Metacyclic groups, related closely to cyclic groups, are perhaps the least "non-abelian" of all non-abelian groups. Define metacyclic group G as

$$G = G(m,n,r) = \{\ x^i y^j : x^m = y^n = 1,\ yx = x^r y\ \} \qquad [4].$$

Absolutely irreducible representations of such groups are well-understood [3, pp. 333-340]; those representations that correspond to two-sided ideals in LG which decompose to left codes are induced from representations of a related cyclic subgroup. Thus, they provide an opportunity to apply Theorem 1 in a straightforward way.

Remark: In the examples below and throughout the remainder of this paper we use the notation <x> to indicate the ideal generated by idempotent x. If x is not in the center of FG, <x> will indicate the left code generated by x.

Example: Let $G = G(7,3,2)$; $L = GF(2^6)$. G has 5 absolutely irreducible representations: $\mathfrak{T}_1$, $\mathfrak{T}_2$, $\mathfrak{T}_3$ of degree one, and $\mathfrak{T}_4$, $\mathfrak{T}_5$ each of degree 3 and induced from representations of $\mathbb{Z}_7$. Each $\mathfrak{T}_i$ corresponds to a minimal two-sided ideal V_i and $LG = \bigoplus_{i=1}^{5} V_i$ where V_i is generated by an idempotent in the center: (ω and δ are primitive third and seventh roots of 1, respectively)

$$V_1 = \langle 1111111\ 1111111\ 1111111 \rangle$$
$$V_2 = \langle 1111111\ \omega^2\omega^2\omega^2\omega^2\omega^2\omega^2\omega^2\ \omega\omega\omega\omega\omega\omega\omega \rangle$$
$$V_3 = \langle 1111111\ \omega\omega\omega\omega\omega\omega\omega\ \omega^2\omega^2\omega^2\omega^2\omega^2\omega^2\omega^2 \rangle\ ;\ V_i \text{ is } (21,1),\ \ 1 \leq i \leq 3.$$

$V_4 = \langle 1110100\ 0000000\ 0000000 \rangle$
$V_5 = \langle 1001011\ 0000000\ 0000000 \rangle$ $\qquad V_i$ is (21,9), $4 \le i \le 5$.

If $\mathcal{T}_4(x) = \begin{bmatrix} \zeta & 0 & 0 \\ 0 & \zeta^2 & 0 \\ 0 & 0 & \zeta^4 \end{bmatrix}$, $\mathcal{T}_4(y) = \begin{bmatrix} 0 & 0 & 1 \\ 1 & 0 & 0 \\ 0 & 1 & 0 \end{bmatrix}$, Theorem 1 yields

$V_4 = \bigoplus_{i=1}^{3} W_i$ with idempotent generators: $1\zeta^6\zeta^5\zeta^4\zeta^3\zeta^2\zeta^1\ 0000000\ 0000000$, $1\zeta^5\zeta^3\zeta^1\zeta^6\zeta^4\zeta^2\ 0000000\ 0000000$, $1\zeta^3\zeta^6\zeta^2\zeta^5\zeta^1\zeta^4\ 0000000\ 0000000$.
$\mathcal{T}_5$ is the contragradient of $\mathcal{T}_4$ and has a similar decomposition.

3.2 Minimal Codes in FG

If F is not a splitting field for G, the representations of G irreducible over F are not, in general, absolutely irreducible. Let K be the "smallest" field with $F \subseteq K$ so that the representations of G irreducible over K are absolutely irreducible, i.e., any such representation is irreducible for any extension of K. Clearly, there exists L, a splitting field for G, such that $K \subseteq L$. A representation irreducible over F is the direct sum of representations irreducible over such a K [7]. The group characters afforded by the absolutely irreducible representations can assist in the identification of the components of the direct sums since such a direct sum group character $\chi : G \to F$.

The representations of G irreducible over F yield the decomposition of FG to two-sided ideals by use of the group characters. As was the case for absolutely irreducible representations,

$$e_i = |G|^{-1} \sum_{g \in G} \chi_i(g^{-1})\, g$$

where χ_i is a character afforded by a representation irreducible over F, is the idempotent generator of a two-sided code in FG [5]. The ideal generated by e_i may decompose to minimal left (or right) ideals. We wish to determine the generators of these codes, but find that Theorem 1 may only be applied if the representation is absolutely irreducible.

Assume that representation $\mathcal{U}$, irreducible over F, is the direct sum of p absolutely irreducible representations over K (as defined above), each of degree n: $\mathcal{U} = \bigoplus_{i=1}^{p} \mathcal{T}_i$, $1 \le i \le p$. (We use the symbol $\oplus$ to indicate the direct sum of group representations as in [7].) Each $\mathcal{T}_i$ corresponds to a minimal two-sided ideal $V_i \subseteq KG$ of dimension n^2, and representation $\mathcal{U}$ corresponds to minimal two-sided ideal $\mathcal{V} \subseteq FG$ of dimension pn^2. The representation space for $\mathcal{T}_i$ is isomorphic to each minimal left ideal $W_{ij} \subseteq V_i$, for $1 \le j \le n$, and, by Theorem 1, the choice of basis for the representation space determines the idempotent generator e_{ij} of each W_{ij}. We claim that it is possible to choose the bases for the p representation spaces so that $e_j = \sum_{i=1}^{p} e_{ij}$, $1 \le j \le n$, generate disjoint minimal left codes, $\mathcal{W}_j \subseteq \mathcal{V}$ where $\mathcal{V} = \bigoplus_{j=1}^{n} \mathcal{W}_j$. Bases should be chosen so that the sums of corresponding basis vectors have coefficients in F.

Theorem 2: Let irreducible representations $\mathcal{U}$, $\mathcal{T}_i$ and ideals $\mathcal{V}$, V_i be as described above with $\mathcal{U} = \bigoplus_{i=1}^{p} \mathcal{T}_i$. Let $\mathcal{T}_i$ have representation space B_i with basis $\{ b_{ij} : 1 \le j \le n \} \subset KG$ with resulting decomposition $V_i = \bigoplus_{j=1}^{n} W_{ij}$ where W_{ij} generated by idempotent e_{ij} as identified by Theorem 1. If $\sum_{i=1}^{p} b_{ij} \in FG$ for all $1 \le j \le n$, then $\mathcal{V} = \bigoplus_{j=1}^{n} \mathcal{W}_j$ where $\mathcal{W}_j \subset FG$ has idempotent generator $e_j = \sum_{i=1}^{n} e_{ij}$ for all $1 \le j \le n$.

Proof: By the structure of the p_i operators of Theorem 1,

$$\mathcal{T}_i(e_{ij}): \begin{cases} b_{ij} \mapsto b_{ij} \\ b_{ik} \mapsto 0 \ \forall\, k \ne j \end{cases} \qquad [10].$$

For every j, $1 \le j \le n$,

$$\mathcal{U}(e_j) = \bigoplus_{i=1}^{p} \mathcal{T}_i(e_{ij}) : \begin{cases} b_{ij} \mapsto b_{ij} \\ b_{ik} \mapsto 0 \ \forall\, k \ne j \end{cases}.$$

Therefore, $\mathcal{U}(e_j) : \sum_{i=1}^{p} b_{ij} \mapsto \sum_{i=1}^{p} b_{ij}$. But $\sum_{i=1}^{p} b_{ij} \in FG$. Since $\mathcal{U}$ is irreducible over F, $e_j \in FG$. Clearly, e_j generates $\mathcal{W}_j$ a subfield subcode of $\bigoplus_{i=1}^{p} W_{ij}$. By the relationship among codes in LG and FG, $\mathcal{V} = \bigoplus_{j=1}^{n} \mathcal{W}_j$ and e_j is idempotent for all j. □

Figure 1 illustrates the decomposition.

Choosing the bases for the constituent absolutely irreducible representations is of great importance. In the examples below, the structure of the metacyclic group and its absolutely irreducible representations simplify the task.

Example 1: As in the Example of 3.1, let G = G(7,3,2), L = GF(2^6). Let $F = \mathbb{F}_2$. $\mathcal{T}_4$ is induced from irreducible representations (characters) of $\mathbb{Z}_7$. As usually derived [3], the representation space for $\mathcal{T}_4$ may be considered to have the basis in $L\mathbb{Z}_7$ (where ζ is a primitive 7-th root of one):

$$b_1 = \zeta^6 + \zeta^5 x + \zeta^4 x^2 + \zeta^3 x^3 + \zeta^2 x^4 + \zeta^1 x^5 + x^6.$$
$$b_2 = \zeta^5 + \zeta^3 x + \zeta^1 x^2 + \zeta^6 x^3 + \zeta^4 x^4 + \zeta^2 x^5 + x^6.$$
$$b_3 = \zeta^3 + \zeta^6 x + \zeta^2 x^2 + \zeta^5 x^3 + \zeta^1 x^4 + \zeta^4 x^5 + x^6.$$

(The representation space is the dual of a familiar binary Hamming (7,4) code.)

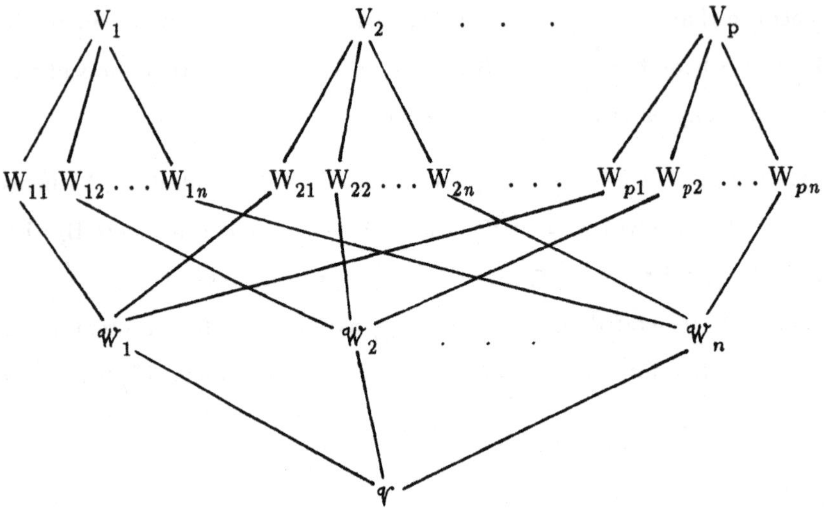

Fig. 1: Decomposition to Minimal Left Ideals in LG and FG

Let $K = \mathbb{F}_2$. Representations $\mathfrak{T}_4$ and $\mathfrak{T}_5$ are equivalent to representations absolutely irreducible over $\mathbb{F}_2$. ($\mathfrak{T}_4$ and $\mathfrak{T}_5$ are realizable in K.) The transformation is accomplished by choosing a new basis $\{ b'_i \}$ in $\mathbb{F}_2\mathbb{Z}_7$:

$$b'_1 = 1 + x + x^2 + x^4 = \zeta^1 b_1 + \zeta^2 b_2 + \zeta^4 b_3$$
$$b'_2 = x + x^2 + x^3 + x^5 = \zeta^2 b_1 + \zeta^4 b_2 + \zeta^1 b_3$$
$$b'_3 = x^2 + x^3 + x^4 + x^6 = \zeta^3 b_1 + \zeta^6 b_2 + \zeta^5 b_3.$$

Then $P = \begin{bmatrix} \zeta^1 & \zeta^2 & \zeta^3 \\ \zeta^2 & \zeta^4 & \zeta^6 \\ \zeta^4 & \zeta^1 & \zeta^5 \end{bmatrix}$ and $P^{-1} = \begin{bmatrix} \zeta^6 & \zeta^5 & \zeta^3 \\ \zeta^1 & \zeta^2 & \zeta^4 \\ 1 & 1 & 1 \end{bmatrix}$.

$$\mathfrak{T}'_4(x) = P^{-1} \cdot \mathfrak{T}_4(x) \cdot P = \begin{bmatrix} 0 & 0 & 1 \\ 1 & 0 & 1 \\ 0 & 1 & 0 \end{bmatrix} \quad \text{and}$$

$$\mathfrak{T}'_4(y) = P^{-1} \cdot \mathfrak{T}'_4(y) \cdot P = \begin{bmatrix} 0 & 1 & 1 \\ 1 & 1 & 1 \\ 0 & 0 & 1 \end{bmatrix}.$$

Applying Theorem 1, $V_3 = \bigoplus_{i=1}^{3} W'_i$, with idempotent generators 1110100 1101001 0100111, 1001110 0100111 1101001, 1001110 1001110 1001110. Each is a (21,3,12) code.

Example 2: Let $G = G(11,5,3)$. Let $L = GF(2^{10})$ and $F = \mathbb{F}_2$. Let $K = GF(2^2)$ and let $\mathcal{V}$ correspond to representation $\mathcal{U} = \mathcal{T}_6 \oplus \mathcal{T}_7$ where (as usually derived)

$$\mathcal{T}_6(x) = \begin{bmatrix} \zeta & 0 & 0 & 0 & 0 \\ 0 & \zeta^3 & 0 & 0 & 0 \\ 0 & 0 & \zeta^9 & 0 & 0 \\ 0 & 0 & 0 & \zeta^5 & 0 \\ 0 & 0 & 0 & 0 & \zeta^4 \end{bmatrix} \qquad \mathcal{T}_6(y) = \begin{bmatrix} 0 & 1 & 0 & 0 & 0 \\ 0 & 0 & 1 & 0 & 0 \\ 0 & 0 & 0 & 1 & 0 \\ 0 & 0 & 0 & 0 & 1 \\ 1 & 0 & 0 & 0 & 0 \end{bmatrix}$$

$$\mathcal{T}_7(x) = \begin{bmatrix} \zeta^{10} & 0 & 0 & 0 & 0 \\ 0 & \zeta^8 & 0 & 0 & 0 \\ 0 & 0 & \zeta^2 & 0 & 0 \\ 0 & 0 & 0 & \zeta^6 & 0 \\ 0 & 0 & 0 & 0 & \zeta^7 \end{bmatrix} \qquad \mathcal{T}_7(y) = \begin{bmatrix} 0 & 1 & 0 & 0 & 0 \\ 0 & 0 & 1 & 0 & 0 \\ 0 & 0 & 0 & 1 & 0 \\ 0 & 0 & 0 & 0 & 1 \\ 1 & 0 & 0 & 0 & 0 \end{bmatrix}$$

Alter $\mathcal{T}_6$ by using $\omega\omega^2 110010011$, $1\omega\omega^2 11001001$, $11\omega\omega^2 1100100$, $011\omega\omega^2 110010$, and $0011\omega\omega^2 11001$ as basis for the representation space. (ω is a primitive third root of 1.) $\mathcal{T}'_6$, irreducible over K, results where

$$\mathcal{T}'_6(x) = \begin{bmatrix} 0 & 0 & 0 & 0 & 1 \\ 1 & 0 & 0 & 0 & \omega \\ 0 & 1 & 0 & 0 & 1 \\ 0 & 0 & 1 & 0 & 1 \\ 0 & 0 & 0 & 1 & \omega^2 \end{bmatrix} \qquad \mathcal{T}'_6(y) = \begin{bmatrix} 1 & 1 & 0 & 1 & 0 \\ 1 & \omega & 1 & \omega & 1 \\ 1 & 1 & 0 & 0 & 1 \\ 0 & 0 & \omega^2 & \omega^2 & 1 \\ 0 & \omega & 1 & \omega^2 & 0 \end{bmatrix}$$

Alter $\mathcal{T}_7$ by using $\omega^2\omega 110010011$, $1\omega^2\omega 11001001$, $11\omega^2\omega 1100100$, $011\omega^2\omega 110010$, and $0011\omega^2\omega 11001$ as bases for the representation space. $\mathcal{T}'_7$ results where

$$\mathcal{T}'_7(x) = \begin{bmatrix} 0 & 0 & 0 & 0 & 1 \\ 1 & 0 & 0 & 0 & \omega^2 \\ 0 & 1 & 0 & 0 & 1 \\ 0 & 0 & 1 & 0 & 1 \\ 0 & 0 & 0 & 1 & \omega \end{bmatrix} \qquad \mathcal{T}'_7(y) = \begin{bmatrix} 1 & 1 & 0 & 1 & 0 \\ 1 & \omega^2 & 1 & \omega^2 & 1 \\ 1 & 1 & 0 & 0 & 1 \\ 0 & 0 & \omega & \omega & 1 \\ 0 & \omega^2 & 1 & \omega & 0 \end{bmatrix}$$

Sums of idempotent generators of minimal left codes in $GF(2^2)G(11,5,3)$ identified by Theorem 1 using $\mathcal{T}'_6$ and $\mathcal{T}'_7$, yield (Theorem 2) idempotent generators of minimal left codes in $\mathcal{V}$: $\mathcal{V} = \bigoplus_{i=1}^{5} \mathcal{W}_i$ where $\mathcal{W}_i$ has idempotent generator e_i and

e_1 = 01101100000 01111000000 00011000011 00000110011 00001010000
e_2 = 00011101000 00110111100 00111100110 01001100001 10000011010
e_3 = 00000110000 11000000000 00000001100 00000011000 00011000000
e_4 = 00001011100 00001111011 11100011001 01010000110 11000001010
e_5 = 00000011011 10000000111 11000110000 00011001100 01010000000.

Each $\mathcal{W}_i$ is of dimension 10, however, they are not equivalent. Minimum distances are 10, 20, 10, 20, and 10 respectively.

Remark: Choosing different bases for $\mathcal{T}_6$ and $\mathcal{T}_7$ could produce different codes. For example, altering $\mathcal{T}_6$ by using $b_{61} = \omega^2 1\omega\omega^2\omega\omega\omega\omega^2\omega^2\omega^2\omega$ and the 4 vectors resulting from shifting b_{11} 1 to 4 places to the right (in a circular fashion) as bases for the representation space. Analogously, use $b_{71} = \omega\omega^2\omega\omega^2\omega^2\omega^2\omega\omega\omega\omega^2$ and its circular shifts as basis for $\mathcal{T}_7$. Then, $\mathcal{V} = \bigoplus_{i=1}^{5} \mathcal{W}'_i$; $\mathcal{W}'_i$ is of dimension 10 for all i; minimum distances are 20, 20, 10, 20, and 20 respectively. Four of these codes are not equivalent to any of the $\mathcal{W}_i$ derived above.

Example 3: G(25,5,6). L = GF(2^{20}). F = $\mathbb{F}_2$. Over L, G has 4 multi-dimensional absolutely irreducible representations, $\mathcal{T}_i$, $1 \le i \le 4$, each of degree 5. (Additionally G has 25 absolutely irreducible representations over L of degree 1.) Each multi-dimensional $\mathcal{T}_i$ is realizable in K = GF(2^4). A direct sum of these 4 representations is irreducible over $\mathbb{F}_2$. Bases may be selected by letting $G' = \mathbb{Z}_{25}$, choosing 5 linearly independent vectors from the cyclic code $\mathcal{A} = <0020410 20_8> \subset \mathbb{F}_2\mathbb{Z}_{25}$ that has non-zero ν^1 (ν is a primitive 25-th root of one). Then, considering the code $\mathcal{A}' \subset L\mathbb{Z}_{25}$ of which $\mathcal{A}$ is a subfield subcode, project each vector onto the cyclic code $C_i \subset GF(2^4)\mathbb{Z}_{25}$, $1 \le i \le 4$, where the C_i have non-zeros ν, ν^4, ν^3, and ν^2 respectively. Use the resulting vectors as the bases $\{ b_{ij} : 1 \le j \le 5 \}$.

4 A Case Study: Metacyclic Codes

The method described above indicates that there are numerous possible decompositions of some minimal two-sided codes in FG where G is non-abelian. Different decompositions need not produce equivalent codes. No metrics have been devised to determine the conditions under which different bases produce decompositions with equivalent codes. The search for minimal codes (and all codes) in LG, even though facilitated by knowledge of the structure of the irreducible representations of G, is a tedious one, involving choosing various bases for the representation spaces involved.

In practice, the process can be avoided if additional information about the structure of LG is available. In the case of metacyclic codes, it has been determined that for many G, all minimal left codes in FG must conform to a particular template [8]. The irreducible representations of G over F are used to derived the minimal two-sided codes. Then application of the template produces all non-equivalent minimal left subcodes. The decomposition of the two-sided code is not determined, however all subcodes are derived and it can then be determined whether any good or "best" codes can be found in FG [9]. We apply this method to Example 3 of 3.2.

Example: G(25,5,6). $L = GF(2^{20})$. $F = \mathbb{F}_2$. Only 1 two-sided ideal $V \subset FG$ contains minimal left codes in FG. (This code corresponds to the sum of the 4 absolutely irreducible representations of G, each of degree 5.) V is (125,100) and $V = \bigoplus_{i=1}^{5} W_i$ where W_i is (125,20). As indicated above, many decompositions of V are possible. Using other information about G, however, assures us all any W_i so derived will be equivalent, i.e., have the same weight distribution, to one of 162 distinct codes. These codes, each minimal in FG, range in minimum distance from 10 to 44. Below are presented the weight polynomials (in z) for a sample of 5 of these codes. (The weight polynomial for code of length n is displayed as $\sum_{i=0}^{n} c_i z^i$ where c_i codewords have weight i.)

Code	Weight Polynomial
W_1	$1 + 50z^{10} + 1025z^{20} + 11000z^{30} + 65250z^{40} + 207500z^{50} + 326250z^{60} + 275000z^{70} + 128125z^{80} + 31250z^{90} + 3125z^{100}$
W_2	$1 + 125z^{36} + 125z^{40} + 2000z^{44} + 10500z^{48} + 47000z^{52} + 151500z^{56} + 277550z^{60} + 290125z^{64} + 180875z^{68} + 68375z^{72} + 17250z^{76} + 3150z^{80}$
W_3	$1 + 375z^{40} + 1000z^{44} + 9875z^{48} + 48625z^{52} + 159125z^{56} + 269675z^{60} + 284500z^{64} + 182625z^{68} + 76000z^{72} + 14625z^{76} + 1900z^{80} + 250z^{84}$
W_4	$1 + 1375z^{44} + 10500z^{48} + 51250z^{52} + 150750z^{56} + 272800z^{60} + 290625z^{64} + 180750z^{68} + 70250z^{72} + 18625z^{76} + 1650z^{80}$
W_5	$1 + 1250z^{44} + 11375z^{48} + 48250z^{52} + 155000z^{56} + 273300z^{60} + 283625z^{64} + 185500z^{68} + 72000z^{72} + 16250z^{76} + 1775z^{80} + 250z^{84}$

5 Conclusion

The method of decomposition described in this paper has yielded new codes that are not equivalent to any abelian code and that equal in minimum distance that of the best linear codes of the same length and dimension [9]. Unfortunately, until those conditions that produce equivalent decompositions are articulated, the process is a tedious one. But the many possible decompositions of non-abelian codes of non-trivial length admit the possibility of discovering new "best" codes that are desirable not only for their high minimum distances but also for their well-defined algebraic structure.

References

1. Camion, P., "Abelian Codes," MRC Tech. Sum. Rep. #1059, University of Wisconsin, Madison, Wis (1971).
2. Cheng, Ying, N. J. A. Sloane, "Codes from Symmetry Groups, and a [32,17,8] Code," *SIAM J. Disc. Math.*, 2 (1989), pp. 28-37.
3. Curtis, Charles W., Irving Reiner, *Representation Theory of Finite Groups and Associative Algebras*, Interscience Publishers, New York (1962).
4. Hall, Marshall, Jr., *The Theory of Groups*, The Macmillan Co., New York (1959).
5. Keown, R., *An Introduction to Group Representation Theory*, Academic Press, New York (1975).
6. MacWilliams, F. J., N. J. A. Sloane, *The Theory of Error-Correcting Codes, North-Holland*, New York (1977).
7. Nagao, Hirosi, Yukio Tsushima, *Representations of Finite Groups*, Academic Press, New York (1987).
8. Sabin, Roberta Evans, "Metacyclic Error-Correcting Codes," PhD dissertation, University of Maryland, Baltimore (1990).
9. Sabin, Roberta Evans, "On Row-Cyclic Codes with Algebraic Structure" (in preparation).
10. Serre, J.-P., *Linear Representations of Finite Groups*, Graduate Text in Math., 42, Springer-Verlag, New York (1977).

On Hyperbolic Cascaded Reed–Solomon codes

Keith Saints
Center for Applied Mathematics
Cornell University, Ithaca NY 14853
email: keith@macomb.tn.cornell.edu

Chris Heegard
School of Electrical Engineering
Cornell University, Ithaca NY 14853
email: heegard@ee.cornell.edu

Abstract

This paper describes a class of two-dimensional codes called cascaded Reed-Solomon (CRS) codes and an algorithm for decoding these codes up to their minimum distance. CRS codes are cascade (or generalized concatenated) codes in which Reed-Solomon codes are used for both the inner and outer codes. We introduce hyperbolic cascaded Reed-Solomon (HCRS) codes, which have maximal rate among CRS codes of a given minimum distance. Our algorithm decodes any CRS code to its minimum distance by calculating a Gröbner basis for an ideal which identifies the error locations. This error location algorithm is based on Sakata's algorithm, but with two significant modifications. First of all, the iterations and terms of polynomials are ordered according to the lexicographic ordering. Secondly, unknown syndromes are calculated as needed, by a simple threshold rule. Once the error locations are known, the error values can be calculated by solving an analog of the key equation for Reed–Solomon codes.

Section 1. Introduction

Reed–Solomon codes are of considerable importance in the theory and practice of error-correcting codes because their structure is well-understood, their minimum distance is optimal, and efficient decoding algorithms are known. However, for a given size q of the alphabet $\mathbf{F}_q$, a (doubly-extended) Reed–Solomon code has length at most $q+1$. Various constructions of 2-D cyclic codes, such as product codes and cascade codes, produce codes of longer blocklength for the same complexity of arithmetic. (Although BCH codes over $\mathbf{F}_q$ can have longer blocklengths, we do not consider them to have the same complexity of arithmetic, since they must be decoded in an extension field of $\mathbf{F}_q$.) This is our motivation for studying the algebraic structure of a class of 2-D cyclic codes we call *hyperbolic cascaded Reed–Solomon* (HCRS) codes, and developing an algebraic method of decoding them.

Section 2. Algebraic Framework

The following notation is used throughout this paper. We assume that q is a prime power, and $\mathbf{F}_q$ denotes the finite field with q elements: the alphabet used to compose codewords and for decoding. We set $n = q-1$, and denote by α a primitive n^{th} root of unity, that is, a primitive generator of $\mathbf{F}_q^* = \mathbf{F}_q \setminus \{0\}$.

This work was supported in part by the U.S. Army Research Office through the Mathematical Sciences Institute of Cornell University, and in part by NSF grants NCR-8903931 and NCR-9207331.

We relate our codes and algorithms to ideals in the ring $\mathbf{F}_q[x, y]$ of bivariate polynomials. For this purpose, it is desirable to introduce the concepts of monomial orders and Gröbner bases. We give but a brief exposition of these ideas. For more details, see [4, 5].

2.1 *Term Orderings*

We denote by $\mathbf{Z}_+^2$ the set of all ordered pairs of non-negative integers. We define a monomial to be a product of powers of the variables, $x^{p_1}y^{p_2}$, for $\mathbf{p} = (p_1, p_2) \in \mathbf{Z}_+^2$. We abbreviate this with the notation $x^{\mathbf{p}} = x^{p_1}y^{p_2}$.

We write $\mathbf{p} \leq \mathbf{q}$ if $p_1 \leq q_1$ and $p_2 \leq q_2$. This indicates that $x^{\mathbf{p}}$ divides $x^{\mathbf{q}}$, and so we refer to $\leq$ as the *divisibility order*. The divisibility order is only a partial order on $\mathbf{Z}_+^2$: for example, $(0, 1)$ and $(1, 0)$ are not comparable under this order.

We define a *monomial order* on $\mathbf{Z}_+^2$ to be a *total* order $\leq_T$ with the property that $\mathbf{p} \leq_T \mathbf{q}$ whenever $\mathbf{p} \leq \mathbf{q}$. We consider monomials to be ordered according to their exponent vectors: thus we say $x^{\mathbf{p}} \leq x^{\mathbf{q}}$ (or $x^{\mathbf{p}} \leq_T x^{\mathbf{q}}$) if and only if $\mathbf{p} \leq \mathbf{q}$ (or $\mathbf{p} \leq_T \mathbf{q}$). The *leading term* of a polynomial $f(x, y) \in \mathbf{F}_q[x, y]$ (with respect to $\leq_T$) is the monomial with nonzero coefficient that is greatest according to $\leq_T$. For a polynomial with leading term $x^{\mathbf{s}}$, we will write either $\text{lead}(f) = x^{\mathbf{s}}$, or $\text{lead}(f) = \mathbf{s}$ depending on the context.

Although many of the results in this paper apply for any monomial order, we wish to emphasize the special properties of the *lexicographic order*. The lexicographic order $\leq_L$ is defined by $\mathbf{p} \leq_L \mathbf{q}$ if either $p_1 < q_1$ or $p_1 = q_1$ and $p_2 \leq q_2$. Thus

$$1 <_L y <_L y^2 <_L \cdots <_L x <_L xy <_L xy^2 <_L \cdots <_L x^2 <_L x^2y <_L x^2y^2 <_L \cdots$$

2.2 *Gröbner bases*

Let I be an ideal in the ring $\mathbf{F}_q[x, y]$. Any monomial which occurs as the leading term of a polynomial in I is called *non-standard*. Any monomial which does *not* occur as the leading term of any polynomial in I is thus *standard*, and we define $\Delta = \{\mathbf{p} : x^{\mathbf{p}} \text{ is standard}\}$. Clearly, the exterior of Δ is closed under $\geq$ and the interior of Δ is closed under $\leq$. In general we will refer to sets with this property as *delta sets*. An *exterior corner* $\mathbf{p}$ of Δ is a point outside Δ which is minimal under $\leq$. An *interior corner* $\mathbf{p}$ of Δ is a point inside Δ which is maximal under $\leq$.

We say that a subset $\mathcal{G}$ of I is a *Gröbner basis* for I (with respect to the chosen monomial order $\leq_T$) if the set $\{\text{lead}(f) : f \in \mathcal{G}\}$ contains all exterior corners of Δ. It follows that the ideal generated by $\{\text{lead}(f) : f \in \mathcal{G}\}$ is the same as the ideal generated by $\{\text{lead}(f) : f \in I\}$, (the usual definition of a Gröbner basis). A Gröbner basis $\mathcal{G}$ for I generates I as an ideal.

For an ideal I, the zero set of I, denoted $Z(I)$, is the set of all zeros common to all of the polynomials in I: $Z(I) = \{(\beta,\gamma) : f(\beta,\gamma) = 0 \text{ for each } f \in I\}$. If $\mathcal{G}$ is any set (in particular a Gröbner basis) which generates I as an ideal, then $Z(I)$ consists of the zeros common to each of the basis polynomials.

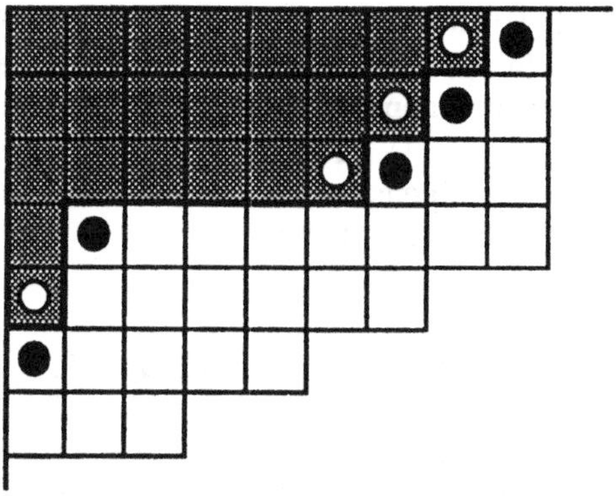

Figure 1. Delta Set showing Interior (white) and Exterior (black) Corners

Proposition. *The following quantities are equal:*

(1) The dimension of $\mathbf{F}_q[x, y]/I$ as a vector space over $\mathbf{F}_q$.

(2) The cardinality of Δ.

(3) The number of zeros of I (with coordinates in the algebraic closure of $\mathbf{F}_q$).

This relationship between the size of Δ and the number of zeros of the ideal will be an important one for the decoding algorithm.

2.3 The Discrete Fourier Transform

We denote by $\mathbf{F}_q^{n\times n}[x, y]$ the set of polynomials in the variables x and y with coefficients in $\mathbf{F}_q$ and with degree strictly less than $n = q - 1$ in both variables. This consists of all polynomials of the form:

$$c(x, y) = \sum_{i=0}^{n-1}\sum_{j=0}^{n-1} c_{ij}x^iy^j.$$

We identify a polynomial $c(x, y) \in \mathbf{F}_q^{n\times n}[x, y]$ with the $n \times n$ array $c = (c_{ij}) \in \mathbf{F}_q^{n\times n}$ of its coefficients. The (two-dimensional) Discrete Fourier transform is defined by the pair:

$$C(X, Y) = \sum_{i=0}^{n-1}\sum_{j=0}^{n-1} c(\alpha^i, \alpha^j)X^iY^j, \qquad c(x, y) = \frac{1}{n^2}\sum_{i=0}^{n-1}\sum_{j=0}^{n-1} C(\alpha^{-i}, \alpha^{-j})x^iy^j.$$

Section 3. Definitions of Codes

All of the codes considered in this paper are sets of polynomials with specified zeros. In the language of the Fourier transform, we consider the set of arrays whose transform has nulls at specified frequencies. Let $\mathbf{Z}_n =$

$\{0, \cdots, n-1\}$, so that the entries of our arrays and their transforms are indexed by $\mathbf{Z}_n^2$. Suppose P is a set of *parity frequencies* $P \subset \mathbf{Z}_n^2$. Then the collection $C(P)$ of polynomials $c(x, y) \in \mathbf{F}_q^{n \times n}[x, y]$ which vanish at each of the points in $Z = (\alpha^i, \alpha^j)$ for each $(i, j) \in P$ forms a *two-dimensional cyclic* code. This code has dimension $n^2 - |P|$, as each equation $c(\alpha^i, \alpha^j) = 0$ is an independent parity check equation.

3.1 CRS codes

For the purposes of this paper, a *cascaded Reed–Solomon* (CRS) code is a 2-D cyclic code $C(P)$ where the parity check region P is a delta set. We refer to [8-11] for a complete description of the construction of $C(P)$ as a cascade code using Reed–Solomon codes as inner and outer codes.

The cascade-code bound for the minimum distance bounds the minimum distance by the products of the distances of the component codes (Reed–Solomon codes, in our case).

$$d[C(P)] \geq \min_{1 \leq k \leq n} d\left[C_k^{\text{in}}\right] d\left[C_k^{\text{out}}\right].$$

Let $\mathrm{p} = (p_1, p_2)$ be an element not in P. This point p lies in a row of P with r consecutive zeros, where $r \leq p_2$. These zeros represent the parity checks for one of the outer Reed–Solomon codes, whose minimum distance is $(r+1)$. In the expression for the cascade-code bound, this distance is multiplied by the distance $(p_1 + 1)$ of one of the inner Reed–Solomon codes. This leads to a simple bound:

$$d[C(P)] \geq \min_{\mathrm{p} \notin P} (p_1 + 1)(p_2 + 1).$$

Note that if g_{p_1} and g_{p_2} are generators of Reed–Solomon codes of distance p_1+1 and p_2+1 respectively, then the product $g_{p_1}(x) g_{p_2}(y)$ has weight $(p_1+1)(p_2+1)$ and lies in the code $C(P)$. Thus the minimum distance of the CRS code is exactly:

$$d[C(P)] = \min_{\mathrm{p} \notin P} (p_1 + 1)(p_2 + 1).$$

3.2 HCRS codes

The above bound makes it easy to design a code to have a minimum distance at least d. Define

$$\mathrm{HCRS}_d = C(P),$$

$$P = \{\mathrm{p} \in \mathbf{Z}_+^2 \ : \ (p_1 + 1)(p_2 + 1) < d\}.$$

We call this code a *hyperbolic cascaded Reed–Solomon* (HCRS) code because P is defined by the graph of a hyperbola. This code has minimum distance $d^* = \min_{\mathrm{p} \notin P} (p_1 + 1)(p_2 + 1) \geq d$. Since $\mathrm{HCRS}_{d^*} = \mathrm{HCRS}_d$, we only consider codes HCRS_d for which $d^* = d$, and refer to d as the designed distance of the HCRS code.

Section 4. Decoding CRS codes

We note that there is a general method proposed by Zinoviev and Zyablov [12], which incorporates decoders for the inner and outer codes into a decoder for the cascade code. Our emphasis is on the algebraic aspects of our code, and we present an algebraic decoding algorithm for the CRS codes, less general than the method of Zinoviev and Zyablov, but in our view, more efficient for CRS codes. We also note that our method of decoding is based on an algorithm for decoding proposed by Sakata in [2]. However, our modifications make it possible to correct the same number of errors using considerably less parity checks. We shall have more to say about this later.

We consider the decoding problem for the t-error correcting HCRS code HCRS_{2t+1} or any of its 2-D cyclic subcodes. So we may assume that if $\mathbf{p}$ is *not* a parity-check frequency, then $(p_1+1)(p_2+1) \geq 2t+1$.

A codeword $\mathbf{c} \in \mathbf{F}_q^{n\times n}$ is sent through the channel where it is corrupted by an error $\mathbf{e}$ of Hamming weight at most t. The decoder receives the array $\mathbf{a} = \mathbf{c}+\mathbf{e}$, from which it must deduce the value of $\mathbf{c}$. In the frequency domain, $\mathbf{A} = \mathbf{C} + \mathbf{E}$, and so syndromes $E_\mathbf{p} = A_\mathbf{p}$ are known at each parity frequency $\mathbf{p} \in P$.

4.1 *Finding the Error Locations*

Let the error pattern array $\mathbf{e}$ have weight $\nu \leq t$, and be given by $e(x, y) = \sum_{k=1}^{\nu} e_k x^{\mathbf{m}^{(k)}}$. For each $k = 1, \cdots, \nu$, we define $(\beta_k, \gamma_k) = (\alpha^{m_1^{(k)}}, \alpha^{m_2^{(k)}})$. Informally, we may refer to either $\mathbf{m}^{(k)}$ or (β_k, γ_k) as *error locations*, since either one determines the terms (though not the coefficients) of the error polynomial. We define the *error locator ideal* L to be the set of all polynomials which vanish at all of the error locations (β_k, γ_k), $k = 1, \cdots, \nu$.

The syndrome array has dimension $n \times n$, but it may be extended in a natural way to an infinite array by defining

$$E_{ij} = e(\alpha^i, \alpha^j), \qquad \text{for all } (i,j) \in \mathbf{Z}_+^2.$$

We say that the (infinite) array E satisfies a *two-dimensional linear recursion* (2-D LR) relation with characteristic polynomial $f \in \mathbf{F}_q[x, y]$ if

$$\sum f_\mathbf{m} E_{\mathbf{p}+\mathbf{m}} = 0, \qquad \text{for all } \mathbf{p} \in \mathbf{Z}_+^2.$$

From now on we will identify 2-D LR relations with their characteristic polynomials, and thus "the 2-D LR relation f" refers to the 2-D LR relation with characteristic polynomial f. We are interested in these relations because of the following theorem [2, 15]:

Error Location Theorem. *The set of characteristic polynomials of 2-D LR relations satisfied by the (infinite) syndrome array* $\mathbf{E}$ *is precisely the error locator ideal* L.

4.2 Finding the Error Values

In order to perform decoding, we will have to compute error values as well as error locations. Before proceeding with the details of the error location algorithm, let us finish our discussion of the decoding process.

We shall see that our error location algorithm produces a Gröbner basis for the error locator ideal. We must now solve for the common zeros of our basis polynomials, for these are the error locations (β_k, γ_k). Our choice of the lexicographic order on the monomials makes it possible to accomplish this easily and efficiently. One member of the Gröbner basis must be a polynomial $g(Y)$ univariate in Y. We solve for the roots $\{\gamma_k\}$ of g using a Chien search. Each root γ indicates a particular column which contains errors. To find the locations of errors within this column, we plug in $Y = \gamma$ for each of the other polynomials in the Gröbner basis to obtain a set of polynomials univariate in X. The common roots β_k of these polynomials give zeros of the Gröbner basis of the form (β_k, γ) which in turn identify locations of errors within the column corresponding to γ. Since we have ordered our terms lexicographically, the Gröbner basis actually gives more detailed information about how the errors are distributed among the columns, and this information allows us to perform the above process (of plugging in γ and finding roots β_k) even more efficiently than described above.

Now we assume that we have completed the above process, so that we know all the (β_k, γ_k). We express $E(X, Y)$ as a rational function in order to develop an analog of the key equation. Note

$$E(X, Y) = \sum_{i=0}^{\infty}\sum_{j=0}^{\infty} e(\alpha^i, \alpha^j)X^iY^j = \sum_{k=1}^{\nu} e_k \sum_{i=0}^{\infty}(\beta_k X)^i \sum_{j=0}^{\infty}(\gamma_k Y)^j,$$

and replacing each geometric series with its sum, we obtain

$$E(X, Y) = \sum_{k=1}^{\nu} \frac{e_k}{(1-\beta_k X)(1-\gamma_k Y)} = \frac{\Omega(X, Y)}{\Lambda(X, Y)}.$$

Here, the numerator and common denominator are given by:

$$\Omega(X, Y) = \sum_k e_k \prod_{\substack{\beta\in\{\beta_l\}\\ \beta\neq\beta_k}} (1-\beta X) \prod_{\substack{\gamma\in\{\gamma_l\}\\ \gamma\neq\gamma_k}} (1-\gamma Y),$$

$$\Lambda(X, Y) = \prod_{\beta\in\{\beta_l\}} (1-\beta X) \prod_{\gamma\in\{\gamma_l\}} (1-\gamma Y),$$

where the products are over the sets $\{\beta_l\}$ (resp. $\{\gamma_l\}$) corresponding to all the rows (resp. columns) which contain an error. Note that we can compute Λ based on our knowledge of the error locations, and that each error value e_k

may be obtained from the following equation, which results from evaluating Ω at a point determined by the corresponding error location,

$$\Omega(\beta_k^{-1}, \gamma_k^{-1}) = e_k \cdot \prod_{\substack{\beta \in \{\beta_l\} \\ \beta \neq \beta_k}} (1 - \beta\beta_k^{-1}) \prod_{\substack{\gamma \in \{\gamma_l\} \\ \gamma \neq \gamma_k}} (1 - \gamma\gamma_k^{-1}).$$

Thus we find the error values by first computing Λ, and then Ω as the product $\Lambda \cdot E$. Then evaluation of Ω yields the error values. Examination of the formula for Ω shows that it has terms $X^i Y^j$, for all i less than the number of rows in error, and all j less than the number of columns in error. Thus to calculate Ω by polynomial multiplication, we will need to know E_{ij} on the same set of indices. In many cases, this calculation involves syndromes which were not known at the outset. But we can show that each syndrome that is needed to make this computation must have been generated by the error location algorithm.

Section 5. The Error Location Algorithm

We now focus our attention on finding the error locations. As indicated by the Error Location Theorem, this may be accomplished by finding the 2-D LR relations which are valid on the syndrome array. Sakata has developed an algorithm to find a minimal set of 2-D LR relations valid on a finite array. We describe here some modifications to his algorithm which allow us to conclude that this minimal set is valid on the infinite syndrome array, and therefore is a Gröbner basis for the error locator ideal. Note that in sections 5.1 and 5.2 we state general results valid for any choice $\leq_T$ of monomial order, but afterwards we specialize to the lexicographic order $\leq_L$ to describe our algorithm.

5.1 Validity of 2-D LR Relations

Although we are ultimately interested in 2-D LR relations valid on the entire array, we build up to this by finding 2-D LR relations which are valid on a portion of the array. To this end, we make the following definitions. A 2-D LR relation f with leading term $x^{\mathbf{s}}$ is *valid* for the entire array $E = (E_{\mathbf{p}})$ if

$$\sum_{\mathbf{m} \leq_T \mathbf{s}} f_{\mathbf{m}} E_{\mathbf{p}+\mathbf{m}} = 0, \qquad \text{for all } \mathbf{p} \geq 0.$$

Solve this equation for its leading term, and adjust subscripts to obtain:

$$E_{\mathbf{p}} = -f_{\mathbf{s}}^{-1} \sum_{\mathbf{m} <_T \mathbf{s}} f_{\mathbf{m}} E_{\mathbf{p}+\mathbf{m}-\mathbf{s}}, \qquad \text{for all } \mathbf{p} \geq \mathbf{s}.$$

We want to interpret this as the equality of the actual value (LHS) with the predicted value (RHS). Accordingly, we define, for $\mathbf{p} \geq \mathbf{s}$,

$$P_{\mathbf{p}}(f) = -f_{\mathbf{s}}^{-1} \sum_{\mathbf{m} <_T \mathbf{s}} f_{\mathbf{m}} E_{\mathbf{p}+\mathbf{m}-\mathbf{s}},$$

to be the value *predicted* by f for E at $\mathbf{p}$ (based on previous elements $E_{\mathbf{q}}$, $\mathbf{q} <_T \mathbf{p}$). We will say that f is *valid* for E at $\mathbf{p}$ if $\mathbf{p} \not\geq \mathbf{s}$ or $E_{\mathbf{p}} = P_{\mathbf{p}}(f)$. Thus f is valid for the entire array E if and only if f is valid for E at each $\mathbf{p} \in \mathbf{Z}_+^2$.

5.2 *Two Fundamental Theorems*

The following two theorems, both due to Sakata [1, 3], are crucial for our method of finding the error locations. The Agreement theorem asserts that under certain conditions, two 2-D LR relations are forced to agree in their predictions. The second theorem, presented here as a simple corollary of the first one, is called the 2-D Massey's theorem, because it is a two-dimensional generalization of a theorem first stated by Massey in [7].

Theorem (Agreement Theorem). *Suppose* $\mathbf{p} \geq \mathrm{lead}(f) + \mathrm{lead}(g)$. *If f and g are valid for E at each* $\mathbf{q} <_T \mathbf{p}$, *then* $P_{\mathbf{p}}(f) = P_{\mathbf{p}}(g)$.

Proof. Let $\mathbf{s} = \mathrm{lead}(f)$, and $\mathbf{t} = \mathrm{lead}(g)$. By hypothesis, $E_{\mathbf{q}} = P_{\mathbf{q}}(g)$, for $\mathbf{t} \leq \mathbf{q} <_T \mathbf{p}$. Note that $\mathbf{p} - \mathbf{s} \geq \mathbf{t}$, and so

$$\mathbf{m} <_T \mathbf{s} \Longrightarrow \mathbf{t} \leq (\mathbf{m} + \mathbf{p} - \mathbf{s}) <_T \mathbf{p} \Longrightarrow E_{\mathbf{m}+\mathbf{p}-\mathbf{s}} = P_{\mathbf{m}+\mathbf{p}-\mathbf{s}}(g),$$

Now,

$$P_{\mathbf{p}}(f) = -f_{\mathbf{s}}^{-1} \sum_{\mathbf{m} <_T \mathbf{s}} f_{\mathbf{m}} E_{\mathbf{m}+\mathbf{p}-\mathbf{s}} = -f_{\mathbf{s}}^{-1} \sum_{\mathbf{m} <_T \mathbf{s}} f_{\mathbf{m}} P_{\mathbf{m}+\mathbf{p}-\mathbf{s}}(g),$$

$$= f_{\mathbf{s}}^{-1} \sum_{\mathbf{m} <_T \mathbf{s}} f_{\mathbf{m}} g_{\mathbf{t}}^{-1} \sum_{\mathbf{r} <_T \mathbf{t}} g_{\mathbf{r}} E_{\mathbf{r}+(\mathbf{m}+\mathbf{p}-\mathbf{s})-\mathbf{t}}.$$

By symmetry,

$$P_{\mathbf{p}}(g) = g_{\mathbf{t}}^{-1} \sum_{\mathbf{r} <_T \mathbf{t}} g_{\mathbf{r}} f_{\mathbf{s}}^{-1} \sum_{\mathbf{m} <_T \mathbf{s}} f_{\mathbf{m}} E_{\mathbf{m}+(\mathbf{r}+\mathbf{p}-\mathbf{t})-\mathbf{s}}.$$

But the two expressions are just rearrangements of the same sum, so $P_{\mathbf{p}}(f) = P_{\mathbf{p}}(g)$. ■

Theorem (Massey's Theorem: 2-D version). *Suppose* $\mathrm{lead}(f) = x^{\mathbf{s}}$, *and suppose f is valid for E at all* $\mathbf{q} <_T \mathbf{p}$, *but f is invalid at* $\mathbf{p}$. *Then* $\mathbf{p} - \mathbf{s} \in \Delta$. *To be more precise, there does not exist g with* $\mathrm{lead}(g) = x^{\mathbf{p}-\mathbf{s}}$ *such that g is valid for E at all* $\mathbf{q} \leq_T \mathbf{p}$.

Proof. Suppose there exists g with $\mathrm{lead}(g) = x^{\mathbf{p}-\mathbf{s}}$ such that g is valid for E at all $\mathbf{q} <_T \mathbf{p}$. Then by the Agreement Theorem, g must agree with f at $\mathbf{p}$, and so

$$P_{\mathbf{p}}(g) = P_{\mathbf{p}}(f) \neq E_{\mathbf{p}},$$

which implies that g is not valid for E at $\mathbf{p}$. ■

This motivates the following definition: Suppose $\mathrm{lead}(f) = x^{\mathbf{s}}$, and let $\mathbf{p}$ be the index which has the property that f is valid for E at all $\mathbf{q} <_T \mathbf{p}$, but f is invalid at $\mathbf{p}$. Then we define the *discrepancy vector* $\mathbf{D}(f) = \mathbf{p} - \mathbf{s}$ and the *discrepancy* $\delta_f = E_{\mathbf{p}} - P_{\mathbf{p}}(f)$. In case f is valid on the infinite array E, we take $\mathbf{D}(f) = (\infty, \infty)$. According to the 2-D Massey's theorem, we can be sure that $\mathbf{q} \in \Delta$ if we have a polynomial g such that $\mathbf{D}(g) = \mathbf{q}$.

5.3 *Overview of the algorithm*

The algorithm maintains four objects $(\Delta_{\mathbf{p}},\ F_{\mathbf{p}},\ G_{\mathbf{p}},\ D_{\mathbf{p}})$, subscripted by the current point $\mathbf{p}$, as $\mathbf{p}$ steps through the syndrome array. We now describe the state of the algorithm before it observes the syndrome $E_{\mathbf{p}}$. The delta set $\Delta_{\mathbf{p}}$ consists of all monomials which do not occur as the leading term of any 2-D LR relation valid for E at all $\mathbf{m} <_L \mathbf{p}$. The set $F_{\mathbf{p}}$ consists of polynomials $f^{(\mathbf{s})}$ with lead term $x^{\mathbf{s}}$ for each exterior corner $\mathbf{s}$ of $\Delta_{\mathbf{p}}$, valid for E at all $\mathbf{m} <_L \mathbf{p}$. The set $G_{\mathbf{p}}$ consists of polynomials $g^{(\mathbf{t})}$, such that $\mathbf{D}(g^{(\mathbf{t})}) = \mathbf{t}$ for each interior corner $\mathbf{t}$ of $\Delta_{\mathbf{p}}$. The values of the discrepancies are stored as the set $D_{\mathbf{p}} = \{\delta_g : g \in G_{\mathbf{p}}\}$.

What follows next depends upon whether the syndrome $E_{\mathbf{p}}$ is known or unknown. These two cases are treated in sections 5.4 and 5.5. In either case, we are able to calculate updated versions of $(\Delta_{\mathbf{p}'},\ F_{\mathbf{p}'},\ G_{\mathbf{p}'},\ D_{\mathbf{p}'})$, where $\mathbf{p}' = (p_1, p_2 + 1)$ is the successor of $\mathbf{p}$ in the lexicographic order, that is, one entry to the right along the same row. However, before proceeding to the next iteration, we check a criterion (described in section 5.6) that would allow us to conclude that each of the $f \in F_{\mathbf{p}'}$ are valid for all subsequent entries in the row. If this criterion holds, we update $\mathbf{p}$ with the value $(p_1+1,\ 0)$, that is, the first entry in the next row. Before beginning a row, we check a criterion (also described in section 5.6) that would allow us to conclude that each $f \in F_{\mathbf{p}}$ is valid for the entire array. When this criterion holds, the algorithm is finished, and at that point we may conclude that F is a Gröbner basis for the ideal L.

5.4 *The Main Step in the Algorithm*

We now describe how the algorithm takes the data $(\Delta_{\mathbf{p}},\ F_{\mathbf{p}},\ G_{\mathbf{p}},\ D_{\mathbf{p}})$, along with the known syndrome $E_{\mathbf{p}}$, and produces the updated version $(\Delta_{\mathbf{p}'},\ F_{\mathbf{p}'},\ G_{\mathbf{p}'},\ D_{\mathbf{p}'})$ of these objects. This step is the same as the main step in Sakata's algorithm [1].

1. The algorithm tests each $f \in F_{\mathbf{p}}$ for validity at $\mathbf{p}$. For each f found to be invalid, $\mathbf{D}(f) = \mathbf{p} - \text{lead}(f)$ is computed, and we append to $\Delta_{\mathbf{p}}$ all $\mathbf{m}$ such that $\mathbf{m} \leq_L \mathbf{D}(f)$, resulting in a possibly larger set $\Delta_{\mathbf{p}'}$. Some of the interior corners of $\Delta_{\mathbf{p}'}$ may be the same as an interior corner of $\Delta_{\mathbf{p}}$. We continue to associate with these corners their respective polynomials from $G_{\mathbf{p}}$. Each newly-created interior corner must be of the form $\mathbf{D}(f)$ for some $f \in F_{\mathbf{p}}$, and so this f is placed in $G_{\mathbf{p}'}$ as the polynomial associated with this corner.

2. Now for each exterior corner $\mathbf{s}$ of $\Delta_{\mathbf{p}'}$ we need to construct a new polynomial with the specified leading term $\mathbf{s}$.

(a) If $\mathbf{s}$ was an exterior corner for $\Delta_{\mathbf{p}}$, and the corresponding polynomial $f^{(\mathbf{s})}$ was valid at $\mathbf{p}$, then f continues to be in $F_{\mathbf{p}'}$.

(b) Otherwise we have the following situation: $\mathbf{s} \geq \mathbf{r}$ for some exterior corner $\mathbf{r}$ for $\Delta_{\mathbf{p}}$, and the polynomial $f^{(\mathbf{r})}$ in $F_{\mathbf{p}}$ with lead term $\mathbf{r}$ is invalid at $\mathbf{p}$, with discrepancy δ_f.

(b1) In the special case that $\mathbf{s} \not\leq \mathbf{p}$, then associate with $\mathbf{s}$ the polynomial $x^{\mathbf{s}-\mathbf{r}} f^{(\mathbf{r})}$.

(b2) It is easy to show that if $\mathbf{s} \leq \mathbf{p}$, then $\mathbf{p}-\mathbf{s} \in \Delta_{\mathbf{p}}$, and hence $\mathbf{p}-\mathbf{s} \leq \mathbf{m}$ for some interior corner $\mathbf{m}$ with associated polynomial $g^{(\mathbf{m})}$. Now we calculate the polynomial

$$h(x,\ y) = x^{\mathbf{s}-\mathbf{r}} f^{(\mathbf{r})}(x,\ y) - \left(\frac{\delta_g}{\delta_f}\right) x^{\mathbf{m}-(\mathbf{p}-\mathbf{s})} g^{(\mathbf{m})}(x,\ y).$$

We can show that h has leading term $x^{\mathbf{s}}$ and is valid for all $\mathbf{q} \leq_L \mathbf{p}$, and so we may place h in $F_{\mathbf{p}'}$ as the polynomial associated with $\mathbf{s}$.

5.5 Generation of Unknown Syndromes

Our method for calculating an unknown syndrome $E_{\mathbf{p}}$ is simple. We take each of the f in $F_{\mathbf{p}}$ in turn, and calculate the value it predicts for $E_{\mathbf{p}}$. This gives a list of candidate values which we try one by one. For each candidate value, we calculate the new delta set $\Delta_{\mathbf{p}'}$ as in the Main Step (part 1). If $|\Delta_{\mathbf{p}'}| \leq t$, then we claim that we have computed the correct value for $E_{\mathbf{p}}$, and thus we finish part 2 of the Main Step, and proceed. But if $|\Delta_{\mathbf{p}'}| \geq t$, then we reject that candidate value, and try the next one instead.

We assert that this procedure always fills in $E_{\mathbf{p}}$ with the correct value. We need to prove two things: first of all, that the actual value always occurs among the predictions of the 2-D LR relations $f \in F$, and secondly, that only one of these candidate values leads to $|\Delta_{\mathbf{p}'}| \leq t$. (Note that the successful candidate may have been predicted by several 2-D LR relations; the point is that only one candidate *value* may pass this test for correctness.)

We introduce the notation $(\mathbf{p}-\Delta) = \{\mathbf{m} \in \mathbf{Z}_+^2 \ :\ \mathbf{m} = \mathbf{p}-\mathbf{q} \text{ for some } \mathbf{q} \in \Delta\}$. Geometrically, the set $\mathbf{p} - \Delta$ is another copy of Δ, originating at $\mathbf{p}$, inverted, and possibly truncated at the coordinate axes to avoid introducing negative coordinates. Thus it is clear that this set has area $|\mathbf{p} - \Delta| \leq |\Delta|$. Let us also introduce the rectangle $R_{\mathbf{p}} = \{\mathbf{m} \in \mathbf{Z}_+^2 \ :\ \mathbf{m} \leq \mathbf{p}\}$, which has area $|R_{\mathbf{p}}| = (p_1+1)\,(p_2+1)$.

Theorem. *Let Δ and Δ^* be the delta sets computed for $\Delta_{\mathbf{p}'}$ according to two different values v and v^* for $E_{\mathbf{p}}$. Then $\Delta \cup (\mathbf{p} - \Delta^*) \supset R_{\mathbf{p}}$.*

Proof. Suppose $\mathbf{s} \leq \mathbf{p}$, and $\mathbf{s} \notin \Delta^*$. Then there exists a polynomial f with $\text{lead}(f) = \mathbf{s}$ such that f is valid for all $\mathbf{q} <_L \mathbf{p}$, and f predicts that $E_{\mathbf{p}} = v$. Under the assumption that v^* is correct, f is invalid at $\mathbf{p}$, and so $\mathbf{D}(f) = \mathbf{p} - \mathbf{s} \in \Delta^*$. In other words, $\mathbf{s} \in (\mathbf{p} - \Delta^*)$. ∎

Corollary. *If t or fewer errors occurred, and $\mathbf{p}$ is outside the parity-check set P for the HCRS code of distance $2t+1$, then the actual value of $E_{\mathbf{p}}$ occurs among the candidate values, and it is the only candidate value which leads to $|\Delta_{\mathbf{p}'}| \leq t$.*

Proof. If the algorithm were given the true value of $E_{\mathbf{p}}$, it would produce a delta set Δ with area at most t. Suppose another value v^* is postulated for $E_{\mathbf{p}}$, causing the algorithm to produce a delta set Δ^*. By the previous theorem, Δ, together with a copy of Δ^*, cover the rectangle $R_{\mathbf{p}}$. But $\mathbf{p}$ is not a parity check, so the rectangle has area $(p_1+1)\,(p_2+1)$ which is at least $2t+1$. This means Δ^* must have area at least $t+1$. Moreover, if a value not on the list of candidates is postulated for $E_{\mathbf{p}}$, then all of the 2-D LR relations in $F_{\mathbf{p}}$ are invalid, and thus the Δ^* produced is as large as possible. Since the true value of $E_{\mathbf{p}}$ leads to a Δ of minimal size, it must occur in the list of candidates. ∎

5.6 Termination Criteria

Note that the 2-D Massey's theorem allows us to determine the consequences of the failure of a 2-D LR relation before we actually test its validity. If $f \in F_{\mathbf{p}}$ is a 2-D LR relation known to be valid so far, we know that if f should fail at p, we would have to append $\mathbf{p}-\mathbf{s}$ to $\Delta_{\mathbf{p}}$ to create $\Delta_{\mathbf{p}'}$. Moreover, other 2-D LR relations from $F_{\mathbf{p}}$ may be forced to agree with f according to the Agreement theorem, and so if f now fails, so do these others, and we will have a whole collection of points to append to $\Delta_{\mathbf{p}}$. It could happen that the number of the points thus appended would cause the size of $\Delta_{\mathbf{p}'}$ to exceed t. In this case, we could be sure that f is valid at p (the decoder is allowed to assume that no more than t errors occurred), *without examining or computing the syndrome* $E_{\mathbf{p}}$. If *all* of the relations in $F_{\mathbf{p}}$ can be certified to be correct in this way, then there is no need to examine or compute the syndrome $E_{\mathbf{p}}$. Moreover, if this condition occurs at p, it will occur at each subsequent $\mathbf{q} \geq \mathbf{p}$, and in particular, all subsequent entries in the same row. This condition is checked before beginning each iteration, and when it holds, we may be sure that the collection $F_{\mathbf{p}}$ of 2-D LR relations correctly generates the infinite sequence of syndromes that make up the current row, and so we skip to the beginning of the next row.

Eventually it happens that each 2-D LR relation is assured to be correct at all subsequent entries in the array to which it could possibly be applied. When this happens, the algorithm terminates and F is a Gröbner basis for the error locator ideal. Note that it can happen that known syndromes are skipped in a given row, or that several rows of known syndromes are ignored.

Section 6. *Conclusions and Further Directions*

We estimate the overall complexity of our decoding algorithm to be $O(t^{5/2} \log t)$ in the worst case, and perhaps $O(t^2 \log^2 t)$ in the average case. These estimates are quite rough, as it is not easy to estimate two quantities: the number of syndromes that need to be generated, and the number of 2-D LR relations in the minimal set.

It is interesting to note that the ability of the decoder to correct up to t errors gives an independent proof that the code has minimum distance at least $2t+1$. Moreover, if we were to design a code around this decoding algorithm, we would arrive at precisely the same set of parity checks, since our last result says that we do not require parity checks p for which $(p_1+1)\,(p_2+1) \geq 2t+1$, as these may be generated by extension. The t-error correcting codes proposed in [2] have a parity check set (called $\Delta(t)$ in Sakata's paper) which contains the parity check set P of our code HCRS_{2t+1}. Thus the HCRS code has higher rate and the same error-correcting capability. The set $\Delta(t)$ has the property that for any $\mathbf{p} \notin \Delta(t)$, the termination rule of our algorithm must be satisfied, and so all 2-D LR relations known at that point are certified to be valid at $\mathbf{p}$. Conversely, for any point $\mathbf{p} \in \Delta(t) \setminus P$, there is at least one error pattern which will cause our algorithm to require the generation of the unknown syndrome $E_{\mathbf{p}}$.

We presented an algorithm for finding the error locations which is a two-dimensional analog of the Berlekamp–Massey algorithm [6, 7] for Reed–Solomon codes. One possible avenue for future investigation would be to find a two-dimensional analog of the Reed–Solomon decoder which uses the Euclidean algorithm [13]. One might expect the calculation of a Gröbner basis to take the place of the Euclidean algorithm in the bivariate case. A first step in this direction was made in [14].

We should point out the definition of CRS code given here is not the most general. For simplicity, we presented cascade codes in which each inner and outer Reed–Solomon code was defined by a spectrum which had nulls in a band of frequencies *beginning at zero*. We could obtain a more general code by forming a cascade code with arbitrary Reed–Solomon codes as outer codes and an arbitrary nested sequence of Reed–Solomon codes as inner codes. We have modified our decoding algorithm to treat these more general codes, but at the expense of some extra computation.

We are also interested in investigating the parameters of subfield-subcodes of HCRS codes. These are codes whose alphabet is a subfield of $\mathbf{F}_q$, but are defined by the same frequency constraints in $\mathbf{F}_q$. Moving to a subfield-subcode will require that the set of zeros of the code will have to satisfy conjugacy conditions, and so there will be some sacrifice in rate, due to "extra" zeros which do not contribute to the cascade-code bound. In this context, it will be worthwhile to consider the more general CRS codes described in the previous paragraph, in order to minimize the number of extra conjugates.

It is interesting to note that the idea of generating unknown syndromes has also proved to be fruitful in decoding BCH codes [16] and algebraic-geometry codes [15, 17]. The common idea in these various contexts is that dependence relations among known syndromes are extrapolated to infer candidate values for unknown syndromes. In the scheme of Feng and Rao, for example, the true value is determined from among the candidate values by a majority rule, whereas in our method the true value is determined by a threshold rule.

References

[1] S. Sakata, "Finding a minimal set of linear recurring relations capable of generating a given finite two-dimensional array," *J. Symbolic Computation*, vol. 5, pp. 321–337, 1988.

[2] S. Sakata, "Decoding binary 2-D cyclic codes by the 2-D Berlekamp–Massey algorithm," *IEEE Trans. Inform. Theory*, vol. 37, pp. 1200–1203, July 1991.

[3] S. Sakata, "A Gröbner basis and a minimal polynomial set of a finite nD array," in *Applied Algebra, Algebraic Algorithms, and Error-correcting Codes: Proceedings of AAECC-8, Tokyo, 1990* (S. Sakata, ed.). Berlin: Springer-Verlag, 1991.

[4] B. Buchberger, "Gröbner bases: An algorithmic method in polynomial ideal theory," in *Multidimensional Systems Theory: Progress, Directions and Open Problems in Multidimensional Systems* (N. K. Bose, ed.). Dordrecht, Holland: D. Reidel, 1985.

[5] D. Cox, J. Little, and D. O'Shea, *Ideals, Varieties, and Algorithms*. New York: Springer-Verlag, 1992.

[6] E. R. Berlekamp, *Algebraic Coding Theory*. New York: McGraw-Hill, 1968.

[7] J. L. Massey, "Shift-register synthesis and BCH decoding," *IEEE Trans. Inform. Theory*, vol. 15, pp. 122–127, Jan. 1969.

[8] E. L. Blokh and V. V. Zyablov, "Coding of generalized cascade codes," *Probl. Info. Trans.*, vol. 10, pp. 45–50, 1974.

[9] J. Wu and D. J. Costello Jr., "New multi-level codes over GF(q)," *IEEE Trans. Inform. Theory*, vol. 38, pp. 933–939, Jan. 1992.

[10] R. Krishnamoorthy and C. Heegard, "Structure and decoding of Reed–Solomon based cascade codes," in *Proc. 25th Ann. Conf. Inform. Sci. Syst.*, pp. 29–33, 1991.

[11] R. Krishnamoorthy, *Algorithms for Capacity Computations and Algebraic Cascade Coding with Applications to Data Storage*. PhD thesis, Cornell University, Aug. 1991.

[12] V. A. Zinoviev and V. V. Zyablov, "Decoding of nonlinear generalized concatenated codes," *Probl. Info. Trans.*, vol. 14, pp. 46–52, 1978.

[13] Y. Sugiyama, M. Kasahara, S. Hirasawa, and T. Namekawa, "A method for solving key equation for decoding Goppa codes," *Inf. and Contr.*, vol. 27, pp. 87–99, 1975.

[14] P. Fitzpatrick and J. Flynn, "A Gröbner basis technique for Padé approximation," *J. Symbolic Computation*, vol. 13, pp. 133–138, 1992.

[15] J. Justesen, K. Larsen, H. Jensen, and T. Høholdt, "Fast decoding of codes from algebraic plane curves," *IEEE Trans. Inform. Theory*, vol. 38, pp. 111–119, Jan. 1992.

[16] G.-L. Feng and K. K. Tzeng, "Decoding cyclic and BCH codes up to actual minimum distance using nonrecurrent syndrome dependence relations," *IEEE Trans. Inform. Theory*, vol. 37, pp. 1716–1723, Nov. 1991.

[17] G.-L. Feng and T.R.N. Rao, "Decoding algebraic-geometric codes up to the designed minimum distance," *IEEE Trans. Inform. Theory*, vol. 39, pp. 37–45, Jan. 1993.

Peak-Shift and Bit Error-Correction with Channel Side Information in Runlength-Limited Sequences*

Yuichi Saitoh[1]**, Ikuyo Ibe[1]***, and Hideki Imai[2]

[1] Division of Electrical and Computer Engineering,
Yokohama National University,
156 Tokiwadai, Hodogaya-ku, Yokohama 240, Japan
[2] Institute of Industrial Science, University of Tokyo,
7-22-1 Roppongi, Minato-ku, Tokyo 106, Japan

Abstract. An error-correcting modulation coding technique is proposed for correcting combinations of multiple peak-shift errors and single bit errors in (d, k)-constrained sequences. The decoding process of the proposed technique uses channel side information in a magnetic recording system. Examples and comparisons are also presented.

1 Introduction

A (d, k) code is well known as a binary *modulation* or *recording code* [1], [2], [3], [4], [7], which is applied to input restricted channels, such as magnetic and optical recording systems. (d, k) means that any two consecutive 1's in the input bit stream are separated by at least d 0's and by at most k 0's. During the NRZI (non-return-to-zero-inverse) precoding step that is usually applied to a (d, k) code, the 1's in the (d, k)-constrained sequences are mapped onto transitions in the corresponding binary channel bit sequences. Then, in the (d, k)-constrained sequences, the minimum separation d and the maximum separation k are imposed in order to reduce intersymbol interference and extract clock control from the received bit stream, respectively. In this paper, we consider error-corrections in (d, k)-constrained sequences.

It has been reported that in magnetic recording channels, most of the errors are *right-* and *left-shift errors*, and *drop-out* and *drop-in errors* also arise due to various physical reasons [2], [3]. A right-shift error shifts a 1 in right direction; a left-shift error shifts a 1 in left direction; a drop-out error turns a 1 into a 0; a drop-in error turns a 0 into a 1. In this paper, right- and left-shift errors are referred to as *peak-shift errors*, and drop-out and drop-in errors are called *bit errors*. In most practical applications, the probability that a 1 is shifted by two

---

* This work was presented in part at the IEICE Technical Meeting on Information Theory, Tokyo, Japan, Nov. 26, 1992.

** Y. Saitoh will be with the Institute of Industrial Science, the University of Tokyo from April 1993.

*** I. Ibe will be with Hitachi, Ltd. from April 1993.

or more positions is very small. In this paper, we assume that a 1 may be shifted by at most one position.

In a conventional recording system, after data are encoded by an error-correction encoder, a modulation encoder encodes the code sequences into (d, k)-constrained channel sequences, as depicted in Fig. 1. Then, errors in the received sequences are extended by the modulation decoder. Therefore, the error-correction decoder must correct the burst error even if only a single error is in the received sequences. It is desirable that the modulation decoding is performed after the error-correction decoding, as illustrated in Fig. 2. In Fig. 2, an error-correcting modulation encoder encodes modulation code sequences into channel sequences with error-correction capabilities. Such a recording system has been proposed in [1] and [2]. In this paper, we consider an error-correcting modulation coding technique in the recording system as shown in Fig. 2.

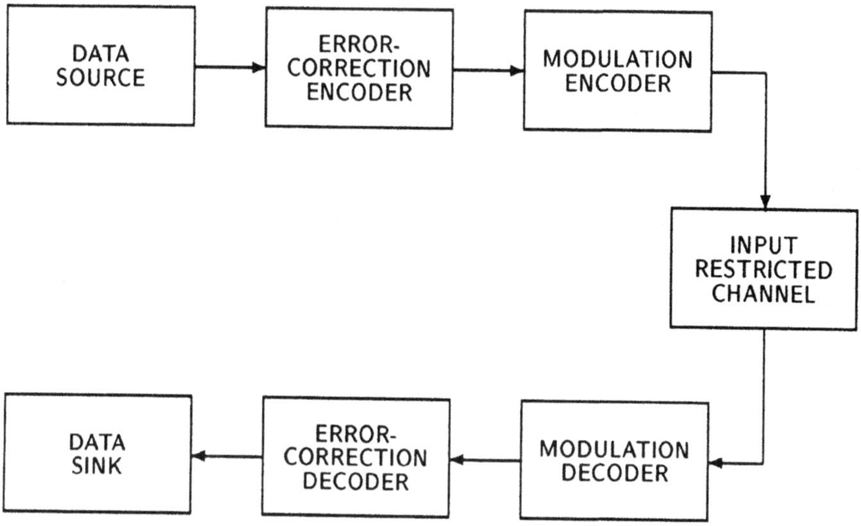

Fig. 1. Block diagram of a conventional recording system

Ferreira and Lin [1], Hilden, Howe, and Weldon [2], Kuznetsov and Vinck [7], and Saitoh, Ohno, and Imai [8] have developed peak-shift error-correcting modulation coding techniques. Ferreira and Lin [1] has also presented bit error-correcting modulation coding techniques. In this paper, we propose an error-correcting modulation coding technique for correcting combinations of multiple peak-shift errors and single bit errors in (d, k)-constrained sequences. The decoding process of the proposed technique uses well known channel side information [1], [5] in a magnetic recording system.

The structure of this paper is as follows: In Sect. 2, a peak-shift error-correction technique is described. Section 3 describes channel side information for determinating whether or not a single bit error is in the received sequence.

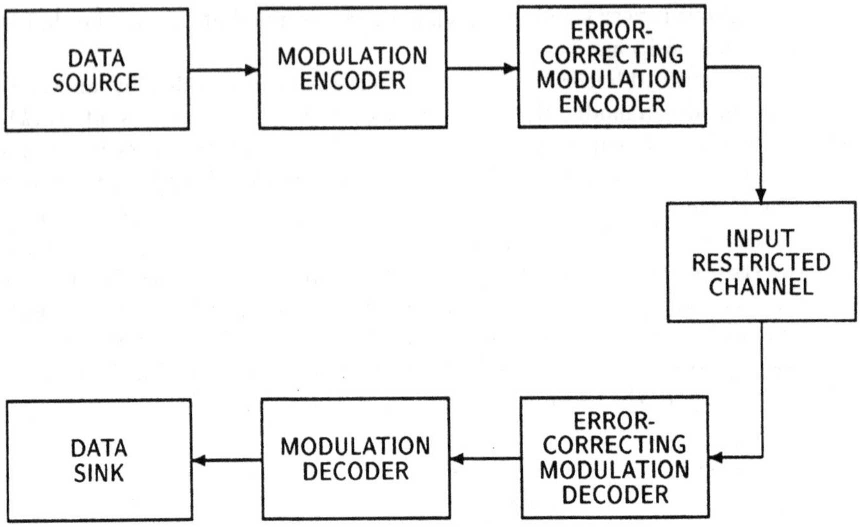

Fig. 2. Block diagram of a recording system considered in this paper

In Sect. 4, the encoding and decoding procedures of the proposed coding technique is strictly described for correction of combinations of multiple peak-shift errors and single bit errors. Examples and comparisons of the proposed coding technique are shown in Sect. 5.

2 Peak-Shift Error-Correction

A peak-shift error-correction technique has been developed in [2]. This technique is a variable length coding technique. In this section, this technique is easily modified to one for fixed length coding.

Consider a (d, k)-constrained sequence X of length L and Hamming weight W that starts with a 1 and terminates with at least d 0's. Data can be encoded into X by the technique described in [4, Ch. 5]. X is represented by

$$\begin{aligned} X &= (x_1, x_2, \cdots, x_L), \quad x_i \in \{0, 1\}, \\ &= (10^d, E_{d,k,k-d,k-d}(z), 0^d), \end{aligned}$$

where z is a data sequence and $E_{d,k,l,r}$ is an encoding function of (d, k)-constrained block sequences that start with at most l 0's and terminate with at most r 0's.

Let s_i be the number indicating the position of the 1 in X for $i = 0, 1, \cdots, W-1$ that satisfies

$$1 = s_0 < s_1 < \cdots < s_{W-1} \leq L - d,$$

i.e.,

$$\begin{aligned} x_{s_i} &= 1 \quad \text{for } i = 0, 1, \cdots, W-1, \\ x_j &= 0 \quad \text{if } j \neq s_i \text{ for } i = 0, 1, \cdots, W-1. \end{aligned}$$

For $K = \lfloor L/(d+1) \rfloor$, a bijection A_L from $\{0,1\}^L$ to K-dimensional vectors is defined as

$$A_L(X) = (s_1, \cdots, s_{W-1}, 0, \cdots, 0),$$
$$A_L^{-1}((s_1, \cdots, s_{W-1}, 0, \cdots, 0)) = X.$$

Let E be an encoding function for a ternary systematic t-fold error-correcting code of length N and dimension K. Suppose that E_2 is an encoding function for the (d,k)-constrained check part of length N_2 and is represented by

$$E_2(S_2) = (10^d, l_{K+1}, 0^d, l_{K+2}, 0^d, \cdots, l_N, 0^d),$$

where S_2 is a ternary $(N-K)$-dimensional vector such that

$$S_2 = (s_{K+1}, s_{K+2}, \cdots, s_N),$$

and l_i is a binary 3-dimensional vector as follows:

$$l_i = (l_{i,0}, l_{i,1}, l_{i,2}),$$

where $l_{i,j} = 1$ if $j = s_i$; $l_{i,j} = 0$ otherwise. We then have $N_2 = (N-K)(d+3) + d+1$. E_2 is used for encoding $N-K$ check symbols of the ternary error-correcting code into the (d,k)-constrained sequence.

A (d,k)-constrained block code of length $L + N_2$ capable of correcting t peak-shift errors can be encoded as follows:

1. From the input data sequence z, calculate

$$X = (10^d, E_{d,k,k-d,k-d}(z), 0^d).$$

2. Calculate

$$T \equiv A_L(X) \pmod 3,$$
$$(T, S_2) = E(T).$$

3. Let the codeword corresponding to X be $(X, E_2(S_2))$.

It is immediately proved from [2] that the code defined by the above encoding procedure has the t-fold peak-shift error-correction capability.

3 Channel Side Information

Channel side information provides some error-detecting capability in magnetic recording channels [1], [5]. In NRZI recording, a 1 is stored as a change in direction of magnetization, while a 0 is stored as no change. In the readback process, the changes in direction of magnetization are read by the peak detector as the peaks of the current or voltage, where the polarity of the peaks may be observed, as illustrated in Fig. 3. Then, the peaks alternate in polarity, i.e., if the readback sequence is

$$(\tau_1, \tau_2, \cdots), \quad \tau_i \in \{-1, 0, 1\}$$

satisfying $\tau_i \neq 0$, $\tau_{i+h} \neq 0$, and $\tau_{i+j} = 0$ for $j = 1, 2, \cdots, h-1$, then

$$\tau_{i+h} = -\tau_i.$$

Therefore, all single bit errors can be detected in $(\tau_1, \tau_2, \cdots)$. Note that combinations of multiple peak-shift errors and any single bit error can be detected, i.e., it can be determined whether or not any single bit error is in the readback sequence if no more than two bit errors other than peak-shift errors are in the readback sequence.

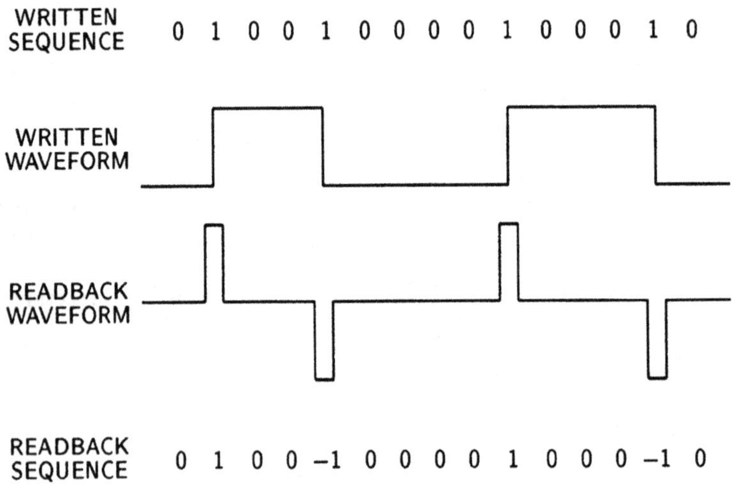

Fig. 3. Writing and reading processes during the NRZI precoding step in a magnetic recording system

By channel side information, single bit errors are not only detected but also the error-locations may almost be estimated if it is determined whether the errors are drop-out or drop-in errors. When the readback sequence is

$$(\cdots, \tau_i, 0^{h-1}, \tau_{i+h}, \cdots)$$

and $\tau_i = \tau_{i+h}$, a single error has arised, and we should

- let $\tau_{i+j} = -\tau_i$ for a j if the error is a drop-out error.
- let $\tau_i = 0$ or $\tau_{i+h} = 0$ if the error is a drop-in error.

A single drop-out or drop-in error adds -1 or $+1$ to the Hamming weight of the sequence. We can determine whether the type of the single bit error is drop-out or drop-in if we obtain W mod 3 where W is the Hamming weight.

4 Encoding and Decoding of the Proposed Coding Technique

In this section, the encoding and decoding procedures for the proposed coding technique are described.

Suppose that $W(*)$ is the Hamming weight of $*$. Let $C_{d,k}^{(q)}$ be a (d,k)-constrained block code with q codewords that start with a 1 and terminate with at least d 0's, let $E_{d,k}^{(q)}$ be an encoding function from $\{0,1,\cdots,q-1\}$ to $C_{d,k}^{(q)}$, let $D_{d,k}^{(q)}$ be the decoding function corresponding to $E_{d,k}^{(q)}$, and let $D_2 = E_2^{-1}$. Also let $C_{d,k,t}^{(3)}$, $E_{d,k,t}^{(3)}$, and $D_{d,k,t}^{(3)}$ be a (d,k)-constrained block code capable of correcting t peak-shift errors having three codewords that start with a 1 and terminate with at least d 0's, encoding and decoding functions for $C_{d,k,t}^{(3)}$, respectively. Assume that E and D are encoding and decoding functions, respectively, for a ternary systematic block code of length N, dimension K, and minimum distance $2t+2$.

Encoding Procedure:

1. Calculate

$$
\begin{aligned}
A_L(X) &= (s_1, s_2, \cdots, s_{W(X)-1}, 0, \cdots, 0),\\
p_1 &= \left\lfloor \frac{1}{3}\left(\left(\sum_{i=1}^{W(X)-1} s_i\right) \bmod (k-d+1)\right)\right\rfloor,\\
X_1 &= E_{d,k}^{(\lceil (k-d+1)/3\rceil)}(p_1),\\
T &\equiv (A_L(X), A_{N_1}(X_1)) \pmod 3,\\
(T, S_2) &= E(T),\\
X_2 &= E_2(S_2),\\
X_3 &= E_{d,k,t}^{(3)}(W((X, X_1, X_2)) \bmod 3),
\end{aligned}
$$

where N_1 is the length of X_1.

2. Let the encoder output sequence be (X, X_1, X_2, X_3).

K must satisfy

$$
K = \left\lfloor \frac{L}{d+1} \right\rfloor + \left\lfloor \frac{N_1}{d+1} \right\rfloor - 2.
$$

Assume that the received sequence is (X', X_1', X_2', X_3'). Hereafter the symbol with a prime ($'$) implies that of the received side. The decoding procedure is sketched as follows. This is illustrated in Fig. 4.

Decoding Procedure: If a single bit error is in the received sequence, then channel side information provides the location of the subsequence that consists of consecutive two peaks with the same polarity, as described in Sect. 3. If no single bit error has arised, then the error-correction procedure is as follows:

1. Calculate

$$\begin{aligned} T' &\equiv (A_L(X'), A_{N_1}(X_1')) \pmod 3, \\ R &= (T', D_2(X_2')), \\ (v_1, \cdots, v_K) &\equiv D(R) - T' \pmod 3, v_i \in \{-1, 0, 1\}, \\ (y_1, \cdots, y_L) &= A_L^{-1}(A_L(X') + (v_1, \cdots, v_M)), y_i \in \{0, 1\}, \end{aligned}$$

where $M = \left\lfloor \frac{L}{d+1} \right\rfloor - 1$.

2. Let the decoder output sequence be $E^{-1}_{d,k,k-d,k-d}((y_{d+2}, \cdots, y_{L-d}))$.

If a single bit error has arised, then the error-correction procedure is as follows:

1. Calculate

$$S = (A_L(X'), A_{N_1}(X_1'), D_2(X_2')).$$

2. It can be found from $D^{(3)}_{d,k,t}(X_3')$ whether the error is a drop-out or drop-in error. If the error is a drop-out error, then an erasure symbol is inserted into the location corresponding to the erroneous subsequence of S; otherwise the symbol corresponding to the erroneous subsequence is set as an erasure symbol in S. Let the new sequence corresponding to S be $S' = (s_1', \cdots, s_N')$, and let the erasure symbol be the Jth symbol s_J'.
3. Calculate

$$\begin{aligned} T' &= (s_1', \cdots, s_K') \pmod 3, \\ R &= S' \pmod 3, \\ (u_1, \cdots, u_K) &= D(R), \\ (v_1, \cdots, v_K) &\equiv (u_1, \cdots, u_K) - T' \pmod 3, v_i \in \{-1, 0, 1\}, \\ (w_1, \cdots, w_K) &= S' + (v_1, \cdots, v_K). \end{aligned}$$

   If $J > M$, where $M = \left\lfloor \frac{L}{d+1} \right\rfloor - 1$, then let the decoding output sequence be $A_L^{-1}((w_1, \cdots, w_M))$, and stop this procedure; otherwise go to the next step.
4. Find j satisfying

$$j \equiv 3 \cdot D_{d,k}^{(\lceil (k-d+1)/3 \rceil)}(A_{N_1}^{-1}((w_{M+1}, \cdots, w_K))) + u_J \pmod{k - d + 1}$$

   and $w_{J-1} + d < j < w_{J+1} - d$.
5. Calculate

$$(y_1, \cdots, y_L) = A_L^{-1}((w_1, \cdots, w_{J-1}, j, w_{J+1}, \cdots, w_M)), y_i \in \{0, 1\}.$$

6. Let the decoder output sequence be $E^{-1}_{d,k,k-d,k-d}((y_{d+2}, \cdots, y_{L-d}))$.

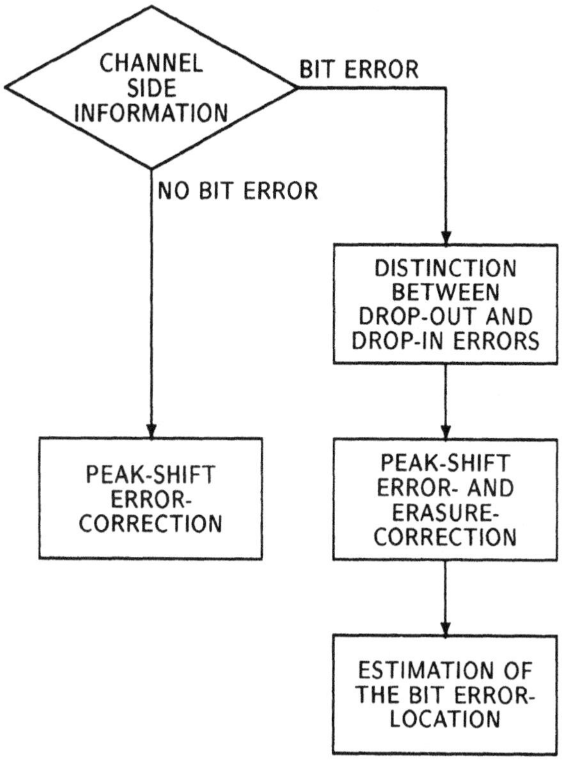

Fig. 4. Block diagram of the error-correction procedure for the proposed coding technique

5 Examples and Comparisons

The length of the check part of a codeword of the proposed coding technique is

$$N_1 + N_2 + N_3,$$

where N_3 is the length of X_3. When d, k, and t are constant, N_1 and N_3 are constant, and N_2 is represented by $N_2 = (N - K)(d+3) + d + 1$. Examples of the parameters are listed in Table 1. Ternary codes in [6] are used for Table 1.

Example 1. Consider the case where $(d, k) = (2, 7)$ and $t = 3$, i.e., combinations of triple peak-shift errors and single bit errors are correctable. From $(d, k) = (2, 7)$ the number of words for X_1 is 2, and thus we use for the encoder of X_1

$$E_{2,7}^{(2)}(0) = 100100,$$
$$E_{2,7}^{(2)}(1) = 100000$$

Table 1. Parameters of error-correcting modulation codes

(a) $(d,k)=(1,7)$

t	L	N_1	N_2	N_3	Code Length
1	30	5	22	7	64
2	34	5	34	7	80
3	24	5	46	7	82
1	84	5	26	7	122
2	76	5	50	7	138
3	58	5	70	7	140

(b) $(d,k)=(2,7)$

t	L	N_1	N_2	N_3	Code Length
1	45	6	28	9	88
2	51	6	43	9	109
3	36	6	58	9	109
1	126	6	33	9	174
2	114	6	63	9	192
3	87	6	88	9	190

of $N_1 = 6$. From $(d,k) = (2,7)$ and $t = 3$ we use for the encoder of X_3

$$E^{(3)}_{2,7,3}(0) = 100100100,$$
$$E^{(3)}_{2,7,3}(1) = 100000100,$$
$$E^{(3)}_{2,7,3}(2) = 100100000$$

of $N_3 = 9$. Suppose that the length of input data sequences is 16 bits. We then have $L = 36$. We select a ternary cyclic code of length 23, dimension 12, and minimum distance 8 [6] for E and D. We then have $(N,K) = (23,12)$ and the following code format:

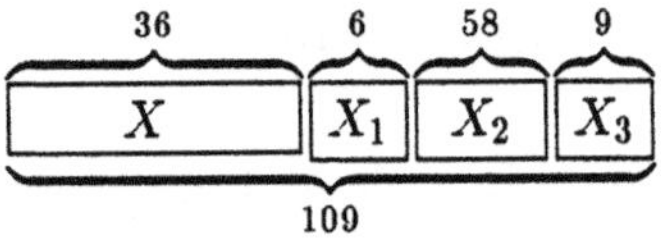

The coding rate is $16/109 = 0.147$. Examples of the encoding and decoding processes are shown in Table 2. We note that by conventional concatenated coding scheme, it is difficult to realize a coding block length of 109 bits with the same error-correcting capability.

Example 2. Consider the case where $(d,k) = (2,7)$ and $t = 1$. Then, combinations of single bit errors and single peak-shift errors are correctable. We use the ternary code of length 48, dimension 42, and minimum distance 4 [6]. Thus, set $N = 48$ and $K = 42$. Then $N_2 = 33$. Let

$$C^{(2)}_{2,7} = \{100100, 100000\},$$
$$C^{(3)}_{2,7,1} = \{100100100, 100000100, 100100000\}.$$

Table 2. Examples of encoding and decoding processes

(a) Encoding

Data sequence z	0101001110100110
$E_{2,7,5,5}(z)$	0001001000100100100000100000001
X	(100,00010010001001001000001000000001,00)
$A_{36}(X)$	$(7, 10, 14, 17, 20, 26, 34, 0, 0, 0, 0)$
p_1	0
X_1	100100
$A_6(X_1)$	4
T	$(1, 1, 2, 2, 2, 2, 1, 0, 0, 0, 0, 1)$
Polynomial of T	$T(x) = x^{11} + x^{10} + 2x^9 + 2x^8 + 2x^7 + 2x^6 + x^5 + 1$
Generator polynomial of E	$G(x) = x^{11} + 2x^8 + 2x^6 + x^4 + x^3 + 2x^2 + 2x + 2$
Polynomial of S_2	$-x^{11}T(x) \bmod G(x)$
S_2	(1,2,0,0,0,0,0,1,2,2,2)
X_2	(100, 010, 00, 001, 00, 100, 00, 100, 00, 100, 00, 100, 00, 100, 00, 010, 00, 001, 00, 001, 00, 001, 00)
$W((X, X_1, X_2))$	22
$W((X, X_1, X_2)) \bmod 3$	1
X_3	100000100
(X, X_1, X_2, X_3)	(1000001001000100100100000100000000100, 100100, 10001000001001000010000100001000010000 1000001000001000010000100, 100000100)

(b) Decoding

(X', X'_1, X'_2, X'_3)	$(\overset{+}{1}\ 00000\ \overset{-}{1}\ 00\ \overset{+}{1}\ 000\ \overset{-}{1}\ 00\ \overset{+}{1}\ 00\underline{0}000000\underline{0\ \overset{+}{1}}000000\ \overset{-}{1}\ 00,$ $\overset{+}{1}\ 00\underline{0\ \overset{-}{1}}0,$ $\overset{+}{1}\ 000\ \overset{-}{1}\ 00000\ \overset{+}{1}\ 00\ \overset{-}{1}\ 0000\ \overset{+}{1}\ 0000\ \overset{-}{1}\ 0000\ \overset{+}{1}\ 0000$ $\overset{-}{1}\ 00000\ \overset{+}{1}\ 00000\ \overset{-}{1}\ 0000\ \overset{+}{1}\ 000\underline{\overset{-}{1}\ 0}00,$ $\overset{+}{1}\ 00000\ \overset{-}{1}\ 00)$
$D^{(3)}_{2,7,3}(X'_3)$	1
$W((X', X'_1, X'_2))$	21
$D^{(3)}_{2,7,3}(X'_3) - W((X', X'_1, X'_2))$	1 (mod 3) → The bit error is a drop-out error.
S'	$(7, 10, 14, 17, *, 27, 34, 0, 0, 0, 0, 5, 1, 2, 0, 0, 0, 0, 0, 1, 2, 2, 1)$
J	5
T'	$(1, 1, 2, 2, *, 0, 1, 0, 0, 0, 0, 2)$
R	$(1, 1, 2, 2, *, 0, 1, 0, 0, 0, 0, 2, 1, 2, 0, 0, 0, 0, 0, 1, 2, 2, 1)$
$D(R) = (u_1, \cdots, u_{12})$	$(1, 1, 2, 2, 2, 2, 1, 0, 0, 0, 0, 1)$
$(v_1, \cdots, v_{12})$	$(0, 0, 0, 0, *, -1, 0, 0, 0, 0, 0, -1)$
$(w_1, \cdots, w_{12})$	$(7, 10, 14, 17, *, 26, 34, 0, 0, 0, 0, 4)$
$A_6^{-1}(w_{12})$	100100
$D^{(2)}_{2,7}(A_6^{-1}(w_{12}))$	0
u_J	2
j	20
$(y_1, \cdots, y_{36})$	1000001001000100100100000100000000100
$E^{-1}_{2,7,5,5}((y_4, \cdots, y_{34}))$	0101001110100110

$*$: Erasure symbol.

Thus, $N_1 = 6$ and $N_3 = 9$. From $K = 42$ and $N_1 = 6$, we have $L = 126$. As the result, we have the following code format:

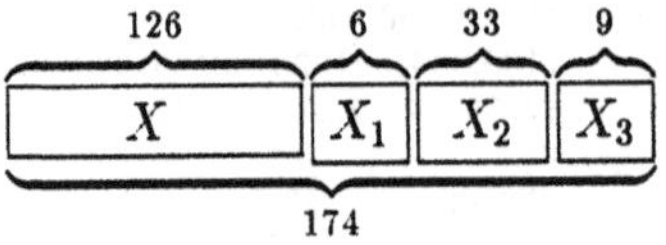

Thus, the coding rate of this error-correcting modulation encoder is $126/174 = 0.724$. The coding rate of the encoding for X approximates 0.5, and thus the coding rate of the coding system employing the proposed error-correcting modulation encoder and the modulation encoder of rate 0.5 is 0.362. Consider the conventional concatenated coding scheme in order to compare the proposed one with it. In the concatenated coding scheme, an error might well become a burst error since the demodulation will have some error-propagation. Let b be the length of the longest burst error that can be caused by a peak-shift or bit error. For the $(2,7)$ code of [4, pp. 125–127] we have rate 1/2 and $b = 5$. When we use this inner code, the error-correcting outer code must correct burst errors of $b = 5$. For example, when we use the code from interleaving the Reed-Solomon code of length 11 and dimension 7 over GF(16) twice as the outer code, the concatenated coding system can correct double errors, and the length of a codeword of the system is 176 bits. Then we have rate $56/176 = 0.318$, and thus the proposed error-correcting modulation encoder is efficient.

6 Conclusions

In this paper, we present an error-correcting modulation code capable of correcting combinations of single bit errors and t peak-shift errors. The remaining research problems include

- finding more efficient representation for the check part.
- generalization to correction of combinations of two or more bit errors and peak-shift errors.

We note that the proposed technique can be easily modified to a variable length version. The variable length codes are efficient, but they might cause catastrophic error-propagation by uncorrectable errors.

Acknowledgement

This work was supported in part by the Grant-in-Aid for Encouragement of Young Scientists of the Ministry of Education, Science and Culture of Japan.

References

1. Ferreira, H. C., Lin, S.: Error and erasure control (d, k) block codes. IEEE Trans. Inform. Theory **37** (1991) 1399–1408
2. Hilden, H. M., Howe, D. G., Weldon, E. J., Jr.: Shift error correcting modulation codes. IEEE Trans. Magn. **27** (1991) 4600–4605
3. Howell, T. D.: Analysis of correctable errors in the IBM 3380 disk file. IBM J. Res. Develop. **28**, 2 (1984) 206–211
4. Immink, K. A. S.: Coding Techniques for Digital Recorders. London, UK: Prentice Hall Int. (1991)
5. Kobayashi, H., Tang, D. T.: Application of partial-response channel coding to magnetic recording systems. IBM J. Res. Develop. **14** (1970) 368–375
6. Kschischang, F. R., Pasupathy, S.: Some ternary and quaternary codes and associated sphere packings. IEEE Trans. Inform. Theory **38** (1992) 227–246
7. Kuznetsov, A. V., Vinck, A. J. H.: Single peak-shift correction in (d,k)-sequences. 1991 IEEE Int. Symp. Inform. Theory (1991) 256.
8. Saitoh, Y., Ohno, T., Imai, H.: Construction techniques for error-control runlength-limited block codes. IEICE Trans. Fundamentals (to appear)

On a Third Order Differential Equation whose Differential Galois Group is the Simple Group of 168 Elements

Michael F. Singer* and Felix Ulmer**

North Carolina State University
Department of Mathematics
Box 8205
Raleigh, N.C. 27695-8205

Abstract. In this paper we compute the differential Galois group of a third order linear differential equation whose existence was predicted by F. Klein [9] and whose construction is due to A. Hurwitz [7]. The aim of this paper is to apply the results of [15] in order to prove, starting only with the equation, that the simple group of 168 elements is the the differential Galois group of this equation.

0 Introduction

Let $L(y) = 0$ be a (homogeneous) linear differentatial equation of degree n whose coefficients belong to a differential field k whose field of constants $\mathcal{C}$ is algebraically closed of characteristic 0 (see e.g. [8, 14, 16] for those notions). Similar to the case of algebraic equations, there is a notion of a "splitting field" for this equation. More specifically, there is a field $K = k < y_1, \ldots, y_n >$, generated (in the differential sense) by a fundamental set of solutions $\{y_1, \ldots, y_n\}$ of $L(y) = 0$ such that K and k have the same set of constants. K is called the *Picard-Vessiot extension of k corresponding to L(y)=0* and is unique up to a differential k-isomorphism. The set of differential automorphisms of the field extension K/k (i.e., the field automorphisms of the field extension K/k which commute with the derivation of K) that leave k elementwise fixed is called the *differential Galois group* $\mathcal{G}(L)$ *of* $L(y) = 0$. Since the solution space of $L(y) = 0$ is an n-dimensional vector space over $\mathcal{C}$ and since the group $\mathcal{G}(L)$ sends a solution of $L(y) = 0$ into another solution of $L(y) = 0$, we get a faithfull representation of $\mathcal{G}(L)$ as a subgroup of $GL(n, \mathcal{C})$. In fact, $\mathcal{G}(L)$ can be shown to be a linear algebraic group. Many properties of the differential equation are mirrored in the group structure of $\mathcal{G}(L)$. For example, irreducibility of the equation is equivalent to irreducibility of the group and solvability in terms of exponentials, integrals

* Partially supported by NSF Grant 90-24624

** Partially supported by Deutsche Forschungsgemeinschaft, while on leave from Universität Karlsruhe. The second author would like to thank North Carolina State University for its hospitality and partial support during the preparation of this paper.

and algebraic functions (the liouvillian functions) is equivalent to the connected component of the identity of $\mathcal{G}(L)$ being solvable. It is therefore an important problem to be able to calculate the Galois group or at least determine various properties of this group. We note that despite much recent work in determining the Galois group of a differential equation, there is, at present, no general algorithm that will calculate the defining polynomial equations of this group or even determine its dimension (see [15] for references to recent work).

In [15], we use the representation theory of groups to give simple necessary and sufficient conditions regarding the structure of the Galois groups of second and third order linear differential equations. These allow us to give simple necessary and sufficient conditions for a second order linear differential equation to have liouvillian solutions and for a third order linear differential equation to have liouvillian solutions or be solvable in terms of second order equations. In this note we show how these results can be applied to calculate the Galois group of the following equation due to Hurwitz[3] (see [7]):

$$\begin{aligned} H(y) = {} & x^2(x-1)^2y''' + (7x-4)x(x-1)y'' + \\ & (\frac{72}{7}(x^2-x) - \frac{20}{9}(x-1) + \frac{3}{4}x)y' + \\ & (\frac{792}{7^3}(x-1) + \frac{5}{8} + \frac{2}{63})y \end{aligned}$$

In the first section we review the relevant results of [15], in the second section we show how these can be used to calculate the example and in the final section we make some concluding remarks.

1 Symmetric powers and Galois groups

In [15] we show how by factoring differential operators, one can decide if the differential Galois group $\mathcal{G}(L)$ is reducible, imprimitive or primitive and, if the $\mathcal{G}(L) \subseteq SL(n, C)$ is a primitive linear group, how to compute this group. The method uses the construction of the differential equation

$$L^{\circledS m}(y) = \overbrace{L(y) \circledS \cdots \circledS L(y)}^{m} = 0,$$

called the *symmetric power* of order m of $L(y) = 0$. The equation $L^{\circledS m}(y) = 0$ is characterized by the property that it is the monic linear differential operator of smallest order whose solution space is spanned by all products of length m of solutions of $L(y) = 0$. In [13, 15] an algorithm to construct the above equation is given. The differential equation $L^{\circledS m}(y)$ is of order at most $\binom{n+m-1}{n-1}$, where n is the degree of $L(y)$. For example, the second symmetric power $H^{\circledS 2}(y)$ of

[3] This equation is related to an equation previously produced by Halphen in a letter to F. Klein (cf. [6])

$H(y)$ is:

$$y^{(6)} + \frac{31x-18}{x(x-1)}y^{(5)} + \frac{\frac{2320}{7}x^2 - \frac{97135}{252}x + \frac{880}{9}}{x^2(x-1)^2}y^{(4)}$$
$$+\frac{\frac{73760}{49}x^3 - \frac{9282095}{3528}x^2 + \frac{526415}{392}x - \frac{1640}{9}}{x^3(x-1)^3}y^{(3)}$$
$$+\frac{\frac{141840}{49}x^3 - \frac{4548685}{1176}x^2 + \frac{21381505}{15876}x - \frac{7840}{81}}{x^4(x-1)^3}y''$$
$$+\frac{\frac{4898160}{2401}x^3 - \frac{54698425}{28812}x^2 + \frac{5780305}{15876}x - \frac{560}{81}}{x^5(x-1)^3}y'$$
$$+\frac{\frac{40930560}{117649}x^2 - \frac{87839125}{470596}x + \frac{40805}{3969}}{x^5(x-1)^3}y$$

In [15] it is shown how the orders of irreducible factors of $L^{\circledS m}(y)$ are related to the decomposition of the character of the symmetric power of the representation of $\mathcal{G} \subseteq SL(n, C)$. This can be used to determine properties of $\mathcal{G}(L)$. For example, we are able to give necessary and sufficient conditions for these linear differential equations to have liouvillian solutions. In particular, we show:

Let $L(y) = y'' + ry = 0$ be a second order linear differential equation with $r \in k$. $L(y) = 0$ has liouvillian solutions if and only if $L^{\circledS 6}(y)$ is reducible.

Factorization properties can also be used to determine Galois groups in many cases. For example (note that the Tetrahedral group is the finite subgroup of SL_2 having a center of order two such that its quotient by this center is isomorphic to the alternating group on 4 letters):

Let $L(y) = y'' + ry = 0$ be a second order linear differential equation with $r \in k$. The Galois group of $L(y) = 0$ is the Tetrahedral Group if and only if $L^{\circledS 2}(y)$ is irreducible and $L^{\circledS 3}(y)$ is reducible.

For third order equations we proved similar results that we now state in more detail. We first recall that a subgroup G of $GL(V)$ is said to *act irreducibly* if the only G-invariant subspaces of V are $\{0\}$ and V, otherwise it is said to act *reducibly*. Let G be a subgroup of $GL(V)$ acting irreducibly. G is called *imprimitive* if, for $k > 1$, there exist non-trivial subspaces $V_1, \cdots, V_k$ such that $V = V_1 \oplus \cdots \oplus V_k$ and, for each $g \in G$, the mapping $V_i \rightarrow g(V_i)$ is a permutation of the set $\mathcal{S} = \{V_1, \ldots, V_k\}$. An irreducible group $G \subseteq GL(n, C)$ which is not imprimitive is called *primitive*. One knows the finite primitive subgroups of $PGL(3, \mathbb{C})$ (c.f., [1]). From this list, one can derive the primitive subgroups of $SL(3, C)$ (c.f., [1]). Any finite primitive group of $SL(3, \mathbb{C})$ is isomorphic to one of the following groups:

1. The Valentiner Group $A_6^{SL_3}$ of order 1080 generated as a transitive permutation group of 18 letters by:

$$(1,2,4)(3,8,13)(5,7,9)(6,10,12)(11,15,14),$$
$$(1,3)(2,6)(4,5)(7,12)(8,9)(10,13),$$
$$(1,4)(3,8)(5,9)(6,11)(10,14)(12,15),$$
$$(1,4,8,3,5,9)(2,7,13)(6,12,10)(11,16,14,17,15,18),$$
$$(1,5,8)(2,7,13)(3,4,9)(6,12,10)(11,15,14)(16,18,17).$$

We have $A_6^{SL_3}/Z(A_6^{SL_3}) \cong A_6$.

2. The simple group G_{168} of order 168 defined by:

$$\{X,Y \mid X^7=(X^4Y)^4=(XY)^3=Y^2=id\}.$$

3. $G_{168} \times C_3$, the direct product of G_{168} with the cyclic group C_3 of order 3.
4. A_5, the alternating group of five letters.
5. $A_5 \times C_3$, the direct product of A_5 with a cyclic group C_3 of order 3.
6. The group $H_{216}^{SL_3}$ of order 648 defined by:

$$\{U,V,S,T \mid U^9=V^4=T^3=S^3=(UV)^3=id, VS=TV$$
$$VT=S^2V,[U^6,V]=[U^6,T]=[U,S]=id,[U,V^2]=S\}.$$

The group $H_{216}^{SL_3}/Z(H_{216}^{SL_3})$ is the *hessian group* of order 216.

7. The group $H_{72}^{SL_3}$ of order 216 generated by the elements S, T, V and UVU^{-1} of $H_{216}^{SL_3}$.
8. The group $F_{36}^{SL_3}$ of order 108 generated by the elements S, T and V of $H_{216}^{SL_3}$.

For each finite primitive subgroup of $SL(3,\mathbb{C})$, using its character table (computed using the group theory system Cayley [2]) and the orthogonality relations of characters, we can decompose the characters of the symmetric product (computed using the computer algebra system AXIOM). The result is summarized in the following table, where the numbers $4, 3^2$ in the column A_5 and row 3 of Figure 1 means that the 3rd symmetric product of the character of any faithfull irreducible representation of A_5 in $SL(3,\mathbb{C})$ has an irreducible summand of degree 4 and two irreducible summands of degree 3.

	PSL_2 / $PSL_2 \times C_3$	A_5 / $A_5 \times C_3$	$F_{36}^{SL_3}$	G_{168} / $G_{168} \times C_3$	$A_6^{SL_3}$	$H_{72}^{SL_3}$	$H_{216}^{SL_3}$
2	$5,1$	$5,1$	$3,3$	6	6	6	6
3	$7,3$	$4,3^2$	$1^2,4^2$	$7,3$	10	$8,2$	$8,2$
4	$9,5,1$	$5^2,4,1$	3^5	$8,6,1$	$9,6$	$6^2,3$	$6^2,3$
5	$11,7,3$	$5,4,3^4$	3^7	$8,7,3^2$	$15,3^2$	$6,3^5$	$9,6,3^2$

Figure 1

The next two results are proved in [15]:

Theorem 1. *Let $L(y) = 0$ be a third order linear differential equation with coefficients in a differential field k with algebraically closed field of constants whose differential Galois $\mathcal{G}(L)$ group is unimodular.*

1. *$L(y) = 0$ is reducible if and only if $L(y) = 0$ has a solution $y \neq 0$ such that $y'/y \in k$ or $L^*(y) = 0$, the adjoint of $L(y) = 0$, has a solution $y \neq 0$ such that $y'/y \in k$ (if $L(y) = \sum_{i=0}^n a_i y^{(i)}$, then $L^*(y) = \sum_{i=0}^n (-1)^i (a_i y)^{(i)}$) .*
2. *Assume $L(y)$ is irreducible. Then $\mathcal{G}(L)$ is imprimitive if and only if $L^{\textcircled{s}3}(y) = 0$ has a solution $y \neq 0$ such that $y^2 \in k$. In this case $\mathcal{G}(L)$ is isomorphic to a subgroup of $C^* \rtimes S_3$, where S_3 is the symmetric group on three letters. If $\mathcal{G}(L)$ is isomorphic to a subgroup of $C^* \rtimes A_3$, where A_3 is the alternating group on three letters, then the above solution y is already in k.*
3. *Assume $L(y)$ is irreducible and 2. does not hold, then $\mathcal{G}(L)$ is a primitive group.*

Theorem 2. *Let $L(y) = 0$ be a third order linear differential equation with coefficients in a differential field k with algebraically closed field of constants, whose differential Galois group $\mathcal{G}(L)$ is unimodular. Assume that $\mathcal{G}(L)$ is primitive.*

1. *If $L^{\textcircled{s}2}(y)$ has order 5 or factors then $\mathcal{G}(L)$ is isomorphic to PSL_2, $PSL_2 \times C_3$, A_5, $A_5 \times C_3$ or $F_{36}^{SL_3}$. In this case one of the following holds*
   - *$\mathcal{G}(L) \cong F_{36}^{SL_3}$ if and only if $L^{\textcircled{s}2}(y)$ has a factor of order 3, or*
   - *$\mathcal{G}(L) \cong A_5$ or $A_5 \times C_3$ if and only if $L^{\textcircled{s}3}(y)$ has a factor of order 3 and a factor of order 4, or*
   - *$\mathcal{G}(L) \cong PSL_2$ or $\mathcal{G}(L) \cong PSL_2 \times C_3$ if and only if the previous two cases do not hold.*
2. *If $L^{\textcircled{s}2}(y)$ has order 6 and is irreducible, then one of the following holds*
   - *$\mathcal{G}(L) \cong G_{168}$ or $G_{168} \times C_3$ if and only if $L^{\textcircled{s}3}(y)$ has a factor of order 3.*
   - *$\mathcal{G}(L) \cong A_6^{SL_3}$ if and only if $L^{\textcircled{s}4}(y)$ is reducible and $L^{\textcircled{s}3}(y)$ is irreducible.*
   - *$\mathcal{G}(L) \cong H_{72}^{SL_3}$ if and only if $L^{\textcircled{s}5}(y)$ has more than 2 factors of order 3.*
   - *$\mathcal{G}(L) \cong H_{216}^{SL_3}$ if and only if $L^{\textcircled{s}5}(y)$ has exactly 2 factors of order 3 and $L^{\textcircled{s}2}(y)$ has a factor of degree 2.*
   - *The Galois group is $SL(3, C)$ if and only if none of the above happen.*

Algorithms for factoring linear differential operators are well known, [11, 12, 5]. Therefore the above results will *in theory* allow one to determine the Galois group of $H(y) = 0$. Nonetheless, we show how simpler calculations allow us to determine these factorization properties.

2 The Galois group of Hurwitz's equation

We start our computations by first showing that $\mathcal{G}(H)$ is unimodular (i.e. a subgroup of $SL(n, C)$). According to [8] p. 41, this follows from the fact that for the rational function $w = x^4(x-1)^3$ we have

$$\frac{w'}{w} = \frac{4}{x} + \frac{3}{x-1} = \frac{7x-4}{x(x-1)}$$

2.1 Reducibility

We must show that $H(y)$ is irreducible. Theorem 1 says that we just need to check if $H(y) = 0$ or $H^*(y) = 0$ have a solution $y \neq 0$ such that $y'/y \in \mathbb{C}(x)$. There is an algorithm to decide this (cf. [16], section 3.2) which has been implemented in the AXIOM system by M. Bronstein. We used this implementation to show that $H(y) = 0$ and $H^*(y) = 0$ have no such solution. Therefore $H(y)$ is irreducible.

2.2 Imprimitivity

Now one must check to see if $\mathcal{G}(H)$ is an imprimitive linear group. Theorem 1 says that we need to check if $H^{\circledS 3}(y) = 0$ has a solution $y \neq 0$ such that $y^2 \in \mathbb{C}(x)$. One can calculate $H^{\circledS 3}(y) = 0$ and use AXIOM to decide this question, but there is an easier way using exponents (c.f., [10] for a definition of exponents and their elementary properties). One can calculate the exponents of $H(y) = 0$ at the singular points 0, 1 and ∞ (which are all regular singular points) and one gets:

- $\{\frac{11}{7}, \frac{9}{7}, \frac{8}{7}\}$ at ∞
- $\{0, -\frac{1}{3}, -\frac{2}{3}\}$ at $x = 0$
- $\{\frac{1}{2}, 0, -\frac{1}{2}\}$ at $x = 1$

Since the exponents at infinity and 0 do not differ by integers, there exist solutions of the form $x^\rho \sum_{i=0}^{\infty} x^{-i}$ for each exponent ρ at infinity and $x^\rho \sum_{i=0}^{\infty} x^i$for each exponent ρ at 0. Calculating further, one can show that despite the fact that the exponents differ by an integer at 1 there exist similar solutions at $x = 1$, i.e., no logarithmic terms. A basis for the solution space of $H^{\circledS 3}(y) = 0$ will be gotten by taking three such solutions (allowing repetitions) and forming their products. Using this fact one sees that the exponents at infinity are $\frac{33}{7}, \frac{31}{7}, \frac{30}{7}, \frac{29}{7}, \frac{28}{7} = 4, \frac{27}{7}, \frac{27}{7} + n$ for some non-negative integer n, $\frac{26}{7}, \frac{25}{7}, \frac{24}{7}$. Similar lists can be made at 0 and 1. Note that for 0 any such exponent is at worst a fraction with denominator 3 and is ≥ -2 and at 1 they are at worst a fraction with denominator 2 and is $\geq -\frac{3}{2}$.

A solution y of $H^{\circledS 3}(y) = 0$ with $y^2 \in \mathbb{C}(x)$ must be of the form (because everything is fuchsian) $y = p(x)x^a(x-1)^b$ where $p(x)$ is a polynomial and a resp. b are exponents of $H^{\circledS 3}(y) = 0$ at 0 resp. 1. We also have that $-a - b$-deg(p) will be an exponent at infinity. by comparing denominators, we see that the only possibility is that $-a - b-$deg$(p) = 4$. Since $a \geq -2$ and $b \geq -\frac{3}{2}$, we have $-a - b$-deg$(p) \leq \frac{7}{2} < 4$. Therefore, $H^{\circledS 3}(y) = 0$ does not have a solution of the required form and so $\mathcal{G}(H)$ is not an imprimitive linear group.

2.3 Primitivity

We now know that the Galois group is a primitive group subgroup of $SL(3, C)$. It therefore must be one of the groups listed above. The computation of $H^{\circledS 4}(y)$ shows that this equation has order 14. This already gives us some information.

If we start with a vector space V of dimension 3 and form its fourth symmetric power $\mathcal{S}^4(V)$, we will get a vector space of dimension 15. If we let V be the solution space of $H(y) = 0$, then there is a $\mathcal{G}(H)$ morphism of $\mathcal{S}^4(V)$ onto the solution space of $H^{\textcircled{s}4}(y) = 0$ (cf. [15], Section 3.2.2). This means that the kernel of this morphism is a $\mathcal{G}(H)$ invariant subspace of $\mathcal{S}^4(V)$, i.e., $\mathcal{S}^4(V)$ will have a one dimensional invariant subspace. Looking at Figure 1, we see that this can only happen for groups corresponding to the first, second and fourth columns. Thus $\mathcal{G}(H)$ must be one of the groups PSL_2, A_5, G_{168} or the direct product of one of those groups with C_3.

We now show that the groups corresponding to the first two columns cannot occur. Since $H^{\textcircled{s}2}(y)$ is of order 6, we get from Figure 1 that if $\mathcal{G}(H)$ is one of the groups PSL_2, A_5, $PSL_2 \times C_3$ or $A_5 \times C_3$, then $H^{\textcircled{s}2}(y)$ must have a factor of order 1. Since the singularities are fuchsian and the exponents are rational, this implies that $H^{\textcircled{s}2}(y) = 0$ would have a solution y such that $y^i \in \mathbb{C}(x)$ for some i. If $\sigma \in \mathcal{G}(H)$ then $\sigma(y) = \chi(\sigma)y$ where χ is a character of $\mathcal{G}(H)$. PSL_2 and A_5 are simple groups and so have no nontrivial characters while if χ is a character of $PSL_2 \times C_3$ or $A_5 \times C_3$ then $\chi^3 = 1$. In all these cases we therefore must have that y^3 is left fixed by the Galois group and so $y^3 \in \mathbb{C}(x)$. Looking at the exponents $\frac{16}{7}, \frac{17}{7}, \frac{18}{7}, \frac{19}{7}, \frac{22}{7}$ at infinity we see that this is impossible. Therefore the Galois group must be one of the groups corresponding to column 4, i.e., G_{168} or $G_{168} \times C_3$. We must now distinguish between G_{168} and $G_{168} \times C_3$.

To do this, we shall use the monodromy group of the above equation. Consider a linear differential equation with coeficients in $\mathbb{C}(x)$. For these equations we can use analytic considerations to define a group called the *monodromy group* that is a subgroup of the Galois group. Let $c_1, \ldots, c_n$ be the singular points of $L(y) = 0$ (including infinity if it is a singular point) and let c_0 be an ordinary point of the equation. We consider these points as lying on the Riemann Sphere S^2. Let $\{y_1, \ldots, y_n\}$ be a fundamental set of solutions of $L(y) = 0$ analytic at c_0 and let γ be a closed path in $S^2 - \{c_1, \ldots, c_n\}$ that begins and ends at c_0. One can analytically continue $\{y_1, \ldots, y_n\}$ along γ and get new fundamental solutions $\{\overline{y_1}, \ldots, \overline{y_n}\}$ analytic at c_0. These two sets must be related via $(\overline{y_1}, \ldots, \overline{y_n})^T = M_\gamma(y_1, \ldots, y_n)^T$ where $M_\gamma \in GL(n, \mathbb{C})$. One can show that M_γ depends only on the homotopy class of γ and that the map $\gamma \mapsto M_\gamma$ defines a group homomorphism from $\pi_1(S^2 - \{c_1, \ldots, c_n\})$ to $\mathcal{G}(L)$. The image of this map depends on the choice of c_0 and $\{y_1, \ldots, y_n\}$ but is unique up to conjugacy and is called the *monodromy group of L(y)*. In general the image of this group will be a proper subgroup of $\mathcal{G}(L)$ but when $L(y)$ is fuchsian, the Zariski closure of this group will be the full Galois group $\mathcal{G}(L)$ (c.f., [17]). In particular if $\mathcal{G}(L)$ is finite (i.e., all solutions of $L(y) = 0$ are algebraic) then the map is surjective and the monodromy and Galois groups coincide.

We now return to $H(y) = 0$. At each singular point α, we have linearly independent solutions $y_i = (x-\alpha)^{\rho_i} \sum a_{ij}(x-\alpha)^j$, $i = 1, 2, 3$ where the ρ_i are the distinct exponents at α. Each ρ_i is a rational number, say $\rho_i = \frac{r_i}{s_i}$, $(r_i, s_i) = 1$. If we analytically continue each y_i around α, we get a new solution $y_i = \zeta_i y_i$ where ζ_i =exp($\frac{2i\pi}{s_i}$). Therefore the local monodromy group of $H(y) = 0$ around each singular point α is a cyclic subgroup generated by one element g_α whose order is the least common multiple of the denominator of the exponents at α. Thus g_∞ is of order 7, g_0 is of order 3 and g_1 is of order 2. The product $g_\infty g_0 g_1$ corresponds to the zero path and thus must be the identity. Since g_0, g_1 and g_∞ generate the monodromy group and $g_0 g_1 = g_\infty^{-1}$, we get that the group $\mathcal{G}(H)$ is generated by an element of order 2 and an element of order 3 whose product is an element of order 7. Using the group theory system CAYLEY (see [2]) one can see that G_{168} has such a set of generators, while $G_{168} \times C_3$ does not have such a set of generators. The group G_{168} is generated by S and T, where $S^7 = (S^4T)^4 = (ST)^3 = T^2 = 1$. A set of generators of the above form is given by the element T of order 2 and TS^{-1} of order 3 whose product is of order 7.

This shows that $\mathcal{G}(H) \cong G_{168}$.

3 Final comments

The techniques of [16] give another way to distinguish between these two groups G_{168} and $G_{168} \times C_3$. In[16], we show (among other things) that when the Galois group of a linear differential equation is a finite primitive group one can use invariant theory to construct the minimal polynomial of a solution of the linear differential equation (this is an idea going back to L. Fuchs [3]). In particular if $\mathcal{G}(H) \cong G_{168}$, then $H(y) = 0$ has an algebraic solution whose minimal polynomial is of degree 42, while if $\mathcal{G}(H) \cong G_{168} \times C_3$, then any solution of $H(y) = 0$ has minimal polynomial of degree at least 126 (cf. [16], Theorem 4.2). Techniques are given in [16] to compute these polynomials and one can use these to distinguish between G_{168} and $G_{168} \times C_3$.

We finally mention the source of $H(y) = 0$. In [9], Klein studied the Riemann suface $\mathcal{S}$ defined by $x^3y+y^3z+z^3x = 0$ in $\mathbf{CP}^2$. This surface has genus 3 and its automophism group $Aut(\mathcal{S})$ is the group G_{168}. The quotient of S under the action of $Aut(\mathcal{S})$ is just the Riemann sphere $\mathbf{CP}^1$. Let $\omega_1 = f_1 dt, \omega_2 = f_2 dt, \omega_3 = f_3 dt$ be a basis for the holomorphic 1-forms on S, where t is a $Aut(\mathcal{S})$ invariant function. The map $\kappa(p) \mapsto (f_1(p), f_2(p), f_3(p))$ defines the canonical embedding of the curve into $\mathbf{CP}^2$. One can show that (after an automorphism of $\mathbf{CP}^2$, if necessary), $\kappa(\mathcal{S}) = \mathcal{S}$. $Aut(\mathcal{S})$ acts linearly on the space of holomorphic 1-forms and leaves $\mathcal{S}$ invariant. Therefore $\{f_1, f_2, f_3\}$ span an $Aut(\mathcal{S})$ invariant vector space of dimension 3. This implies that $\{f_1, f_2, f_3\}$ span the solution space of a linear differential equation $H(y) = Wr(y, f_1, f_2, f_3)/Wr(f_1, f_2, f_3) = 0$ having coefficients that are rational functions. This equation has Galois group G_{168} and it has solutions parameterizing the Riemann surface $\mathcal{S}$. Referring to this curve, Klein says in footnote 21 of [9]: *"Sie muss sich auch durch eine lineare Differentialgleichung dritter Ordnung lösen lassen; wie hat man dieselbe aufzustellen?"*. Hurwitz [7] used the above reasoning to find this equation (c.f., [4], p. 232, 390).

References

1. Blichfeld, H. F., *Finite collineation groups*, University of Chicago Press, 1917
2. Cannon, J. J., *An introduction to the group theory language Cayley.* In Computational Group Theory, Atkinson, M.D. (ed), New York: Academic Press 1984
3. Fuchs, L., *Ueber die linearen Differentialgleichungen zweiter Ordnung, welche algebraische Integrale besitzen, zweite Abhandlung*, J. für Math., **85** (1878)
4. Gray, J., *Linear differential equations and group theory from Riemann to Poincaré*, Boston, Basel, Stuttgart: Birkhäuser 1986
5. Grigor'ev, D. Yu., *Complexity of factoring and calculating the GCD of linear ordinary differential operators*, J. Symb. Comp., **10** (1990)
6. Halphen, G., *Sur une équation différentielle linéaire du troisième ordre*, Math. Ann., **24** (1884)
7. Hurwitz, A., *Ueber einege besondere homogene lineare Differentialgleichungen*, Math. Ann., **26** (1886)
8. Kaplansky, I., *Introduction to differential algebra*, Paris: Hermann 1957
9. Klein, F., *Über die Transformation siebenter Ordnung der elliptischen Funktionen* Math. Ann, **14**, (1878/79)
10. Poole, E. G. C., *Introduction to the Theory of Linear Differential Equations*, Dover Publications, Newy York, 1960
11. Schlesinger, L., *Handbuch der Theorie der linearen Differentialgleichungen*, Leipzig: Teubner 1895
12. Schwarz, F., *A Factorization algorithm for linear ordinary differential equations*, Proceedings of the 1989 Symposium on Symbolic and Algebraic Computation, ACM 1989
13. Singer, M. F., *Liouvillian solutions of nth order linear differential equations*, Amer. J. Math., **103** (1981)
14. Singer, M. F., *An outline of differential Galois theory*, in *Computer Algebra and Differential Equations*, Ed. E. Tournier, New York: Academic Press 1990
15. Singer, M. F., Ulmer, F., *Galois group of second and third order Linear differential equations*, Preprint, North Carolina State University (1992)
16. Singer, M. F., Ulmer, F., *Liouvillian and algebraic solutions of second and third order Linear differential equations*, Preprint, North Carolina State University (1992)
17. Tretkoff, C., Tretkoff, M., *Solution of the inverse problem of differential Galois theory in the classical case*, Amer. J. Math., **101** (1979)

Approximating the number of error locations within a constant ratio is $\mathcal{NP}$-complete

Jacques Stern

Laboratoire d'Informatique, Ecole Normale Supérieure

Abstract. Using recent results from complexity theory, we show that, under the assumption $\mathcal{P} \neq \mathcal{NP}$, no polynomial time algorithm can compute an upper bound for the number of error locations of a word y with respect to a code C, which is guaranteed to be within a constant ratio of the true (Hamming) distance of y to C.
Thus the barrier which prevents the design of very general decoding algorithms that would apply to unstructured codes is even more solid than was thought before.
We also give an analogous result for integer lattices.

1 Introduction

The notion of $\mathcal{NP}$-completeness is by now classical and has produced in the last twenty years an elegant body of results(see [5]). Recently, progress in complexity theory using the notion of *interactive proofs* has produced various results, culminating in a paper by Arora, Lund, Motwani, Sudan and Szegedy ([1]). These authors constructed proofs of membership for any $\mathcal{NP}$ language that can be checked probabilistically by querying only a *constant* number of bits in the proof.

This new achievement has startling consequences for the approximation of $\mathcal{NP}$-complete problems. For example, it implies that, unless $\mathcal{P} = \mathcal{NP}$, there exists a constant ϵ such that no polynomial time algorithm can approximate within $(1+\epsilon)$ the maximal number of 3-clauses of a 3-SAT instance which can be simultaneously satisfiable: we say that the underlying optimization problem (called $MAX3SAT$) does not have a polynomial approximation scheme. Actually, $MAX3SAT$ is a member of a larger family of problems, called $MAXSNP$-complete and whose definition is due to Papadimitriou and Yannakakis ([7]). In [1], it is shown that, unless $\mathcal{P} = \mathcal{NP}$, none of these problems can have an approximation scheme. Other results rule out the possibility of having good approximation algorithms for specific problems: for example, the maximal size of a clique in a graph with n vertices cannot be approximated within n^{ϵ}, for some constant ϵ.

The present paper investigates an optimization problem that appears in the theory of error-correcting codes, namely the computation of the (Hamming) distance of a vector y from the nearest codeword of a given code (also called the number of error locations of y). Shortly after the appearence of the notion of

$\mathcal{NP}$-completeness, the following decision problem was shown to be $\mathcal{NP}$-complete (see [3]):

INSTANCE: a linear binary code C of length n, a word y of length n, an integer k.
QUESTION: Is there a codeword x of C such that

$$d(x, y) \leq k$$

where d denotes the Hamming distance.

Thus, unless $\mathcal{P} = \mathcal{NP}$, no polynomial time algorithm can compute the value $\inf\{d(x, y) : x \in C\}$. We now discuss the possibility of obtaining an approximation of this value. Rather than giving general definitions, for which we refer to [7], we restrict ourselves to the specific problem under dicussion:

Definition: Let α, β be real numbers. An algorithm *approximates the number of error locations* with performance (α, β) if, when given as input a linear binary error-correcting code C of length n and a word y, it computes an integer t such that

$$\alpha(\inf\{d(x, y) : x \in C\}) \leq t \leq \beta(\inf\{d(x, y) : x \in C\})$$

Remarks.
1. The use of real numbers in the above is quite artificial and, by weakening the performances of an algorithm, we can restrict ourselves to rational numbers.
2. If an algorithm approximates the number of error locations with performance (α, β), then by outputing $t\alpha^{-1}$, we get another algorithm approximating the number of error locations with performance $(1, \alpha^{-1}\beta)$. Thus we will restrict our attention to performances of type $(1, \rho)$, and write performance ρ to simplify notation.

Our main result reads as follows.

Theorem 1 *Assume $\mathcal{P} \neq \mathcal{NP}$. Then, given any constants α, β, there is no polynomial time algorithm which approximates the number of error locations with performance (α, β).*

As oberved above, it will be enough to rule out approximation algorithms with performance ρ. Our starting point will be the following result which we quote from reference [1]

Theorem 2 *Let L be a language in $\mathcal{NP}$. There exists a polynomial time algorithm A and a rational constant γ, $0 < \gamma < 1$ such that, given an input x, algorithm A produces an instance Φ_x of 3-SAT consisting of polynomially many clauses satisfying the following*

1. *if $x \in L$ then Φ_x is satisfiable.*
2. *if $x \notin L$, then at most a fraction $1 - \gamma$ of the clauses in Φ_x are satisfiable simultaneously.*

The proof of theorem 1 will be based on a polynomial time reduction from MAX3SAT: taking as an input a set of clauses Φ, this reduction outputs the generating matrix of a binary code C_Φ and an integer K_Φ such that, denoting by u the word whose bits are all one, the following holds:

1. if Φ_x is satisfiable, then the distance of u to C_Φ is K_Φ.
2. if at most a fraction $1-\gamma$ of the clauses in Φ_x are satisfiable simultaneously, then the distance of u to C_Φ is at least λK_Φ, where λ is a constant depending only on γ.

Such a reduction clearly rules out the existence of a polynomial time algorithm approximating the number of error locations with performance $(1, \lambda-\epsilon)$, for any fix ϵ such that $\lambda - \epsilon > 0$: by combining the reduction of theorem 2, the reduction just defined and finally applying such a polynomial algorithm, we could solve any problem in $\mathcal{NP}$ in polynomial time.

Thus if we can build an infinite sequence of reductions whose corresponding constants form a sequence λ_q such that $\lim_{q\to\infty} \lambda_q(\gamma) = \infty$, our result will follow.

2 A simplified result.

We first prove a weaker version of theorem 1

Theorem 3 *Assume $\mathcal{P} \neq \mathcal{NP}$. Then, there exists a constant ρ such that no polynomial time algorithm can approximate the number of error locations with performance ρ.*

In other terms, there is no polynomial time approximation scheme for the number of error locations. As was pointed out by one of the referees, this result was somehow implicit in the literature: a paper of Bruck and Naor ([4]) mentions that the set of cuts in a graph form a linear space (when viewed as vectors of length the number of edges representing the characteristic function of the set of edges crossing the cut). This yields a polynomial time reduction of the problem *MAXCUT* to the maximum likelihood decoding problem. Furthermore, this reduction preserves aprroximations. Hence, we can apply the results from [1].

We give a different reduction which uses, as a building block, of the $(7,3,4)$ simplex code whose generating matrix is

$$\begin{pmatrix} 0\,1\,1\,0\,1\,1\,0 \\ 1\,0\,1\,1\,0\,1\,0 \\ 0\,0\,0\,1\,1\,1\,1 \end{pmatrix}$$

As is well known, all non-zero codewords of this code have weight 4.

We start from an instance of 3-SAT Φ consisting of m clauses $c_1, \cdots, c_m$, each having three literals taken from $v_1, \cdots, v_r; \overline{v_1} \cdots \overline{v_r}$. The generating matrix of the code C_Φ has the following form

$$M = \left(\begin{array}{|c|c|c|c|c|c|} B_1 & \cdots & B_m & A_1 & \cdots & A_r \end{array} \right)$$

Matrices A_i and B_j have $2r$ rows, which we will number by the literals $v_1, \overline{v_1}, \cdots, v_r, \overline{v_r}$.

Each of the matrices A_i corresponds to one propositional variable v_i and has $8m$ columns. All its rows are zero except for the two rows corresponding to $v_i, \overline{v_i}$ which consist of ones.

Each of the matrices B_j corresponds to one clause of Φ say c_j and has seven columns. All its rows are zero except the three rows labelled by the three literals which appear in c_j. These three rows are filled in such a way that by discarding the zero rows one obtains a copy of the generating matrix of the simplex code described above.

Now, we observe that M generates a $(7m + 8mn, 2r)$ code. Any codeword is produced by a linear combination of the rows, where the coefficients are obtained by assigning values $\tau(v_i), \tau(\overline{v_i})$ to the literals. We denote by $w(\tau)$ the codeword built from τ. We say that an assignment is *legal* if $\tau(v_i) = 1 - \tau(\overline{v_i})$ and we make the following easy remarks (recall that u is the all-ones word).

Remarks

1. For any codeword $w(\tau)$ built from an illegal assignment τ the Hamming weight of $u \oplus w(\tau)$ is at least $8m$.
2. For a codeword $w(\tau)$ built from a legal assignment satisfying the set of clauses Φ, the Hamming weight of $u \oplus w(\tau)$ is exactly $3m$.
3. For a codeword $w(\tau)$ built from a legal assignment satisfying at most a proportion $(1 - \gamma)$ of the set of clauses Φ, the Hamming weight of $u \oplus w(\tau)$ is at least $(3 + 4\gamma)m$ and at most $7m$.

The proofs of the first remark is easy: an illegal τ uses the two non-zero rows of one of the matrices A_i and therefore cannot cancel the $8m$ corresponding ones in u. As for other remarks, they follow from the fact that legal τ's are proper truth assignments and cancel 4 ones of u at positions corresponding to a matrix B_j, where τ satisfies c_j and do not cancel anything at other positions because the non-zero rows of B_j are not hit in this case.

From the remarks, it follows that we have built a reduction of the kind described at the end of section 1 with $\lambda(\gamma) = \frac{3+4\gamma}{3}$. As observed in section 1, this rules out approximation algorithms with performance $1 + \frac{4}{3}\gamma$ and proves theorem 3.

3 Proof of the main theorem.

In order to prove theorem 1, we will build a sequence of reductions similar to the one that was described at the end of the previous section but with $\lambda(\gamma)$ being replaced by $\lambda_q(\gamma) = (1 + \frac{4}{3}\gamma)^q$. As mentioned in section 1, this yields the result.

Actually, our construction is recursive and we will simply explain how to go from the case $q = 1$ (covered by the previous section) to $q = 2$. The general inductive step is similar and will be only sketched very briefly. We start from the generating matrix M of the code constructed in section 2. M is a $(2r, p)$ matrix with $p = 7m + 8mr$, following previous notations. We let $\tilde{M}$ be the $(2r, p^2)$ matrix obtained from M by replacing each one by a block of p ones and each zero by a block of p zeros (this is a concatenation construction). The generating matrix of the code C^2_Φ has the form

$$M_2 = \left(\begin{array}{c} \tilde{M} \\ \hline M' \end{array} \right)$$

where M' is built as shown on the diagram displayed on the next page. In this diagram, each matrix P_k has $2r$ rows and p columns and is a copy of M. Thus, we have built a matrix M_2 with $2r(p+1)$ rows and p^2 columns through a construction that can clearly be achieved in polynomial time.

We now make the following remarks where u stands for the all-ones word of suitable length.

Remarks

1. If Φ (the given set of clauses) is satisfiable, then there is a codeword w built as a linear combination of the first $2n$ rows of M_2 such that $u \oplus w$ has exactly $3m$ non-zero blocks. In each of these blocks, ones can be further cancelled by using the corresponding matrix P_k so as to leave exactly $3m$ ones. Thus, the distance of u to C^2_Φ is exactly $(3m)^2$.
2. If at most a proportion $(1 - \gamma)$ of the set of clauses Φ can be satisfied simultaneously, then any codeword w built as a linear combination of the first $2n$ rows of M_2 will leave at least $(3 + 4\gamma)m$ non-zero blocks in $u \oplus w$. In

each of these blocks, use of the corresponding matrix will cancel more ones but will leave at least $(3+4\gamma)m$ ones. Therefore the distance of u to C_{Φ}^{2} is at least $((3+4\gamma)m)^2$.

$$M' = \begin{pmatrix} P_1 & \cdots & 0 & 0 \\ 0 & P_2 & \cdots & 0 \\ \cdots & \cdots & \cdots & \cdots \\ 0 & 0 & \cdots & P_p \end{pmatrix}$$

Thus, as announced, we have built a reduction with $\lambda_2(\gamma) = (1+\frac{4}{3}\gamma)^2$. The same construction can be repeated so as to get $\lambda_q(\gamma) = (1+\frac{4}{3}\gamma)^q$. The construction involves replacing the generating matrix M_q obtained at step q by the matrix $\tilde{M}_q$ obtained from M_q by replacing each one by a block of p ones and each zero by a block of p zeros. The generating matrix of the code C_{Φ}^{q+1} is constructed by adding below $\tilde{M}_q$ a diagonal matrix, each element of the diagonal being a copy of the initial matrix M, defined in section 2. This proves theorem 1.

4 Approximating the closest lattice vector.

A *lattice* is an additive subgroup of Z^n. Lattices have received a lot of attention recently because of their use in algorithmic and cryptographic problems. From this point of view, the main achievement is the celebrated LLL algorithm (see [6]), which produces in polynomial time a so-called *reduced basis*. Using such a basis, it is possible to approximate the length of the shortest non-zero vector of a lattice with performance 2^{n-1}, where n is the dimension of the lattice (the reader will easily supply a definition of performance modelled after those in section 1). It is also possible to approximate the distance of a given point to the closest lattice point but the performance is worse (see [2]).

It turns out that a lattice can be attached to any linear binary code C of length n by letting

$$L(C) = \{x \in Z^n : x \bmod 2 \in C\}$$

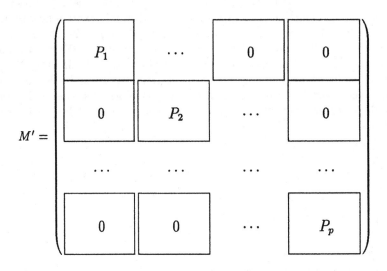

Now, it is easily seen that the distance of the all-ones vector u to $L(C)$ is precisely the square root of the (Hamming) distance of u to L. From this, we get the following:

Theorem 4 *Assume $\mathcal{P} \neq \mathcal{NP}$. Then, there is no polynomial time algorithm which, given a lattice L and a vector y approximates the distance of y to the closest point of L within a constant ratio.*

Acknowledgements. The author wishes to thank Adi Shamir for drawing his attention to reference [1], for going through this paper with him and for discussions on the present work. The author is also grateful to the anonymous referees for their positive suggestions.

References

1. S. Arora, C. Lund, R. Motwani, M. Sudan and M. Szegedy. Proof verification and intractability of approximation problems. *Proc. 31st IEEE Symp. on Foundations of Computer Science* (1992), to appear.
2. L. Babai. On Lovàsz lattice reduction and the nearest point problem, *Proc STACS 85*, Lecture Notes in Computer Science **182**, Springer, Berlin, 13-20.
3. E. R. Berlekamp, R. J. Mc Eliece and H. C. A. Van Tilborg. On the inherent intractability of certain coding problems, *IEEE Trans. Inform. Theory,* (1978) 384-386.
4. J. Bruck and M. Naor. The hardness of decoding linear codes with preprocessing, *IEEE Trans. Inform. Theory,* (1990), 381-385.
5. M. R. Garey, D. S. Johnson. *Computers and Intractability: a Guide to the Theory of NP-completeness*, Freeman, San Francisco (1979).
6. A. K. Lenstra, H. W. Lenstra and L. Lovàsz. Factoring polynomials with rational coefficients. *Math. Annalen* 261 (1982) 515-534.
7. C. H. Papadimitriou and M. Yannakakis. Optimization, approximation and complexity classes. *J. Comp. Syst. Sc.* **43** (1990), 425-440.

Two Chosen-Plaintext Attacks on the Li-Wang Joint Authentication and Encryption Scheme

Johan van Tilburg*

Abstract. In [LW91], Li and Wang proposed a joint authentication and encryption scheme based on algebraic coding theory. They claimed that their scheme is as secure as the Rao-Nam scheme [RN89]. However, in contrast with their claim, it will be shown that this joint authentication and encryption scheme appears to be less secure. In this paper two inherently different chosen-plaintext attacks are presented.
The first attack is based on the linearity of the bit selection function, and obtains a $k \times n$ matrix equivalent to the encryption matrix in $O(k)$ encryptions. If the set of error vectors $\mathcal{Z}$ is randomly chosen, about $|\mathcal{Z}|$ encryptions are necessary to obtain a corresponding set of error vectors. With knowledge of only r error vectors one can always encrypt and decrypt $r2^{2k-n}$ messages.
The second attack makes use of the non-linearity of the error function, and always obtains $2k - n$ rows of the encryption matrix after $O(k^2)$ encryptions. Hereafter $|\mathcal{Z}|$ encryptions are required to create a cryptosystem equivalent to the Li-Wang scheme.
Some extensions of the scheme are discussed, and a general question raised by Brickell and Odlyzko [BO88] related to the Rao-Nam scheme is settled in a negative way.

1 Introduction

In [LW91], Li and Wang introduced a joint authentication and encryption scheme based on algebraic coding theory. They claim that their secret-key cryptosystem, which we call LW-scheme, can be used to authenticate and encrypt messages simultaneously. Under the assumption that the LW-scheme is secure, this means that two users sharing the same secret key are able to exchange secret messages over an insecure communication channel and are able to detect if the received ciphertext was tampered with. We refer to [Simm88] and [Mass88] for a detailed treatment of this subject.

In this paper, the LW-scheme is described for code rates $R \geq 1/2$. However, the results obtained are also valid for the less interesting case $R < 1/2$. For the original description we refer to [LW91].

The LW-scheme is based on a secret-key cryptosystem, called RN-scheme, proposed by Rao and Nam [RN89]. The RN-scheme is a secret-key variant of the McEliece public-key cryptosystem [McEl78]. All these schemes are based on

* PTT Research, Dept. of Applied Mathematics and Signal Processing, P.O. Box 421, 2260 AK Leidschendam, the Netherlands.

algebraic coding theory and allow high-speed implementations. Without details, the RN-scheme can be summarized as follows. A message $\underline{x}$ is encrypted by randomly selecting an error vector $\underline{z}$ from a predefined set of error vectors $\mathcal{Z}$ and forming the ciphertext $\underline{y} = \underline{x}E + \underline{z}$, where E is the encryption matrix. As a consequence of the special way the set $\mathcal{Z}$ was generated, unique decoding is possible. For more details, we refer the reader to [RN89].

In [Mass88], Massey remarks that most cryptosystems in use today are intended by their designers to be secure against at least a chosen-plaintext attack, even if it is hoped that the enemy cryptanalyst will never have the opportunity to mount more than a cipher-text-only attack. In [LW91], Li and Wang conclude that their secret-key cryptosystem is secure under a chosen-plaintext attack by a work factor of about $|\mathcal{Z}|^k$, where $|\mathcal{Z}|$ denotes the cardinality of the set of error vectors. It should be noted that the set $\mathcal{Z}$ is part of the secret key. In the RN-scheme each error vector can be used to encrypt any given message, but in the LW-scheme each message corresponds to only one error vector. Besides, each error vector corresponds to several messages. It will be shown in Sections 4 and 5 that this is a serious weakness of the LW-scheme and enables us to mount two attacks with a work factor of about $|\mathcal{Z}|$ chosen-plaintexts. Because the RN-scheme does not have this weakness, the proposed attacks do not apply to the RN-scheme.

In Section 6, some extensions of the LW-scheme will be discussed. Moreover, we will settle a general question raised by Brickell and Odlyzko in their survey article [BO88]. Their question was related to the security of the RN-scheme: Could the security of the RN-scheme be improved if the scheme is slightly modified by using a pseudo-random function f, and letting $\underline{z} = f(\underline{x})$ so that there is only one encryption for each message $\underline{x}$?

Li and Wang suggest to use the set $\mathcal{Z}$, if not used for authentication, for purpose of encryption and improving the code rate of their scheme. And in their conclusion, they state that a similar modification of the McEliece scheme [McEl78] is possible to get a new public-key cryptosystem for purpose of both encryption and authentication. These two situations will be discussed in Section 6.

Finally, conclusions can be found in Section 7.

2 Notations and Definitions

In this section notations and definitions are introduced, necessary to describe the LW-scheme as presented in [LW91] more strictly.

Let C be a linear $[n, k]$ code over $\mathbb{F}_2$ with code length n and dimension k. Let G be a $k \times n$ generator matrix of the code C and let H be an $(n-k) \times n$ parity check matrix, so $GH^T = O_{k,(n-k)}$. The code C has exactly 2^k codewords, all are linear combinations of the rows of G. Let $\underline{y} \in \mathbb{F}_2^n$. The set $\underline{y}+C = \{\underline{y}+\underline{c} \mid \underline{c} \in C\}$ is called a coset of C. Note that two vectors $\underline{y}$ and $\underline{z}$ are in the same coset if and only if $\underline{y} - \underline{z} \in C$. Each coset contains 2^k vectors, and there are exactly 2^{n-k} different cosets. The vector $\underline{s} = \underline{y}H^T$ is called the syndrome of $\underline{y}$ with respect to the parity check matrix H of C. Two vectors $\underline{x}$ and $\underline{y}$ are in the same coset if and only if their syndromes are equal. The Hamming weight $w(\underline{y})$ gives the number of nonzero coordinates of $\underline{y}$.

Definition 2.1 *Let C be a linear $[n,k]$-code over $\mathbb{F}_2$ with parity check matrix H. The coset $\mathcal{Z}_{\underline{s}}$ with syndrome $\underline{s} \in \mathbb{F}_2^{n-k}$ is defined as*

$$\mathcal{Z}_{\underline{s}} = \{\underline{z} \in \mathbb{F}_2^n \mid \underline{z}H^T = \underline{s}\}.$$

A set of error vectors $\mathcal{Z}$ is obtained by randomly selecting one vector from each coset.

Definition 2.2 (Set of error vectors) *The set of error vectors $\mathcal{Z}$ is a 'random' subset of $\mathbb{F}_2^n$ such that*

$$|\mathcal{Z} \cap \mathcal{Z}_{\underline{s}}| = 1 \quad \textit{for all} \quad \underline{s} \in \mathbb{F}_2^{n-k}.$$

In [LW91], the set of error vectors satisfy also the weight constraint: $w(\underline{z}) \approx n/2$. As has been shown in [MT91b] this restriction is risky. Besides, the fraction f_w of vectors in $\mathbb{F}_2^n$ of weight smaller than $w < n/2$ is upper bounded by

$$f_w = 2^{-n} \sum_{i=0}^{w-1} \binom{n}{i} < 2^{-n(1-H(\frac{w}{n}))}.$$

For $n = 256$ about 99% of the vectors in $\mathbb{F}_2^{256}$ has weight in the interval $[128-15, 128+15]$. Hence, the probability that one randomly selects a set of error vectors with low Hamming weight is negligible.

Next, define a selection function as follows:

Definition 2.3 (Selection function) *The function $\sigma_A(\underline{x})$ selects an $(n-k)$-bit subvector from $\underline{x} \in \mathbb{F}_2^k$, i.e.,*

$$\sigma_A(\underline{x}) = (x_{a_1}, \ldots, x_{a_{n-k}}),$$

where $A = \{a_1, \ldots, a_{n-k}\}$ is a subset of the set $\{1, \ldots, k\}$.

In the LW-scheme, the set A is part of the secret key. The function $\sigma_A(\underline{x})$ extracts $|A| = n-k$ bits from the k-bit vector $\underline{x}$.

Introduce an error function $f_A: \mathbb{F}_2^{n-k} \to \mathbb{F}_2^n$, which assigns to each message $\underline{x}$ one error vector $\underline{z}$ as follows:

Definition 2.4 (Error function) *The error function $f_A(\underline{x})$ selects a vector $\underline{z}$ from the set of error vectors $\mathcal{Z}$:*

$$f_A(\underline{x}) \in \{\mathcal{Z} \cap \mathcal{Z}_{\sigma_A(\underline{x})}\}.$$

It should be observed that two different messages might be mapped on the same error vector as $\underline{z}$ does not uniquely determine $\underline{x}$. From the above definitions it follows that $f_A(\underline{x})H^T = \underline{z}H^T = \sigma_A(\underline{x})$.

3 The Li-Wang Scheme

In this section the LW-scheme as proposed in [LW91] will be described in a different but equivalent manner.

Let $\underline{x} \in \mathbb{F}_2^k$ be the message to be encrypted into a ciphertext $\underline{y} \in \mathbb{F}_2^n$. The encryption function in the LW-scheme can be defined as follows

$$\underline{y} = (\underline{x}SG + f_A(\underline{x}))P,$$

where S is a $k \times k$ random non-singular matrix over $\mathbb{F}_2$ and P a random $n \times n$ permutation matrix. The LW-scheme is a secret-key cryptosystem. Hence the S and P matrices do not conceal a code or a decoding algorithm. It can be shown that the two matrices are redundant and can be eliminated:

$$\underline{y} = (\underline{x}SG + f_A(\underline{x}))P = \underline{x}SGP + f_A(\underline{x})P = \underline{x}E + g_A(\underline{x}),$$

where $E = SGP$ is a full rank $k \times n$ encryption matrix over $\mathbb{F}_2$ and $g_A(\underline{x}) = f_A(\underline{x})P$ is the new error function. To avoid the introduction of a new notation and new Definitions 2.1 and 2.4, for E and $g_A(\underline{x})$, we will ignore the S and P matrices without loss of generality.

The following secret-key cryptosystem, equivalent to the LW-scheme, will be used in the cryptanalysis of the LW-scheme. A message $\underline{x} \in \mathbb{F}_2^k$ is *encrypted* into a ciphertext $\underline{y} \in \mathbb{F}_2^n$ by

$$\underline{y} = \underline{x}G + f_A(\underline{x}).$$

A ciphertext $\underline{y} \in \mathbb{F}_2^n$ is *decrypted* into a message $\underline{x} \in \mathbb{F}_2^k$ as follows.

- Calculate $\underline{y}H^T = \sigma_A(\underline{x})$, which uniquely determines $\underline{z}$;
- Subtract $\underline{z}$ from $\underline{y}$ to obtain $\underline{x}G$;
- Recover $\underline{x}$ as follows

$$(\underline{y} - \underline{z})G^{-R} = \underline{x}GG^{-R} = \underline{x},$$

where G^{-R} is a right inverse of G, i.e. $GG^{-R} = I_k$, the $k \times k$ identity matrix.

The matrix G (H) and the error function $f_A(\underline{x})$ form the secret key.

4 Chosen-Plaintext Attack A

In this section an efficient chosen-plaintext attack on the LW-scheme will be described. For this attack to be successful with overwhelming probability, $O(k)$ chosen-plaintexts are required to obtain an $k \times n$ matrix $\widehat{G}$ equivalent to the encryption matrix G. Depending on the way the set of error vectors $\mathcal{Z}$ was generated, about $|\mathcal{Z}|$ chosen-plaintexts are required to construct a corresponding error function $\widehat{f_A(\underline{x})}$. The matrix $\widehat{G}$ and the error function $\widehat{f_A(\underline{x})}$ together define a secret-key cryptosystem equivalent to the LW-scheme.

The attack is based on the following theorem, which states that the sum of three ciphertexts obtained from two randomly chosen messages yields a codeword.

Theorem 4.1 *Let G be a generator matrix for code C. If $\underline{y}_i, \underline{y}_j, \underline{y}_{ij} \in \mathbb{F}_2^n$ are ciphertexts satisfying*

$$\underline{y}_i = \underline{x}_i G + f_A(\underline{x}_i), \quad \underline{y}_j = \underline{x}_j G + f_A(\underline{x}_j) \quad \textit{and} \quad \underline{y}_{ij} = (\underline{x}_i + \underline{x}_j)G + f_A(\underline{x}_i + \underline{x}_j),$$

where $\underline{x}_i, \underline{x}_j \in \mathbb{F}_2^k$, then it holds that

$$\underline{y}_i + \underline{y}_j + \underline{y}_{ij} \in C.$$

Proof: Observe that $(\underline{y}_i + \underline{y}_j + \underline{y}_{ij})H^T = f_A(\underline{x}_i)H^T + f_A(\underline{x}_j)H^T + f_A(\underline{x}_i + \underline{x}_j)H^T = \sigma_A(\underline{x}_i) + \sigma_A(\underline{x}_j) + \sigma_A(\underline{x}_i + \underline{x}_j) = \underline{0}$, as σ_A is a linear function. A vector $\underline{y}$ is in C if and only if $\underline{y}H^T = \underline{0}$, hence $\underline{y}_i + \underline{y}_j + \underline{y}_{ij}$ is a codeword. □

To obtain a generator matrix $\widehat{G}$ for C any set of k linearly independent codewords from C can be used. If Algorithm 4.2 terminates then a $k \times n$ matrix $\widehat{G}$ that generates the same code C is returned.

Algorithm 4.2

1. *Select at random two different messages $\underline{x}_i$ and $\underline{x}_j$ and obtain the ciphertexts: $\underline{y}_i, \underline{y}_j$ and $\underline{y}_{ij}$.*
2. *Compute $\underline{c} := \underline{y}_i + \underline{y}_j + \underline{y}_{ij}$.*
3. *Repeat Steps 1 and 2 until k linearly independent codewords $\underline{c}$ are obtained, say, $\underline{c}_1, \ldots, \underline{c}_k$.*
4. *Let $\widehat{G} := (\underline{c}_1^T, \ldots, \underline{c}_k^T)^T$.*

The next lemma states that Algorithm 4.2 obtains with overwhelming probability a generator matrix.

Lemma 4.3 *Algorithm 4.2 terminates with overwhelming probability in $O(k)$ iterations.*

Before we prove Lemma 4.3 we will pay attention to the idea behind the proof. We need to estimate how many iterations in Algorithm 4.2 are necessary to obtain k linearly independent codewords.

Consider k randomly selected vectors $\underline{x}_i \in \mathbb{F}_2^k$ and obtain k corresponding vectors $\underline{c}_i \in \mathbb{F}_2^k$ such that $\underline{x}_i E = \underline{c}_i$, $1 \leq i \leq k$. Let $X = (\underline{x}_1^T, \ldots, \underline{x}_k^T)^T$ and $C = (\underline{c}_1^T, \ldots, \underline{c}_k^T)^T$. If matrix X is non-singular, then the full rank $k \times n$ matrix E is recovered from $E = X^{-1}C$.

The probability q_k that X has full rank given k randomly chosen vectors $\underline{x}_i$ is equal to $\prod_{i=0}^{k-1}(1 - 2^{i-k})$ [MS77]. Clearly, C has full rank if and only if X has full rank. Thus if k codewords are randomly selected, then with probability $q_k \approx 0.289$ ($k > 10$) a linearly independent set of k codewords is obtained. This probability can be improved by picking more than k vectors. (See also [KZYR89].) Clearly, the expected number of vectors to choose before their rank equals k is less than $q_k^{-1} \times k = O(k)$.

Now we return to our problem and prove Lemma 4.3.

Proof: In Step 1, two different messages $\underline{x}_i$ and $\underline{x}_j$ are randomly selected. The set $\mathcal{Z}$ is randomly chosen, therefore from Theorem 4.1 it follows that Step 2 yields a random codeword. If Steps 1 and 2 are repeated $O(k)$ times, then with overwhelming probability ($|\mathcal{Z}| \gg k$) k linearly independent code words are obtained. As a result Algorithm 4.2 terminates. □

Costs: In Step 1 exactly 3 encryptions are required, so for Algorithm 4.2 to be successful with overwhelming probability we need $3 \times O(k) = O(k)$ encryptions. Step 2 requires $O(kn^2)$ bit operations and $O(kn)$ bits of memory.

In the literature no chosen-plaintext attack for the RN-scheme is known that recovers a matrix $\widehat{G}$ equivalent to the encryption matrix G within $O(k)$ chosen encryptions. As stated in [RN89] this is really the hard part of the RN-scheme. In the RN-scheme we are not able to choose three encryptions that will certainly yield a codeword, hence this attack will not be successful.

To recover the secret set A proceed as follows. Solving $\widehat{G}\widehat{H}^T = O_{k,(n-k)}$ yields a parity check matrix $\widehat{H}$ for C, hence $G\widehat{H}^T = O_{k,(n-k)}$. Let $\underline{u}_i \in \mathbb{F}_2^k$ be the i-the unit vector. The following algorithm recovers not only the set A but also recovers $2k - n$ rows of the encryption matrix G.

Algorithm 4.4

1. *Let $\tilde{A} = \emptyset$. Select at random a message $\underline{x}$, obtain the ciphertext $\underline{y} = \underline{x}G + f_A(\underline{x})$ and compute $\widehat{\underline{s}} := \underline{y}\widehat{H}^T$.*
2. *For $1 \leq i \leq k$, obtain the ciphertext $\underline{y}_i = (\underline{x}+\underline{u}_i)G + f_A(\underline{x}+\underline{u}_i)$ and compute $\widehat{\underline{s}}_i := \underline{y}_i\widehat{H}^T$. If $\widehat{\underline{s}}_i = \underline{s}$ then $\underline{g}_i := \underline{y} + \underline{y}_i$ else $\tilde{A} := \tilde{A} \cup \{i\}$.*

Note that if $\widehat{\underline{s}}_i = \underline{s}$ then $f_A(\underline{x}) = f_A(\underline{x} + \underline{u}_i)$ and hence $\underline{y} + \underline{y}_i = \underline{u}_i G = \underline{g}_i$, the i-th row of the encryption matrix G.

Update matrix $\widehat{G}$ such that the rows $\widehat{\underline{g}}_i$ with $i \in \tilde{A}$ are equal to the rows $\underline{g}_i$ obtained in Algorithm 4.4. Next, with knowledge of set $\tilde{A}$ it is straightforward to obtain the corresponding set of error vectors within at most $|\mathcal{Z}|$ chosen-plaintexts:

$$\underline{y} = \underline{x}G + f_A(\underline{x}), \quad \text{so that} \quad \widehat{\underline{z}} = \widehat{f_A(\underline{x})} = \underline{y} - \underline{x}\widehat{G}.$$

Note that with only r error vectors, one can always encrypt and decrypt $r2^{2k-n}$ messages.

5 Chosen-Plaintext Attack B

In this section an efficient chosen-plaintext attack is presented. This attack is not based on the linearity of the σ_A-function: $\sigma_A(\underline{x}_i) + \sigma_A(\underline{x}_j) + \sigma_A(\underline{x}_i + \underline{x}_j) = \underline{0}$, but on the non-linearity of the error function: $f_A(\underline{x}_i + \underline{x}_j) \neq f_A(\underline{x}_i) + f_A(\underline{x}_j)$.

An important feature of this attack is that $2k-n$ rows of the $k \times n$ matrix G can be recovered and verified independently. The attack always obtains $2k-n$ rows of the encryption matrix after $O(k^2)$ chosen-plaintexts. Hereafter $|\mathcal{Z}|$ chosen-plaintexts are required to create a cryptosystem equivalent to the LW-scheme. For reasons of clarity, the first part of the attack is presented with unit vectors. However, this part can be extended to arbitrary messages.

Lemma 5.1 *Let $\underline{u}_i, \underline{u}_j \in \mathbb{F}_2^k$ indicate the i-th and j-th unit vectors and let $\underline{g}_j$ be the j-th row of the full rank $k \times n$ matrix G. If $\underline{y}_{ij}$ and $\underline{y}_i$ are ciphertexts satisfying $\underline{y}_{ij} = (\underline{u}_i + \underline{u}_j)G + f_A(\underline{u}_i + \underline{u}_j)$ and $\underline{y}_i = \underline{u}_i G + f_A(\underline{u}_i)$, then it holds that*

$$\forall_{i,1 \leq i \leq k}\ [\underline{y}_{ij} + \underline{y}_i = \underline{g}_j] \quad \text{if and only if} \quad \sigma_A(\underline{u}_j) = \underline{0}.$$

There are exactly $2k-n$ vectors $\underline{u}_j$ that satisfy $\sigma_A(\underline{u}_j) = \underline{0}$.

Proof: Compute $\underline{y}_{ij} + \underline{y}_i = \underline{u}_j G + f_A(\underline{u}_i + \underline{u}_j) + f_A(\underline{u}_i)$. Because for all i $(1 \leq i \leq k)$: $f_A(\underline{u}_i + \underline{u}_j) + f_A(\underline{u}_i) = \underline{0}$ if and only if $\sigma_A(\underline{u}_j) = \underline{0}$, we have $\forall_{i,1 \leq i \leq k}$ $[\underline{y}_{ij} + \underline{y}_i = \underline{u}_j G = \underline{g}_j]$ if and only if $\sigma_A(\underline{u}_j) = \underline{0}$.

The function $\sigma_A(\underline{u}_j)$ results in the $(n-k)$-bit all zero vector if and only if the selected $n-k$ bits from $\underline{u}_j$ are all equal to zero. Therefore $k-(n-k) = 2k-n$ unit vectors $\underline{u}_j$ exist that satisfy $\sigma_A(\underline{u}_j) = \underline{0}$. □

The next theorem provides an efficient method to recover $2k-n$ rows of the encryption matrix G independently.

Theorem 5.2 *Let $\underline{u}_i, \underline{u}_j \in \mathbb{F}_2^k$ be unit vectors and let $f_A: \mathbb{F}_2^k \to \mathbb{F}_2^n$ be a secret non-linear function. Let $\underline{y}_{ij}$ and $\underline{y}_i$ be ciphertexts satisfying $\underline{y}_{ij} = (\underline{u}_i + \underline{u}_j)G + f_A(\underline{u}_i + \underline{u}_j)$ and $\underline{y}_i = \underline{u}_i G + f_A(\underline{u}_i)$. Let $\widehat{\underline{g}}_j = \underline{y}_{1j} + \underline{y}_1$. A sufficient condition to recover the j-th row $\underline{g}_j$ of the $k \times n$ encryption matrix G is:*

$$\text{If} \quad \forall_{i,2 \leq i \leq k} \quad [\underline{y}_{ij} + \underline{y}_i = \widehat{\underline{g}}_j] \quad \text{then} \quad \widehat{\underline{g}}_j = \underline{g}_j. \tag{1}$$

There are exactly $2k-n$ rows $\underline{g}_j$ that can be obtained this way.

Proof: If $\sigma_A(\underline{u}_j) = \underline{0}$, then for all unit vectors $\underline{u}_i$, $1 \leq i \leq k$, it holds (Lemma 5.1) that $\underline{y}_i + \underline{y}_{ij} = \underline{g}_j + f_A(\underline{u}_i) + f_A(\underline{u}_i + \underline{u}_j) = \underline{g}_j$. Hence $\widehat{\underline{g}}_j = \underline{g}_j$.

Suppose $\underline{u}_j$ is a unit vector such that $\sigma_A(\underline{u}_j) \neq \underline{0}$. According to Lemma 5.1, exactly $2k-n$ vectors $\underline{u}_i$ can be selected such that $\underline{y}_i + \underline{y}_{ij} = \underline{g}_j + f_A(\underline{0}) + f_A(\underline{u}_j) = \widehat{\underline{g}}_j$. All other unit vectors, say $\underline{u}_s$, satisfy $\underline{y}_s + \underline{y}_{sj} = \underline{g}_j + f_A(\underline{u}_s) + f_A(\underline{u}_s + \underline{u}_j)$. Because f_A is a non-linear function $f_A(\underline{0}) + f_A(\underline{u}_j)$ cannot match $f_A(\underline{u}_s) + f_A(\underline{u}_s + \underline{u}_j)$ for all values of s.

In accordance with Lemma 5.1, exactly $2k-n$ unit vectors have $\sigma_A(\underline{u}_j) = \underline{0}$. So exactly $2k-n$ rows of G can be obtained this way. □

Theorem 5.2 states that if a row estimate $\widehat{\underline{g}}_j$ satisfies (1) then $\widehat{\underline{g}}_j$ is the j-th row of the encryption matrix G. Algorithm 5.3 obtains k row estimates and uses

(1) as verification condition. As soon as Algorithm 5.3 terminates, the $2k-n$ rows for which $\hat{\underline{g}}_j = \underline{g}_j$ are returned.

Algorithm 5.3

1. *Let* $r := 1$ *and* $j := 0$.
2. *Let* $j := j+1$ *and* $\hat{\underline{g}}_j := \underline{y}_{1j} + \underline{y}_1$.
   If $\forall_{i, 2 \leq i \leq k}\ \ \underline{y}_{ij} + \underline{y}_i = \hat{\underline{g}}_j$ *then let* $b_r := j$, $\underline{g}_{b_r} := \hat{\underline{g}}_j$ *and* $r := r+1$.
3. *Repeat Step 2 until the* $2k-n$ *different vectors* $\underline{g}_{b_r}$ *of* G *are recovered.*
4. *Return the* $(2k-n) \times n$ *matrix* $\tilde{G} := (\underline{g}^T_{b_1}, \ldots, \underline{g}^T_{b_{2k-n}})^T$ *and the corresponding set* $B := \{b_1, \ldots, b_{2k-n}\}$.

Note that $|A|+|B| = k$ and $A \cap B = \emptyset$. As stated in Lemma 5.1 and Theorem 5.2, there are exactly $2k-n$ rows of the full rank $k \times n$ matrix G that satisfy the verification condition (1). Therefore at most all k rows estimates $\hat{\underline{g}}_j$ have to be considered. The following lemma is now obvious.

Lemma 5.4 *Algorithm 5.3 terminates after at most k iterations and returns a* $(2k-n) \times n$ *sub-matrix* $\tilde{G}$ *of the generator matrix* G.

Costs: In the worst case all k row estimates $\hat{\underline{g}}_j$ have to be considered, i.e., Step 2 is repeated k times. Hence in the worst case $O(k^2)$ chosen-plaintexts, $O(k^2 n)$ bit operations and $O(kn)$ bits of memory are required.

To obtain encryption and decryption procedures equivalent to the LW-scheme, proceed as follows. Split $\underline{x} \in \mathbb{F}_2^k$ into an $(n-k)$-bit subvector $\underline{x}^{(A)} = \sigma_A(\underline{x}) = (x_{a_1}, \ldots, x_{a_{n-k}})$ and a $(2k-n)$-bit subvector $\underline{x}^{(B)} = \sigma_B(\underline{x}) = (x_{b_1}, \ldots, x_{b_{2k-n}})$ such that the concatenation $\underline{x}^{(A)} || \underline{x}^{(B)}$ is a permutation of the k-bit vector $\underline{x}$. Without loss of generality, assume that the set A is the same as in Definition 2.3.

Use a chosen-plaintext attack to obtain all $|\mathcal{Z}|$ ciphertexts $\underline{y} = (\underline{x}^{(A)} || \underline{0})G + f_A(\underline{x}^{(A)} || \underline{0})$, and let

$$\mathcal{T} = \{(\underline{x}^{(A)}, \underline{y}) \mid \underline{y} = \underline{x}G + f_A(\underline{x}),\ \underline{x}^{(B)} = \underline{0} \text{ and } \underline{x}^{(A)} \in \mathbb{F}_2^{n-k}\}.$$

Note that $\underline{x}^{(A)}$ uniquely determines $f_A(\underline{x})$, and given $\underline{x}^{(B)} = \underline{0}$ also $\underline{y}$. Rewrite the set $\mathcal{T}$ as a substitution table: $\underline{y} = \mathcal{T}(\underline{x}^{(A)})$.

Split the matrix G into two sub-matrices G_A and G_B such that

$$\underline{y} = \underline{x}G + f_A(\underline{x}) = \sigma_A(\underline{x})G_A + \sigma_B(\underline{x})G_B + f_A(\underline{x}).$$

Replace G_B by matrix $\tilde{G}$, and replace G_A together with the error function $f_A(\underline{x})$ by the substitution table $\mathcal{T}$. The following encryption scheme is equivalent to the LW-scheme.

A message $\underline{x}$ is *encrypted* into a ciphertext $\underline{y}$ by computing

$$\underline{y} = \sigma_B(\underline{x})\tilde{G} + \mathcal{T}(\sigma_A(\underline{x})).$$

To obtain an equivalent decoding procedure proceed as follows. Compute $\tilde{H}^T$ such that $\tilde{G}\tilde{H}^T = O$ and let

$$\tilde{T} = \{(\underline{x}^{(A)}, \underline{\tilde{s}}) \mid \underline{\tilde{s}} = T(\underline{x}^{(A)})\tilde{H}^T,\ \underline{x}^{(A)} \in \mathbb{F}_2^{n-k}\}.$$

It can be easily verified that $\underline{\tilde{s}}$ uniquely determines $\underline{x}^{(A)}$. Rewrite the set $\tilde{T}$ as a substitution table: $\underline{x}^{(A)} = \tilde{T}(\underline{\tilde{s}})$. The following decryption procedure recovers the message $\underline{x}$ from the ciphertext $\underline{y}$ by the following three steps:

- Compute $\underline{\tilde{s}} = \underline{y}\tilde{H}^T$ and obtain $\underline{x}^{(A)} := \tilde{T}(\underline{\tilde{s}})$;
- Compute $(\underline{y} - T(\underline{x}^{(A)}))\tilde{G}^{-R} = \underline{x}^{(B)}$;
- Let $\underline{x} = \sigma_A^{-1}(\underline{x}^{(A)}) + \sigma_B^{-1}(\underline{x}^{(B)})$.

Again this attack will not apply to the RN-scheme for the obvious reason that in the RN-scheme a message $\underline{x}$ does not uniquely determine the error vector $\underline{z}$ used.

6 Discussion

As stated before, attack A is based on the linearity of the bit selection function σ_A: $\sigma_A(\underline{x}_i) + \sigma_A(\underline{x}_j) + \sigma_A(\underline{x}_i + \underline{x}_j) = \underline{0}$. To prevent cryptanalysts using this linearity, the LW-scheme can be modified as follows:

$$\underline{y} = \underline{x}G + \underline{z} \quad \text{with} \quad \underline{z}H^T = h(\sigma_A(\underline{x})),$$

where $h : \mathbb{F}_2^{n-k} \rightarrow \mathbb{F}_2^{n-k}$ is a nonlinear function. Although attack A is averted, attack B is still valid as it is based on non-linearity of the f_A-function.

Another possible modification, let the set A in the selection function $\sigma_A(\underline{x}_i)$ also depend on the message $\underline{x}_i$ used, i.e., $\sigma_{A(\underline{x}_i)}(\underline{x}_i)$. In this case $\sigma_{A(\underline{x}_i)}(\underline{x}_i)$ equals $\sigma_{A(\underline{x}_j)}(\underline{x}_j)$ with high probability when $\underline{x}_i$ and $\underline{x}_j$ are chosen to be of low Hamming weight. Hence, an efficient probabilistic chosen-plaintext attack (with independent row verification) is possible.

A better extension would be to use a nonlinear function $h : \mathbb{F}_2^k \rightarrow \mathbb{F}_2^{n-k}$ such that $\underline{z}H^T = h(\underline{x})$. As mentioned in [Mass88], the almost universal assumption of cryptography (due to Aug. Kerckhoffs, 1835-1903) is that the security of a secret-key cryptosystem must reside entirely in the secret-key. Equivalently, Kerckhoffs' assumption is that the entire mechanism of encipherment, except for the value of the secret-key, is known to the enemy cryptanalyst. As a consequence, the h-function should be made key-dependent, say h_K. In this way the encryption scheme becomes

$$\underline{y}_i = \underline{x}_i G + \underline{z}_i \quad \text{such that} \quad \underline{z}_i H^T = h_K(\underline{x}_i),$$

where K is a secret-key.

Unfortunately, an efficient chosen-plaintext attack exists for this extension. It is outside the scope of this paper to discuss this attack in detail. The attack is a modification of the one presented in [MT91a], with the following observation:

$$\text{If} \quad \underline{u}_s = \underline{x}_i + \underline{x}_j \quad \text{then} \quad \widehat{\underline{g}}_s = \underline{y}_i + \underline{y}_j = (\underline{x}_i G + \underline{z}_i) + (\underline{x}_j G + \underline{z}_j) = \underline{g}_s + \underline{z}_i + \underline{z}_j .$$

So instead of encrypting the same message $O(|\mathcal{Z}|^{1/2})$ time as in [MT91a], we encrypt $O(|\mathcal{Z}|^{1/2})$ different messages $\underline{x}_i$ and $\underline{x}_j$ for which it holds that $\underline{x}_i + \underline{x}_j = \underline{u}_s$. It can be shown that a $k \times n$ matrix equivalent to the encryption matrix G can be obtained within $O(k \times |\mathcal{Z}|^{1/2})$ chosen encryptions, with high probability. To obtain the full set of error vectors, we need $O(|\mathcal{Z}|)$ chosen encryptions. The remaining part is to find the key K used. Therefore we may conclude that the security of this modification is actually based on the cryptographic strength of the function h_K used. Note that a complex h_K-function effects the efficiency of the scheme. As a consequence this is not a good extension of the LW-scheme.

The previous discussion shows that the security of the RN-scheme is not improved if the scheme is slightly modified by using a pseudo-random function f, and letting $\underline{z} = f(\underline{x})$ so that there is only one encryption for each message $\underline{x}$. This settles the question raised by Brickell and Odlyzko [BO88] in a negative way.

Li and Wang suggest to use the set $\mathcal{Z}$, if not used for authentication, for purposes of encryption and improving the code rate of their scheme. However, their scheme is not based on code $\mathcal{C}$'s error correcting capability to recover random errors, but is based on finding an error vector in a predefined set $\mathcal{Z}$. Hence, the cryptanalyst obtains additional information about the used set of error vectors. As a result, this is not a good extension of this type of secret-key cryptosystem.

In their conclusion, Li and Wang state that a similar modification of the McEliece scheme [McEl78] is possible to obtain a new public-key cryptosystem for purposes of both encryption and authentication. To the best of our knowledge, we do not see how such an extension can be made possible without rendering the *public-key* cryptosystem insecure. Note that part of the security of the McEliece cryptosystem is related to the general decoding problem for linear codes.

7 Conclusion

In this paper we proposed two chosen-plaintext attacks on the Li-Wang joint authentication and encryption scheme [LW91].

The first attack is based on the linearity of the bit selection function σ_A. Due to this linearity, the sum of three ciphertexts obtained from two randomly chosen messages yields a codeword. A $k \times n$ matrix equivalent to the encryption matrix is obtained in $O(k)$ chosen encryptions. If the set of error vectors $\mathcal{Z}$ is randomly chosen, about $|\mathcal{Z}|$ chosen encryptions are necessary to obtain a corresponding set of error vectors. In the literature no chosen-plaintext attack for the RN-scheme is known that recovers an equivalent $k \times n$ encryption matrix within $O(k)$ chosen encryptions. As stated in [RN89] this is really the hard part of the RN-scheme. Moreover, in the RN-scheme we are not able to choose three encryptions that will certainly yield a codeword.

The second attack is inherently different and is based on non-linearity of the error function f_A. The attack always yields $2k - n$ rows of the $k \times n$ encryption matrix after $O(k^2)$ chosen encryptions. Hereafter $|\mathcal{Z}|$ chosen-plaintexts are required to create a cryptosystem equivalent to the LW-scheme. The attack does not apply to the RN-scheme as in the RN-scheme a message $\underline{x}$ not uniquely determine the error vector $\underline{z}$ used.

In the RN-scheme, each error vector can be used to encrypt any given message, but in the LW-scheme each message corresponds to only one error vector. Besides, each error vector corresponds to several messages, which is a serious weakness of the LW-scheme and made the two attacks with a work factor of about $|\mathcal{Z}|$ chosen-plaintexts possible. Because the RN-scheme does not have this weakness, the proposed attacks do not apply. Therefore, we may conclude that in contrast with their claim, the LW-scheme is less secure than the RN-scheme.

We have considered several generalizations of the LW-scheme, and reached the conclusion that none of them is a good extension of the RN-scheme. Therefore, the security of the RN-scheme is not improved when the scheme is slightly modified by using a pseudo-random function f, and letting $\underline{z} = f(\underline{x})$ so that there is only one encryption for each message $\underline{x}$. This settles the question raised by Brickell and Odlyzko [BO88] in a negative way.

Acknowledgements

We are grateful to Henk van Tilborg for very helpful discussions concerning this work and to Chung-Huang Yang and Peter de Rooij for their comments on a previous draft.

References

[BO88] E.F. Brickell and A.M. Odlyzko, Cryptanalysis: A Survey of Recent Results, *Proceedings of the IEEE*, vol. 76, no. 5, pp. 578–593, May 1988.

[KZYR89] Kencheng Zeng, C.H. Yang and T.R.N. Rao, On the Linear Consistency Test (LCT) in Cryptanalysis with Applications, *Advances in Cryptology-CRYPTO'89*, Lecture Notes in Computer Science **435**, Springer-Verlag, pp. 164–174, 1989.

[LW91] Y. Li and X. Wang, A Joint Authentication and Encryption Scheme Based on Algebraic Coding Theory, *Applied Algebra, Algebraic Algorithms and Error-Correcting Codes* (AAECC'9), Lecture Notes in Computer Science **539**, Springer-Verlag, pp. 241–245, October 1991.

[MS77] F.J. MacWilliams and N.J.A. Sloane, *The Theory of Error-Correcting Codes*, North-Holland Mathematical Library, Vol. 16, North-Holland, Amsterdam, 1977.

[McEl78] R.J. McEliece, A Public-Key Cryptosystem Based on Algebraic Coding Theory, DSN Progress Report 42–44, Jet Propulsion Laboratory, Pasadena, pp. 114–116, January 1978.

[Mass88] J.L. Massey, An Introduction to Contemporary Cryptology, *Proceedings of the IEEE*, vol. 76, no. 5, pp. 533–549, May 1988.

[MT91a] J. Meijers and J. van Tilburg, On the Rao-Nam Private-Key Cryptosystem using Linear Codes, *Proceedings 1991 IEEE-ISIT*, p. 126, Budapest, Hungary, June 1991.

[MT91b] J. Meijers and J. van Tilburg, Extended Majority Voting and Private-Key Algebraic-Code Encryptions, ASIACRYPT'91, Fujiyoshida, Japan, November 1991.

[RN89] T.R.N. Rao and K.H. Nam, Private-Key Algebraic-Code Encryptions, *IEEE Trans. Inform. Theory*, vol. IT-35, no. 4, pp. 829–833, July 1989.

[Simm88] G.J. Simmons, A Survey of Information Authentication, *Proceedings of the IEEE*, vol. 76, no. 5, pp. 603–620, May 1988.

Some Constructions of Perfect Binary Codes

Alexander Vardy[*1] and **Tuvi Etzion**[**2]

[1] IBM Research Division
Almaden Research Center
650 Harry Road, San Jose, CA 95120
e-mail: vardy@almaden.ibm.com

[2] Technion — Israel Institute of Technology
Department of Computer Science
Haifa 32000, Israel
e-mail: etzion@cs.technion.ac.il

Abstract. A construction of perfect binary codes is presented. It is shown that this construction gives rise to perfect codes that are nonequivalent to any of the previously known perfect codes. Furthermore, perfect codes C_1 and C_2 are constructed such that their intersection $C_1 \cap C_2$ has the maximum possible cardinality. The latter result is then employed to explicitly construct $2^{2^{cn}}$ nonequivalent perfect codes of length n, for sufficiently large n and some constant c slightly less than 0.5.

1. Introduction

Let $\mathbb{F}_2^n$ be a vector space of dimension n over $GF(2)$. A subset of $\mathbb{F}_2^n$ is a binary code of length n. For any code $C \subset \mathbb{F}_2^n$ we denote by $C^\perp$ the subspace of $\mathbb{F}_2^n$ consisting of those vectors that are orthogonal to all the codewords of C. Two codes $C_1, C_2 \subset \mathbb{F}_2^n$ are said to be isomorphic if there exists a permutation π, such that $C_2 = \{\pi(c) : c \in C_1\}$. They are said to be equivalent if there exist a vector a and a permutation π, such that $C_2 = \{a + \pi(c) : c \in C_1\}$. The Hamming distance between vectors $x, y \in \mathbb{F}_2^n$, denoted $d(x,y)$, is the number of coordinates in which x and y differ. The Hamming weight of x is given by $wt(x) = d(x, \mathbf{0})$, where $\mathbf{0}$ denotes the all-zero vector.

A code C of length $n = 2^m - 1$ is *perfect* if the minimum Hamming distance between the codewords is 3, and any $x \in \mathbb{F}_2^n$ is within distance 1 from some codeword. Without loss of generality we shall always assume $\mathbf{0} \in C$. The linear perfect codes are unique, — these are the well-known Hamming codes [3]. Nonlinear perfect codes were first constructed by Vasil'ev [6]. Other constructions of nonlinear perfect codes have been subsequently presented by Phelps [4], Phelps [5], and Bauer et al. [1]. These constructions are reviewed in the next section.

* Research supported in part by the Rothschild Fellowship.

** Research supported in part by the Technion V.P.R. fund and in part by the fund for the promotion of research at the Technion.

In Section 3 we present a new construction of perfect binary codes. The codes resulting from this construction have the property of being "interlaced". This property will be elaborated upon in the sequel. Further, the construction can be employed to produce perfect codes of any rank in the range of $(n-m)$ to $(n-1)$. Using these two facts we show that, at least for length 15, our construction gives rise to codes that are nonequivalent to any perfect code obtained through the constructions of [1,4,5,6].

In Section 4 we consider the following question. Let C_1, C_2 be two distinct perfect codes of length $n = 2^m-1$. What is the maximum possible cardinality of their intersection $C_1 \cap C_2$? We prove that

$$|C_1 \cap C_2| \leq 2^{n-m} - 2^{\frac{n-1}{2}} \tag{1}$$

and then present a construction that attains the upper bound of (1). This construction does not yield new perfect codes. In fact, it can be shown that it is a special case of the Vasil'ev construction of [6]. Yet, using this construction, the space $\mathbb{F}_2^n$ may be partitioned into disjoint sets, such that each set contains two different perfect coverings of itself. The number of such sets is $|C|$, where C is the Hamming code of length $\nu = (n-1)/2$. Taking all the possible unions of these coverings we explicitly construct $2^{|C|}$ distinct perfect codes of length n.

2. Preliminaries

In this section we briefly outline the known constructions of nonlinear perfect codes, and give upper bounds on the rank of a code obtained using these constructions.

For $v \in \mathbb{F}_2^n$, let $p(v) = wt(v) \bmod 2$. Let C_n be a perfect binary code of length $n = 2^m - 1$. Let $f : C_n \to \{0,1\}$ be an arbitrary mapping, such that $f(0) = 0$ and $f(c_1) + f(c_2) \neq f(c_1 + c_2)$ for some $c_1, c_2, c_1 + c_2 \in C_n$.

Proposition 2.1. (Vasil'ev [6]). *The code C_{2n+1} defined by*

$$C_{2n+1} = \{(\, v \mid v + c \mid p(v) + f(c)\,) \; : \; v \in \mathbb{F}_2^n, c \in C_n\},$$

where $(\cdot|\cdot)$ denotes concatenation, is perfect. ∎

Let $V(n)$ be the set of all the perfect codes of length n that may be obtained using the foregoing construction. We define the rank of a set of vectors as the maximum number of linearly independent vectors in the set.

Lemma 2.2. *For $C_{2n+1} \in V(2n+1)$,*

$$\mathsf{rank}(C_{2n+1}) = \mathsf{rank}(C_n) + n + 1 \,.$$

Proof. Denote $r = \mathsf{rank}(C_n)$. Let w be a vector in $C_n^{\perp}$. Then $(w|w|0)$ is obviously in $C_{2n+1}^{\perp}$. Hence $\mathsf{rank}(C_{2n+1}^{\perp}) \geq \mathsf{rank}(C_n^{\perp})$ and therefore

$$\mathsf{rank}(C_{2n+1}) \leq (2n+1) - (n-r) .$$

Let $c_1, c_2, \ldots c_r$ be some r linearly independent vectors in C_n. Let $u_1, u_2, \ldots u_n$ be the set of vectors of weight 1 in $\mathbb{F}_2^n$. Then the union of $\mathcal{U}$ and $\mathcal{C}$, where

$$\mathcal{C} = \{ \ (0|c_i|f(c_i)) \ : \ i = 1, 2, \ldots r \ \}$$

$$\mathcal{U} = \{ \ (u_i|u_i|1) \ : \ i = 1, 2, \ldots n \ \} ,$$

belongs to C_{2n+1} and is linearly independent. Now assume that any codeword of C_{2n+1} can be represented as a sum of elements of $\mathcal{U} \cup \mathcal{C}$. But then for any $c \in C_n$, the vector $(0|c|f(c))$ can be represented as a sum of elements of $\mathcal{C}$, and therefore the mapping f is linear. ∎

A code C of length $n{+}1 = 2^m$ is an extended perfect code iff $|C| = 2^{n-m}$ and the minimum distance of C is 4, that is $d(c_1, c_2) \geq 4$ for any distinct $c_1, c_2 \in C$. Let $\mathbb{E}_2^n$ denote the set of all even weight vectors of $\mathbb{F}_2^n$. Let $C_0^0, C_1^0, \ldots C_n^0$ and $C_0^1, C_1^1, \ldots C_n^1$ be partitions of $\mathbb{E}_2^{n+1}$ and $\mathbb{F}_2^{n+1} \backslash \mathbb{E}_2^{n+1}$, respectively, into extended perfect codes. Let π be a permutation on the set $\{0, 1, \ldots n\}$.

Proposition 2.3. (Phelps [4]). *The code C defined by*

$$C \ = \ \{ \ (\ c_0 \ | \ c_1 \ ) \ : \ c_0 \in C_i^0, \ c_1 \in C_j^1, \ \pi(i) = j \ \}$$

is an extended perfect code. ∎

Puncturing any coordinate of C gives a perfect code of length $2n+1$. We let $P_2(2n+1)$ denote the set of all such perfect codes (the meaning of the subscript will become clear immediately).

Now let R be an extended perfect code of length k, where $k = 2^{\kappa}$ for some κ. Let Q be a minimum distance 2 code of length k over an alphabet of $(n+1)$ symbols, with $|Q| = (n+1)^{k-1}$.

Proposition 2.4. (Phelps [5]). *The code P defined by*

$$P \ = \ \{ \ (\ c_1|c_2|\cdots|c_k \ ) \ : \ c_i \in C_{j_i}^{r_i}, \ (r_1, r_2, \ldots r_k) \in R, \ (j_1, j_2, \ldots j_k) \in Q \ \}$$

is an extended perfect code of length $k(n+1)$. ∎

The set of all the perfect codes obtained by puncturing the codes of Proposition 2.4 is denoted $P_k(k(n{+}1) - 1)$. Note that for $k = 2$, the code Q is in effect a permutation on the set $\{0, 1, \ldots n\}$ and R is a "perfect" code consisting of a single vector. Thus Proposition 2.3 is a special case of Proposition 2.4, and the two definitions of $P_2(n)$ coincide.

Lemma 2.5. *For $C \in P_k(n)$,*

$$\mathsf{rank}(C) \leq n + \mathsf{rank}(R) - k + 1 .$$

Proof. Let $w = (w_1, w_2, \ldots w_k)$ be a vector orthogonal to R. Let $\mathbf{1}$ denote the all-ones vector. Then the vector $(\, u_1 | u_2 | \cdots | u_k \,) \in \mathbb{F}_2^{k(n+1)}$ given by

$$u_i = \begin{cases} \mathbf{0} & w_i = 0 \\ \mathbf{1} & w_i = 1 \end{cases}$$

is orthogonal to P, for any Q and any partition of the space. Therefore $(n+1) - \mathsf{rank}(P) \geq k - \mathsf{rank}(R)$. Yet $\mathsf{rank}(C) \leq \mathsf{rank}(P)$. ∎

In addition to the constructions discussed above, there are three "non-Vasil'ev" perfect codes of length 15 constructed by Bauer et al. [1]. The study of [1] provides a necessary and sufficient condition for a systematic code to be perfect. The authors then employ this condition to construct the three perfect codes of length 15. In general it is not clear, however, how to find codes which satisfy the condition of Bauer et al. [1].

3. New perfect codes

Lemma 3.1. *If C is a perfect code of length $n = 2^m - 1$ then all the vectors in $C^\perp \backslash \{0\}$ have weight 2^{m-1}.*

Proof. Since C is an orthogonal array of strength $2^{m-1} - 1$ (cf. [2]), any vector of weight less than 2^{m-1} is orthogonal to exactly half the codewords of C. To see that the same applies to a vector x of weight greater than 2^{m-1} extend C by a parity check coordinate. The resulting code C^* is again an orthogonal array of strength $2^{m-1} - 1$ and the vector $\mathbf{1}$ is orthogonal to C^*. The vector $\mathbf{1} + (x|0)$ has weight less than 2^{m-1} and is therefore orthogonal to exactly half the codewords of C^*. Hence x is orthogonal to exactly half the codewords of C^*, and consequently also of C. ∎

We now give a necessary and sufficient condition for a code C of length $n = 2^m - 1$ and rank less than n to be perfect. Clearly for such a code $C^\perp \backslash \{\mathbf{0}\} \neq \emptyset$. It follows from the foregoing lemma that if C is to be perfect then there should exist a vector $w \in C^\perp$ of weight $\nu + 1 = 2^{m-1}$. W.l.o.g. assume that the nonzero entries of w are in the first $\nu + 1$ positions. Define

$$T(u) = \{\, v \in \mathbb{F}_2^\nu \;:\; (u|v) \in C \,\}$$
$$H(v) = \{\, u \in \mathbb{E}_2^{\nu+1} \;:\; (u|v) \in C \,\}$$

Lemma 3.2. *The code C is perfect if and only if the following conditions hold.*

1. $\forall u \in \mathbb{E}_2^{\nu+1}$, *$T(u)$ is a perfect code of length ν.*
2. $\forall v \in \mathbb{F}_2^{\nu}$, *$H(v)$ is an extended perfect code of length $\nu+1$.*

Proof. ($\Rightarrow$) Let $c_1 = (x_1|y_1)$ and $c_2 = (x_2|y_2)$ be two distinct codewords of C, where $x_1, x_2 \in \mathbb{E}_2^{\nu+1}$ and $y_1, y_2 \in \mathbb{F}_2^{\nu}$. If $x_1 = x_2$ then $y_1, y_2 \in T(x_1)$ and hence $d(c_1, c_2) \geq 3$. Similarly if $y_1 = y_2$ then $x_1, x_2 \in H(y_1)$ and $d(c_1, c_2) \geq 4$. Now if $x_1 \neq x_2$ and $y_1 \neq y_2$ then $d(c_1, c_2) = d(x_1, x_2) + d(y_1, y_2) \geq 3$. Clearly,

$$|C| = \sum_{u \in E_2^{\nu+1}} |T(u)| = 2^{\nu} \cdot 2^{\nu-(m-1)} = 2^{n-m} .$$

Hence C is perfect. ($\Leftarrow$) Obviously $d(T(u)) = \min_{v_1, v_2 \in T(u)} d(v_1, v_2) \geq 3$. Hence by the sphere packing bound $|T(u)| \leq 2^{\nu-(m-1)}$. Yet if $|T(u^*)| < 2^{\nu-(m-1)}$ for some u^*, then

$$|C| = \sum_{u \in E_2^{\nu+1}} |T(u)| \leq (2^{\nu} - 1) \cdot 2^{\nu-(m-1)} + |T(u^*)| < 2^{n-m} .$$

Therefore $|T(u)| = 2^{\nu-(m-1)}$ and $T(u)$ is perfect. A similar argument shows that $H(v)$ is extended perfect. ∎

The following construction of perfect codes is in a sense a generalization of the first construction of Phelps [4]. Let V be a subset of $\mathbb{F}_2^n$. Let $\mathcal{A} = \{A_1, A_2, \ldots A_k\}$ and $\mathcal{B} = \{B_1, B_2, \ldots B_k\}$ be two ordered sets of subsets of V. For $v \in V$, define

$$\Lambda_{\mathcal{A}}(v) = \{ i \ : \ v \in A_i \}$$
$$\Lambda_{\mathcal{B}}(v) = \{ i \ : \ v \in B_i \}$$

where $A_i \in \mathcal{A}$ and $B_i \in \mathcal{B}$. We say that $\mathcal{A}$ and $\mathcal{B}$ form a *perfect segmentation* of order k of the set V, if both $\cup_{i \in \Lambda_{\mathcal{B}}(v)} A_i$ and $\cup_{i \in \Lambda_{\mathcal{A}}(v)} B_i$ are perfect codes of length n, for all $v \in V$.

Proposition 3.3. *Let $\mathcal{A}$ and $\mathcal{B}$ be a perfect segmentation of $\mathbb{F}_2^n$. The code C defined by*

$$C = \{ (u \,|\, v) \ : \ u \in A_i^*, \ v \in B_i \},$$

where the superscript * denotes extension by a adding a parity check, is a perfect code of length $2n+1$.

Proof. Let the sets $T(u)$ and $H(v)$ be defined in the obvious way on the first n+1 coordinates of C. The set of all the elements of $\mathcal{B}$ that contain a given vector $v \in \mathbb{F}_2^n$ is clearly $\Lambda_{\mathcal{B}}(v)$. Hence $H(v) = \cup_{i \in \Lambda_{\mathcal{B}}(v)} A_i^* = (\cup_{i \in \Lambda_{\mathcal{B}}(v)} A_i)^*$. Similarly $T(u) = \cup_{i \in \Lambda_{\mathcal{A}}(v)} B_i$. Thus C is perfect by Lemma 3.2. ∎

To see that Proposition 2.3 is a special case of Proposition 3.3 note that any two partitions of $\mathbb{F}_2^n$ into $(n+1)$ perfect codes form a perfect segmentation of $\mathbb{F}_2^n$. In fact, it is evident that $(n+1)$ is the minimum order of any perfect segmentation of $\mathbb{F}_2^n$. We presently show that perfect segmentations of higher order exist. These can be easily constructed from the Hamming codes. Let C_1 and C_2 be two isomorphic Hamming codes of length $n = 2^m - 1$, such that $C' = C_1 \cap C_2$ has cardinality 2^{n-m-1}. Then $C_1 = C' \cup (c_1 + C')$ and $C_2 = C' \cup (c_2 + C')$ for some $c_1 \in C_1 \backslash C'$ and $c_2 \in C_2 \backslash C'$. Define

$$\begin{array}{llll} A_1 = C' & A_2 = c_1 + C' & A_3 = c_2 + C' & A_4 = c_1 + c_2 + C' \\ B_1 = C_1 & B_2 = C_2 & B_3 = c_1 + C_2 & B_4 = c_2 + C_1 \end{array} \quad (2)$$

It can be readily verified that $\{A_1, A_2, A_3, A_4\}$ and $\{B_1, B_2, B_3, B_4\}$ form a perfect segmentation of the set $V = C_1 \cup (c_2 + C_1)$. Furthermore, any two partitions $A_5, A_6, \ldots A_{n+3}$ and $B_5, B_6, \ldots B_{n+3}$ of the set $\mathbb{F}_2^n \backslash V$ into perfect codes (say, the cosets of C_1 or C_2) complete $\{A_1, A_2, A_3, A_4\}$ and $\{B_1, B_2, B_3, B_4\}$ to a perfect segmentation of $\mathbb{F}_2^n$. Alternatively, let $a \in \mathbb{F}_2^n \backslash V$ and define, for instance,

$$\begin{array}{llll} B_5 = a + C' & B_6 = a + c_1 + C' & B_7 = a + c_2 + C' & B_8 = a + c_1 + c_2 + C' \\ A_5 = a + C_1 & A_6 = a + C_2 & A_7 = a + c_1 + C_2 & A_8 = a + c_2 + C_1 \end{array} \quad (3)$$

Then again $\{A_5, A_6, A_7, A_8\}$ and $\{B_5, B_6, B_7, B_8\}$ is a perfect segmentation of the set $a + V$. Continuing in this manner one can construct perfect segmentations of any even order up to $2(n+1)$.

Let $V_1, V_2, \ldots V_k$ be an arbitrary collection of sets. We say that these sets are *non-interlaced* if for $1 \le i, j \le k$ either $V_i = V_j$ or $V_i \cap V_j = \emptyset$.

Lemma 3.4. *For any perfect code of length $n = 2\nu + 1 = 2^m - 1$, the following four statements are equivalent:*

a. *The 2^ν sets $T(u)$ are non-interlaced.*

b. *The 2^ν sets $H(v)$ are non-interlaced.*

c. *There are exactly $\nu + 1$ distinct sets $T(u)$.*

d. *There are exactly $\nu + 1$ distinct sets $H(v)$.*

Proof. **(c)⇒(a)** Let $T(u_1) \neq T(u_2)$ and $T(u_1) \cap T(u_2) \neq \emptyset$. Then $|\bigcup_{u \in E_2^{\nu+1}} T(u)| \le (\nu+1) \cdot 2^{\nu-(m-1)} - |T(u_1) \cap T(u_2)| \le 2^\nu - 1$. It follows that there exist $v \in \mathbb{F}_2^\nu$ such that $\bigcup_{u \in E_2^{\nu+1}} T(u) \not\ni v$. But then $H(v) = \emptyset$, a contradiction to Lemma 3.2. **(a)⇒(c)** Let there be k distinct sets $T(u)$. Obviously if the 2^ν sets $T(u)$ are non-interlaced then the k distinct sets are disjoint. Hence $2^\nu = |\mathbb{F}_2^\nu| = |\bigcup_{u \in E_2^{\nu+1}} T(u)| = k \cdot 2^{\nu-(m-1)}$ and $k = 2^{m-1} = \nu + 1$. Thus **(a)⇔(c)**. By a similar argument **(b)⇔(d)**. It remains to show **(a)⇔(b)**. For $v_1 \neq v_2$ let $H(v_1) = \{x_1, x_2, \ldots x_s\}$ and $H(v_2) = \{y_1, y_2, \ldots y_s\}$, where $s = 2^{\nu-(m-1)}$. Then for $i = 1, 2, \ldots s$, $T(x_i) \ni v_1$ and $T(y_i) \ni v_2$. Hence by **(a)**, $T(x_1) = T(x_2) = \cdots = T(x_s)$ and $T(y_1) = T(y_2) = \cdots = T(y_s)$. Assume $T(x_i) \neq T(y_i)$. But then obviously $\{x_1, x_2, \ldots x_s\} \cap \{y_1, y_2, \ldots y_s\} = \emptyset$, that is $H(v_1) \cap H(v_2) = \emptyset$. Now assume $T(x_i) = T(y_i)$. Yet $y_i \notin H(v_1)$ implies $v_1 \notin T(y_i)$ and hence $T(y_i) \neq T(x_i)$. Therefore $y_i \in H(v_1)$ for all i, and $H(v_1) = H(v_2)$. Thus **(a)⇒(b)**. A similar argument shows **(b)⇒(a)**. ∎

We say that a perfect code is "non-interlaced" if it satisfies, **(a)**–**(d)**, otherwise we say that it is "interlaced". The set of all the non-interlaced perfect codes is produced by the construction of Phelps [4] (Proposition 2.3).

Corollary 3.5. *A perfect code C of length $n = 2\nu + 1$ is in $P_2(n)$ if and only if it is non-interlaced for some vector of weight $\nu + 1$ in $C^\perp$.*

Proof. ($\Leftarrow$) As the construction of Proposition 2.3 employs partition of $\mathbb{F}_2^\nu$ into $(\nu + 1)$ perfect codes, (**a**) and (**c**) hold. ($\Rightarrow$) It follows from the proof of (**c**)$\Rightarrow$(**a**) in Lemma 3.4 that for any perfect code

$$\cup_{u \in E_2^{\nu+1}} T(u) = \mathbb{F}_2^\nu \tag{4}$$

$$\cup_{v \in F_2^\nu} H(v) = \mathbb{E}_2^{\nu+1} \tag{5}$$

By (**a**), (**c**) and (4), the $(\nu + 1)$ distinct sets $T(u)$ constitute a partition of $\mathbb{F}_2^\nu$. By Lemma 3.2, each of the codes in the partition is a perfect code of length ν. Similarly by (**b**), (**d**), (5) and Lemma 3.2, the $(\nu+1)$ distinct sets $H(v)$ constitute a partition of $\mathbb{E}_2^{\nu+1}$ into extended perfect codes. ∎

Using Proposition 3.3 we now construct interlaced perfect codes of length $n = 2\nu + 1 = 2^m - 1$ and any rank in the range of $(n - m + 1)$ to $(n - 1)$. As before let $V = C_1 \cup (c_2 + C_1) = C_2 \cup (c_1 + C_2)$, where C_1 and C_2 are isomorphic Hamming codes of length ν and rank $\nu - (m-1)$, and let $\{A_1, A_2, A_3, A_4\}$ and $\{B_1, B_2, B_3, B_4\}$ be a perfect segmentation of V as in (2) or (3). Let $(a_0 + C_1), (a_1+C_1), \ldots (a_\nu+C_1)$ and $(b_0+C_1), (b_1+C_1), \ldots (b_\nu+C_1)$ be two partitions of $\mathbb{F}_2^\nu$ into cosets of C_1, such that $a_0 = b_0 = 0$ and $a_1 = b_1 = c_2$. Define a perfect segmentation of $\mathbb{F}_2^\nu$ by completing $\{A_1, A_2, A_3, A_4\}$ and $\{B_1, B_2, B_3, B_4\}$ with $A_{i+3} = a_i + C_1$ and $B_{i+3} = b_i + C_1$, for $i = 2, 3, \ldots \nu$. The rank of a perfect code C, constructed by applying Proposition 3.3 to this perfect segmentation, is given by

$$\mathsf{rank}(C) = 2\mathsf{rank}(C_1) + \mathsf{rank}(\Gamma),$$

where $\Gamma = \{(0|0), (c_2|0), (0|c_2)\} \cup \{(a_i|b_i) : i = 2, 3, \ldots \nu\}$. Obviously,

$$m \leq \mathsf{rank}(\Gamma) \leq 2(m-1).$$

and the vectors $a_2, a_3, \ldots a_\nu$ and $b_2, b_3, \ldots b_\nu$ can be always chosen such as to make Γ have any rank in the above range. Since $\mathsf{rank}(C_1) = \nu - (m-1)$, the rank of C can be made to attain any value in the range of $(n-m+1)$ to $(n-1)$. Furthermore, since we employed a perfect segmentation of order $\nu + 3$ in the construction of C, it follows from Lemma 3.4 that C is interlaced.

Now assume that the rank of C is $(n - 1)$. Then C cannot be equivalent to any non-interlaced code of length n, since there is only one nonzero codeword in $C^\perp$ and the definition of $T(u)$ and $H(v)$ for C is unique. Hence, in view of Corollary 3.5, $C \notin P_2(n)$ for all n.

As a specific example of an interlaced code C of length 15 and rank 14 consider the following construction. Let C_1 and C_2 be two $(7, 4, 3)$ Hamming codes, generated by

$$G_1 = \begin{bmatrix} 1\,0\,0\,0\,0\,1\,1 \\ 0\,1\,0\,0\,1\,0\,1 \\ 0\,0\,1\,0\,1\,1\,0 \\ 0\,0\,0\,1\,1\,1\,1 \end{bmatrix} \qquad G_2 = \begin{bmatrix} 0\,1\,0\,0\,0\,1\,1 \\ 1\,0\,0\,0\,1\,0\,1 \\ 0\,0\,1\,0\,1\,1\,0 \\ 0\,0\,0\,1\,1\,1\,1 \end{bmatrix}$$

respectively. Thus C_2 is the image of C_1 under the permuatation $(1,2)$. Clearly we have $c_1 = (1000011)$, $c_2 = (0100011)$, and $C' = C_1 \cap C_2$ is generated by

$$G' = \begin{bmatrix} 1\,1\,0\,0\,1\,1\,0 \\ 0\,0\,1\,0\,1\,1\,0 \\ 0\,0\,0\,1\,1\,1\,1 \end{bmatrix}$$

The vectors $a_2, a_3, \ldots a_7$ and $b_2, b_3, \ldots b_7$ could be chosen, for instance, as follows. Let

$$\begin{aligned} e_1 &= (0100011) \\ e_2 &= (0100000) \\ e_3 &= (0000010) \end{aligned}$$

Then we may define

$$\begin{aligned} a_2 &= e_2 & \qquad b_2 &= e_2 \\ a_3 &= e_1 + e_2 & b_3 &= e_3 \\ a_4 &= e_3 & b_4 &= e_1 + e_2 + e_3 \\ a_5 &= e_1 + e_3 & b_5 &= e_1 + e_2 \\ a_6 &= e_2 + e_3 & b_6 &= e_2 + e_3 \\ a_7 &= e_1 + e_2 + e_3 & b_7 &= e_1 + e_3 \end{aligned}$$

Since $c_2 = e_1$ we may write Γ as

$$\Gamma = \left\{ \begin{array}{c} (\,0 \mid 0\,) \\ (\,e_1 \mid 0\,) \\ (\,0 \mid e_1\,) \\ (\,e_2 \mid e_2\,) \\ (\,e_1 + e_2 \mid e_3\,) \\ (\,e_3 \mid e_1 + e_2 + e_3\,) \\ (\,e_1 + e_3 \mid e_1 + e_2\,) \\ (\,e_2 + e_3 \mid e_2 + e_3\,) \\ (\,e_1 + e_2 + e_3 \mid e_1 + e_3\,) \end{array} \right\}$$

Using the fact that e_1, e_2, e_3 are linearly independent, it is now easy to verify that $\mathsf{rank}(\Gamma) = 6$ and therefore $\mathsf{rank}(C) = 2{\cdot}4 + 6 = 14$. Hence, in view of the foregoing argument $C \notin P_2(15)$.

Establishing non-equivalence between two families of perfect codes is in general a difficult task. This task is greatly simplified for codes of length 15, by virtue of the well-known fact that the only perfect code of length 7 is the linear Hamming code (cf. [3]). Thus by Lemma 2.2 the rank of all the codes in $V(15)$ is 12. Hence $C \notin V(15)$. The set of all the Phelps codes of length 15 is given by the union of $P_2(15)$, $P_4(15)$ and $P_8(15)$. We have already established $C \notin P_2(15)$. Yet by Lemma 2.5 the rank of all the codes in $P_4(15)$ and $P_8(15)$ is at most 13. Thus $C \notin \cup_{k=2,4,8} P_k(15)$. Finally, it is a matter of straightforward verification that C is not equivalent to any of the three perfect codes of length 15 in Bauer et al. [1].

4. Intersections of perfect codes

Let C_1 and C_2 be two distinct perfect codes of length $n = 2^m - 1$. What is the maximum possible cardinality of their intersection $|C_1 \cap C_2|$? In the following we give a complete answer to this question. In fact, we consider a slightly more general situation. For a positive integer $n = 2\nu+1$, let V be the smallest subset of $\mathbb{F}_2^n$ such that there exist two distinct perfect coverings of V. Namely, let $\mathcal{A}$ and $\mathcal{B}$ be two distinct subcodes of V, such that any vector in V is within distance 1 from a unique codeword of $\mathcal{A}$ *and* a unique codeword of $\mathcal{B}$.

Lemma 4.1. *The cardinality of V is at least $(\nu+1)2^{\nu+1}$.*

Proof. Without loss of generality we assume $\mathbf{0} \in \mathcal{A}$. Otherwise let $a \in \mathcal{A}$. Then $(a + V)$ is a subset of $\mathbb{F}_2^n$ and $(a + \mathcal{A})$, $(a + \mathcal{B})$ are two distinct perfect coverings of $(a + V)$, with $\mathbf{0} \in (a + \mathcal{A})$. It is also clear that $\mathcal{A} \cap \mathcal{B} = \emptyset$, since otherwise V is not minimal. Hence $\mathbf{0} \notin \mathcal{B}$. It follows that $\mathcal{B}$ contains a unique codeword of weight 1. Indeed, since $\mathbf{0} \in \mathcal{A}$, there are no codewords of weight 1 in $\mathcal{A}$. Now let A_i and B_i denote the number of codewords of weight i in $\mathcal{A}$ and $\mathcal{B}$, respectively. Counting the number of vectors of weight i in V, we have for $1 \le i \le 2\nu$,

$$(n-i+1)A_{i-1} + A_i + (i+1)A_{i+1} = (n-i+1)B_{i-1} + B_i + (i+1)B_{i+1} \quad (6)$$

Let $\Delta_i = A_i - B_i$. Then (6) implies the following recurrence

$$i\Delta_i = -\Delta_{i-1} - (n-i+2)\Delta_{i-2} \quad (7)$$

We have already established $A_0 = 1, A_1 = 0$ and $B_0 = 0, B_1 = 1$. Hence $\Delta_0 = 1$ and $\Delta_1 = -1$. The unique solution of the recurrence (7) with these initial conditions is

$$\Delta_i = \begin{cases} \binom{\nu}{\lfloor \frac{i}{2} \rfloor} & i \equiv 0,3 \mod 4 \\ -\binom{\nu}{\lfloor \frac{i}{2} \rfloor} & i \equiv 1,2 \mod 4 \end{cases}$$

Obviously, $|V| = (1+n)|\mathcal{A}| = (1+n)\sum_{i=0}^{n} A_i$. Hence

$$\frac{|V|}{1+n} = \sum_{\substack{i=0 \\ i\equiv 0,3}}^{n} A_i + \sum_{\substack{i=0 \\ i\equiv 1,2}}^{n} A_i \ge \sum_{\substack{i=0 \\ i\equiv 0,3}}^{n} A_i = \sum_{\substack{i=0 \\ i\equiv 0,3}}^{n} [\Delta_i + B_i] \ge \sum_{\substack{i=0 \\ i\equiv 0,3}}^{n} \Delta_i = \sum_{\substack{i=0 \\ i\equiv 0,3}}^{n} \binom{\nu}{\lfloor \frac{i}{2} \rfloor}$$

where all the equivalences are modulo 4. Substituting $j = \lfloor i/2 \rfloor$ in the above inequality, yields $|V| \ge (1+n)\sum_{j=0}^{\nu} \binom{\nu}{j} = (\nu+1)2^{\nu+1}$. ∎

Corollary 4.2. *Let C_1 and C_2 be two distinct perfect codes of length $n = 2^m - 1$. Then*

$$|C_1 \cap C_2| \leq 2^{n-m} - 2^{\nu}$$

where $\nu = (n-1)/2$.

Proof. Since for distinct perfect codes we may always set $\mathcal{A} = C_1 \backslash (C_1 \cap C_2)$ and $\mathcal{B} = C_2 \backslash (C_1 \cap C_2)$, the upper bound follows immediately by Lemma 4.1. ∎

We now construct C_1 and C_2, such that the cardinality of $C_1 \cap C_2$ attains the upper bound. Let $\mathcal{H}_n$ be the Hamming code of length $n = 2\nu + 1 = 2^m - 1$, and let H be its parity check matrix. Further, assume that the columns of H, — $h_0, h_1, \ldots h_{n-1}$, are arranged such that for some fixed column vector $z = h_{n-1}$ and for all $i = 0, 1, \ldots \nu - 1$,

$$h_i + h_{i+\nu} = z .$$

Let C_1 be a coset of $\mathcal{H}_n$, such that the syndrome $s(c) = Hc^t$ is z for all $c \in C_1$. Define

$$\mathcal{A} = \{ (x \mid x \mid p(x)) \ : \ x \in \mathbb{F}_2^{\nu} \} \tag{8}$$

$$\mathcal{B} = \{ (x \mid x \mid p(x)+1) \ : \ x \in \mathbb{F}_2^{\nu} \} \tag{9}$$

where $p(x)$, the parity of x, is given by $wt(x)$ mod 2. Let $C_2 = (C_1 \backslash \mathcal{B}) \cup \mathcal{A}$.

Proposition 4.3. *The codes C_1 and C_2 are perfect, and $C_1 \cap C_2$ has the maximum possible cardinality.*

Proof. Obviously $\mathcal{A} \subset \mathcal{H}_n$, and $\mathcal{B} \subset C_1$ is a coset of $\mathcal{A}$ disjoint with $\mathcal{H}_n$. Hence $|C_2| = |C_1| - |\mathcal{B}| + |\mathcal{A}| = 2^{n-m}$. Indeed, $d(\mathcal{A}) = 3$ and $d(C_1 \backslash \mathcal{B}) = 3$. Now let $v = a + c$, where $a \in \mathcal{A}$ and $c \in C_1$. Clearly $s(v) = s(a) + s(c) = z$. Hence if $wt(v) \leq 2$ then either $v = (\mathbf{0}|\mathbf{0}|1)$ or $v = (u|u|0)$, where u is a vector of weight 1. But then $c = a + v$ is either $(x|x|p(x)+1)$ or $(x+u \mid x+u \mid p(x))$. Since $p(x) = p(x+u) + 1$, in both cases $c \in \mathcal{B}$. Thus $d(\mathcal{A}, C_1 \backslash \mathcal{B}) = \min_{a \in \mathcal{A}, c \in C_1 \backslash \mathcal{B}} d(a, c) \geq 3$, and C_2 is perfect. Furthermore, $|C_1 \cap C_2| = |C_1| - |\mathcal{B}| = 2^{n-m} - 2^{\nu}$, which is the maximum possible by Lemma 4.1. ∎

Note that the set $\mathcal{A}$ as defined in (8) is a linear subcode of the Hamming code $\mathcal{H}_n$. Let us rearrange the columns of H, the parity check matrix of $\mathcal{H}_n$, such that $h_{2i} + h_{2i-1} = h_0 = z$ for $i = 1, 2, \ldots \nu$. Then the code $\mathcal{A}$ is generated by the set of ν vectors of weight 3 in $\mathbb{F}_2^n$, hereafter called triples, which all concur in the first (i.e. the leftmost) coordinate. There is an essentially unique set of coset representatives for $\mathcal{A}$ in $\mathcal{H}_n$. These may be characterized as follows. Let $\mathcal{C}$ be a Hamming code of length ν. Then the coset representatives of $\mathcal{A}$ are obtained from the codewords of $\mathcal{C}$ by writing 0 as 00, writing 1 as 01, and then appending 0 in the first position. With a slight abuse of notation we shall write the coset of $\mathcal{A}$ corresponding to the codeword $c \in \mathcal{C}$ as $(c + \mathcal{A})$. Thus $\mathcal{H}_n = \cup_{c \in \mathcal{C}} (c + \mathcal{A})$. Note that the set $\mathcal{B}$ is also a coset of $\mathcal{A}$ corresponding to the coset leader $(1|\mathbf{0})$. Furthermore, for any $c \in \mathcal{C}$ the cosets $(c + \mathcal{A})$ and $(c + \mathcal{B})$ perfectly cover the same subset of $\mathbb{F}_2^n$ of cardinality $(n+1)2^{\nu}$. Hence choosing for each $c \in \mathcal{C}$ either

$(c+\mathcal{A})$ or $(c+\mathcal{B})$, and taking the union of the $|\mathcal{C}|$ cosets, we obtain a set Ω of distinct perfect codes of length n. Let $N = 2^{\nu-m+1}$ denote the cardinality of $\mathcal{C}$. A code $C \in \Omega$ may be written uniquely as

$$C = \bigcup_{i=0}^{N-1} \Big((x_i|0) + c_i + \mathcal{A} \Big) \tag{10}$$

where $c_0 = 0, c_1, \ldots c_{N-1}$ are the codewords of $\mathcal{C}$, and $x = (x_0, x_1, \ldots x_{N-1})$ is an arbitrary vector of $\mathbb{F}_2^N$. Thus Ω can be identified with the space $\mathbb{F}_2^N$ whose coordinates are indexed by the codewords of $\mathcal{C}$.

Clearly $|\Omega| = 2^N = 2^{2^{0.5(n+1)-\log(n+1)}}$. Hence in order to see that the number of nonequivalent codes in Ω approaches $2^{2^{0.5(n+1)}}$ as $n \to \infty$, it is sufficient to observe that the number of codes equivalent to a given $C \in \Omega$ is at most $2n!|C| = O(2^{n \log n})$.

Acknowledgement. The authors are indebted to Gilles Zémor for stimulating and helpful discussions. Alexander Vardy also wishes to thank Hagit Itzkowitz.

References

1. H. Bauer, B. Ganter, and F. Hergert, Algebraic techniques for nonlinear codes, *Combinatorica*, **3** (1983), 21–33.
2. P. Delsarte, Four fundamental parameters of a code and their combinatorial significance, *Info. and Control*, **23** (1973), 407–438.
3. F.J. MacWilliams and N.J.A. Sloane, *The Theory of Error-Correcting Codes*, New York: North-Holland, 1977.
4. K.T. Phelps, A combinatorial construction of perfect codes, *SIAM J. Alg. Disc. Meth.*, 4 (1983), 398–403.
5. K.T. Phelps, A general product construction for error-correcting codes, *SIAM J. Alg. Disc. Meth.*, 5 (1984), 224–228.
6. J.L. Vasil'ev, On nongroup close-packed codes, *Probl. Kibernet.*, 8 (1962), 375–378, (in Russian).

Authors' Index

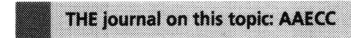

Lecture Notes in Computer Science

For information about Vols. 1–595
please contact your bookseller or Springer-Verlag

Vol. 596: L.-H. Eriksson, L. Hallnäs, P. Schroeder-Heister (Eds.), Extensions of Logic Programming. Proceedings, 1991. VII, 369 pages. 1992. (Subseries LNAI).

Vol. 597: H. W. Guesgen, J. Hertzberg, A Perspective of Constraint-Based Reasoning. VIII, 123 pages. 1992. (Subseries LNAI).

Vol. 598: S. Brookes, M. Main, A. Melton, M. Mislove, D. Schmidt (Eds.), Mathematical Foundations of Programming Semantics. Proceedings, 1991. VIII, 506 pages. 1992.

Vol. 599: Th. Wetter, K.-D. Althoff, J. Boose, B. R. Gaines, M. Linster, F. Schmalhofer (Eds.), Current Developments in Knowledge Acquisition - EKAW '92. Proceedings. XIII, 444 pages. 1992. (Subseries LNAI).

Vol. 600: J. W. de Bakker, C. Huizing, W. P. de Roever, G. Rozenberg (Eds.), Real-Time: Theory in Practice. Proceedings, 1991. VIII, 723 pages. 1992.

Vol. 601: D. Dolev, Z. Galil, M. Rodeh (Eds.), Theory of Computing and Systems. Proceedings, 1992. VIII, 220 pages. 1992.

Vol. 602: I. Tomek (Ed.), Computer Assisted Learning. Proceedings, 1992. X, 615 pages. 1992.

Vol. 603: J. van Katwijk (Ed.), Ada: Moving Towards 2000. Proceedings, 1992. VIII, 324 pages. 1992.

Vol. 604: F. Belli, F.-J. Radermacher (Eds.), Industrial and Engineering Applications of Artificial Intelligence and Expert Systems. Proceedings, 1992. XV, 702 pages. 1992. (Subseries LNAI).

Vol. 605: D. Etiemble, J.-C. Syre (Eds.), PARLE '92. Parallel Architectures and Languages Europe. Proceedings, 1992. XVII, 984 pages. 1992.

Vol. 606: D. E. Knuth, Axioms and Hulls. IX, 109 pages. 1992.

Vol. 607: D. Kapur (Ed.), Automated Deduction – CADE-11. Proceedings, 1992. XV, 793 pages. 1992. (Subseries LNAI).

Vol. 608: C. Frasson, G. Gauthier, G. I. McCalla (Eds.), Intelligent Tutoring Systems. Proceedings, 1992. XIV, 686 pages. 1992.

Vol. 609: G. Rozenberg (Ed.), Advances in Petri Nets 1992. VIII, 472 pages. 1992.

Vol. 610: F. von Martial, Coordinating Plans of Autonomous Agents. XII, 246 pages. 1992. (Subseries LNAI).

Vol. 611: M. P. Papazoglou, J. Zeleznikow (Eds.), The Next Generation of Information Systems: From Data to Knowledge. VIII, 310 pages. 1992. (Subseries LNAI).

Vol. 612: M. Tokoro, O. Nierstrasz, P. Wegner (Eds.), Object-Based Concurrent Computing. Proceedings, 1991. X, 265 pages. 1992.

Vol. 613: J. P. Myers, Jr., M. J. O'Donnell (Eds.), Constructivity in Computer Science. Proceedings, 1991. X, 247 pages. 1992.

Vol. 614: R. G. Herrtwich (Ed.), Network and Operating System Support for Digital Audio and Video. Proceedings, 1991. XII, 403 pages. 1992.

Vol. 615: O. Lehrmann Madsen (Ed.), ECOOP '92. European Conference on Object Oriented Programming. Proceedings. X, 426 pages. 1992.

Vol. 616: K. Jensen (Ed.), Application and Theory of Petri Nets 1992. Proceedings, 1992. VIII, 398 pages. 1992.

Vol. 617: V. Mařík, O. Štěpánková, R. Trappl (Eds.), Advanced Topics in Artificial Intelligence. Proceedings, 1992. IX, 484 pages. 1992. (Subseries LNAI).

Vol. 618: P. M. D. Gray, R. J. Lucas (Eds.), Advanced Database Systems. Proceedings, 1992. X, 260 pages. 1992.

Vol. 619: D. Pearce, H. Wansing (Eds.), Nonclassical Logics and Information Proceedings. Proceedings, 1990. VII, 171 pages. 1992. (Subseries LNAI).

Vol. 620: A. Nerode, M. Taitslin (Eds.), Logical Foundations of Computer Science – Tver '92. Proceedings. IX, 514 pages. 1992.

Vol. 621: O. Nurmi, E. Ukkonen (Eds.), Algorithm Theory – SWAT '92. Proceedings. VIII, 434 pages. 1992.

Vol. 622: F. Schmalhofer, G. Strube, Th. Wetter (Eds.), Contemporary Knowledge Engineering and Cognition. Proceedings, 1991. XII, 258 pages. 1992. (Subseries LNAI).

Vol. 623: W. Kuich (Ed.), Automata, Languages and Programming. Proceedings, 1992. XII, 721 pages. 1992.

Vol. 624: A. Voronkov (Ed.), Logic Programming and Automated Reasoning. Proceedings, 1992. XIV, 509 pages. 1992. (Subseries LNAI).

Vol. 625: W. Vogler, Modular Construction and Partial Order Semantics of Petri Nets. IX, 252 pages. 1992.

Vol. 626: E. Börger, G. Jäger, H. Kleine Büning, M. M. Richter (Eds.), Computer Science Logic. Proceedings, 1991. VIII, 428 pages. 1992.

Vol. 628: G. Vosselman, Relational Matching. IX, 190 pages. 1992.

Vol. 629: I. M. Havel, V. Koubek (Eds.), Mathematical Foundations of Computer Science 1992. Proceedings. IX, 521 pages. 1992.

Vol. 630: W. R. Cleaveland (Ed.), CONCUR '92. Proceedings. X, 580 pages. 1992.

Vol. 631: M. Bruynooghe, M. Wirsing (Eds.), Programming Language Implementation and Logic Programming. Proceedings, 1992. XI, 492 pages. 1992.

Vol. 632: H. Kirchner, G. Levi (Eds.), Algebraic and Logic Programming. Proceedings, 1992. IX, 457 pages. 1992.

Vol. 633: D. Pearce, G. Wagner (Eds.), Logics in AI. Proceedings. VIII, 410 pages. 1992. (Subseries LNAI).

Vol. 634: L. Bougé, M. Cosnard, Y. Robert, D. Trystram (Eds.), Parallel Processing: CONPAR 92 – VAPP V. Proceedings. XVII, 853 pages. 1992.

Vol. 635: J. C. Derniame (Ed.), Software Process Technology. Proceedings, 1992. VIII, 253 pages. 1992.

Vol. 636: G. Comyn, N. E. Fuchs, M. J. Ratcliffe (Eds.), Logic Programming in Action. Proceedings, 1992. X, 324 pages. 1992. (Subseries LNAI).

Vol. 637: Y. Bekkers, J. Cohen (Eds.), Memory Management. Proceedings, 1992. XI, 525 pages. 1992.

Vol. 639: A. U. Frank, I. Campari, U. Formentini (Eds.), Theories and Methods of Spatio-Temporal Reasoning in Geographic Space. Proceedings, 1992. XI, 431 pages. 1992.

Vol. 640: C. Sledge (Ed.), Software Engineering Education. Proceedings, 1992. X, 451 pages. 1992.

Vol. 641: U. Kastens, P. Pfahler (Eds.), Compiler Construction. Proceedings, 1992. VIII, 320 pages. 1992.

Vol. 642: K. P. Jantke (Ed.), Analogical and Inductive Inference. Proceedings, 1992. VIII, 319 pages. 1992. (Subseries LNAI).

Vol. 643: A. Habel, Hyperedge Replacement: Grammars and Languages. X, 214 pages. 1992.

Vol. 644: A. Apostolico, M. Crochemore, Z. Galil, U. Manber (Eds.), Combinatorial Pattern Matching. Proceedings, 1992. X, 287 pages. 1992.

Vol. 645: G. Pernul, A M. Tjoa (Eds.), Entity-Relationship Approach – ER '92. Proceedings, 1992. XI, 439 pages, 1992.

Vol. 646: J. Biskup, R. Hull (Eds.), Database Theory – ICDT '92. Proceedings, 1992. IX, 449 pages. 1992.

Vol. 647: A. Segall, S. Zaks (Eds.), Distributed Algorithms. X, 380 pages. 1992.

Vol. 648: Y. Deswarte, G. Eizenberg, J.-J. Quisquater (Eds.), Computer Security – ESORICS 92. Proceedings. XI, 451 pages. 1992.

Vol. 649: A. Pettorossi (Ed.), Meta-Programming in Logic. Proceedings, 1992. XII, 535 pages. 1992.

Vol. 650: T. Ibaraki, Y. Inagaki, K. Iwama, T. Nishizeki, M. Yamashita (Eds.), Algorithms and Computation. Proceedings, 1992. XI, 510 pages. 1992.

Vol. 651: R. Koymans, Specifying Message Passing and Time-Critical Systems with Temporal Logic. IX, 164 pages. 1992.

Vol. 652: R. Shyamasundar (Ed.), Foundations of Software Technology and Theoretical Computer Science. Proceedings, 1992. XIII, 405 pages. 1992.

Vol. 653: A. Bensoussan, J.-P. Verjus (Eds.), Future Tendencies in Computer Science, Control and Applied Mathematics. Proceedings, 1992. XV, 371 pages. 1992.

Vol. 654: A. Nakamura, M. Nivat, A. Saoudi, P. S. P. Wang, K. Inoue (Eds.), Prallel Image Analysis. Proceedings, 1992. VIII, 312 pages. 1992.

Vol. 655: M. Bidoit, C. Choppy (Eds.), Recent Trends in Data Type Specification. X, 344 pages. 1993.

Vol. 656: M. Rusinowitch, J. L. Rémy (Eds.), Conditional Term Rewriting Systems. Proceedings, 1992. XI, 501 pages. 1993.

Vol. 657: E. W. Mayr (Ed.), Graph-Theoretic Concepts in Computer Science. Proceedings, 1992. VIII, 350 pages. 1993.

Vol. 658: R. A. Rueppel (Ed.), Advances in Cryptology – EUROCRYPT '92. Proceedings, 1992. X, 493 pages. 1993.

Vol. 659: G. Brewka, K. P. Jantke, P. H. Schmitt (Eds.), Nonmonotonic and Inductive Logic. Proceedings, 1991. VIII, 332 pages. 1993. (Subseries LNAI).

Vol. 660: E. Lamma, P. Mello (Eds.), Extensions of Logic Programming. Proceedings, 1992. VIII, 417 pages. 1993. (Subseries LNAI).

Vol. 661: S. J. Hanson, W. Remmele, R. L. Rivest (Eds.), Machine Learning: From Theory to Applications. VIII, 271 pages. 1993.

Vol. 662: M. Nitzberg, D. Mumford, T. Shiota, Filtering, Segmentation and Depth. VIII, 143 pages. 1993.

Vol. 663: G. v. Bochmann, D. K. Probst (Eds.), Computer Aided Verification. Proceedings, 1992. IX, 422 pages. 1993.

Vol. 664: M. Bezem, J. F. Groote (Eds.), Typed Lambda Calculi and Applications. Proceedings, 1993. VIII, 433 pages. 1993.

Vol. 665: P. Enjalbert, A. Finkel, K. W. Wagner (Eds.), STACS 93. Proceedings, 1993. XIV, 724 pages. 1993.

Vol. 666: J. W. de Bakker, W.-P. de Roever, G. Rozenberg (Eds.), Semantics: Foundations and Applications. Proceedings, 1992. VIII, 659 pages. 1993.

Vol. 667: P. B. Brazdil (Ed.), Machine Learning: ECML – 93. Proceedings, 1993. XII, 471 pages. 1993. (Subseries LNAI).

Vol. 668: M.-C. Gaudel, J.-P. Jouannaud (Eds.), TAPSOFT '93: Theory and Practice of Software Development. Proceedings, 1993. XII, 762 pages. 1993.

Vol. 669: R. S. Bird, C. C. Morgan, J. C. P. Woodcock (Eds.), Mathematics of Program Construction. Proceedings, 1992. VIII, 378 pages. 1993.

Vol. 670: J. C. P. Woodcock, P. G. Larsen (Eds.), FME '93: Industrial-Strength Formal Methods. Proceedings, 1993. XI, 689 pages. 1993.

Vol. 671: H. J. Ohlbach (Ed.), GWAI-92: Advances in Artificial Intelligence. Proceedings, 1992. XI, 397 pages. 1993. (Subseries LNAI).

Vol. 672: A. Barak, S. Guday, R. G. Wheeler, The MOSIX Distributed Operating System. X, 221 pages. 1993.

Vol. 673: G. Cohen, T. Mora, O. Moreno (Eds.), AAECC-10: Applied Algebra, Algebraic Algorithms and Error-Correcting Codes. Proceedings, 1993. X, 355 pages. 1993.

Vol. 674: G. Rozenberg (Ed.), Advances in Petri Nets 1993. VII, 457 pages. 1993.

Vol. 675: A. Mulkers, Live Data Structures in Logic Programs. VIII, 220 pages. 1993.